Lecture Notes in Computer Science 3769

Commenced Publication in 1973
Founding and Former Series Editors:
Gerhard Goos, Juris Hartmanis, and Jan van Leeuwen

David A. Bader   Manish Parashar
Varadarajan Sridhar   Viktor K. Prasanna (Eds.)

High Performance Computing – HiPC 2005

12th International Conference
Goa, India, December 18-21, 2005
Proceedings

 Springer

Volume Editors

David A. Bader
Georgia Institute of Technology
College of Computing
Atlanta, GA 30332, USA
E-mail: bader@cc.gatech.edu

Manish Parashar
University of New Jersey
Department of Electrical and Computer Engineering
94 Brett Road, Piscataway, NJ 08854, USA
E-mail: parashar@caip.rutgers.edu

Varadarajan Sridhar
Satyam Computer Services Ltd.
Entrepreneurship Centre, SID Block, Indian Institute of Science Campus
Bangalore - 560 012, India
E-mail: Sridhar@satyam.com

Viktor K. Prasanna
University of Southern California
Department of Electrical Engineering
Los Angeles, California, 90089-2562, USA
E-mail: prasanna@usc.edu

Library of Congress Control Number: 2005937125

CR Subject Classification (1998): D.1-4, C.1-4, F.1-2, G.1-2

ISSN   0302-9743
ISBN-10   3-540-30936-5 Springer Berlin Heidelberg New York
ISBN-13   978-3-540-30936-9 Springer Berlin Heidelberg New York

Springer is a part of Springer Science+Business Media

springeronline.com

© Springer-Verlag Berlin Heidelberg 2005

Typesetting: Camera-ready by author, data conversion by Scientific Publishing Services, Chennai, India
Printed on acid-free paper   SPIN: 11602569   06/3142   5 4 3 2 1 0

Message from the Program Chair

Welcome to Goa and the 12th International Conference on High Performance Computing!

This year, we were delighted to receive 362 submissions to this conference from more than 30 different countries, including (besides India!) countries in North and South America, Europe, Asia, Africa, and Australia. This is a major increase compared with last year (253 submissions from 25 countries). Eventually, 50 submissions from 12 different countries were selected for presentation at the conference and publication in the conference proceedings.

This sharp increase in the number of submissions meant we had to adapt the regular selection process used in previous years. First, all submitted papers were carefully considered by the Program Chair and Vice-Chairs to check their consistency with the minimal syntactic requirements for acceptance. At the end of this first stage, we were left with 271 submissions, which were further considered by the Program Committee. Each of these papers was reviewed by three Program Committee members. As many as 785 reviews were collected (2.90 per paper on average) and each paper was discussed at the online Program Committee meeting. Finally, 50 out of 271 (18.5%) were accepted for presentation and publication in the proceedings.

Among them, two outstanding papers were selected as "Best Papers"; one focusing on the Systems Software area ("Preemption Adaptivity in Time-Published Queue-Based Spin Locks," by Bijun He, William N. Scherer III, and Michael L. Scott) and the other focusing on the Architecture area ("Criticality Driven Energy Aware Speculation for Speculative Multithreaded Processors," by Rahul Nagpal and Anasua Bhowmik). They will be presented in a separate plenary session, and each paper will be awarded a prize sponsored by InfoSys.

Here is a general summary of the results with respect to the origin of the submissions:

Submission origin	Reviewed	Accepted	Acceptance rate
Overall	362	50	13.8%
India	56.4%	28%	6.9%
Asia except India	19.3%	10%	7.1%
North America (mainly USA)	14.4%	46%	44%
Elsewhere (mainly Europe)	9.9%	16%	22%
Total	100%	100%	

These figures show that the selection process was highly competitive. We are pleased to accommodate ten (parallel) technical sessions of high-quality contributed papers, plus the special plenary "Best Papers" session. In addition, this year's conference also features a poster session, industrial exhibits, five keynote addresses, five tutorials and four workshops.

It has been a pleasure putting together this program with the help of five excellent Program Vice-Chairs and their 70 Program Committee members. The hard work of all the Program Committee members has been deeply appreciated. I especially wish to acknowledge the dedicated effort put forth by the Vice-Chairs: Michael A. Bender (Algorithms), Zhiwei Xu (Applications), José Duato (Architecture), M. Cristina Pinotti (Communication Networks), and Satoshi Matsuoka (System Software). Without their help and timely work, the quality of this program would not be as high nor would the process have run so smoothly.

I thank the other organizers who have contributed to assembling this program, including those who organized the keynotes, tutorials, workshops, awards, poster session, industry exhibits, and those who performed the administrative functions that have been essential to the success of this conference. The work of Sushil K. Prasad in putting together the conference proceedings is also acknowledged, as well as the support provided by Kamesh Madduri and Vaddadi Chandu, Ph.D. students at Georgia Institute of Technology, and Vipin Sachdeva, M.S. student at the University of New Mexico, in assisting with the EDAS online paper submission and evaluation software. Last, but certainly not least, I express heart-felt thanks to our General Co-chairs, Manish Parashar and V. Sridhar; Steering Chair, Viktor Prasanna; and to the Vice-General Chair, Rajendra V. Boppana; for all their useful advice.

Lastly, I thank the Conference General Co-chairs for allowing me to serve our community as the Program Chair of this high-quality international conference. It has been my pleasure to correspond with so many of you, and I personally welcome you to Goa. As you can see from these proceedings, we have made considerable effort to select and assemble the highest-quality technical program for this year's meeting. Please enjoy the informative and stimulating presentations and your entire experience at HiPC 2005 including the food and beautiful coastal scenery in this culturally-rich location of Goa, India!

December 2005   David A. Bader

Message from the General Co-chairs and the Vice General Chair

On behalf of the organizers of the 12th International Conference on High Performance Computing (HiPC), it is our pleasure to welcome you to the paradise state of Goa.

The HiPC 2005 technical program includes technical paper sessions interspersed with keynotes from leading HPC researchers, a poster paper session, an industry session with presentations from leading HPC companies, a HPC user community meeting, an exhibition, tutorials on hot topics in computing and networking, and several workshops focusing on emerging areas. We do hope you find the conference and these proceedings exciting and rewarding.

This year, the HiPC call for papers, once again, received an overwhelming response with a record number of submissions from across the globe. For this, we would like to specially thank David Bader, Program Chair, who, with remarkable dedication, put together an outstanding technical program consisting of the papers that appear in these proceedings. We would also like to thank the program committee for their efforts in assembling such an excellent program and the authors who submitted the high quality material from which that program was selected. We would like to especially thank the presenters of the keynotes, posters and tutorials, the organizers of the workshops, and all the participants, who complete the program.

Arranging an exciting meeting with a high quality technical program is easy when one is working with an excellent and dedicated team and can build on the practices and levels of excellence established by a quality research community. HiPC 2005 would not have been possible without the tremendous efforts of the many volunteers. We would like to acknowledge the critical contributions of each one. We would specially like to thank Viktor Prasanna, Chair of the HiPC Steering Committee, for his leadership, sage guidance, and untiring dedication. This year we were lucky to have a number of new volunteers joining us. We would like to welcome you to the team and thank you – your efforts are critical to the continued success of this conference.

We would like to gratefully acknowledge our academic and industrial sponsors including IEEE Computer Society, ACM SIGARCH, EATCS, IFIP, BHU, Infosys, Satyam, HP, IBM, Cray, Dell, and Sun. We would also like to acknowledge the local support we have received from Cidade De Goa, Goa Chamber of Commerce and Industry, National Institute of Oceanography, and Goa University.

Once again, we welcome you to Goa and HiPC 2005.

December 2005

Manish Parashar
V. Sridhar
Rajendra V Boppana

Message from the Steering Chair

It is my pleasure to welcome you to the 12th International Conference on High Performance Computing and to Goa, a unique city with its blend of Indian and Portugese culture. First, I would like to single out the contribution of David Bader, Program Chair, for his enthusiasm, commitment and attention to details in putting together an excellent technical program. We received a record number of submissions this year, surpassing our previous high set last year. I am grateful to David for his efforts and thoughtful inputs in putting together the program.

2005 marks a year in transition. We have several new volunteers who continue the rich tradition set by HiPC over the years. Manish Parashar, General Co-chair, took responsibility for the overall meeting organization. He identified several new volunteers as well as handling local arrangments in Goa. I would like to welcome Rajendra Boppana, Vice General Chair, Rajeev Thakur, Poster/Presentation Chair, Anu Bourgeois, Student Scholarships Co-chair, Rajeev Raje, Publicity Co-chair, and Viraj Bhat, Cyber Chair, to the "HiPC family." The continued efforts of Rajeev Muralidhar of Intel India and Ramamurthy Badrinath of HP India are gratefully acknowledged.

The technical program was expanded through committed efforts from several volunteers. Ramamurthy Badrinath, with assistance from Venkat Ramana of Hinditron Infosystems, put together the first HPC Users' Group meeting. Frank Baetke of HP organized a panel entitled "Processors, Instruction Sets, Operating Systems – Challengers and Survivors".

I would like to thank all our volunteers for their tireless efforts. The meeting would not have been possible without the enthusiastic commitment of these individuals.

Major financial support for the meeting was provided by several leading IT companies in India. I would like to acknowledge the following individuals and their organizations for their support: N.R. Narayana Murthy, Infosys, India; Kris Gopalakrishnan, Infosys, India; Harish Grama, IBM India; P. Gopalakrishnan, IBM Solutions Research Center, India; Venkat Ramana, Hinditron Infosystems; Dinkar Sitaram and Faisal Paul, HP India; V. Sridhar, Satyam; Raghuram Tupuri, AMD; Raj Yavatkar and Kumar Ranganathan, Intel.

Finally, I would like to thank Animesh Pathak at USC for his continued assistance and enthusiasm in organizing the meeting.

December 2005   Viktor K. Prasanna

Conference Organization

General Co-chairs
Manish Parashar, Rutgers University, USA
V. Sridhar, Satyam Computer Services Ltd., India

Vice General Chair
Rajendra V Boppana, University of Texas at San Antonio, USA

Program Chair
David A. Bader, Georgia Institute of Technology, USA

Program Vice-Chairs
Algorithms
Michael A. Bender, SUNY Stony Brook, USA
Applications
Zhiwei Xu, Chinese Academy of Sciences, China
Architecture
Jose Duato, Technical University of Valencia, Spain
Communication Networks
Cristina M. Pinotti, University of Perugia, Italy
Systems Software
Satoshi Matsuoka, Tokyo Institute of Technology, Japan

Steering Chair
Viktor K. Prasanna, University of Southern California, USA

Workshops Chair
C.P. Ravikumar, Texas Instruments, India

Poster/Presentation Chair
Rajeev Thakur, Argonne National Laboratory, USA

Scholarships Co-chairs
Anu G. Bourgeois, Georgia State University, USA
Animesh Pathak, University of Southern California, USA

Finance Co-chairs
Ajay Gupta, Western Michigan University, USA
B.V. Ramachandran, Software Technology Park, Bangalore, India

Tutorials Chair
Srinivas Aluru, Iowa State University, USA

Industry Liaison Chair
Sudheendra Hangal, Sun Microsystems, India

Publicity Co-chairs
Ramamurthy Badrinath, HP, India
Rajeev R. Raje, Indiana University, Purdue University, Indianapolis, USA

Publications Chair
Sushil K. Prasad, Georgia State University, USA

Cyber Chair
Viray Bhat, Rutgers University, USA
Nikesh Patel, Indiana University, Purdue University, Indianapolis, USA

Local Arrangements Chair
Rajeev D. Muralidhar, Intel, India

Registration Chair
Susamma Barua, California State University, Fullerton, USA

Goa Coordination Committee
Air Cmd PK Pinto, Director General
Mr. Sandeep Verenkar, Chairman IT Committee
Mr. Manguirish Raiker, Hon. Treasurer
Dr. V Kamat, Goa University, Member IT Committee
Mr. K D Kulkarni, VP Controlnet India, Member IT Committee
Ms. Shilpa Mhatapat, Event Coordinator

Steering Committee
Ramamurthy Badrinath, HP, India
Jose Duato , Universidad Politécnica de Valencia, Spain
Harish Grama, IBM, India
N.S. Nagaraj, Infosys, India
N. Radhakrishnan, North Carolina A&T State University, USA
Viktor K. Prasanna, University of Southern California, USA
Venkat Ramana, Cray-Hinditron, India
Shubhra Roy, Intel, India
Hari Rao, Oracle, India
Dheeraj Sanghi, Indian Institute of Technology, Kanpur, India
Sartaj Sahni, University of Florida, USA
Assaf Schuster, Technion, Israel Institute of Technology, Israel
V. Sridhar, Satyam, India

Program Committee

Algorithms

Srinivas Aluru, Iowa State University, USA
Frank Dehne, Griffith University, Australia
Martin Farach-Colton, Rutgers University, USA
Cyril Gavoille, University of Bordeaux, France
Seth Gilbert, MIT, USA
Ananth Grama, Purdue University, USA
Bradley C. Kuszmaul, MIT, USA
Cynthia A. Phillips, Sandia National Laboratories, USA
Ali Pinar, Lawrence Berkeley National Laboratory, USA
Rajmohan Rajaraman, Northeastern University, USA
Peter Sanders, University of Karlsruhe, Germany
Christian Scheideler, Johns Hopkins University, USA
Denis Trystram, Grenoble, France
Ramachandran Vaidyanathan, Louisiana State University, USA

Applications

Bill Appelbe, VPAC, Australia
Rudolf Eigenmann, Purdue University, USA
Yike Guo, Imperial College, UK
Jens Gustedt, Loria/INRIA, France
Naoki Hirose, Asian Technology Information Program, Japan
Chung-Ta King, TsingHua University, Taiwan
Sharad C. Purohit, Centre for Development of Advanced Computing (CDAC), India
Hong Shen, Japan Advanced Institute of Science & Technology (JAIST), Japan
Peter Sloot, University of Amsterdam, Netherlands
Xian-He Sun, Illinois Institute of Technology, USA
Cho-Li Wang, University of Hong Kong, Hong Kong, China
Vladimir L. Yakushev, Russian Academy of Sciences, Russia
Yao Zheng, Zhejiang University, China

Architecture

Ricardo Bianchini, Rutgers University, USA
Angelos Bilas, University of Crete, Greece
Antonio Gonzalez, UPC/Intel Labs, Barcelona, Spain
Jose Gonzalez, Intel Labs, Barcelona, Spain
David Kaeli, Northeastern University, USA
Stefanos Kaxiras, University of Patras, Greece
Olav Lysne, Simula Research Laboratory, Norway
Li-Shiuan Peh, Princeton University, USA

Timothy Mark Pinkston, University of Southern California, USA
Anand Sivasubramaniam, Pennsylvania State University, USA
Evan Speight, IBM, USA
Per Stenstrom, Chalmers University of Technology, Sweden
Sudhakar Yalamanchili, Georgia Institute of Technology, USA

Communication Networks

Alan A. Bertossi, University of Bologna, Italy
Amiya Bhattacharya, New Mexico State University, USA
Azzedine Boukerche, University of Ottawa, Canada
Mainak Chatterjee, University of Central Florida, USA
Ajoy K. Datta, University of Nevada, Las Vegas, USA
Evangelos Kranakis, Carleton University, Canada
Bhaskhar Krishnamachari, University of Southern California, USA
Anurag Kumar, IISc Bangalore, India
Alessandro Mei, University of Rome "La Sapienza", Italy
Loren Schwiebert, Wayne State University, USA
Ivan Stojmenovic, University of Ottawa, Canada
Violet R. Syrotiuk, Arizona State University, USA
Roger Wattenhoher, ETH Zurich, Switzerland
Jie Wu, Florida Atlantic University, USA
Taieb Znati , University of Pittsburgh, USA

Systems Software

Franck Cappello, Universitè Paris Sud, France
Denis Caromel, Univ. of Nice, CNRS/I3S, INRIA, IUF, France
Dennis Gannon, Indiana University, USA
Manish Gupta, IBM Research, USA
Laxmikant V. Kale, University of Illinois at Urbana-Champaign, USA
Vijay Karamcheti, New York University, USA
Craig Lee, Aerospace Corp., USA
David Lowenthal, University of Georgia, USA
Andrew Lumsdane, Indiana University, USA
Dhabaleswar K. (DK) Panda, The Ohio State University, USA
Steven Parker, University of Utah, USA
Ramendra Sahoo, IBM T.J. Watson Research Center, USA
Joel Saltz, The Ohio State University, USA
Mitsuhisa Sato, Tsukuba University, Japan
Alan Sussman, University of Maryland, USA

Workshop Organizers

Workshop on Cutting Edge Computing

Co-chairs
   Harish K Grama, IBM Software Lab, India
   Albee Jhoney, IBM Software Lab, India

Trusted Internet Workshop

Chair
   Suresh Subramaniam, The George Washington University, USA

Workshop on Next Generation Wireless Networks

Co-chairs
   C. Siva Ram Murthy, IIT Madras, India
   B.S. Manoj, University of California, San Diego, USA

Workshop on New Horizons in Compilers

Co-chairs
   R. Govindarajan, IISc, Bangalore, India
   Bhagi Narahari, GWU, Washington, USA

Tutorials

Smart Environments: Technology, Protocols and Applications
Sajal Das, University of Texas at Arlington, USA

The Gridbus Toolkit: Creating and Managing Utility Grids for eScience and eBusiness Applications
Rajkumar Buyya, University of Melbourne, Australia

Scheduling Algorithms for Heterogeneous Platforms
Yves Robert, Lyon

State of InfiniBand in Designing Next Generation Clusters, File/Storage Systems and Datacenters
Dhabaleswar Panda, Ohio State University, USA

Building and Deploying Wireless Sensor Network Applications: Tools, Techniques and Challenges
Raju Pandey, University of California, Davis, USA

List of Reviewers

Michelle Ackermann, ETH Zurich
Kunal Agrawal, MIT
Keno Albrecht, ETH Zurich
Srinivas Aluru, Iowa State University
Khaled Alzoubi, Illinois Institute of Technology
Gabriel Antoniu, IRISA
William Appelbe, VPAC
Juan Aragon, University of Murcia
Asad Awan, Purdue University
Shahaan Ayyub, Monash University
David A. Bader, Georgia Institute of Technology
Amnon Barak, The Hebrew University of Jerusalem
Doina Bein, University of Nevada, Las Vegas
Nadia Bel Hadj Aissa, LIFL
Frank Bellosa, University of Karlsruhe
Michael Bender, SUNY Stony Brook
Alan Albert Bertossi, University of Bologna
Shalabh Bhatnagar, Indian Institute of Science
Amiya Bhattacharya, New Mexico State University
Ricardo Bianchini, Rutgers University
Angelos Bilas, FORTH University of Crete
Deepak Bobbarjung, Purdue University
Ladislau Boloni, University of Central Florida
Maurizio Angelo Bonuccelli, Università di Pisa
Azzedine Boukerche, Univ. of Ottawa
Aurelien Bouteiller, Universitè Paris Sud (LRI)
Mauro Brunato, University of Trento
Nicolas Burri, ETH Zurich
Surendra Byna, Illinois Institute of Technology
Ramon Canal, UPC
Franck Cappello, INRIA Futurs
Denis Caromel, Univ. of Nice
Antonio Caruso, CNR
Lakshmi Chakrapani, Georgia Institute of Technology
Sayantan Chakravorty, University of Illinois
Philip Chan, Monash University
Pedro Chaparro, Intel-UPC Barcelona Research Center
François Charoy, Université de Nancy
François Charpillet, LORIA
Mainak Chatterjee, University of Central Florida

Arun Chauhan, Indiana University
Bruno Checcucci, INFN, University of Perugia
Stefano Chessa, University of Pisa
Yogesh Chobe, Georgia Institute of Technology
Yeh-Ching Chung, National Tsing Hua University
Josep M. Codina, Intel-UPC Barcelona Research Center
Reuven Cohen, Technion
Marco Conti, CNR - Istituto CNUCE
Gilberto Contreras, Princeton University
Richard Copeland, Georgia Institute of Technology
Jesus Corbal, Intel-UPC Laboratories
Llorenc Cruz, UPC
Kostadin Damevski, University of Utah
Saikat Dan, NEC Labs, America
Sajal Das, University of Texas at Arlington
Ajoy Datta, UNLV
Mark Davis, Intel
Frank Dehne, Griffith University
Alex Delis, University of Athens
Roberto Di Pietro, University of Rome "La Sapienza"
Chyi-Ren Dow, Feng Chia University
Carole Dulong, Intel
Pierre-François Dutot, ID-IMAG Laboratory
Rudolf Eigenmann, Purdue University
Mohamed Eltoweissy, Virginia Tech
Jacob Engel, Univ. of Central Florida
Colin Enticott, Monash University
Oguz Ergin, Intel Laboratories Barcelona
Mohamed Essaïdi, INRIA Sophia-Antipolis
Lionel Eyraud, ID-IMAG
Martin Farach-Colton, Rutgers University
Rohan Fernandes, Rutgers University
Ronaldo Ferreira, Purdue University
Roland Flury, ETH Zurich
Andrea Formisano, University of L'Aquila
Dennis Gannon, Indiana University
Carlos Garcia, Intel Laboratories Barcelona - UPC
Slavisa Garic, Monash University
Chris Gauthier-Dickey, University of Oregon
Cyril Gavoille, University of Bordeaux
Enric Gibert, UPC
Seth Gilbert, MIT CSAIL
Antonio Gonzalez, Universitat Politècnica de Catalunya
Jose Gonzalez, Intel Laboratories Barcelona
Lawrence Gordon, University of Maryland

Jin Li, ETH Zurich
Wei Li, Institute of Computing Technology, Chinese Academy of Sciences
Haitao Lin, Nortel
Hwa-Chun Lin, National Tsing Hua University
Thomas Locher, ETH Zurich
Pedro Lopez, Universidad Politécnica de Valencia
David Lowenthal, University of Georgia
Gang Lu, University of Southern California
Yingping Lu, University of Minnesota
Andrew Lumsdaine, Indiana University
Olav Lysne, Simula Research Laboratory
C.E. Veni Madhavan, Indian Institute of Science
Praveen Madiraju, Georgia State University
Carlos Madriles, Intel-UPC Barcelona Research Center
Grigorios Magklis, Intel-UPC Barcelona Research Center
Srilaxmi Malladi, Georgia State University
Vittorio Maniezzo, University of Bologna
D. Manjunath, Indian Institute of Technology, Bombay
Pedro Marcuello, Intel Laboratories Barcelona
José Martinez, Univ. Nantes
Alessandro Mei, University of Rome "La Sapienza"
Carlos Molina, UPC
Edoardo Mollona, University of Bologna
Vincent Mooney, Georgia Institute of Technology
David Morano, Northeastern University
Tangui Morlier, INRIA Futurs
Thomas Moscibroda, ETH Zurich
Feryal Moulai, ID-IMAG
Nicolas Navet, INRIA Lorraine LORIA
Vincent Neri, CNRS
Regina O'Dell, ETH Zurich
Andreas Ottiger, ETH Zurich
Sourav Pal, University of Texas, Arlington
Dhabaleswar Panda, The Ohio State University
Symeon Papavassiliou, NJIT
Steven Parker, University of Utah
Tom Peachey, Monash University
Johnatan Pecero, ID-IMAG
Li-Shiuan Peh, Princeton University
Salvador Petit, Universidad Politècnica de Valencia
Chiara Petrioli, University of Rome "La Sapienza"
Cynthia Phillips, Sandia National Lab.
Alejandro Piñeiro, Intel-UPC Barcelona Research Center
Ali Pinar, Lawrence Berkeley National Laboratory
Timothy Pinkston, University of Southern California

Cristina M. Pinotti, University of Perugia
Viktor Prasanna, University of Southern California
Antonio Puliafito, University of Messina
Sharad Purohit, Centre for Development of Advanced Computing (CDAC)
Francisco Quiles, Universidad de Castilla La Mancha
Patrice Quinton, ENS Cachan Bretagne
Rajmohan Rajaraman, Northeastern University
Subramanium Ramaswamy, Georgia Institute of Technology
Prathima Rao, Purdue University
Yves Robert, École Normale Supérieure de Lyon
Krzysztof Rzadca, ID-IMAG
Narayanan Sadagopan, USC
Ramendra Sahoo, IBM Research, Yorktown Heights
Joel Saltz, Ohio State University
Peter Sanders, Universität Karlsruhe
Paolo Santi, CNR
Mitsuhisa Sato, University of Tsukuba
Christian Scheideler, Johns Hopkins University
Stefan Schmid, ETH Zurich
Loren Schwiebert, Wayne State University
Hong Shen, Japan Advanced Institute of Science and Technology
Huaping Shen, The University of Texas at Arlington
Anand Sivasubramaniam, Penn State
Peter Sloot, University of Amsterdam
Evan Speight, IBM Austin Research Lab
Jeffrey Squyres, Indiana University
Per Stenstrom, Chalmers University of Technology
Ivan Stojmenovic, University of Ottawa
Xian-He Sun, Illinois Institute of Technology
Alan Sussman, University of Maryland
Frederic Suter, LORIA
Violet Syrotiuk, Arizona State University
Jeff Tan, Monash University
David Taylor, University of Waterloo
David Taylor, Washington University in St. Louis
Anita Thomas, EMC
Umut Topkara, Purdue University
Denis Trystram, Univ. of Grenoble
Osman Unsal, Intel Laboratories Barcelona
Kiran Vadde, Arizona State University
Ramachandran Vaidyanathan, Louisiana State University
Todd Veldhuizen, Indiana University
Andrea Vitaletti, University of Rome La Sapienza
Pascal von Rickenbach, ETH Zurich
Cho-Li Wang, The University of Hong Kong

Table of Contents

Session II - Applications

Session III - Architecture

Session IV - Applications

Session V - Systems Software

Session VI - Communication Networks

Session VII - Architecture

Session VIII - Communication Networks

Session IX - Algorithms

Session X - Systems and Networks

Data Confidentiality in Collaborative Computing

Mikhail Atallah

Purdue University, CERIAS, Recitation Bldg,
West Lafayette, IN 47907, USA
mja@cs.purdue.edu

Abstract. Even though collaborative computing can yield substantial economic, social, and scientific benefits, a serious impediment to fully achieving that potential is a reluctance to share data, for fear of losing control over its subsequent dissemination and usage. An organization's most valuable and useful data is often proprietary/confidential, or the law may forbid its disclosure or regulate the form of that disclosure. We survey security technologies that mitigate this problem, and discuss research directions towards enforcing the data owner's approved purposes on the data used in grid computing. These include techniques for cooperatively computing answers without revealing any private data, even though the computed answers depend on all the participants' private data. They also include computational outsourcing, where computationally weak entities use computationally powerful entities to carry out intensive computing tasks without revealing to them either their inputs or the computed outputs.

Biography. Mikhail ("Mike") Atallah obtained the Ph.D. in 1982 from the Johns Hopkins University and joined the Computer Science Department at Purdue University, where he currently holds the rank of Distinguished Professor. A Fellow of the IEEE, he served on the editorial boards of many top journals (including SIAM Journal on Computing, JPDC, IEETC, etc), and on the Program Committees of many top conferences and workshops (including PODS, SODA, SoCG, WWW, PET, DRM, SACMAT, etc). He was Keynote and Invited Speaker at many national and international meetings, and a speaker in the Distinguished Colloquium Series of six top Computer Science Departments. He was selected in 1999 as one of the best teachers in the history of Purdue University and included in Purdue's Book of Great Teachers, a permanent wall display of Purdue's best teachers past and present. He is a co-founder of Arxan Technologies Inc. See http://www.cs.purdue.edu/people/mja for further information.

D.A. Bader et al. (Eds.): HiPC 2005, LNCS 3769, p. 1, 2005.
© Springer-Verlag Berlin Heidelberg 2005

Productivity in High Performance Computing

James C. Browne

The University of Texas at Austin,
Department of Computer Sciences,
Austin, Texas 78712, USA
browne@cs.utexas.edu

Abstract. Productivity gains in development of software systems has been modest over the 45-50 years that I have been involved in writing programs. Productivity gains in high performance computing in particular have been even more modest than across the industry in general. This talk will rationalize the relative failure in HPC and sketch an approach which could enable orders of magnitude improvement in productivity for development of HPC software systems for all types of execution environments. The technical challenges and opportunities implicit in the approach will be discussed. Barriers to productivity gains in HPC include: (i) The dramatic increase in complexity of HPC algorithms and applications coupled with a dramatic increase in complexity, diversity and scale of HPC execution environments, (ii) Little CS/CE research on productivity directly addresses HPC-specific problems such as parallelism, macro-locality and distributed state. The conceptual basis for dramatically increasing productivity in HPC already exists and includes: (i) Attention to HPC-specific design, evaluation and execution issues, (ii) Component-based development coupled with automation and tools, (iii) Coherent unification of design time, compile time and runtime languages mechanisms. Implementation of these concepts in HPC raises major technical challenges: (i) development environments must be extended to address HPC-specific issues including diverse and scalable parallelism, macro-locality and distributed state, (ii) Compilation processes must become more semantically complex and (iii) Compilation and runtime systems must become more knowledge-based and be more effectively unified.

Biography. James C. Browne. Browne is Professor of Computer Science and Physics and holds the Regents Chair #2 in Computer Sciences at The University of Texas at Austin. Browne earned his Ph.D. in Chemical Physics at The University of Texas in 1960. He taught in the Physics Department at The University of Texas from 1960 through 1964. He was, from 1965 through 1968, Professor of Computer Science at Queens University in Belfast and directed the Computer Laboratory. Browne joined The University of Texas in 1968 as Professor of Physics AND Computer Science. He served as Department Chair for Computer Science in 1968–69, 1971–75 and 1984–87. Browne's current research interests span parallel programming, distributed and grid computing methods, performance measurement and analysis, software engineering and formal methods including model checking of software systems. He has

D.A. Bader et al. (Eds.): HiPC 2005, LNCS 3769, pp. 2–3, 2005.

recently begun a research program on integration of methods for verification and validation of software systems He is a Fellow of the Association for Computing Machinery, of the British Computer Society, the American Physical Society and of the American Association for the Advancement of Science. He was Chairman of the ACM Special Interest Group on Operating Systems 1973–75. Browne has published approximately 100 papers in computational physics and 250 papers in Computer Science.

A New Approach to Programming and Prototyping Parallel Systems

Kunle Olukotun

Stanford University,
Department of Electrical Engineering,
Stanford, CA 94305-9030 USA
kunle@ogun.stanford.edu

Abstract. Writing parallel programs to run on them. Transactional Coherence and Consistency (TCC) is a new way to implement cache coherency in shared-memory parallel systems by using programmer defined transactions as the fundamental unit of parallel work, communication, coherence, consistency and failure atomicity. TCC has the potential to simplify parallel programming by providing a smooth transition from sequential programs to parallel programs. In this talk, I will describe TCC and explain how to develop parallel applications using the TCC programming model. I will also briefly describe the architecture of the Stanford Flexible Architecture Research Machine (FARM). FARM is a scalable system based on commercial high-density blade server technology. FARM is designed to improve the capability of architects and applications developers to experiment with large-scale parallel systems.

Biography. Kunle Olukotun is a Professor of Electrical Engineering and Computer Science. Olukotun received his Ph.D. from The University of Michigan. Olukotun led the Stanford Hydra single-chip multiprocessor research project which was the first microprocessor with multiple processors on a single silicon chip. Olukotun founded Afara Websystems to develop commercial server systems with chip multiprocessor technology. Afara was acquired by Sun Microsystems; the Afara microprocessor technology, called Niagara, is the core of Sun's "Throughput Computing" initiative. Olukotun is actively involved in research in computer architecture, parallel programming environments and scalable parallel systems.

D.A. Bader et al. (Eds.): HiPC 2005, LNCS 3769, p. 4, 2005.

The Changing Challenges
of Collaborative Algorithmics

Arnold L. Rosenberg

University of Massachusetts Amherst,
Department of Computer Science,
Amherst, MA 01003, USA
rsnbrg@cs.umass.edu

Abstract. It seems to be a truism that one can gain computational efficiency by enlisting more computers in the solution of a single computational problem. (We refer to such use of multiple computers as "collaborative computing.") In order to realize the promise of collaborative computing, however, one must know how to exploit the strengths of the technology used to build the computing platform - the multiple computers and the networks that interconnect them - and how to avoid the weaknesses of the technology. Changes in technology - even apparently modest ones - often call for dramatic changes in algorithmic strategy. In this talk, I describe some of the challenges that the algorithm designer has faced as the dominant collaborative computing platforms have changed.

Biography. Arnold L. Rosenberg is a Distinguished University Professor of Computer Science at the University of Massachusetts Amherst, where he co-directs the Theoretical Aspects of Parallel and Distributed Systems (TAPADS) Laboratory. Prior to joining UMass, he was a Professor of Computer Science at Duke University from 1981 to 1986, and a Research Staff Member at the IBM Watson Research Center from 1965 to 1981. He has held visiting positions at Yale University and the University of Toronto; he was a Lady Davis Visiting Professor at the Technion (Israel Institute of Technology) in 1994, and a Fulbright Senior Research Scholar at the University of Paris-South in 2000. Dr. Rosenberg's research focuses on developing algorithmic models and techniques to deal with the new modalities of "collaborative computing" that result from emerging technologies. He is the author of more than 150 technical papers on these and other topics in theoretical computer science and discrete mathematics and is the coauthor of the book "Graph Separators, with Applications." Dr. Rosenberg is a Fellow of the ACM, a Fellow of the IEEE, and a Golden Core member of the IEEE Computer Society. See http://www.cs.umass.edu/~rsnbrg/ for further information.

D.A. Bader et al. (Eds.): HiPC 2005, LNCS 3769, p. 5, 2005.
© Springer-Verlag Berlin Heidelberg 2005

Quantum Physics and the Nature of Computation

Umesh Vazirani

U.C. Berkeley,
Computer Science Division,
Berkeley, CA 94720, USA
vazirani@eecs.berkeley.edu

Abstract. Quantum physics is a fascinating area from a computational viewpoint. The features that make quantum systems prohibitively hard to simulate classically are precisely the aspects exploited by quantum computation to obtain exponential speedups over classical computers. In this talk I will survey our current understanding of the power of quantum computers and prospects for experimentally realizing them in the near future. I will also touch upon insights from quantum comuptation that have resulted in new classical algorithms for efficient simulation of certain important quantum systems.

Biography. Umesh Vazirani received his B.Tech in computer science from M.I.T. in 1981 and his PhD in computer science from U.C. Berkeley in 1985. He is currently professor of computer science at U.C. Berkeley and director of BQIC - the Berkeley Center for Quantum Information and Computation. Prof. Vazirani is a theoretician with broad interests in novel models of computation. He has done seminal work in quantum computation and on the computational foundations of randomness. He is the author of a leading textbook in computational learning theory (*An Introduction to Computational Learning Theory*, MIT Press, 1995, with Michael Kearns). See www.cs.berkeley.edu/~vazirani for further information.

D.A. Bader et al. (Eds.): HiPC 2005, LNCS 3769, p. 6, 2005.

Preemption Adaptivity in Time-Published Queue-Based Spin Locks*

Bijun He, William N. Scherer III, and Michael L. Scott

Department of Computer Science,
University of Rochester,
Rochester, NY 14627-0226, USA
{bijun, scherer, scott}@cs.rochester.edu

Abstract. The proliferation of multiprocessor servers and multithreaded applications has increased the demand for high-performance synchronization. Traditional scheduler-based locks incur the overhead of a full context switch between threads and are thus unacceptably slow for many applications. Spin locks offer low overhead, but they either scale poorly (test-and-set style locks) or handle preemption badly (queue-based locks). Previous work has shown how to build preemption-tolerant locks using an extended kernel interface, but such locks are neither portable to nor even compatible with most operating systems.

In this work, we propose a *time-publishing* heuristic in which each thread periodically records its current timestamp to a shared memory location. Given the high resolution, roughly synchronized clocks of modern processors, this convention allows threads to guess accurately which peers are active based on the currency of their timestamps. We implement two queue-based locks, MCS-TP and CLH-TP, and evaluate their performance relative to traditional spin locks on a 32-processor IBM p690 multiprocessor. These results show that time-published locks make it feasible, for the first time, to use queue-based spin locks on multi-programmed systems with a standard kernel interface.

1 Introduction

Historically, spin locks have found most of their use in operating systems and dedicated servers, where the entire machine is dedicated to whatever task the locks are protecting. This is fortunate, because spin locks typically don't handle preemption very well: if the thread that holds a lock is suspended before releasing it, any processor time given to waiting threads will be wasted on fruitless spinning.

Recent years, however, have seen a marked trend toward multithreaded user-level programs, such as databases and on-line servers. Further, large multiprocessors are increasingly shared among multiple multithreaded programs. As a result, modern applications cannot in general count on any specific number of processors; spawning one thread per processor does not suffice to avoid preemption.

* This work was supported in part by NSF grants numbers CCR-9988361, EIA-0080124, CCR-0204344, and CNS-0411127, by DARPA/ITO under AFRL contract F29601-00-K-0182, by financial and equipment grants from Sun Microsystems Laboratories, and by an IBM Shared University Research grant.

D.A. Bader et al. (Eds.): HiPC 2005, LNCS 3769, pp. 7–18, 2005.

For multithreaded servers, the high cost of context switches makes scheduler-based locking unattractive, so implementors are increasingly turning to spin locks to gain performance. Unfortunately, this solution comes with hidden drawbacks: queue-based locks are highly vulnerable to preemption, but test-and-set locks do not scale beyond a modest number of processors. Although several heuristic strategies can reduce wasted spinning time [12, 15], multiprogrammed systems usually rely on non-queue-based locks [19]. Our goal is to combine the efficiency and scalability of queue-based spin locks with the preemption tolerance of the scheduler-based approach.

1.1   Related Work

One approach to avoiding excessive wait times can be found in *abortable locks* (sometimes called *try locks*), in which a thread "times out" if it fails to acquire the lock within a specified *patience* interval [11, 26, 27]. Although timeout prevents a thread from being blocked behind a preempted peer, it does nothing to improve system-wide throughput if the lock is squarely in the application's critical path. Further, any timeout sequence that requires cooperation with neighboring threads in a queue opens yet another window of preemption vulnerability. Known approaches to avoiding this window result in unbounded worst-case space overhead [26] or very high base time overhead [11].

An alternative approach is to adopt *nonblocking* synchronization, eliminating the use of locks [8]. Unfortunately, while excellent nonblocking implementations exist for many important data structures (only a few of which we have room to cite here [20, 22, 23, 28, 29]), general-purpose mechanisms remain elusive. Several groups (including our own) are working on this topic [6, 10, 17, 25], but it still seems unlikely that nonblocking synchronization will displace locks entirely soon.

Finally, several researchers have suggested operating system mechanisms that provide user applications with a limited degree of control over scheduling, allowing them to avoid [4, 5, 13, 18, 24] or recover from [1, 2, 30, 32] inopportune preemption. Commercial support for such mechanisms, however, is neither universal nor consistent.

Assuming, then, that locks will remain important, and that many systems will not provide an OS-level solution, how can we hope to leverage the fairness and scalability of queue-based spin locks in multithreaded user-level programs?

In this work, we answer this question with two new abortable queue-based spin locks that combine fair and scalable performance with good preemption tolerance: the MCS time-published lock (MCS-TP) and the CLH time-published (CLH-TP) lock. In this context, we use the term *time-published* to mean that contending threads periodically write their wall clock timestamp to shared memory in order to be able to estimate each other's run-time states. In particular, given a low-overhead hardware timer with bounded skew across processors and a memory bus that handles requests in bounded time[1] we can guess with high accuracy that another thread is preempted if the current system time exceeds the thread's latest timestamp by some appropriate threshold. We can then selectively pass a lock only between active threads. Although this doesn't solve

---

[1] Our requirements are modest: While it must be possible to read the clock within, say, $100ns$, clock skew or remote access times of tens of microseconds would be tolerable. Most modern machines do much better than that.

the preemption problem completely (threads can be preempted while holding the lock, and our heuristic suffers from a race condition in which we read a value that has just been written by a thread immediately before it was preempted), experimental results (Sections 4 and 5) confirm that our approach suffices to make the locks *preemption adaptive*: free, in practice, from virtually all preemption-induced performance loss.

2   Algorithms

We begin this section by presenting common features of our two time-published (TP) locks; Sections 2.1 and 2.2 cover algorithm-specific details.

Our TP locks are abortable variants of the well-known MCS [19] and CLH [3, 16] queue-based spin locks. Their `acquire` functions return `success` if the thread acquired the lock within a supplied *patience* interval parameter, and `failure` otherwise. In both locks, the thread owning the head node of a linked-list queue holds the lock.

With abortable queue-based locks, there are three ways in which preemption can interfere with throughput. First, as with any lock, a thread that is preempted in its critical section will block all competitors. Second, a thread preempted while waiting in the queue will block others once it reaches the head; strict FIFO ordering is a disadvantage in the face of preemption. Third, any timeout protocol that requires explicit handshaking among neighboring threads will block a timed-out thread if its neighbors are not active.

The third case can be avoided with *nonblocking* timeout protocols, which guarantee a waiting thread can abandon an acquisition attempt in a bounded number of its own time steps [26]. To address the remaining cases, we use a timestamp-based heuristic. Each waiting thread periodically writes the current system time to a shared location. If a thread A finds a stale timestamp for another thread B, A assumes that B has been preempted and removes B's node from the queue. Further, any time A fails to acquire the lock, it checks the critical section entry time recorded by the current lock holder. If this time is sufficiently far in the past, A yields the processor in the hope that a suspended lock holder might resume.

There is a wide design space for time-published locks, which we have only begun to explore. Our initial algorithms, described in the two subsections below, are designed to be fast in the common case, where timeout is uncommon. They reflect our attempt to adopt straightforward strategies consistent with the head-to-tail and tail-to-head linking of the MCS and CLH locks, respectively. These strategies are summarized in Table 1. Time and space bounds are considered in the technical report version of this paper [7].

2.1   MCS Time-Published Lock

Our first algorithm is adapted from Mellor-Crummey and Scott's MCS lock [19]. In the original MCS algorithm, a contending thread A atomically swaps a pointer to its queue node α into the queue's tail. If the swap returns `nil`, A has acquired the lock; otherwise the return value is A's predecessor B. A then updates B's `next` pointer to α and spins until B explicitly changes the A's state from `waiting` to `available`. To release the lock, B reads its `next` pointer to find a successor node. If it has no successor, B atomically updates the queue's tail pointer to `nil`.

Table 1. Comparison between MCS and CLH time-published locks

Lock	MCS-TP	CLH-TP
Link Structure	Queue linked head to tail	Queue linked tail to head
Lock handoff	Lock holder explicitly grants the lock to a waiter	Lock holder marks lock available and leaves; next-in-queue claims lock
Timeout precision	Strict adherence to patience	Bounded delay from removing timed-out and preempted predecessors
Queue management	Only the lock holder removes timed-out or preempted nodes (at handoff)	Concurrent removal by all waiting threads
Space management	Semi-dynamic allocation: waiters may reinhabit abandoned nodes until removed from the queue	Dynamic allocation: separate node per acquisition attempt

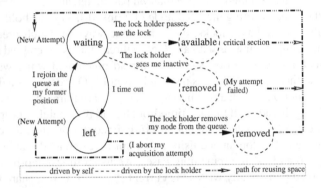

Fig. 1. State transitions for MCS-TP queue nodes

The MCS-TP lock uses the same head-to-tail linking as MCS, but adds two additional states: `left` and `removed`. When a waiting thread times out before acquiring the lock, it marks its node `left` and returns, leaving the node in the queue. When a node reaches the head of the queue but is either marked `left` or appears to be owned by a preempted thread (i.e., has a stale timestamp), the lock holder marks it `removed`, and follows its `next` pointer to find a new candidate lock recipient, repeating as necessary. Figure 1 shows the state transitions for MCS-TP queue nodes. Source code can be found in the technical report version of this paper [7]. It runs to about 3 pages.

The MCS-TP algorithm allows each thread at most one node per lock. If a thread that calls `acquire` finds its node marked `left`, it reverts the state to `waiting`, resuming its former place in line. Otherwise, it begins a fresh attempt from the tail of the queue. To all other threads, timeout and retry are indistinguishable from an execution in which the thread was waiting all along.

2.2   CLH Time-Published Lock

Our second time-published lock is based on the CLH lock of Craig [3] and Landin and Hagersten [16]. In CLH, a contending thread A atomically swaps a pointer to its queue node α into the queue's tail. This swap always returns a pointer to the node β inserted by A's predecessor B (or, the very first time, to a dummy node, marked `available`, created at initialization time). A updates α's `prev` pointer to β and spins until β's state is `available`. Note that, in contrast to MCS, links point from the tail of the queue toward the head, and a thread spins on the node inserted by its predecessor. To release the lock, a thread marks the node it inserted `available`; it then takes the node inserted by its predecessor for use in its next acquisition attempt. Because a thread cannot choose the location on which it is going to spin, the CLH lock requires cache-coherent hardware in order to bound contention-inducing remote memory operations.

CLH-TP retains the link structure of the CLH lock, but adds both non-blocking time-out and removal of nodes inserted by preempted threads. Unlike MCS-TP, CLH-TP allows any thread to remove the node inserted by a preempted predecessor; removal is not reserved to the lock holder. Middle-of-the-queue removal adds significant complexity to CLH-TP; experience with earlier abortable locks [26, 27] suggests that it would be very difficult to add to MCS-TP. Source code for the CLH-TP lock can be found in the technical report version of this paper [7]. It runs to about 5 pages.

We use low-order bits in a CLH-TP node's `prev` pointer to store the node's state, allowing us to atomically modify the state and the pointer together. If `prev` is a valid pointer, its two lowest-order bits specify one of three states: `waiting`, `transient`, and `left`. Alternatively, `prev` can be a `nil` pointer with low-order bits set to indicate three more states: `available`, `holding`, and `removed`. Figure 2 shows the state transition diagram for CLH-TP queue nodes.

In each lock acquisition attempt, thread B dynamically allocates a new node β and links it to predecessor α as before. While waiting, B handles three events. The simplest occurs when α's state changes to `available`; B atomically updates β's state to `holding` to claim the lock.

The second event occurs when B believes A to be preempted or timed out. Here, B performs a three-step *removal sequence* to unlink A's node from the queue. First, B atomically changes α's state from `waiting` to `transient`, to prevent A from acquiring the lock or from reclaiming and reusing α if it is removed from the queue by some successor of B (more on this below). Second, B removes α from the queue,

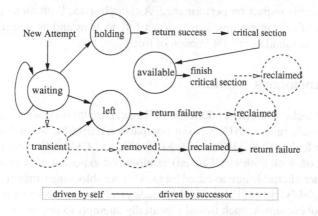

Fig. 2. State transitions for CLH-TP queue nodes

simultaneously verifying that B's own state is still waiting (since β's prev pointer and state share a word, this is a single *compare-and-swap*). Hereafter, α is no longer visible to other threads in the queue, and B spins on A's predecessor's node. Finally, B marks α as safe for reclamation by changing its state from transient to removed.

The third event occurs when B times out or when it notices that β is transient. In either case, it attempts to change β's state atomically from transient or waiting to left. If the attempt succeeds, B has delegated responsibility for reclamation of β to a successor. Otherwise, B has been removed from the queue and must reclaim its own node. Either way, whichever of B and its successor notices *last* that β has been removed from the queue handles the memory reclamation.

A corner case occurs when, after B marks α transient, β is marked transient by some successor thread C before B removes α from the queue. In this case, B leaves α for C to clean up; C recognizes this case by finding α already transient.

The need for the transient state derives from a race condition in which B decides to remove α from the queue but is preempted before actually doing so. While B is not running, successor C may remove both β and α from the queue, and A may reuse its node in this or another queue. When B resumes running, we must ensure that it does not modify (the new instance of) A. The transient state allows us to so, if we can update α's state and verify that β is still waiting as a single atomic operation. A custom atomic construction (ommitted here but shown in the TR version [7]) implements this operation using *load-linked/store-conditional*. Alternative solutions might rely on a tracing garbage collector (which would decline to recycle α as long as B has a reference) or on manual reference-tracking methodologies [9, 21].

3   Scheduling and Preemption

TP locks publish timestamps to enable a heuristic that guesses whether the lock holder or a waiting thread is preempted. This heuristic admits a race condition wherein a thread's timestamp is polled just before it is descheduled. In this case, the poller will mistakenly assume the thread to be active. In practice, the timing window is too narrow to have a noticeable impact on performance. Although space limitations preclude further discussion of scheduler-conscious locks, a full analysis and an empirical study of the matter may be found in the TR version of this paper [7].

4   Microbenchmark Results

We test our TP locks on an IBM pSeries 690 (Regatta) multiprocessor with 32 1.3 GHz Power4 processors, running AIX 5.2. For comparison purposes, we include a range of user-level spin locks: TAS, MCS, CLH, MCS-NB, and CLH-NB. TAS is a test-and-test-and-set lock with (well tuned) randomized exponential backoff. MCS-NB and CLH-NB are abortable queue-based locks with non-blocking timeout [26]. We also test *spin-then-yield* variants [12] in which threads yield after exceeding a wait threshold.

In our microbenchmark, each thread repeatedly attempts to acquire a lock. We simulate critical sections (CS) by updating a variable number of cache lines; we simulate non-critical sections (NCS) by varying the time spent spinning in an idle loop between acquisitions. We measure the total throughput of lock acquisitions and we count successful and unsuccessful acquisition attempts, across all threads for one second, averaging results of 6 runs. For abortable locks, we retry unsuccessful acquisitions immediately, without executing a non-critical section. We use a fixed patience of 50 μs.

4.1 Single Thread Performance

Because low overhead is crucial for locks in real systems, we assess it by measuring throughput absent contention with one thread and empty critical and non-critical sections. We present results in Figure 3.

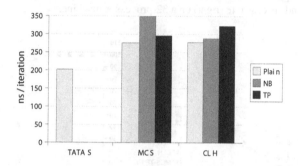

Fig. 3. Single-thread performance results

As expected, the TAS variants are the most efficient for one thread, absent contention. MCS-NB has one *compare-and-swap* more than the base MCS lock; this appears in its single-thread overhead. Similarly, other differences between locks trace back to the operations in their `acquire` and `release` methods. We note that time-publishing functionality adds little overhead to locks.

A single-thread atomic update on our p690 takes about 60 ns. Adding additional threads introduces contention from memory and processor interconnect bus traffic and adds cache coherence overhead when transferring a cache line between processors. We have measured atomic update overheads of 120 and 420 ns with 2 and 32 threads.

4.2 Multi-thread Performance

Under high contention, serialization of critical sections causes application performance to depend primarily on the overhead of handing a lock from one thread to the next; other overheads are typically subsumed by waiting. We examine two typical configurations.

Our first configuration simulates contention for a small critical section with a 2-cache-line-update critical section and a 1 μs non-critical section. Figure 4 plots lock performance for this configuration. Up through 32 threads (our machine size), queue-based locks outperform TAS; however, only the TP and TAS locks maintain throughput

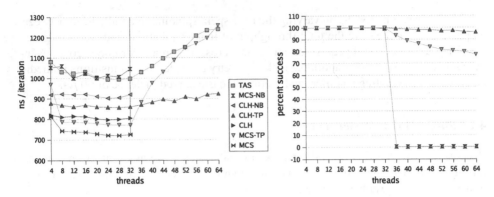

Fig. 4. 2 cache line-update critical section (CS). 1 μs non-critical section (NCS). Critical section service time (left) and success rate (right) on a 32-processor machine.

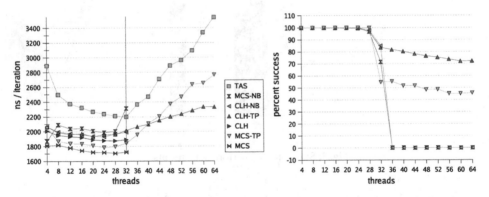

Fig. 5. 40 cache line CS; 4 μs NCS. CS service time (left) and success rate (right).

in the presence of preemption. MCS-TP's overhead increases with the number of preempted threads because it relies on the lock holder to remove nodes. By contrast, CLH-TP distributes cleanup work across active threads and keeps throughput more steady. The right-hand graph in Figure 4 shows the percentage of successful lock acquisition attempts for the abortable locks. MCS-TP's increasing handoff time forces its success rate below that of CLH-TP as the thread count increases. CLH-NB and MCS-NB drop to nearly zero due to preemption while waiting in the queue.

Our second configuration uses 40-cache-line-update critical sections 4 μs non-critical sections. It models larger longer operations in which preemption of the lock holder is more likely. Figure 5 shows lock performance for this configuration. That the TP locks outperform TAS demonstrates the utility of cooperative yielding for preemption recovery. Moreover, the CLH-TP–MCS-TP performance gap is smaller here than in our first configuration since the relative importance of removing inactive queue nodes goes down compared to that of recovering from preemption in the critical section.

In Figure 5, the success rates for abortable locks drop off beyond 24 threads. Since each critical section takes about 2 μs, our 50 μs patience is just enough for a thread to sit through 25 predecessors. We note that TP locks adapt better to insufficient patience.

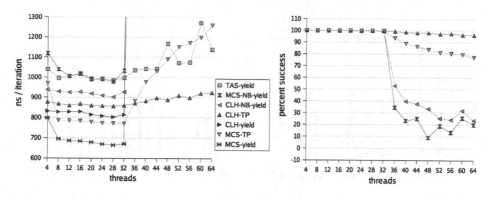

Fig. 6. Spin-then-yield variants; 2 cache line CS; $1\,\mu s$ NCS

One might expect a spin-then-yield policy [12] to allow other locks to match TP locks in preemption adaptivity. In Figure 6 we test this hypothesis with a $50\,\mu s$ spinning time threshold and a 2 cache line critical section. (Other settings produce similar results.) Although yielding improves the throughput of non-TP queue-based locks, they still run off the top of the graph. TAS benefits enough to become competitive with MCS-TP, but CLH-TP still outperforms it. These results confirm that targeted removal of inactive queue nodes is much more valuable than simple yielding of the processor.

4.3 Time and Space Bounds

Finally, we measure the time overhead for removing an inactive node. On our Power4 p690, we calculate that the MCS-TP lock holder needs about $200\text{–}350\,ns$ to delete each node. Similarly, a waiting thread in CLH-TP needs about $250\text{–}350\,ns$ to delete a predecessor node. By combining these values with our worst-case analysis for the number of inactive nodes in the lock queues [7], one can estimate an upper bound on delay for lock handoff when the holder is not preempted.

In our analysis of the CLH-TP lock's space bounds [7] we show a worst-case bound quadratic in the number of threads, but claim an expected linear value. In tests designed to maximize space contention (full details available in the TR version [7]), we find space consumption to be very stable over time. Even with very short patience, we obtain results far closer to the expected linear than the worst-case quadratic space bound.

5 Application Results

In this section we measure the performance of our TP locks on the Raytrace and Barnes benchmarks from the SPLASH-2 suite [31].

Application Features: Raytrace and Barnes are heavily synchronized [14, 31]. Raytrace uses no barriers but features high contention on a small number of locks. Barnes uses few barriers but numerous locks. Both offer reasonable parallel speedup.

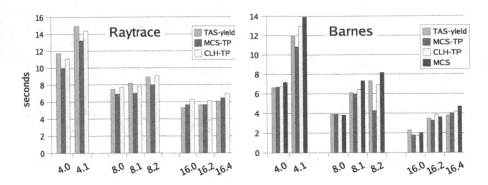

Fig. 7. Parallel execution times for Raytrace and Barnes on a 32-processor IBM p690 Regatta. Test $M.N$ uses M application threads and $(32 - M) + N$ external threads.

Experimental Setup: We test the locks from Section 4 and the native `pthread_mutex` on our p690, averaging results over 6 runs. We choose inputs large enough to execute for several seconds: 800×800 for Raytrace and 60K particles for Barnes. We limit testing to 16 threads due to the applications' limited scalability. External threads running idle loops generate load and force preemption.

Raytrace: The left side of Figure 7 shows results for three preemption adaptive locks: TAS-yield, MCS-TP and CLH-TP. Other spin locks give similar performance absent preemption; when preemption is present, non-TP queue-based locks yield horrible performance (Figures 4, 5, and 6). The `pthread_mutex` lock also scales very badly; with high lock contention, it can spend 80% of its time in kernel mode. Running Raytrace with our input size took several hours for 4 threads.

Barnes: Preemption adaptivity is less important here than in Raytrace because Barnes distributes synchronization over a very large number of locks, greatly reducing the impact of preemption. This can be confirmed by noting that a highly preemption-sensitive lock, MCS, "only" doubles its execution time given heavy preemption. We therefore attribute Barnes' relatively severe preemption-induced performance degradation to the barriers it uses.

With both benchmarks, we find that our TP locks maintain good throughput and adapt well to preemption. With Raytrace, MCS-TP in particular yields 8-18% improvement over a yielding TAS lock with 4 or 8 threads. Barnes is less dependent on lock performance in that different locks have similar performance. Overall, MCS-TP outperforms CLH-TP; this is consistent with our microbenchmark results. We speculate that this disparity is due to lower base-case overhead in the MCS-TP algorithm combined with short-lived lock acquisitions in these applications.

6 Conclusions and Future Work

In this work we have demonstrated that published timestamps provide an effective heuristic by which a thread can accurately guess the running state of its peers, without support from a nonstandard scheduler API. We have used this published-time heuristic to implement preemption adaptive versions of standard MCS and CLH queue-based locks. Empirical tests confirm that these locks combine scalability, strong tolerance for preemption, and low observed space overhead with throughput as high as that of the best previously known solutions. Given the existence of a low-overhead time-of-day register with low system-wide skew, our results make it feasible, for the first time, to use queue-based locks on multiprogrammed systems with a standard kernel interface.

For cache-coherent machines, we recommend CLH-TP when preemption is frequent and strong worst-case performance is needed. MCS-TP gives better performance in the common case. With unbounded clock skew, slow system clock access, or a small number of processors, we recommend a TAS-style lock with exponential backoff combined with a spin-then-yield mechanism. Finally, for non-cache-coherent (e.g. Cray) machines, we recommend MCS-TP if clock registers support it; otherwise the best choice is the abortable MCS-NB try lock.

As future work, we conjecture that time can be used to improve thread interaction in other areas, such as preemption-tolerant barriers, priority-based lock queueing, dynamic adjustment of the worker pool for bag-of-task applications, and contention management for nonblocking concurrent algorithms. Further, we note that we have examined only two points in the design space of TP locks; other variations may merit consideration.

References

[1] T. E. Anderson, B. N. Bershad, E. D. Lazowska, and H. M. Levy. Scheduler Activations: Effective Kernel Support for the User-Level Management of Parallelism. *ACM Trans. on Computer Systems*, 10(1):53–79, Feb. 1992.

[2] D. L. Black. Scheduling support for concurrency and parallelism in the Mach operating system. *IEEE Computer*, 23(5):35–43, 1990.

[3] T. S. Craig. Building FIFO and priority-queueing spin locks from atomic swap. Technical Report TR 93-02-02, Department of Computer Science, Univ. of Washington, Feb. 1993.

[4] J. Edler, J. Lipkis, and E. Schonberg. Process management for highly parallel UNIX systems. *USENIX Workshop on Unix and Supercomputers*, Sep. 1988.

[5] H. Franke, R. Russell, and M. Kirkwood. Fuss, futexes and furwocks: Fast userlevel locking in Linux. *Ottawa Linux Symp.*, June 2002.

[6] T. Harris and K. Fraser. Language support for lightweight transactions. *18th Conf. on Object-Oriented Programming, Systems, Languages, and Applications*, Oct. 2003.

[7] B. He, W. N. Scherer III, and M. L. Scott. Preemption adaptivity in time-published queue-based spin locks. Technical Report URCS-867, University of Rochester, May 2005.

[8] M. Herlihy. Wait-free synchronization. *ACM Trans. on Programming Languages and Systems*, 13(1):124–149, Jan. 1991.

[9] M. Herlihy, V. Luchangco, and M. Moir. The repeat offender problem: A mechanism for supporting dynamic-sized, lock-free data structures. *16th Conf. on Distributed Computing*, Oct. 2002.

[10] M. Herlihy, V. Luchangco, M. Moir, and W. N. Scherer III. Software transactional memory for dynamic-sized data structures. *22nd ACM Symp. on Principles of Distributed Computing*, July 2003.

[11] P. Jayanti. Adaptive and efficient abortable mutual exclusion. *22nd ACM Symp. on Principles of Distributed Computing*, July 2003.

[12] A. R. Karlin, K. Li, M. S. Manasse, and S. Owicki. Empirical studies of competitive spinning for a shared-memory multiprocessor. *13th SIGOPS Symp. on Operating Systems Principles*, Oct. 1991.

[13] L. I. Kontothanassis, R. W. Wisniewski, and M. L. Scott. Scheduler-conscious synchronization. *ACM Trans. on Computer Systems*, 15(1):3–40, Feb. 1997.

[14] S. Kumar, D. Jiang, R. Chandra, and J. P. Singh. Evaluating synchronization on shared address space multiprocessors: Methodology and performance. *SIGMETRICS Conf. on Measurement and Modeling of Computer Systems*, May 1999.

[15] B.-H. Lim and A. Agarwal. Reactive synchronization algorithms for multiprocessors. *6th Intl. Conf. on Architectural Support for Programming Languages and Operating Systems*, Oct. 1994.

[16] P. S. Magnusson, A. Landin, and E. Hagersten. Queue locks on cache coherent multiprocessors. *8th Intl. Parallel Processing Symp.*, Apr. 1994.

[17] V. J. Marathe, W. N. Scherer III, and M. L. Scott. Design Tradeoffs in Modern Software Transactional Memory Systems. *7th Workshop on Languages, Compilers, and Run-time Support for Scalable Systems*, Oct. 2004.

[18] B. D. Marsh, M. L. Scott, T. J. LeBlanc, and E. P. Markatos. First-class user-level threads. *13th SIGPOPS Symp. on Operating Systems Principles*, Oct. 1991.

[19] J. M. Mellor-Crummey and M. L. Scott. Algorithms for scalable synchronization on shared-memory multiprocessors. *ACM Trans. on Computer Systems*, 9(1):21–65, 1991.

[20] M. M. Michael. High performance dynamic lock-free hash tables and list-based sets. *14th Symp. on Parallel Algorithms and Architectures*, Aug. 2002.

[21] M. M. Michael. Hazard pointers: Safe memory reclamation for lock-free objects. *IEEE Trans. on Parallel and Distributed Systems*, 15(8)491–504, 2004.

[22] M. M. Michael. Scalable lock-free dynamic memory allocation. *SIGPLAN Symp. on Programming Language Design and Implementation*, June 2004.

[23] M. M. Michael and M. L. Scott. Nonblocking algorithms and preemption-safe locking on multiprogrammed shared memory multiprocessors. *Journal of Parallel and Distributed Computing*, 51:1–26, 1998.

[24] J. K. Ousterhout. Scheduling techniques for concurrent systems. *3rd Intl. Conf. on Distributed Computing Systems*, Oct. 1982.

[25] W. N. Scherer III and M. L. Scott. Advanced Contention Management for Dynamic Software Transactional Memory. *24th ACM Symp. on Principles of Distributed Computing*, July 2005.

[26] M. L. Scott. Non-blocking timeout in scalable queue-based spin locks. *21st ACM Symp. on Principles of Distributed Computing*, July 2002.

[27] M. L. Scott and W. N. Scherer III. Scalable queue-based spin locks with timeout. *8th ACM Symp. on Principles and Practice of Parallel Programming*, June 2001.

[28] H. Sundell and P. Tsigas. NOBLE: A non-blocking inter-process communication library. *6th Workshop on Languages, Compilers, and Run-time Support for Scalable Systems*, Mar. 2002.

[29] H. Sundell and P. Tsigas. Fast and lock-free concurrent priority queues for multi-thread systems. *Intl. Parallel and Distributed Processing Symp.*, Apr. 2003.

[30] H. Takada and K. Sakamura. A novel approach to multiprogrammed multiprocessor synchronization for real-time kernels. *18th Real-Time Systems Symp.*, Dec. 1997.

[31] S. C. Woo, M. Ohara, E. Torrie, J. P. Singh, and A. Gupta. The Splash-2 programs: Characterization and methodological considerations. *22nd Intl. Symp. on Computer Architecture*, Jun. 1995.

[32] H. Zheng and J. Nieh. SWAP: A scheduler with automatic process dependency detection. *Networked Systems Design and Implementation*, Mar. 2004.

Criticality Driven Energy Aware Speculation for Speculative Multithreaded Processors

Rahul Nagpal[1] and Anasua Bhowmik[2]

[1] Department of Computer Science and Automation,
Indian Institute of Science, Bangalore, India
rahul@csa.iisc.ernet.in
[2] Microprocessor Solutions Sector,
AMD-India Engineering Center, Bangalore, India
anasua.bhowmik@amd.com

Abstract. Speculative multithreaded architecture (SpMT) philosophy relies on aggressive speculative execution for improved performance. Aggressive speculative execution results in a significant wastage of dynamic energy due to useless computation in the event of mis-speculation. As energy consumption is becoming an important constraint in microprocessor design, it is extremely important to reduce such wastage of dynamic energy in SpMT processors in order to achieve a better performance to power ratio. Dynamic instruction criticality information can be effectively applied to control aggressive speculation in SpMT processors. In this paper, we present a model of micro-execution for SpMT processors to determine dynamic instruction criticality. We also present two novel techniques utilizing criticality information, namely delaying non-critical loads and criticality based thread-prediction for reducing useless computation and energy consumption. Our experiments show 17.71% and 11.63% reduction in dynamic energy for criticality based thread prediction and criticality based delayed load scheme respectively while the corresponding improvements in dynamic energy delay products are 13.93% and 5.54%.

1 Introduction

Speculative multithreaded (SpMT) execution paradigm [2] executes in parallel multiple threads obtained from a sequential program. In order to extract parallelism from sequential programs, SpMT processors use aggressive control and data speculation. The hardware speculates that dependencies do not exist among the threads running in parallel and then recovers whenever dependencies occur at run-time. Though aggressive speculation achieves higher speedup when the speculations are correct, the mis-speculations lead to squashing and re-execution of the threads and thus wasting a large amount of work and energy. For example, a thread mis-prediction in SpMT would result in the squashing of the mis-predicted thread and all the subsequent threads and thereby wasting dynamic energy of executing several hundreds of instructions. We observe that

D.A. Bader et al. (Eds.): HiPC 2005, LNCS 3769, pp. 19–28, 2005.

in an 8 PE SpMT processor, nearly 40% of the time is spent in useless computation due to mis-speculations and this results in a significant wastage of dynamic energy. As energy consumption is fast becoming an important constraint in microprocessor design, it is extremely important to reduce such a wastage of dynamic energy in order to achieve a better performance to power ratio in SpMT processors.

In this paper, we focus on reducing the dynamic energy consumption while maintaining the speedup achieved by the SpMT processors using the dynamic instruction criticality information. First, we propose an analytical model for determining the dynamic critical path of a program for SpMT execution paradigm. Our model is based on the critical path model proposed by Fields et. al. [1] for finding critical path in the superscalar processors. Notably, modeling and determining critical paths in SpMT processors are quite challenging since there exist many parallel paths of execution through different threads at the same time. We have analyzed the dynamic critical path of the programs and then used that knowledge to devise useful run-time speculation control techniques. We have used the criticality information successfully in guiding both the control speculation and the data speculation in an aggressive SpMT processor. Our experiments with criticality based thread prediction and criticality based delayed load execution have resulted in 17.71% and 11.63% reduction in dynamic energy consumptions respectively, while the corresponding dynamic energy-delay products are improved by 13.93% and 5.54%. In summary our main contributions are as follows:

- Development and implementation of analytical model for finding critical instructions in SpMT execution framework.
- Guiding speculation control based on criticality information to reduce useless computation and energy wastage.
- A detailed performance evaluation of proposed techniques demonstrating the net savings in useless computation and energy consumption without much performance degradation.

Although, a lot of work has been done on determining and exploiting dynamic critical paths in superscalar processor models [1] [3], ours is the first work to determine the critical paths in the SpMT execution paradigm. Tune et al. [3] have proposed various heuristics, based on micro-execution events, to predict the dynamic instruction criticality. Their heuristic based approach have been effective for driving various optimization techniques such as instruction steering in clustered architectures, value prediction, and reducing power consumption. Fields et al. [1] follow a modeling based approach for predicting critical instructions. They proposed a graph model to capture constraints such as finite reorder buffer and branch mis-prediction apart from traditional data dependencies. They have developed a token passing based criticality predictor and used it for selective value prediction and instruction steering for clustered architectures. They have shown that the model based approach is generic and more accurate than a heuristic based approach.

Reducing energy in microprocessors is an active area of research and many power optimization techniques have been proposed. The works that are closest to our work are [4], [7], and [9]. Manne et. al. [4] introduced the idea of pipeline gating and used branch confidence estimator to reduce wrong-path instructions in the pipeline to save energy. In [7] processor power dissipation is reduced by throttling the different pipeline stages selectively based on branch confidence estimation. In order to save energy, critical path prediction is used in [9] to send the non-critical instructions to slower functional units.

The rest of the paper is organized as follows. Section 2 gives a brief overview of SpMT architectures and presents our graph model for determining the dynamic critical path. In section 3, we present and evaluate the speculation control techniques using the critical path information. Section 4 contains the conclusions and future work.

2   Modeling the Critical Path for SpMT Processor

The central idea behind SpMT is to execute multiple threads obtained from a sequential program in parallel. SpMT architectures support both control speculation and data speculation to enable program parallelization despite any compile-time uncertainty about (control or data) dependencies between the threads running in parallel. This execution model is particularly suitable for exploiting parallelism from non-numeric applications which are difficult to parallelize for traditional parallel processors.

A typical SpMT processor consists of a collection of simpler superscalar out-of-order processing elements (PEs) that are connected by an interconnection network. Each PE has its own fetch, decode, and execution units. A thread spawns the next successor thread speculatively based on thread level prediction. A spawned thread becomes active after a PE becomes available. The threads are spawned and activated in the program order and at any point of time only one thread executes non-speculatively. Although instructions are executed out-of-order, they are committed in program order. After the thread finishes execution it waits to become non-speculative before committing. A speculative thread (and all the subsequent threads) can be squashed before committing because of misprediction or dependence violation.

To capture the micro-execution of the SpMT processor model we build a dynamic dependence graph (DDG) at run-time. The critical path is the longest path in the DDG and the nodes lying in that critical path constitute the critical instructions. Each instruction in the DDG is represented by three nodes namely **F**, **E**, and **C** nodes. The **F** node represents the event of an instruction being fetched, decoded, and put in the ROB. The **E** node represents the execution event and the **C** node represents the commit event of an instruction. A directed edge between two nodes depict the dependence between them and the edge weight represent the resultant delay. There are inter-thread as well as intra thread edges showing the interaction between the threads running in parallel. All the edges in the DDG are listed in Table 1 along with a brief description. A detailed description and an example of the DDG model can be found in [10].

Although our critical path model is based on the model proposed by Fields et. al. [1], there are significant differences between the two models. First of all, in their model, three nodes, namely *dispatch, execute,* and *commit* were sufficient to model the micro-execution of the processor. But in our model, a separate *fetch* node is needed to capture the fetch bottlenecks. In SpMT processor, instruction fetch is affected by various inter-thread and intra-thread constraints and the *fetch* nodes are major components in the critical path. Although we do not model the *dispatch* node explicitly like Fields et. al., we keep track of the dispatch constraints separately during the simulation. We have also extended the graph model to capture the inter-thread data dependencies, structural constraints, and thread squashing.

Table 1. List of edges in DDG, $I_{j,i}$ refers to the ith instruction in j^{th} thread and $F_{j,i}, E_{j,i}$, and $C_{j,i}$ are the corresponding nodes in DDG

Edge type	Constraint modeled	Name	Edge description
Intra-thread	In-order Fetch	FF_i	$F_{j,i-1} \to F_{j,i}$; $I_{j,i}$ fetched after $I_{j,i-1}$.
	Finite Size ROB	CF_i	$C_{j,i-R} \to F_{j,i}$; R: size of ROB. $I_{j,i}$ fetched after $I_{j,i-R}$ committed.
	Control Dependence	EF_i	$E_{j,i-1} \to F_{j,i}$; $I_{j,i-1}$: mis-predicted branch. $I_{j,i}$ not fetched till $I_{j,i-1}$ executed.
Inter-thread	In-order thread activation	FF_I	$F_{j-1,1} \to F_{j,1}$; A thread is activated and starts fetching only after its immediate predecessor is activated and started fetching the instructions.
	Finite PE	CF_I	$C_{j-N,l} \to F_{j,i}$; N:number of PEs. Thread T_j is activated and starts fetching only after thread T_{j-N} commits and PE becomes free.
	Control/data mis-speculation	EF_I	$E_{j-1,i} \to F_{j,1}$; $I_{j-1,i}$ caused control/data mis-speculation; Thread T_j activated and starts fetching after $I_{j-1,i}$ executed.
Intra-thread	Execute Follow Fetch	FE_i	$F_{j,i} \to E_{j,i}$;
	Data Dependency	EE_i	$E_{j,m} \to E_{j,n}, \quad m < n$; $I_{j,n}$ dependent on $I_{j,m}$.
Inter-thread	Data Dependency	EE_I	$E_{k,m} \to E_{j,n}, \quad k < j$; $I_{j,n}$ dependent on $I_{k,m}$.
Intra-thread	Commit Follow Execution	EC_i	$E_{j,i} \to C_{j,i}$;
	In-order Commit	CC_i	$C_{j,i-1} \to C_{j,i}$; $I_{j,i}$ committed after $I_{j,i-1}$
Inter-thread	In-order Commit	CC_I	$C_{j-1,l} \to C_{j,1}$; Thread T_j can start committing after the last instruction of T_{j-1} committed.

3 Reducing Dynamic Energy Through Criticality Based Speculation Control

In this section, we present our instruction criticality based schemes for dynamic energy reduction in SpMT processors. We first present an overview of our dynamic critical path detection methodology and analyze behavior of the critical path. Analysis of the critical path is necessary in order to evaluate the potential for speculation control in SpMT processors and to devise effective mechanisms.

We have implemented the dynamic dependence graph model in a cycle accurate simulator of the multiscalar processor [2]. We build the dependence graph during the simulation and find the critical path. The parameters of the simulated processor are given in Table 2. All the results presented here are for 8 PE configuration. We have obtained similar results for 4 and 12 PE configurations as well.

From Figure 1, we see that on the average 18.77% of load instructions, 12.76% of store instructions, and 33% of mispredicted branch instructions lie on the critical path. To gain insight into the cause of dynamic energy wastage, we have measured the time spent in useless computation due to thread squashing. Figure 2 shows the *squash loss time(SLT)* due to the memory dependence violation and branch mis-prediction. The instructions causing memory dependence

Table 2. Hardware parameters used in experimental evaluation

Component	Description
PEs	2-way issue, 32-entry ROB, 2 integer, 1 FP, 1 branch, 1 memory
Prediction	Intra-task (Inter-task) : gshare (path-based) with 16-bit history, 64K-entry table of 2-bit counters
Register Ring	2 values per cycle, bypass same cycle between adjacent PEs
Memory Buffer	32 entries/PE, 32 x No. of PE bytes/entry, fully associative, 2 cycle hit
L1 I-cache	16 * No. of PE KB , 2-way associative, 32 byte blocks, 1cycle hit
L1 D-cache	16 * No. of PE KB, 2-way associative, 32 byte blocks, 2 cycle hit

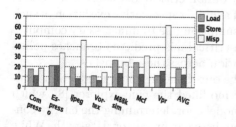

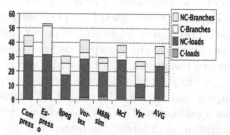

Fig. 1. % of load, store, and mispredicted branches found critical

Fig. 2. % breakup of squash loss time

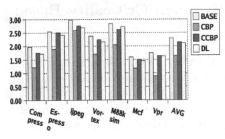

Fig. 3. IPC

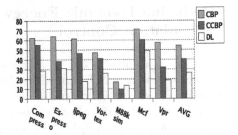

Fig. 4. % reduction in squash loss time

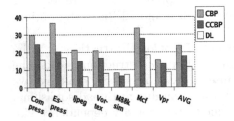

Fig. 5. % reduction in dynamic energy

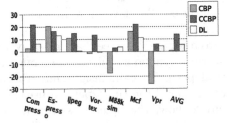

Fig. 6. % improvement in dynamic energy delay product

violation (i.e. load/store instructions) and branch mis-prediction are again classified into critical and non-critical instructions. From Figure 2, we see that nearly 40% of the time is spent in doing useless computation which results in wastage of dynamic energy. This is due to the aggressive speculation techniques employed by the processor to extract thread level and instruction level parallelism. From Figure 2, we see that on the average, critical load-store contributes for only 3.03% of the squashing where as 59.42% of squashing is caused due to non-critical loads. The figure also shows that on the average 24.0% of squashing is caused due to critical branches that are mis-predicted and 13.52% of squashing is attributed to non-critical branches. The high percentage of non-critical load-store instructions in Figure 2 points toward the possibility of avoiding speculative execution of such instructions in order to reduce power consumption due to unnecessary computation. Similarly the high percentage of non-critical mis-predicted branches points toward the possibility of avoiding speculation across these branches.

In the next two subsections we describe our dynamic energy reduction techniques and analyze the results. We have modified the epic-explorer [8], which has a collection of activity based energy models, for determining the energy consumption in different components of the multiscalar processor data-path. While adapting the energy model we have ensured that it faithfully captures all the microarchitectural features of the multiscalar processor implementation. Figure 3 shows the IPC for the base execution model and the IPC with our dynamic energy reduction techniques. In Figure 4, we present the percentage reduction in squash loss time over the base model for the various dynamic energy reduction techniques which gives an indirect measure of reduction in useless activity. Figure 5 and 6 present the reduction in dynamic energy and the improvement in dynamic energy delay product over the base model for all the techniques.

3.1   Reducing Dynamic Energy by Delaying Non-critical Loads

In SpMT processors a load instruction is speculatively executed assuming that the store on which the load is dependent has already taken place. The processor loads the value either from the memory or the intermediate buffer and continues execution. If an earlier store to the same location is executed after the speculative load, the processor detects a memory dependence violation and squashes the violating load and its dependent instructions and this causes dynamic energy wastage.

Although, speculative loads may lead to dynamic energy wastage, they are necessary to speedup the execution as we can see from Figure 1 that nearly 18% of all load instructions lie on the critical path. Therefore speculative loads can not be removed altogether. However, from Figure 2, we see that in all the benchmarks a significant percentage (avg. 59.42%) of squashing is due to non-critical loads. Since the total execution time is not likely to depend on the non-critical loads, we delay the non-critical loads (DL) in order to reduce dynamic energy wastage without affecting the speedup. We determine the average time between the issue of a load and the resultant squashing (average time to squash or ATS) and the non-critical loads are delayed by this time period.

From Figures 4 and 3 we see that the average reduction of squash loss time is 26.43% with IPC reduction of 8.11% for delayed load scheme whereas Figure 5 and Figure 6 depicts the average improvement of 13.93% and 5.54% in dynamic energy and the dynamic energy delay product respectively. The degradation in IPC is happening mainly because at present we delay all the non-critical loads in a benchmark by a fixed ATS. IPC degradation can be reduced, if different loads can be delayed by different amount.

From Figure 4 we see that mcf has the highest reduction in SLT and dynamic energy usage. This is as expected since from Figure 2 we see that it has considerably high squash loss time and a very high percentage of this SLT is due to non-critical loads. Since the IPC degradation of mcf is also very low, it shows a high improvement in dynamic energy delay product. In ijpeg the improvement in energy-delay product is negligible while in vortex there is a little decrease in the energy-delay product. Both vortex and ijpeg has a considerable amount of non-critical loads, however the squash loss time due to the non-critical loads are less compared to the other benchmarks. In particular, ijpeg suffers a large IPC degradation in delayed load scheme compared to the base case leading to only marginal benefits in terms of dynamic energy delay product. Apart from ijpeg and vortex all other benchmarks show considerable improvements in the dynamic energy consumption as well as dynamic energy-delay product.

3.2   Reducing Dynamic Energy Due to Thread Mis-prediction

Thread misprediction is the major source of useless computation in SpMT processors. In SpMT processor a thread mis-prediction leads to the squashing of the mis-speculated thread and all the subsequent threads thus causing huge wastage of dynamic energy. In most cases, the prediction of the next thread is linked to the prediction of the branch on which the existence of the thread depends. From Figure 2 we can see that a significant portion of squash loss is due to branch mis-prediction.

Earlier works have used confidence estimator for speculation control [5, 7]. However, our experiments show that confidence based prediction alone is not sufficient for speculation control in SpMT thread prediction. Unlike superscalar processors, in SpMT processors the performance penalty of not starting a thread in the critical path is also very high. However, in order to do speculation control with the confidence based prediction in the SpMT processor, the confidence estimator needs to be very accurate, i.e. it should give high confidence to the correctly predicted branches while giving low confidence to the wrongly predicted branches. If it gives low confidence for most of the branches, irrespective of whether the prediction is correct or not, we could lose parallelism by not speculating on the correctly predicted branches. On the other hand, if confidence estimator gives high confidence to most of the branches then the processor will perform useless computation by speculating on the wrongly predicted branches as well. Experiments [5] show that it is difficult to achieve desired accuracy from the confidence estimator alone. Therefore we have combined criticality information with the confidence estimation to perform speculation control. In confidence based prediction mechanism, speculation is done only for branches with high confidence value whereas the combined criticality and confidence based prediction mechanism speculates across branches with low confidence as well if it is found on the critical path of the program. We have implemented a 5 bit JRS [6] predictor with resetting counter for confidence estimation. We experiment with both purely confidence based predictor(CBP) and a combined confidence and criticality based predictor(CCBP) and we compare the results with the base line performance (i.e., no confidence estimator).

From Figure 4 we see that a purely confidence based predictor is able to get a maximum reduction in squash loss time (on the average 55%) but suffers from intolerably high performance penalty (29.81% reduction in IPC). This is because the confidence estimator identifies many correct predictions as low confidence and by not speculating on those predictions the processor misses parallelism. This result is in agreement with the earlier studies [5] done in the context of superscalar processors. On the other hand, our combined criticality and confidence based predictor is able to reap most of the benefits of purely confidence based predictor in terms of reducing SLT (on the average, reduction is 40.44% for CCBP compared to 54.44% of CBP) with much less performance penalty (6.52% for CCBP compared to 29.81% of CBP).

From Figure 5 and Figure 6 we see that the average savings in dynamic energy for CBP and CCBP are 23.73% and 17.71% respectively. However, because of large degradation of IPC in case of CBP, the average improvement in dynamic energy delay product is less than 1%. Moreover, as seen in Figure 6, in some programs like `vpr` and `m88ksim` there is a large degradation in dynamic energy delay product attributed to large reduction in performance in these benchmarks for purely confidence based predictor scheme as shown in Figure 3. In case of CCBP the dynamic energy delay product is improved by 13.93% on the average, which is a significant improvement. Only in case of `espresso` we find that the improvement in dynamic energy-delay product is more for CBP than the CCBP. From Figure 2, we see that `espresso` has the highest amount of squash loss time and only 2.69% of squash loss time happens due to non-critical branches which is lowest among all benchmarks. Thus CCBP technique finds fewer opportunities to avoid speculation. Hence in `espresso` dynamic energy reduction is more in case of CBP than in CCBP and also the energy-delay product improvement is higher for CBP. However, for all other benchmarks, the CCBP scheme is able to perform much better than CBP scheme. These results clearly demonstrate the capability of a combined criticality and confidence based prediction scheme in improving the performance to power ratio for SpMT processors.

Our experiments show that the overall energy (i.e. the sum of dynamic energy and leakage energy) reduction is less compared to the reduction in dynamic energy in absence of any leakage energy management scheme. We observe 1.71%, 5.54%, and 2.62% reduction in overall energy for CBP, CCBP, and DL respectively. This is because the speculation control mechanisms we have proposed, increases the idleness in various processor components (apart from reducing the activity) for longer period of time which in turn increases the leakage energy. However, this also creates an opportunity to use leakage energy management techniques more aggressively. With such a scheme in place, we expect to see encouraging gains in terms of overall energy consumption and energy delay product.

4   Conclusions and Future Work

Whereas the earlier work on critical path analysis of program is limited to out-of-order superscalar processors, we have developed a model to identify the dynamic critical instructions in SpMT execution. We proposed two novel techniques that use the criticality information to reduce the useless computation and hence the dynamic energy wastage in SpMT processors. Criticality based load delaying scheme reduces useless computation due to mis-speculation on non-critical loads. It reduces the dynamic energy consumption and dynamic energy-delay product by 11.63% and 5.54% respectively. Criticality based thread predictor makes use of criticality information to speculate across low confidence branches. Our experiments show that confidence based thread prediction alone is not effective to improve the performance to power ratio in SpMT processors. Our combined criticality and confidence based thread predictor improves the dynamic energy by 17.71% and dynamic energy delay product by 13.93%. The significant improvement obtained by the criticality based speculation control, validates both the accuracy of our dynamic critical path model and the effectiveness of our speculation control schemes. Although our experiments are based on the multiscalar processor model, our analytical model and the schemes are generic and can be applied to other SpMT processors as well.

The future extensions to this work involve the development of on-line criticality predictor for determining instruction criticality during program execution and utilizing the criticality information generated offline during a profiling run to guide the speculation control. Development of leakage energy management techniques for SpMT processors also remains as another important future direction.

Acknowledgments

The authors like to thank Prof. M. Franklin for providing the Multiscalar simulator and Prof. R. Bodik and B. Fields for providing their critical path framework for superscalar processors.

References

1. B. Fields, S. Rubin, and R. Bodik. Focusing Processor Policies via Critical-path Prediction. In *Proc. of Intl. Symp. on Computer Architecture*, 2001.
2. M. Franklin. Multiscalar Processors. *Kluwer Academic Publishers*, 2002.
3. E. Tune, D. M. Tullsen, and B. Calder. Quantifying Instruction Criticality. In *Proc. of Intl. Conf. on Parallel Architectures and Compilation Techniques*, 2002.
4. S. Manne, A. Klauser, and D. Grunwald. Pipeline Gating: Speculation Control For Energy Reduction. In *Proc. of Intl. Symp. on Computer Architecture*, 1998.
5. D. Grunwald, A. Klauser, S. Manne, and A. Pleszkun. Confidence Estimation for Speculation Control. In *Proc. of Intl. Symp. on Computer Architecture*, 1998.
6. E. Jacobsen, E. Rotenberg, and J. E. Smith. Assigning Confidence to Conditional Branch Predictions. In *Proc. of Intl. Symp. on Microarchitecture*, 1996.
7. J. L. Aragon, J. Gonzalez, and A. Gonzalez. Power-Aware Control Speculation Through Selective Throttling. In *Proc. of Intl. Symp. on High Performance Computer Architecture*, 2003.
8. G. Ascia, V. Catania, M.Palesi, and D. Patti. A System-level Framework for Evaluating Area/Performance/Power Trade-offs of VLIW-based Embedded Systems. In *Asia and South Pacific Design Automation Conference*, 2005.
9. J. S. Seng, E. S. Tune, and D. M. Tullsen. Reducing Power with Dynamic Critical Path Information. In *Proc. of Intl. Symp. of Microarchitecture*, 2001.
10. R. Nagpal, and A. Bhowmik. Criticality Based Speculation Control for Speculative Multithreaded Architectures. In *Proc. of Intl. Workshop on Advanced Parallel Processing Technologies*, 2005.

Search-Optimized Suffix-Tree Storage for Biological Applications

Srikanta J. Bedathur and Jayant R. Haritsa

Database Systems Lab, SERC,
Indian Institute of Science,
Bangalore 560012, India
{srikanta, haritsa}@dsl.serc.iisc.ernet.in

Abstract. Suffix-trees are popular indexing structures for various sequence processing problems in biological data management. We investigate here the possibility of enhancing the search efficiency of *disk-resident* suffix-trees through customized layouts of tree-nodes to disk-pages. Specifically, we propose a new layout strategy, called *Stellar*, that provides significantly improved search performance on a representative set of real genomic sequences. Further, Stellar supports both the standard root-to-leaf lookup queries as well as sophisticated sequence-search algorithms that exploit the suffix-links of suffix-trees. Our results are encouraging with regard to the ultimate objective of seamlessly integrating sequence processing in database engines.

1 Introduction

The *suffix-tree* is a highly popular mechanism for indexing exponentially growing biological sequence repositories [12,13]. Its appeal lies in its *linear* (in the size of the sequence) time and space complexity of construction, and its *linear* (in the size of the query) search complexity. A unique aspect of suffix-trees is that, unlike traditional database indexes whose size is typically a fraction of the database contents, their size is usually much larger than the underlying sequence data. In fact, standard implementations of suffix-trees require in excess of *an order of magnitude* more space than the indexed data! As a case in point, the entire 3 Gbp of Human Genome is fully representable in about 1 GB memory (with each DNA symbol represented with 2-bits), whereas the corresponding most space-economical suffix-tree occupies close to 25 GB. That is, it is often straightforward to host the sequence data in main memory, but the suffix-tree itself needs to be disk-resident.

This piquant size situation is rendered even worse due to suffix-trees *not being disk-friendly*, as a consequence of the random traversals across tree-nodes induced by the standard construction and search algorithms. Accordingly, there has been significant recent research activity to address this problem and design high-performance disk-resident suffix-trees [5,15,18,19]. However, these efforts have mainly focused on the *construction aspect*, that is, on how to build the tree efficiently on disk.[1] In this paper, we take the next logical step of exploring the *search aspect*, and investigating the

---

[1] Search performance is reported in [14,18], but not analyzed in detail.

D.A. Bader et al. (Eds.): HiPC 2005, LNCS 3769, pp. 29–39, 2005.
© Springer-Verlag Berlin Heidelberg 2005

associated efficiency concerns. Specifically, our focus is on whether it is possible to optimize the *layout* of the suffix-tree with regard to the assignment of tree-nodes to disk-pages, such that the search efficiency is improved. While layout strategies have been well-studied for a variety of data-structures [1,3,10,11,17,20], we are not aware of any work focusing on suffix-trees. Further, carrying out this study for suffix-trees poses *new* problems arising out of the following:

- The patterns of search traversals over suffix-trees are much more complex than those found in traditional index structures, since both tree-edges and special lateral connectors called suffix-links are involved.
- The presence of suffix-links turns suffix-trees into *cyclic* structures.
- Suffix-trees are not inherently balanced, unlike typical disk-resident index structures (e.g. B^+-trees).

Our experiments with a variety of real genomic sequences against representative query workloads demonstrate that the currently available layout choices are *extreme* – they either optimize "vertical" traversal through the tree-edges, or optimize "horizontal" traversal through the suffix-links. But, sequence search algorithms typically need to traverse *both edges and links* – for example, to find all maximal matching substrings between the database sequence and a query, tree-edges are used to walk down the tree matching the query sequence along the way, and the subsequent matches are found by following the suffix-links [7,9].

Given the above motivation for designing a holistic algorithm that optimizes the layout for both kinds of traversals, we present in this paper **Stellar** (Suffix-Tree Edge and Link Locality AmplifieR), an algorithm that attempts to achieve this goal. Stellar is a linear-time, top-down strategy that utilizes the structural relationships between the suffix-links and the tree-edges under associated subtrees, to achieve high locality for both suffix-links and tree-edges. We quantify its effectiveness with a detailed performance study on a variety of real genomic sequences.

In summary, the contributions of this paper are as follows:

1. Demonstrating that standard layouts of suffix-trees optimize only either edge traversals or link traversals, resulting in slow searches of genomic sequences;
2. Presenting Stellar, a new suffix-tree layout that optimizes both kinds of traversals, thereby providing significantly improved search performance.

2   Sequence Search Using Suffix-Trees

A suffix-tree of a string is a compacted trie over all the suffixes of the string. For example, consider the suffix-tree constructed over a DNA fragment, S = "GTTAATTACTGAAT$" shown in Figure 1 (the internal nodes of the tree are filled in dark and the leaf nodes are lightly shaded). The solid edges between nodes represent tree-edges, while the directed dashed lines indicate *suffix-links*. The links play an important role in linear time construction of suffix-trees [16,22], and also in many search algorithms over suffix-trees [7,8,21]. Table 1 summarizes the terminology associated with suffix-trees used in the rest of the paper.

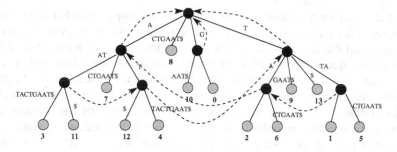

Fig. 1. Suffix-tree for the DNA fragment GTTAATTACTGAAT$

Table 1. Notation

S	Sequence of length n
Σ	Finite alphabet of symbols
$\$$	Delimiter symbol such that $\$ \notin \Sigma$
$S[i]$	Symbol at position i in S, drawn from Σ
$S[i \ldots j]$	Substring of S starting at position i and length $(j - i + 1)$
S_i	Suffix of the sequence S starting at position i
$sl(v)$	Suffix-link starting from the internal node v

Suffix-trees are useful in a large number of sequence search tasks [12], such as exact matching of pattern strings, identification of prefix-suffix pairs over a collection of sequences, common sub-string locations, and so on. A particularly critical use of suffix-trees is in pre-processing a large genomics data repository and subsequently utilizing the index to efficiently answer similarity searches. In these searches, the suffix-tree index is used to quickly locate all common substrings between the database and the given query string. These matching substrings are then used to generate *local alignments*, the regions of similarity between the sequences, through the use of various domain-specific heuristics.

In this paper, we use the *maximal common-substring search*, proposed in [7], as a representative search task over disk-resident suffix-trees. This task is defined as follows:

Definition 1 (Maximal Common-substring Search). *Given a database sequence S, and a query sequence Q, locate $Q[i \ldots i+j]$ and $S[k \ldots k+j]$, such that, $1 \leq i \leq |Q|$, $1 \leq k \leq |S|$, $Q[i \ldots i+j] = S[k \ldots k+j]$ and $Q[i+j+1] \neq S[k+j+1]$. In practice, it is desired that only matches that satisfy a user-defined minimum threshold length, λ, are reported (that is, $j \geq \lambda$).* □

3 Suffix-Tree Layout

Suffix-trees, unlike popular index structures such as B-Trees [4], are not inherently balanced – their structure depends entirely on the combinatorial characteristics of the indexed sequence. Consider, for example, the suffix-tree shown in Figure 1 – here, leaf-node 8 is an immediate child of the root, whereas leaf-node 1 is at depth 3. In the worst-case, the tree can degenerate into a linear chain of internal nodes.

The fan-out of each internal node of a suffix-tree is upper-bounded by the size of the alphabet of the indexed sequence. Therefore, the common strategy of customizing the fanout to suit the disk-page size cannot be adopted here. This means that multiple nodes of a suffix-tree will be stored on a page, with nodes connected both within as well as across pages – it therefore becomes critical to choose the nodes that will be placed in the same disk-page in order to minimize the disk I/O cost incurred during search.

Earlier research on the layouts of disk-resident indexes [10] has considered the problem of packing trees in order to minimize the total disk accesses given a access distribution on the leaf nodes – that is, *average path-length minimization*, following the terminology of [10]. It has been shown that a heuristic-based linear-time algorithm, henceforth called SBFS, that does recursive localized breadth-first layout of the tree, not only outperforms classic tree-layout methods such as Breadth-first and Depth-first strategies, but also results in an I/O-cost that is within a small factor of an *optimal* quadratic-time layout algorithm.

The basic idea behind the SBFS packing strategy is to recursively perform many local breadth-first traversals, beginning from the root of the tree, packing nodes in visit-order into disk pages. Once enough nodes have been visited to fill a page, or there are no more nodes to be visited, the nodes visited so far are assigned to a page. Each of the remaining nodes in the BFS queue then becomes the root of a separate SBFS traversal. The recursion terminates when all nodes have been visited.

3.1   Issues in Suffix-Tree Layout

The general problem of optimal graph layout is known to be NP-complete [11]. Even from a heuristic viewpoint, the storage layout of disk-resident suffix-trees introduces a variety of novel issues:

Structural Complexity: Suffix-trees exhibit greater inherent structural complexity than typical tree index structures due to the presence of *cyclic substructures*. Specifically, the collection of tree-edges as well as the collection of suffix-links in a suffix-tree form two separate tree structures, albeit with a common root. Also note that in the tree structure induced by the collection of suffix-links, the traversal direction between nodes are *reversed* from the natural "parent-to-leaf" direction. That is, there exists a directed path starting at any internal node to the root of the suffix-tree, via a chain of suffix-links. And, from the root node, any of the internal nodes are reachable through a chain of tree-edges, thus completing a cyclic path.

Complex Traversal Patterns: In typical index structures, the queries are mostly lookup searches involving root-to-leaf traversals. But search algorithms over suffix-trees exhibit complex traversal patterns, involving simultaneous use of tree-edges and suffix-links. Thus, the layout strategy has to take into account the two "orthogonal" traversal paths during search.

Due to these complexities, previously proposed layout strategies that are designed to work with either tree or DAG structures are not directly applicable in the context of suffix-trees. Nevertheless, to serve as a comparative yardstick, we investigate the efficacy of the SBFS strategy outlined above for laying out a suffix-tree on disk, by *ignoring* the suffix-links during the layout process.

3.2   Comparing the Quality of Layouts

The overall metric we use to evaluate the quality of layouts obtained using different storage strategies is to execute a representative set of queries over the suffix-trees laid out using these strategies and measure the number of disk accesses incurred. To gain more insight into the observed behavior, we also additionally measure the percentage of tree-edges and suffix-links whose source and target are both present in the same disk page – that is, the *structural localities* of the suffix tree layouts, discussed next.

Table 2 presents the structural locality results for suffix-trees built on a representative 25 Mbp long sequence drawn from Human Chromosome 2, hereafter referred to as HC2/25, with disk pagesize set to 4KB. The storage layouts evaluated here are: (1) **CO** (Creation Order), which corresponds to ordering the nodes as they are created during the construction (Ukkonen's construction algorithm [22] was used here); (2) **SBFS** layout discussed earlier; and (3) our new **Stellar** layout, described in detail in the next section.

Table 2. Structural Edge and Link Localities

Dataset	Storage	Suffix-Links	Tree Edges
	CO	41.8%	0.2%
Human Chromosome 2	SBFS	0.1%	77.5%
	Stellar	**40.0%**	**62.6%**

From the results, we first see that the CO-layout provides practically *no tree-edge locality* – only 0.2% of tree-edges are intra-page, while suffix-link locality is comparatively high – 42%. The SBFS-layout, on the other hand, represents the opposite extreme in structural locality, with 75-80% of tree-edges being intra-page, but less than 0.1% of suffix-links being local! Overall, these results indicate, as also confirmed by our other experiments, that the CO and SBFS layouts represent (negative) extremes in suffix-tree layout. The reasons for this behavior are explained in the extended version of this paper [6].

Finally, note that the structural localities for the Stellar layout in Table 2 indicate that its suffix-link locality (**40.0%**) is close to that of CO, while its tree-edge locality (**62.6%**) is comparable to that of SBFS – clearly *simultaneously* optimizing the locality of both connectors.

4   Design of Stellar

The design of Stellar is based upon the relationship between nodes connected through a suffix-link and the tree-edges under them. This relationship can be derived easily from well-known structural properties of suffix-trees [12]. Specifically, the property we use is as follows:

Property 1. If $v_2 = sl(v_1)$, then all the suffix-links originating from the nodes under v_1 point only to nodes under v_2.

In other words, if two nodes are related through a suffix-link, then *all* the nodes under the source of this suffix-link have their suffix-link targets *only* in the subtree of the

```
Stellar (r,B)
Input
r : Root of the subtree to be traversed
B : Capacity of the disk-page in terms of no. of nodes
Output
An ordering of the suffix-tree under r
```

$$queue \longleftarrow r; \{\text{push root into the BFS queue}\}$$
$$nodecount \leftarrow 0; \{\text{initialize the counter}\}$$
while $queue$ not $\emptyset$ **do**
$\quad$ $r' \longleftarrow queue;$ {remove head of the $queue$}
$\quad$ **if** r' not visited **then**
$\quad\quad$ mark r' as visited and increment $nodecount;$
$\quad$ **for all** c such that c is a child of r' **do**
$\quad\quad$ $s \leftarrow sl(c); \{s$ is the suffix-link of $c\}$
$\quad\quad$ **if** c not visited AND $nodecount < B$ **then**
$\quad\quad\quad$ mark c as visited and increment $nodecount;$
$\quad\quad\quad$ $queue \longleftarrow c;$
$\quad\quad$ **if** s not visited AND $nodecount < B$ **then**
$\quad\quad\quad$ mark s as visited and increment $nodecount;$
$\quad\quad\quad$ $queue \longleftarrow s;$
$\quad$ **if** $nodecount \geq B$ **then**
$\quad\quad$ **while** $queue$ not $\emptyset$ **do**
$\quad\quad\quad$ $m \longleftarrow queue;$
$\quad\quad\quad$ Stellar$(m,B);$

Fig. 2. Stellar Algorithm

target. This property gives us a way to reconcile between the tree-edge and suffix-link localities in the suffix-tree.

The pseudocode of the Stellar algorithm, utilizing the above structural relationship, is presented in Figure 2. The algorithm starts the suffix-tree traversal at the root of the suffix-tree, and recursively traverses the subtree below. When a node is visited, the suffix-link target of the node is visited next, if not already visited through the tree-edges. Thus an internal node and its suffix-link target are treated as a "buddy" pair, and are scheduled for recursive traversal in sequence. This results in the subtree under a node and the subtree under the corresponding suffix-link target to be recursively processed in succession – resulting in a large fraction of suffix-links that span these two subtrees to be intra-page, in addition to the tree-edges of each subtree. When enough nodes have been visited to fill a page, each node in the queue is scheduled for a separate recursive Stellar traversal, until all the nodes have been processed.

It is easy to observe that Stellar's complexity is linear in the size of the suffix-tree being processed – a node is visited only once during the top-down traversal of the tree. Additionally, it does not impose inordinate space overheads, as the only transient data structures required during the layout process are a queue of node ids, and a bit flag for each node of the tree indicating whether it has been visited or not. In our experiments we found that the queue never needs to hold ids of more than 100 nodes, even over DNA sequences exceeding 25Mbp.

4.1   Level-Wise Locality Variation

In addition to the overall locality of tree-edges and suffix-links obtained by the layout schemes, it is also critical to consider the *distribution* of such locality improvements in the suffix-tree.

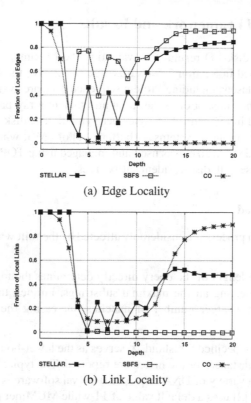

(a) Edge Locality

(b) Link Locality

Fig. 3. Depth-based Structural Localities

Figures 3(a) and (b) illustrate the locality distributions of tree-edges and suffix-links for the suffix-tree over the HC2/25 sequence under the three layout schemes. These values represent the number of intra-page tree-edges (resp. suffix-links) at every level in the suffix-tree as a fraction of all the tree-edges (resp. suffix-links) going out from that level. For example, there are a total of 2,417,879 outgoing edges from level 10, of which approximately 40% become intra-page under a Stellar layout.

As these graphs indicate, the tree-edge and suffix-link locality of all three layouts are comparable at the top portion of the suffix tree. However, as the depth of the suffix-tree increases, the suffix-link locality of CO layout outperforms SBFS significantly, while at the same time SBFS shows significantly better tree-edge locality over CO. On the other hand, the Stellar algorithm shows a steady locality comparable to the best within the tree-edge or suffix-link locality metric. In the middle portion of the suffix-tree, due to the large number of tree nodes, the locality fraction (of both suffix-links as well as tree-edges) is lower than in the top and bottom parts of the tree under all the layouts.

While the above graphs were obtained with a pagesize of 4 KB, our experiments with larger page sizes such as 16 KB, also showed similar trends – details in [6].

5   Experimental Framework and Results

We now present the disk I/O results for evaluating the Maximal Common-substring Search query described in Section 2 for suffix-trees built over the HC2/25 sequence. Results over other datasets, including Protein sequence data, are available in [6].

Our suffix-tree implementation is based on an efficient array-based tree node representation suggested in [5], with 22.5 bytes per symbol. The disk page-size is set to 4KB – a typical value in most systems. A buffer pool of 8MB, which forms approximately 5% of the total index size, was used and managed using TOP-Q [5], a buffering policy designed for use with disk-resident suffix-trees.

5.1   Query Workload

The cost of the search process is considerably affected by the following query workload characteristics:

Query Length: The length of the query directly determines the total number of iterations required for locating all the maximal substrings. Further, the increased query length may result in a larger number of matches, increasing the cost of reporting results.

Value of λ: The user-specified threshold, λ, serves as the lower-bound on the length of the match before all instances of the match are reported. The typical operational range of this parameter in a variety of DNA sequence retrieval software is between 9 and 50. Specifically, BLAST [2] uses a default value of 11 while MUMmer [9] sets it to 50.

We generated our query workload based on a collection of sequences from Expressed Sequence Tag (EST) database of GenBank. The EST-database contains 856,008 sequences with average sequence length of 357.6 basepairs. Using this base collection, we generated 3 length-restricted query collections, with lengths 50, 100, and 200, by randomly sampling fixed-length subsequences from each entry of the EST-database. In order to remove any remaining bias in the ordering of EST fragments, we sampled 10,000 sequences from each length-restricted query set to form three query collections, **hEST50, hEST100** and **hEST200**, used in our evaluation.

5.2   Utility of Disk Layout

The relative performance of maximal substring search over disk-resident suffix-tree laid out using Stellar, normalized to that with the CO layout, is shown in Figure 4. As these results indicate, the Stellar layout results in substantially reduced search costs as compared to CO. For example, at $\lambda = 11$, Stellar requires only 30–45% of the disk I/Os incurred by CO. Although this performance differential reduces with increasing value of λ, Stellar never incurs more than 75% of CO's disk accesses.

When λ values are in the lower end of operational spectrum, e.g. set to 9, the overall I/O cost of search is dominated by the overhead due to producing a large result set. As a result, the Stellar layout, with its larger fraction of intra-page tree-edges, clearly outperforms the CO layout which provides very little tree-edge locality.

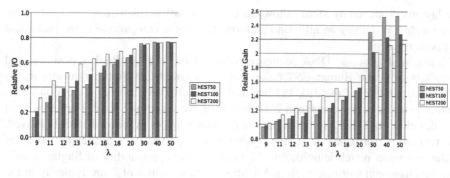

Fig. 4. Stellar Vs. CO   **Fig. 5.** Stellar Vs. SBFS

5.3   Relative Performance of Stellar and SBFS

We now turn our attention towards comparing the disk costs of Stellar and SBFS. In order to provide a normalized measure of performance, we measure their *relative performance gains* over that of the baseline CO layout. These statistics are shown in Figure 5, as a function of λ, and demonstrate that Stellar provides steadily increasing I/O gains with increasing values of λ. For example, at $\lambda = 11$, the performance gain of Stellar over SBFS is close to 20%, which increases to more than 50% at $\lambda = 16$.

In addition to these results, we also performed experiments to show that the suffix-link based searching over Stellar layouts require less than 50% disk I/O as compared to that required for searching *without suffix-links* over SBFS layouts. Details of these experiments are available in [6]. Note that the search performance of disk-resident suffix-trees constructed by the techniques of [15,18,19] is lower-bounded by the performance of the SBFS layout. As a consequence, Stellar-organized suffix-trees outperform the storage organizations produced by all these prior techniques.

6   Conclusions

Developing suffix-trees as a disk-resident sequence index structure has been an active research area in recent times, and many techniques have been proposed to significantly improve the construction time. However, there has been virtually no research on evaluating and optimizing the search performance of these disk-resident suffix-trees, the topic addressed in this paper.

Specifically, we have evaluated the impact of the suffix-tree's disk layout on the I/O performance of common genomic search tasks, and shown through detailed empirical evidence that existing index layouts, such as Creation-Order (CO) and SBFS, are not effective. They provide locality for only one of the two traversal paths, tree-edges and suffix-links, used during suffix-tree searches, and practically zero locality for the other path.

To address this unsatisfactory state of affairs, we presented a layout strategy called Stellar that optimizes the locality of both tree-edges and suffix-links in the suffix-tree.

The layouts produced by Stellar show close to 40% suffix-link locality, and 60% tree-edge locality, providing an all-round performance that is comparable to the individual best performances.

Using real genomic DNA sequences drawn from the GenBank repository, and query-sets from the Human-EST collection, we showed that Stellar typically incurs only about 30-40% of the disk I/O incurred by a suffix-tree stored in creation order. Even in extreme cases, more than 25% disk costs are saved by Stellar. Furthermore, it provides close to *2-fold improvement* over the SBFS layout in terms of disk I/O saved. The relative performance of Stellar significantly improves with increasing values of λ (the minimum match length), thus highlighting the applicability of Stellar in full-genome alignment software such as MUMmer, where values of λ are typically in the range 20–50.

References

1. S. Alstrup et al. Efficient tree layout in a multilevel memory hierarchy. Technical Report arXiv:cs.DS/0211010v1, 2002.
2. S. Altschul et al. A Basic Local Alignment Search Tool. *Journal of Molecular Biology*, 215(3), 1990.
3. S. Baswana and S. Sen. Planar Graph Blocking for External Searching. *Algorithmica*, 34(3), 2002.
4. R. Bayer and E. M. McCreight. Organization and Maintenance of Large Ordered Indexes. *Acta Informatica*, 1(3), 1972.
5. S. Bedathur and J. Haritsa. Engineering a Fast Online Persistent Suffix Tree Construction. In *Proc. of the IEEE Intl. Conf. on Data Engg. (ICDE)*, 2004.
6. S. Bedathur and J. Haritsa. Search-Optimized Persistent Suffix-tree Storage for Biological Applications. Technical Report TR-2004-04, Database Systems Lab, Indian Institute of Science, 2004.
7. W. I. Chang and E. L. Lawler. Approximate String Matching in Sublinear Expected Time. In *Proc. of the IEEE Symp. on Found. of Comp. Sci. (FOCS)*, 1990.
8. A. L. Cobbs. Fast Approximate Matching using Suffix Trees. In *Proc. of the 6th Annual Symp. on Combinatorial Pattern Matching (CPM)*, 1995.
9. A. L. Delcher et al. Alignment of Whole Genomes. *Nucleic Acids Research*, 27(11), 1999.
10. A. A. Diwan et al. Clustering Techniques for Minimizing External Path Length. In *Proc. of the 22nd Intl. Conf. on Very Large Databases (VLDB)*, 1996.
11. J. Gil and A. Itai. How to Pack Trees. *Journal of Algorithms*, 32(2), 1999.
12. D. Gusfield. *Algorithms on Strings, Trees and Sequences: Computer Science and Computational Biology*. Cambridge University Press, Cambridge, 1997.
13. D. Gusfield. Suffix Trees Come of Age in Bioinformatics (Invited Talk). In *IEEE Bioinformatics Conference (CSB)*, 2002.
14. E. Hunt, M. P. Atkinson, and R. W. Irving. Database Indexing for Large DNA and Protein Sequence Collections. *VLDB Journal*, 7(3), 2001.
15. E. Hunt, M. P. Atkinson, and R. W. Irving. A Database Index to Large Biological Sequences. In *Proc. of the 27th Intl. Conf. on Very Large Databases (VLDB)*, 2001.
16. E. M. McCreight. A Space-Efficient Suffix Tree Construction Algorithm. *Jl. of the ACM (JACM)*, 23(2), 1976.
17. M. Nodine, M. Goodrich, and J. Vitter. Blocking for External Graph Searching. In *Proc. of the 12th ACM Symp. on Principles of Database Systems (PODS)*, 1993.

18. K.-B. Schürman and J. Stoye. Suffix Tree Construction and Storage with Limited Main Memory. Technical Report 2003-06, Universität Bielefeld, 2003.
19. S. Tata, R. A. Hankins, and J. M. Patel. Practical Suffix Tree Construction. In *Proc. of the 30th Intl. Conf. on Very Large Databases (VLDB)*, 2004.
20. S. Thite. Optimum Binary Search Trees on the Hierarchical Memory Model. Master's thesis, Dept. of Computer Science, Univ. of Illinois at Urbana-Champaign, 2001.
21. E. Ukkonen. Approximate String Matching over Suffix Trees. In *Proc. of the 4th Annual Symp. on Combinatorial Pattern Matching (CPM)*, 1993.
22. E. Ukkonen. Online Construction of Suffix-trees. *Algorithmica*, 14(3), 1995.

Cost-Optimal Job Allocation Schemes for Bandwidth-Constrained Distributed Computing Systems

Preetam Ghosh, Kalyan Basu, and Sajal K. Das

Center for Research in Wireless Mobility and Networking (CReWMaN),
The University of Texas at Arlington
{ghosh, basu, das}@cse.uta.edu

Abstract. This paper formulates the job allocation problem in distributed systems with bandwidth-constrained nodes. The bandwidth limitations of the nodes play an important role in the design of cost-optimal job allocation schemes. In this paper, we present a pricing strategy for generalized distributed systems by formulating an incomplete information bargaining game on two variables (price and percentage of bandwidth allocated for distributed computing jobs at each node). Next, we present a cost-optimal job allocation scheme for single class jobs that involve the communication delay and hence link bandwidth. We show that our algorithms are comparable to existing job allocation algorithms in minimizing the expected system response time.

1   Introduction

A big challenge in a distributed system consisting of heterogeneous computers (including switches and routers) is the job allocation problem. A distributed system can be viewed as a collection of computing and communication resources shared by active users. When the demand for computing power increases, the job allocation problem becomes important. There are three typical approaches to the job allocation problem: 1) *Global approach*: There is only one decision maker that optimizes the expected response time of the entire system over all jobs and the operating point is called social optimum [7],[3],[6]; 2) *Cooperative approach*: There are several decision makers (e.g. jobs, computers) that cooperate in making the decisions such that each of them will operate at its optimum [2]; 3) *Non-cooperative approach*: Here, each of infinitely many jobs optimize its own response time independently and they all eventually reach an equilibrium [8].

Most of the previous works on static job scheduling considered the minimization of overall expected response time as their main objective. The fairness of allocation, which is also an important issue for modern distributed systems, has received relatively little attention. This problem is tackled in [5] by implementing an incomplete information non-cooperative, alternating offers bargaining game [4] between the wireless access point and the different mobile clients under it. A few other game theoretic resource management models [1] introduce

D.A. Bader et al. (Eds.): HiPC 2005, LNCS 3769, pp. 40–50, 2005.

somewhat different pricing strategies. Our goal in this paper is to find a formal framework for characterization of a fair pricing strategy between the job negotiator and the nodes (i.e, computers). Next, we formulate the cost-optimal job scheduling problem in single class job distributed systems.

Our Contributions: We use our game theoretic framework proposed in a previous work [5] to implement the pricing model. The two players, namely the Job Allocator (JA) and the node, play an incomplete information, alternating-offers, non-cooperative, bargaining game to compute the *price per unit resource* charged by that node and the percentage of bandwidth that can be used for distributed computing. In [5], we assumed that the average bandwidth of the system was fixed and is always in excess to what was required to accomplish the jobs. But in reality, the communication overhead might go higher than the processing overhead as in DSL/volunteer computing. So, we introduce a new variable: *percentage of bandwidth to be used for distributed computing jobs at node i* to characterize this problem. The concept of incomplete information ensures that the two players have no idea of each other's reserved valuations, i.e., the maximum offered price by JA (acting as the buyer of resources) and minimum expected price by the node (acting as the seller of resources); and also the maximum bandwidth percentage (information with JA) and the minimum bandwidth percentage (information with the node). Assuming there are n nodes under a single JA, the JA has to play n such games with the corresponding nodes to form the price per unit resource vector, p_i and the bandwidth percentage vector per_i ($i = 1, ..., n$).

Our second endeavor is the job allocation scheme based on the pricing model. We formulate the job scheduling problem as a *constrained minimization* problem that maximizes the revenue (i.e., minimizes the cost) for the JA.

2   The Pricing Model

We follow the same pricing strategy as in [5] and model the one-to-one relationship between a particular node and its current JA as an incomplete information, alternating-offers bargaining game. The reserved valuation of a node denotes the minimum selling price of its resources and the maximum bandwidth percentage that can be used for the jobs. The JA's reserved valuation is its maximum buying price of a node's resources and minimum bandwidth percentage required. Thus, both the JA and the node can make a surplus and try to reach a mutually beneficial agreement. The bargaining game is characterized by the rules in [5]. Every offer comprises 2-tuples of offered price and percentage of bandwidth.

2.1   Attributes of JA and Node

We use similar real-life parameters as discussed in [5] and concisely present them in Table 1. The two negotiators are denoted by $x \in \{w, m\}$: the Job/Work allocator (denoted by w), and the node/machine (denoted by m). Also, the following attributes are introduced to incorporate the bandwidth in our bargaining game:

Table 1. Different Parameters of JA and node

Parameters	Meaning of the Symbols ($x \in \{w, m\}$)
R_x	Reserved valuation (with respect to *price*) of bargainer x
M_x	Market price calculated by bargainer x
$p_x^{M_x}$	Probability that bargainer x will accept M_x i.e., Market Price
O_x	Offered price by bargainer x
$p_x^{O_x, per_x}(acc)$	Predicted probability by x that opponent accepts (O_x, per_x)
O_{x_y}	Counter offered price of bargainer x predicted by opponent y
$p_x^{O_x, per_x}(rco)$	Probability that bargainer x will reject (O_x, per_x) and counteroffer
$p_x^{O_x, per_x}(rbd)$	Probability that bargainer x will reject (O_x, per_x) and breakdown
$e^{-z_x t}$	Discount factor of bargainer x
ϑ	Resource constraints $\vartheta \in (0, 1)$
per_x	percentage of bandwidth offered/demanded by bargainer x
per_{max}	maximum percentage of bandwidth contributed by node
per_{min}	minimum percentage of bandwidth acceptable to JA
B_x	Market bandwidth calculated by bargainer x
$p_x^{B_x}$	Probability that bargainer x will accept B_x i.e., Market bandwidth
per_{x_y}	Counter-offered bandwidth percentage of bargainer x predicted by y

Percentage of Bandwidth Usage (per_x), $x \in \{w, m\}$: Denotes the percentage of bandwidth available for distributed computing jobs at the corresponding node. Here, per_{max}, per_{min} are respectively the highest and lowest percentage values that represent reserved valuations of node and JA, respectively. These values can change based on the dynamic scenario and also the player's discretion. Also, note that the more per_x is, the lesser is the communication time and larger is the revenue for the JA. Thus, the JA tries to maximize per_x while the nodes try to reduce it, leading to a conflict of interest.

Market Bandwidth Percentage (B_x), $x \in \{w, m\}$: Signifies the market value of percentage of bandwidth offered and is calculated as M_x through statistics [5].

Probability that Bargainers will Accept B_x, ($p_x^{B_x}$): Similarly signifies the probability that bargainer x will accept the market bandwidth percentage similar to $p_x^{M_x}$.

Expected Counter-Offered Bandwidth Percentage of Bargainer x ***Predicted by Opponent*** y (per_{x_y}): This will be determined by an intelligent guess of the opponent's reserved valuation quite like O_{x_y}.

2.2   JA's Utility Functions

All the probabilities are predicted by the JA based on the node's next possible action depending on the game state.

a. If the JA accepts the current offer, its expected utility is given by:

$$Utility\ (Acceptance\ by\ w) = [(R_w - O_m) + (M_w - O_m) - (per_{min} - per_m) - (B_w - per_m)] \times \vartheta$$

The term per_m increases with time (node offers higher percentage of bandwidth), resulting in an increase in the surplus for the JA thereby increasing the chances of an acceptance. Similarly, O_m decreases over time (node asks for lesser price) resulting in increase in the surplus. The gain in surplus due to bandwidth percentage follows the same rules as those for the offered price.

b. If the JA rejects the opponent's offer and breaks down from the game, we get:

$$Utility(Break\text{-}down\ by\ w) = (R_w - M_w) \times p_w^{M_w} - (per_{min} - B_w) \times p_w^{B_w}$$

The second term signifies the profit from the market from bandwidth contributed to distributed computing jobs. The JA will break down finding that there are other nodes providing higher bandwidth.

c. In case of a counter-offer, we get:

$$Utility\ (Counter\ offer\ by\ w) = [[(R_w - O_w) + (M_w - O_w) - (per_{min} - per_w) - (B_w - per_w)] \times p_w^{O_w, per_w}(acc) + [Utility\ (Break\text{-}down\ by\ w) \times p_w^{O_w, per_w}(rbd)] + [(R_w - O_{m_w} - per_{min} + per_{m_w}) \times p_w^{O_w, per_w}(rco)]] \times e^{-z_w t} \times \vartheta$$

Again, when $t = deadline$, $Utility\ (Counter\ offer\ by\ w) = 0$ and the game converges. Also, we have $p_w^{O_w, per_w}(acc) + p_w^{O_w, per_w}(rbd) + p_w^{O_w, per_w}(rco) = 1$ as discussed in [5].

2.3   Node's Utility Functions

All the probabilities herein are predicted by the node based on the JA's next possible action.

$$Utility(Acceptance\ by\ m) = [(O_w - R_m) + (O_w - M_m) - (per_w - per_{max}) - (per_w - B_m)] \times \vartheta$$

$$Utility(Break\text{-}down\ by\ m) = (M_m - R_m) \times p_m^{M_m} - (B_m - per_{max}) \times p_m^{B_m}$$

$$Utility\ (Counter\ offer\ by\ m) = [[(O_m - R_m) + (O_m - M_m) - (per_m - per_{max}) - (per_m - B_m)] \times p_m^{O_m, per_m}(acc) + [Utility\ (Break\text{-}down\ by\ m) \times p_m^{O_m, per_m}(rbd)] + [(O_{w_m} - R_m - per_{w_m} + per_{max}) \times p_m^{O_m, per_m}(rco)]] \times e^{-z_m t} \times \vartheta$$

The price and bandwidth percentage variables both contribute to the surplus and hence have to be normalized to the $(0, 1)$ interval. We can exactly use the same model of interdependent attributes formulation as in [5] to calculate the predicted probabilities for our bargaining game.

3   Job Allocation Scheme

We consider a *single* job class distributed system consisting of n nodes. Because the bargaining game is played *offline*, the Job Allocator (JA) knows of p_i and per_i, for the ith node before the job allocation starts. We assume that the nodes deal with *three* resources: idle CPU cycles, buffer size (required for queueing up jobs), and bandwidth each with equal weight. Thus, p_i = price per unit CPU cycle = price per unit buffer size = price per unit bandwidth consumed at the ith node. Unlike [5], we model each node as an M/G/1 queue with preemptive resume priority because there can be many high priority jobs at the node that can preempt the distributed computing jobs. The work already done for an ongoing class at a node that has been interrupted by the arrival of a higher priority job is remembered, i.e., we assume a *work-conserving discipline* at the nodes. We assume that there are P classes of jobs at a node (class 1 having the lowest priority and class P having the highest priority) where jobs belonging to class g are distributed computing jobs ($1 \leq g \leq P$). For simplicity of notations (and without loss of generality), we have assumed that all the n nodes have the same P classes of jobs. Service time for class l at node i has mean $\overline{X_l^i}$ and second moment $\overline{X_l^{i2}}$. Arrival process for class l jobs at node i is assumed to be Poisson with rate λ_l^i. Also, load of priority class l at node i is given by $\rho_l^i = \lambda_l^i \overline{X_l^i}$, for $l = 1, .., P$ and $i = 1, .., n$. Let, β_i denotes the average arrival rate of distributed computing jobs at node i and Φ the total job arrival rate at the JA. The expected execution time for Class g jobs at a node is given by:

$$W_g^i = \frac{\overline{X_g^i}(1-\rho_P^i - ... - \rho_g^i) + R_g^i}{\Phi(1 - ... - \rho_{g+1}^i)(1-\rho_P^i - ... - \rho_g^i)} = \frac{\overline{X_g^i}(1-\rho_P^i - ... - \rho_{g+1}^i) - \beta_i \overline{X_g^i}^2 + R_g^i}{\Phi(1 - ... - \rho_{g+1}^i)(1-\rho_P^i - ... - \rho_{g+1}^i - \beta_i \overline{X_g^i})}$$

where, R_g^i is the residual lifetime of jobs of Class g at node i and is given by: $R_g^i = \frac{1}{2}\beta_i \overline{X_g^{i2}} + \sum_{l=g+1}^{P} \frac{1}{2}\lambda_l^i \overline{X_l^{i2}}$. The execution time at every node comprises a queueing delay and an actual processing delay. Let, k_i^1 be a constant mapping the execution time to the amount of resources (both CPU cycles and buffer size) consumed at node i. Also, the communication delay can be expressed in terms of b_i (defined as the average bandwidth available to node i) as $\frac{MSB_i}{\Phi per_i b_i}$, where M = average number of messages transferred for one job unit and S = average size of the message in bits. Also k_i^2 is a constant mapping the communication delay to amount of bandwidth resources consumed at node i. Thus, the price to get β_i amount of work done at node i is:

$$\mathcal{C}(\beta_i) = \frac{k_i^1 p_i \{\overline{X_g^i}(1 - \rho_P^i - ... - \rho_{g+1}^i) - \beta_i \overline{X_g^i}^2 + R_g^i\}}{\Phi(1 - ... - \rho_{g+1}^i)(1 - \rho_P^i - ... - \rho_{g+1}^i - \beta_i \overline{X_g^i})} + \frac{MSk_i^2 \beta_i p_i}{\Phi per_i b_i} \quad (1)$$

and the overall cost of the system is given by:

$$C = \sum_{i=1}^{n} \mathcal{C}(\beta_i) = \sum_{i=1}^{n} \left(\frac{k_i^1 p_i \{\overline{X_g^i}(1-\rho_P^i - ... - \rho_{g+1}^i) - \beta_i \overline{X_g^i}^2 + R_g^i\}}{\Phi(1 - ... - \rho_{g+1}^i)(1-\rho_P^i - ... - \rho_{g+1}^i - \beta_i \overline{X_g^i})} + \frac{MSk_i^2 \beta_i p_i}{\Phi per_i b_i} \right)$$

Our objective is to find an efficient job allocation scheme $\{\beta_1, \beta_2, ..., \beta_n\}$ that optimizes the revenue of the JA, by minimizing the cost C that should obey the following conditions:

$$Positivity: \quad \beta_i \geq 0, \quad i = 1, ..., n \tag{2}$$

$$Conservation: \quad \sum_{i=1}^{n} \beta_i = \Phi \tag{3}$$

$$Stability: \beta_i \overline{X_g^i} < (1 - \sum_{l=1}^{g-1} \rho_l^i - \sum_{l=g+1}^{P} \rho_l^i), \quad i = 1, ..., n \tag{4}$$

$$Communicability: \quad MS\beta_i < per_i.b_i, \quad i = 1, ..., n \tag{5}$$

Because we are considering a preemptive priority queue with a single server, the jobs of the different classes $1, .., P$ are kept in different queues, and hence the stability condition needs to verify that the server occupancy is less than 1 (and this condition has to be checked at all the nodes). The positivity and conservation constraints are straightforward as discussed in [2]. The communicability constraint takes care of the bandwidth limitation of node i.

Definition 1. *The optimization problem is denoted by $\{\min_{\beta_i} C\}$, subject to constraints given by Eqns 2-5.*

We first solve the load balancing problem without requiring β_i $(i = 1, ..., n)$ to be non-negative. Scaling down the problem by substituting the expression for R_g^i in Eqn 1 we get:

$$C(\beta_i) = \frac{k_i^1 p_i \overline{X_g^i}}{\Phi(1 - \rho_P^i - ... - \rho_{g+1}^i)} - \frac{k_i^1 p_i \overline{X_g^{i2}}}{2\Phi(1 - \rho_P^i - ... - \rho_{g+1}^i)\overline{X_g^i}}$$
$$+ \frac{k_i^1 p_i [\frac{\overline{X_g^{i2}}(1 - \rho_P^i - ... - \rho_{g+1}^i)}{2\overline{X_g^i}} + \sum_{l=g+1}^{P} \frac{1}{2}\lambda_l^i \overline{X_l^{i2}}]}{\Phi(1 - ... - \rho_{g+1}^i)(1 - \rho_P^i - ... - \rho_{g+1}^i - \beta_i \overline{X_g^i})} + \frac{MSk_i^2 \beta_i p_i}{\Phi per_i b_i}$$

The first two terms in the above expression being constants can be left out of the optimization problem which can be expressed by $C'(\beta_i)$ as follows:

$$C'(\beta_i) = \frac{a_i}{c_i - d_i\beta_i} + e_i\beta_i \tag{6}$$

where, $a_i = \dfrac{k_i^1 p_i [\frac{\overline{X_g^{i2}}(1 - \rho_P^i - ... - \rho_{g+1}^i)}{2\overline{X_g^i}} + \sum_{l=g+1}^{P} \frac{1}{2}\lambda_l^i \overline{X_l^{i2}}]}{\Phi(1 - ... - \rho_{g+1}^i)}$, $c_i = (1 - \rho_P^i - ... - \rho_{g+1}^i)$,

$d_i = \overline{X_g^i}$ and $e_i = \frac{MSk_i^2 p_i}{\Phi per_i b_i}$. Also each of a_i, c_i, d_i and e_i are constants dependent on $i = 1, ..., n$. The following theorem solves this non-linear program:

Theorem 1. *The solution of the optimization problem given in Definition 1 without the constraint $\beta_i \geq 0$, $i = 1, ..., n$ is given by:*

$$\beta_i = \frac{1}{d_i}(c_i - \sqrt{\frac{a_i d_i}{\alpha - e_i}}), \quad i = 1, ..., n \tag{7}$$

where, α is the Lagrange multiplier.

The term α can be calculated iteratively by solving the equation:

$$\sum_{i=1}^{n} \beta_i = \Phi = \sum_{i=1}^{n} \frac{1}{d_i}\left(c_i - \sqrt{\frac{a_i d_i}{\alpha - e_i}}\right).$$

In practice, this solution cannot be used because it can make β_i ($i = 1, ..., n$) negative. Note that β_i becomes negative when $c_i < \sqrt{\frac{a_i d_i}{\alpha - e_i}}$ which means that node i has either very low processing power left to execute distributed computing jobs or it has very low bandwidth available to be able to transfer these jobs. So, we assign $\beta_i = 0$ and thus eliminate node i from consideration and then allocate jobs to the remaining $n - 1$ nodes. The job allocation algorithm, PRIMANGLE, is shown in Fig 1, and its validity is proven by the following theorem:

Theorem 2. *If for an integer i ($1 \leq i \leq n$), $c_i < \sqrt{\frac{a_i d_i}{\alpha - e_i}}$, then C is minimized by setting $\beta_i = 0$, subject to the extra constraint $\beta_i \geq 0$ in addition to the three constraints stated in Definition 1.*

In other words, the job allocation $\{\beta_1, \beta_2, ...\beta_n\}$ by PRIMANGLE is an optimal solution for the minimization problem stated in Definition 1. Recalculating α iteratively inside the while loop is achieved by $Subalgorithm - 1$ (Fig 2) in $O(n)$

PRIMANGLE (dc: abbreviation for distributed computing)
Input: The average service time of P job classes of the nodes:
$\{\overline{X_1^1}, \overline{X_2^1}.., \overline{X_P^1}, \overline{X_1^2}, .., \overline{X_P^2}, .., \overline{X_1^n}, .., \overline{X_P^n}\}$.
Average second moment of P job classes of the nodes:
$\{\overline{X_1^{12}}, \overline{X_2^{12}}.., \overline{X_P^{12}}, \overline{X_1^{22}}, .., \overline{X_P^{22}}, .., \overline{X_1^{n2}}, .., \overline{X_P^{n2}}\}$.
Arrival rate of non-dc jobs at each node:
$\{\lambda_1^1, .., \lambda_{g-1}^1, \lambda_{g+1}^1, .., \lambda_P^1, .., \lambda_1^n, .., \lambda_{g-1}^n, \lambda_{g+1}^n, .., \lambda_P^n\}$.
Total job arrival rate Φ.
The price per unit resource vector: $\{p_1, p_2, ..., p_n\}$.
The processing constant vector: $\{k_1^1, k_2^1, ..., k_n^1\}$.
The communication constant vector: $\{k_1^2, k_2^2, ..., k_n^2\}$.
Number of messages for one unit of dc job, M
and size of each message in bits, S.
Output: The optimal job allocation $\{\beta_1, \beta_2, ...\beta_n\}$.
1. Calculate a_i, c_i, d_i and e_i for $i = 1, ..., n$;
2. Calculate initial α;
3. Sort the nodes in decreasing order of:
$\left(\frac{c_1}{\sqrt{a_1 d_1}} \geq \frac{c_2}{\sqrt{a_2 d_2}} \geq ... \geq \frac{c_n}{\sqrt{a_n d_n}}\right)$;
4. $\Gamma \leftarrow \frac{1}{\sqrt{\alpha - e_n}}$;
5. while $\left(\Gamma > \frac{c_n}{\sqrt{a_n d_n}}\right)$ do
$\quad \beta_n \leftarrow 0$;
$\quad n \leftarrow n - 1$;
$\quad$ Recalculate α iteratively;
$\quad \Gamma \leftarrow \frac{1}{\sqrt{\alpha - e_n}}$;
6. for $i = 1, ..., n$ do
$\quad \beta_i \leftarrow \frac{1}{d_i}\left(c_i - \sqrt{\frac{a_i d_i}{\alpha - e_i}}\right)$;

Subalgorithm-1
1. Set initial
$\quad \alpha = \max_i\{e_i\} + \delta_1$;
2. for $r = 1, ..r_{max}$,
$\quad$//(r: iteration number)
3. $\quad$ for $i = 1, ..., n$
4. $\quad\quad temp+ =$
$\quad\quad\quad \frac{1}{d_i}\left(c_i - \sqrt{\frac{a_i d_i}{\alpha - e_i}}\right)$;
5. $\quad$ end of for i loop
6. $\quad$ if($\phi - temp < \epsilon$)
7. $\quad\quad$ break;
8. $\quad$ else
9. $\quad\quad \alpha_j = \alpha_j + \delta_2$;
10. end of for r loop

Fig. 1. PRIMANGLE: PRIce based optiMAl workload allocation scheme for siNGLE distributed computing class jobs

Fig. 2. Sub-algorithm to calculate α

time. The idea is to start with a very small value for α and then iteratively increase it until we reach Φ. The precision of the algorithm depends on the chosen values of the constants $\delta_1, \delta_2, \epsilon$ and r_{max}. So PRIMANGLE has run-time complexity of $O(n^2 + nP) \approx O(n^2)$, if $n > P$.

4 Performance Analysis

We first demonstrate the performance of the bargaining algorithm based on the following metrics (details can be found in [5]):

Each player draws its reserved valuation independently using a random number generator and uses the same to guess the opponent's reserved valuation. If the JA starts the game, $O_w[0] = \alpha \times minimum(guess(R_m), M_w)$, $per_w[0] = \beta \times maximum(guess(per_{max}), B_w)$. If the node starts the game, $O_m[0] = \beta \times maximum(guess(R_w), M_m)$, $per_m[0] = \alpha \times minimum(guess(per_{min}), B_m)$ where, $O_x[0] =$ Initial offered price from bargainer x; $per_x[0] =$ Initial offered bandwidth percentage by x; $guess(R_x) =$ guess of opponent x's reserved valuation (in terms of price); $guess(per_{min/max}) =$ guess of opponent x's reserved valuation (in terms of bandwidth percentage); $x \in \{w, m\}$; $\alpha = 0.5$; $\beta = 1.5$. The offered prices list is given by $O_w[i] = O_w[0] + (R_w - O_w[0]) \times (1 - e^{-(i \times 6.9/\delta)})$, and $O_m[i] = R_m + (O_m[0] - R_m) \times e^{-(i \times 6.9/\delta)}$ where, $\delta =$ total number of offered prices; $O_x[i] = i^{th}$ offered price of player x. Similarly, $per_m[i] = per_m[0] + (per_{max} - per_m[0]) \times (1 - e^{-(i \times 6.9/\delta)})$ and $per_w[i] = per_{min} + (per_w[0] - per_{min}) \times (1 - e^{-(i \times 6.9/\delta)})$, where $per_x[i] = i^{th}$ offered percentage of bandwidth by player x. Thus the offered prices vector for the JA monotonically increases, and that for the node monotonically decreases. Similarly, the percentage of bandwidth usage vector for JA monotonically decreases, and that for node monotonically increases.

Fig 4 plots the expected revenue of the JA against different offered prices and bandwidth percentage values. With increase in offered prices, i.e., with decrease in bandwidth percentage, the expected revenue decreases, so that the game converges quickly. This behavior is because of the monotonically increas-

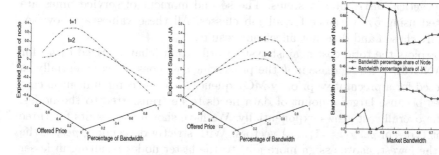

Fig. 3. Expected revenue of Node vs. Offered prices and per_i

Fig. 4. Expected revenue of JA vs. Offered prices, per_i

Fig. 5. Expected bandwidth percentage share of JA and node

ing offered prices and decreasing bandwidth percentage vectors chosen initially. The two plots correspond to the first and second counter-offers ($t = 1, 2$) and we find that the revenue decreases with time (because of the discount factor). Fig 3 plots the revenue of the nodes against the same parameters and we get just the opposite characteristics as in the case of the JA because the expected revenue decreases for the node, as the offered prices decrease and corresponding bandwidth percentage values increase. Also, the revenue decreases with time. Fig 5 shows the variation of the bandwidth percentage of the JA and the node against the market bandwidth. When the market bandwidth is large, the JA will be getting a higher share of revenue and bandwidth percentage than the node. With decrease in market bandwidth, the JA's share decreases as it fails to find nodes that will charge lesser price and also offer higher bandwidth. Similarly, the expected share for the node decreases with the market bandwidth, as all the other nodes will also offer higher bandwidth and the JA will get a higher share.

We next simulate a heterogeneous distributed system with 10 nodes under a single JA with three different types of processing rates depicted in Table 2. The

Table 2. System Configuration

Relative processing rate	1	2	5
Number of nodes	5	3	2
Processing rate (jobs/sec)	0.013	0.026	0.065
k_i^1	1	2	3
k_i^2	1	15	30
b_i (kbps)	15	17	20

first row signifies the relative processing rates of the nodes, and the third row gives the actual processing rate values. The second row signifies the number of nodes of a particular type. The last row shows the average available bandwidth for the nodes. The constants k_i^1 and k_i^2 are chosen such that the faster nodes and those with higher available bandwidth values should charge a higher price to perform distributed computing jobs. The k_i^1 values can simply be chosen to be the relative processing rates, and the k_i^2 values the relative bandwidth for all practical distributed systems. The second moment of service times are calculated using 5% variance for all job classes. All these values are however arbitrarily chosen and does not affect our results.

Fig 8 plots the total cost against system utilization (characterized by Φ) for PRIMANGLE. With increase in Φ the processing cost goes up exponentially as is expected for a preemptive priority MG1 queue. Also, the communication cost increases because larger amount of data needs to be transferred to the nodes. Thus the overall cost grows exponentially. We have shown the results for three different types of per_i ($i = 1, .., n$) vectors. With strictly descending per_i's, the cost is the lowest, as we assign more jobs to the faster nodes resulting in lesser communication cost. The random per_i vector gives better performance than a strictly ascending per_i vector, because of obvious reasons.

Fig 6 plots the response time against Φ for COOP [2], OPTIM [3] and PRIMANGLE in low communication delay systems ($b_i = 1000$ kbps, $\forall i$). Fig 7 plots

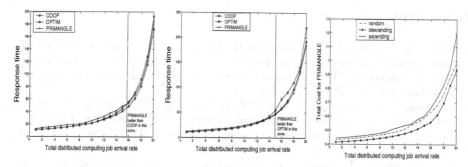

Fig. 6. Response Time vs. Φ plot for high bandwidth systems

Fig. 7. Response Time vs. Φ plot for low bandwidth systems

Fig. 8. Cost vs. Φ plot for PRIMANGLE

the same for high communication delay systems ($b_i = 100$ kbps, $\forall i$). COOP and OPTIM were originally designed with an M/M/1 queueing model at the nodes and we formulated the corresponding linear programming models with preemptive priority M/G/1 queueing models to get the plots. This allows us to incorporate the internal job arrival rate at the nodes. Also, the communication delay is not taken into account in these problem formulations. In low communication delay systems, OPTIM gives the best response time values, but PRIMANGLE catches up with COOP at high values for total distributed computing job arrival rates (region to the right of the straight line in Fig 6). This is expected as the communication delay being low, does not contribute largely to the total system response time, and OPTIM (which is supposed to keep the processing delay to a minimum) performs better than the others. A different characteristic is seen in Fig 7 where the communication delay contributes substantially to the total system response time, and PRIMANGLE performs better than both COOP and OPTIM (region to the right of the straight line in Fig 7) at higher values for total distributed computing job arrival rate. This characteristic however depends on the constants k_i^1 and k_i^2 and the price vector, because the optimization problem for PRIMANGLE minimizes the cost and not the total response time. Lower values for k_i^1, k_i^2 and p_i's should give even better performance for PRIMANGLE.

5   Conclusion

In this paper, we have tackled the job allocation problem in distributed systems considering single-class jobs and communication delay. The bandwidth constraint of the nodes presents a major bottleneck and has to be considered by any job allocation algorithm. The offline bargaining game now incorporates the percentage of bandwidth allotted to distributed computing jobs. We plan to extend this work to $(n + 1)$-player games to reduce the bargaining message overhead.

References

1. R. Buyya, D. Abramson, J. Giddy, and H. Stockinger, *"Economic Models for Resource Management and Scheduling in Grid Computing"*,Special Issue on Grid Computing Environments, The Journal of Concurrency and Computation: Practice and Experience (CCPE), Wiley Press, May 2002.
2. D. Grosu, A. T. Chronopoulos and M.Y. Leung, *"Load Balancing in Distributed Systems: An Approach Using Cooperative Games"*, Proc. of the 16th IEEE International Parallel and Distributed Processing Symposium (IPDPS 2002), Ft Lauderdale, Florida, pp. 501-510.
3. A. N. Tantawi and D. Towsley, *"Optimal static load balancing in distributed computer systems"*, Journal of the ACM, pages 373-382, 32(2):445-465, April 1985.
4. P. Winoto, G. McCalla and J. Vassileva. *"An Extended Alternating-Offers Bargaining Protocol for Automated Negotiation in Multi-Agent Systems"*, Proceedings of the 10th International Conference on Cooperative Information Systems (CoopIS'2002), Irvine, CA, Nov 2002. Springer LNCS vol. 2519, 179-194.
5. P. Ghosh, N. Roy, K. Basu and S.K. Das *"A Game Theory based Pricing Strategy for Job Allocation in Mobile Grids"*, Proc. of the 18th IEEE International Parallel and Distributed Processing Symposium (IPDPS 2004), Santa Fe, New Mexico.
6. X. Tang and S. T. Chanson *"Optimizing static job scheduling in a network of heterogeneous computers"*, In Proc. of the Intl. Conf. on Parallel Processing, pages 373382, August 2000.
7. K. W. Ross and D. D. Yao, *"Optimal load balancing and scheduling in a distributed computer system"*, Journal of the ACM, 38(3):676690, July 1991.
8. H. Kameda, J. Li, C. Kim and Y. Zhang, *"Optimal Load Balancing in Distributed Computer Systems"*, Springer Verlag, 1997.

A Fault Recovery Scheme for P2P Metacomputers*

Keith Power and John P. Morrison

National University of Ireland, Cork, Ireland
{k.power, j.morrison}@cs.ucc.ie

Abstract. Despite the leaps and bounds made by the P2P research field in the last few years, the benefit of this innovation has been constrained to a few areas; search and file-sharing and storage to name a few. In particular, this innovation has had little significant impact in the field of distributed computing.

There are several obstacles to be overcome in the development of any distributed computer, most notably: scalability, fault tolerance, security and load balancing. The difficulty of these is compounded in the dynamic, decentralized environment which characterizes the P2P arena. This paper presents a method of recovering from faults which exploits the distributed hash table functionality provided by modern overlay networks. Its effectiveness is evaluated experimentally using a proof of concept P2P distributed computer.

It is hoped that by providing a solution to one of the obstacles, global, decentralized, dependable distributed computers will be one step closer to reality.

Keywords: peer-to-peer, fault, tolerance, recovery, decentralized, distributed computing, condensed graphs.

1 Introduction

The P2P research field has seen significant advances in technology in the last few years. The advent of Structured Overlay Networks, such as Chord[15], Pastry[14], CAN[13] and Kademlia[9], have had a huge impact on the architecture of P2P applications; the majority are now constructed atop these overlays, inheriting scalability and Distributed Hash Table (DHT) functionality, which enables such diverse facilities as distributed directories, advertising and subscription of services and reliable distributed storage.

However, the field of distributed computing, which should benefit greatly from the addition of scalability and decentralization, has not made any real progress in this direction. The fundamental challenges in the development of any distributed computer are scalability, fault tolerance, security and load balancing. The provision of additional services, for example resource management and brokerage, introduces further challenges. The difficulty of solving these problems is usually compounded in a P2P environment. For instance, in a centralized distributed computer, a single master machine can make decisions on which computers should execute tasks, since it can obtain total knowledge of the system. In a P2P system, on the other hand, where each machine does not have knowledge of every other machine, such knowledge is not easily available. Similarly, a master can decide whether or not it trusts a machine to execute a task. There's an

---

* This work is supported by Science Foundation Ireland.

implicit assumption that the machine trusts the master. In a P2P environment no machine enjoys the unique status afforded the master in a centralized system, no peer is implicitly trustworthy.

Fault tolerance, in particular, is made more difficult in a P2P environment. The mainstays of centralized fault-tolerance policies are redundancy and checkpointing. Neither of these approaches is sufficient in a P2P environment.

Redundancy, in the computational situation, means running more than one copy of a program, or program component, so that if one copy fails, there is a remaining copy to carry on the computation. This is usually used to provide services that are required to be reliable, for instance DNS, or web serving, but is rarely used for distributed computation, since at a minimum it doubles the amount of resources used to produce a result. In a centralized environment, a master must send each task out twice, to two different connected machines, or slaves, just to receive the minimum protection from a fault. In a P2P distributed computer, the cost of providing redundancy can be much greater. For instance, a task may be capable of being decomposed into two parts, then each of those may be decomposed further. If a peer, A, receives the first task, decomposes it into 2 and passes it on, it must do so twice to provide redundancy, resulting in 4 subtasks. If the peers which receive these subtasks do the same, each producing 4, the total subtasks uncovered and executed is 16, when there were only 4 to start. Each level of decomposition brings a doubling in the amount of work. The alternative is to decompose the tasks on a single peer, which is equivalent to the master-slave approach, thereby foregoing the benefits of the P2P approach.

Checkpointing refers to taking a snapshot of a process's state and storing it, so that if the process fails, it can be restarted from the stored state rather than the very beginning. Even ignoring the acknowledged complexities of consistent check-pointing and roll-back in distributed systems[16], this scheme is unsuitable for P2P applications. Traditionally, the checkpoint state is stored on permanent storage, such as a hard disc, or stable-storage in situations where fast fault recovery is critical. This information can then be read when the machine reboots, or if the storage is on a network, it can be read by a remote machine, and used to continue the computation. In a P2P environment, if a peer participating in a computation suddenly departs the system, this constitutes a failure. Storing the state of its process is useless, since the peer might never reconnect. There is no fixed set of machines in a typical P2P environment. In any P2P system, there are often a set of peers which stay connected for considerable lengths of time[7], but it is undesirable that this fact should be exploited, since it a) would introduce a dependency on those peers, and b) would result in a large increase in the volume of data transferred to and from those peers. An option is to checkpoint each process and store the state on another peer, so that if a peer departs an idle or joining peer can be introduced to the computation and begin where the failed peer left off. However, this introduces a costly computational and communication overhead into every process, whether a failure occurs or not.

This paper presents a method of fault tolerance suitable for the dynamic P2P environment. It is fully decentralized and incurs a low overhead. The rest of this paper is structured as follows: a short overview of a decentralized model of computation is given. The fault-recovery algorithm is then described. This algorithm is evaluated in the context of the preceding model, and experimental results are presented. To conclude, the results, and the applicability of the system to other methods of computation are discussed.

2 Model of Computing

Compeer[10], a P2P distributed computer, will be used to illustrate the fault recovery mechanism described in this paper, and to gather results. Compeer is constructed using a structured overlay network. The current implementation uses Chord, but it can use any overlay that assigns peers unique IDs and provides DHT functionality. The overlay is responsible for handling the network topology, and informing Compeer of any changes. The DHT is exploited to provide directory services, but is also indispensable when it comes to fault recovery.

Compeer executes programs expressed as Condensed Graphs ($\mathcal{CG}$s), which are similar to Dataflow, but with added expressiveness[11]. Nodes in $\mathcal{CG}$s can be *primitive*, meaning they represent a piece of code to execute, or *condensed*, meaning they represent another graph. When a condensed node is executed it is effectively replaced by the graph it represents. Writing programs in this hierarchical way lends itself to P2P; a condensed node can be passed to a peer, which can unpack it and pass on more and so on, allowing the three stages of distributed computing, decomposition, distribution and execution of nodes to be carried out by all peers. In centralized master-slave type systems decomposition and distribution is the responsibility of the master, and slaves only execute tasks. The model of computation used by Compeer is referred to as fully decentralized since the responsibility to perform each stage is shared by all peers. Using a fully decentralized model of computation brings numerous benefits, including the potential to distribute work more rapidly than a centralized system, thereby improving scalability[12].

One interesting outcome of using a fully decentralized model is that the results of tasks do not need to return to the peer that distributed them. This is a restriction of centralized models only. Lifting this restriction spreads the impact of a fault among peers, so the failure of any peer is not catastrophic. This is in contrast to a centralized system where the failure of the master renders the system unusable, since slaves can neither receive more tasks, nor return their results. Figure 1 shows an example condensed node, G, which is executed on Peer 1 to produce the graph composed of nodes E, A, B, C and X. Each of these nodes can be placed on a separate peer, and the results can flow directly to their destinations without ever returning to Peer 1.

2.1 Node IDs

Each node is assigned a unique ID. The first, or root, $\mathcal{CG}$ node in an application is given the ID 0. The root executes to produce the program graph. The ID of a child node is created by combining its parent's ID with a number representing how many child nodes have been created for its parent, separated by a period. Graph definitions are expressed

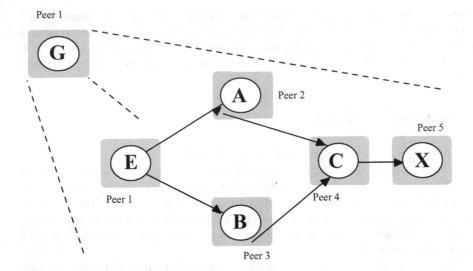

Fig. 1. Node *G* is decomposed into nodes E, A, B, C and X. The nodes are all placed on different peers. In the *bag-of-tasks* approach, a master would have distributed A and B, then waited for their results to return before distributing C.

so that the order of creation of child nodes is deterministic, so the nodes have a preset order of creation. Basically, the child nodes of the root node are numbered 0.0, 0.1 and so on, in the order they are created.

As a result, the *lineage* of a node can be determine from it's ID. The ID of its parent, grand parent and so on can be determined by removing the rightmost period, and those digits to the right of it.

3   Fault Recovery Procedure

The fault recovery procedure outlined assumes the following:

- Only fail-stop failures occur. That is, peers either are responding correctly or they are failed, no incorrect responses are generated.
- Nodes do not fail, except as a result of the peer on which they are hosted failing. It is not possible for nodes on a correctly functioning peer to fail.
- Failures are random. That is, peer failures are unrelated to each other. Thus, faults in the underlying network which might partition a Compeer network are not considered.
- The underlying network is logically fully connected. Any peer can communicate with any peer that has not failed.

The overlay used is responsible for detecting peer disconnection, and rearranging topology. So, the overlay handles *fault survival*. Compeer must handle *fault recovery*, which in this case must ensure all computations terminate with the correct result. In

Compeer this is analogous to graph repairing. The nodes hosted on failed peers must be recreated, if required, and relinked to the nodes that require them. This should be done in a way that prevents work being redone whenever possible. For simplicity, nodes on failed peers are also referred to as having failed.

To facilitate graph rebuilding, recreated condensed nodes must be able to locate their child nodes. To prevent work being redone, recreated nodes must be able to locate any results their former incarnation produced. To this end, peers keep a copy of the results they distributed. This data redundancy, ensures there's at least two copies of any result transmitted.

The *orphan registry* is used to enable recreated nodes to locate nodes they produce. Nodes whose parent has failed register as orphans. Condensed nodes whose children have failed enter *recovery mode*. They execute again, but this time they only create the failed child nodes and relink them to their surviving siblings. Recreated nodes also enter recovery mode. A recreated primitive node searches the registry for a copy of its result before executing, thus preventing any unnecessary computation from being repeated.

The orphan registry is implemented using the overlay's DHT, and thus is fully decentralized. The core of the functionality is provided by two method: registerOrphan and getOrphan, implemented by each Compeer peer. To register as an orphan, a node hashes it's ID to locate a peer, and calls the registerOrphan method on that peer. This method returns the value of its recreated parent, if it already consulted the registry for it. In this way any nodes that register late, that is, after their parent seeks them, can contact the parent node directly and be integrated back into the computation. If an orphaned condensed node registers late, it immediately instructs its children to register as orphans. This prevents two branches of the computation being repeated concurrently. Recreated nodes call the getOrphan method when they execute. They hash the ID of the child they wish to create to identify the peer on which it would have registered if it survived the fault. If no node is found, a reference to the parent is stored, to be given to the node if it subsequently calls registerOrphan.

This recovery procedure is robust in the face of further failures. Orphaned nodes consider the peer on which they register to be their temporary parent, so if it fails, they repeat the registration process. Since the overlay can survive faults, any node can always locate any other node with a simple lookup, if it's registered.

3.1 Special Case: Failure of Root Node

All peers that participate in a computation are given a copy of the program graph definition, and each computation has a unique process identifier (*PID*). If the root node fails, a unique situation arises: there is no node to recreate it. In this case a peer is nominated as responsible for this task. In the analogous situation in parallel processing electing a responsible peer would usually involve a flurry of inter-peer communication to carry out some voting scheme. In contrast, Compeer can once again leverage the overlay to simply hash the *PID* to produce an ID for which a single peer is responsible. Thus, in typically $log_2 N$ steps (where N is the number of overlay nodes), a peer responsible for the root node has been unambiguously located with no broadcasting or voting. This peer recreates the root node, and since it's recreated it will relink to any surviving child nodes via the orphan registry.

Full details of this recovery method are available in [12]. The effectiveness of this method is now investigated.

4 Experimental Results

4.1 Methodology

To measure the characteristics of the presented scheme, the Compeer system was deployed over a cluster of 32 1.5Ghz Pentium 4s, each containing 1GB of RAM, linked via Gigabit ethernet. The author's implementation of the Chord overlay was used, with a neighbourhood size of 5. An "embarrassingly parallel" application (the ubiquitous Mandelbrot Set generator) was used in each case, and the problem size for each run, that is, the number of tasks executed, was 512. The application was written as a hierarchically structured $\mathcal{CG}$ that allows work to propagate rapidly among peers. Each task takes 7 seconds to complete on an unloaded machine. Each result, that is, each plotted point, is the average of 50 executions.

The effectiveness of the fault recovery procedure can be measured in a very simple way. If the procedure does indeed recover orphaned work, then an application submitted

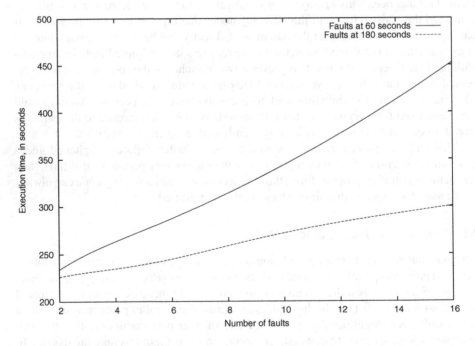

Fig. 2. This graph illustrates the effect of timing on fault recovery. Two sets of experiments were carried out, on a Compeer network initially of size 32. In the first set peers failed 60 seconds after the computation began. In the second set, they failed after 180 seconds. For both cases the experiments were repeated varying the number of faulty peers. The graph illustrates that the earlier a fault happens, the greater its effect.

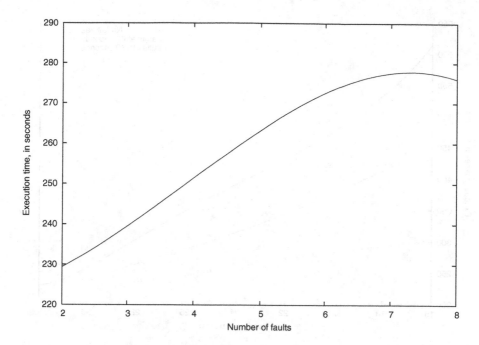

Fig. 3. This graph illustrates the effect of regularly spaced failures on fault recovery, on a Compeer network initially of size 32. In each case, the first failure occurred 30 seconds after the computation. Failures took place 30 seconds apart.

to a Compeer network of N peers, F of which fail during the computation, should take less time to execute than if it were executed solely on $N - F$ peers without faults. Letting T_N^A represent the time taken to execute an application, A, on N peers with no failures, then the time taken for an application run on N peers, F of which fail, should be less than T_{N-F}^A, and will most likely be greater than T_N^A. Several factors can affect the amount of work recoverable after a fault, and therefore the measured performance. One of these factors is the time a fault occurs.

First, some results are presented which demonstrate the effect that the timing of faults has on recovery. These illustrate that work is being carried out, and recovered, since otherwise the timing of faults would yield no difference. These are followed by some results which demonstrate the effectiveness of the method.

4.2 Effect of Timing on Fault Recovery

The time a fault occurs has a clear effect on the amount of work which can be recovered. For instance, a peer which fails directly after a computation begins is likely to have contributed less work than a peer which fails directly before a computation is about to terminate, and the amount of work contributed by a peer directly affects the amount that can be recovered, and therefore the amount that must be redone.

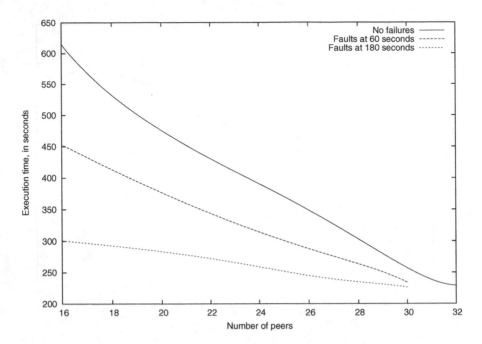

Fig. 4. This graph shows the execution time measured from executions of an application on a Compeer network initially of size 32. After a time a number of peers failed. One line shows the effect of peers failing after 60 seconds, and one line shows the effects of failing at 180 seconds. The other line shows the execution time resulting from an execution with no failures. The data is plotted using the execution time and the number of peers remaining at the end of the computation.

Figure 2 illustrates the effect that timing has on the measured performance. Executions in which multiple peers simultaneous fail 60 seconds after the computation begins finish later than executions in which the failures occur at 180 seconds. This follows since the failed peers had 120 seconds longer in which to execute work and distribute results than in the first case, and therefore left more recoverable work. Figure 3 illustrates the effect of multiple faults regularly spaced at intervals of 30 seconds. In this case, some peers fail having completed very little work, while some have completed a lot. Thus, the resulting graph lies between those in Figure 2.

4.3   Effectiveness of Fault Recovery

Figures 4 and 5 demonstrate the effectiveness of the fault recovery procedure. These graphs were produced by plotting the execution time against the number of peers remaining after all failures. This allows an easy comparison between the T_N^A and T_{N-F}^A measurements. If the the graph representing the execution times with failures is lower than the graph representing execution without failures, then the fault recovery procedure can be considered effective, since it signifies that not all of the work carried out by failed peers is being redone.

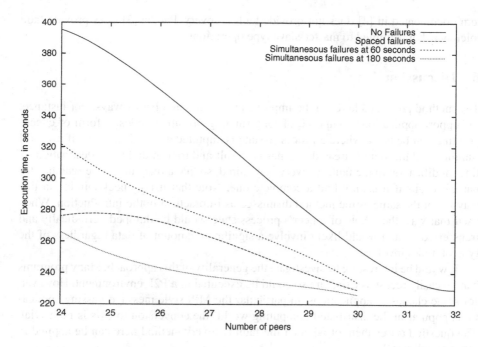

Fig. 5. This graph is similar to Figure 4, but includes an extra line which shows the effect of regularly spaced peer failures. Again, the data is plotted using the execution time and the number of peers remaining at the end of the computation.

Figure 4 shows that even in the face of catastrophic failure, where half of the peers fail simultaneously, the procedure is effective.

Figure 5 shows that, as expected, the performance is better when failures occur at different times over the course of the computation than when they all occur early in the computation. Similarly, the performance is worse than the case where all failures occur late in the computation. Thus, the graph representing spaced failures lies between those representing simultaneous failures occurring early and late in the computation.

5   Related Work

There aren't many distributed computers that qualify as P2P, and of those none provide both fault tolerance with a non-master-slave approach. Systems such SETI@Home[1], while referred to as P2P are centralized in implementation, and so are not comparable to the work presented here. The HYDRA[5] project is decentralized but does not provide fault tolerance. The DREAM[8] system can tolerate faults, but does not recover from them; it is designed to execute only applications that do not require every task to complete. A nameless system[17] based on the JXTA framework[2], which uses a modified version of the master-slave approach, a measure of fault-tolerance but appears to be vulnerable to a single point of failure at the Monitor. The P2P cycle sharing system documented in [4] does not provide fault recovery. ParCop[3] does provide fault tolerance, but is limited to master-slave type operations.

6 Discussion

The method presented here can be improved upon in a number of ways. For instance, to support applications composed of long-running primitive nodes, a form of check-pointing can be used, where a process's state is captured every X minutes, if it is still running, and the state is treated as a partial result and transmitted to its destination. A tiny modification to the fault recovery is required, so that a recreated node searches for partial results if it cannot find a complete one. Note that using checkpointing in this way is not the same as the method dismissed as impractical in the introduction. When used that way, the whole of a peer's process state would be dumped periodically and transferred, which would likely involve a significant amount of data regardless of the type of nodes involved.

It would be interesting to investigate the generality of this approach. Many programs can be easily recast as CGs, and so could be executed in a P2P environment. However, for some classes of applications, in particular the MPI style message-passing applications popular in the distributed computing world, the conversion to CGs is not trivial. The question arises then, of whether the fault recovery method here can be applied in other paradigms ?

In one sense, the messages passed by processes can be viewed as results. In the event of a process failure, the code that initialized the process could re-execute in a recovery mode, such that it only recreates any failed processes. Messages would need to be given an identifier which would relate them to the task that produced them. For example, in a message passing implementation of Conway's Life, each message could be given an identifier which would tie it to the area processed, and the step which produced it. When a recreated process executes, it would examine the orphan registry for results before executing the code. In this way it would be able to transmit its messages without having to recompute anything. The surviving processes would retransmit theirs, if required, from a local cache. Either message passing library, or the processes themselves, would need to be extend the concept of a message, so that it can be associated with a parent, and can be alerted when its parent dies. Since each process would typically be passing messages directly to a small number of other processes, this could be optimized to use separate caches of messages, segregated based on the originating or receiving process. Then information on all messages in a cache can be stored in the orphan registry if the parent associated with that cache fails.

7 Conclusion and Discussion

Traditional approaches to fault tolerance are not suited to the dynamicity of the P2P environment. This paper presented a method which exploits the DHT aspect of structured overlay networks to provide robust, fully decentralized fault recovery, in a manner appropriate for P2P systems. It is experimentally tested and shown to be effective, even in the face of catastrophic failures.

This method provides a solution to one of the major obstacles to P2P distributed computing. It is hoped that this is a step on the road to the (perhaps inevitable [6]) development of world scale, decentralized, dependable P2P computers.

References

1. The SETI@Home home page. http://setiathome.ssl.berkerly.edu, 2002.
2. The JXTA Project. http://www.jxta.org/, 2003.
3. Nidal Al-Dmour and W.J. Teahan. ParCop: A Decentralized Peer-to-Peer Computing System. *Third International Symposium on Parallel and Distributed Computing*, pages 162–168, July 2004.
4. Y. Charlie Hu Ali Raza Butt, Xing Fang and Samuel Midkiff. Java, Peer-to-Peer, and Accountability: Building Blocks for Distributed Cycle Sharing. *Proceedings of the 3rd USENIX Virtual Machines Research and Technology Syposium*, May 2004.
5. Marian Bubak and Pawel Paszczak. HYDRA - Decentralised and Adaptive Approach to Distributed Computing. In *Applied Parallel Computing, New Paradigms for HPC in Industry and Academia, 5th International Workshop, PARA 2000*, volume 1947 of *Lecture Notes in Computer Science*, pages 242–249. Springer, 2000.
6. Ian Foster and Adriana Iamnitchi. On Death, Taxes, and the Convergence of Peer-to-Peer and Grid Computing. In *2nd International Workshop on Peer-to-Peer Systems (IPTPS'03)*. Linkping University, Sweden, 2003.
7. K. Labonte J. Chu and B. Levine. Availability and Locality Measurements of Peer-to-Peer File Systems. *Scalability and Traffic Control in IP Networks, ITCom*, July 2002.
8. M. Jelasity, M. Preu, and B. Paechter. Maintaining Connectivity in a Scaleable and Robust Distributed Environment. In *Proceedings of the IEEE International Symposium on Cluster Computing and the Grid*, pages 389–394, Berlin, Germany, May 2002.
9. P. Maymounkov and D. Mazieres. Kademlia: A peer-to-peer information system based on the xor metric. In *In Proceedings of IPTPS02, Cambridge, USA, March*, 2002.
10. John P. Morrison and Keith Power. Compeer: Peer-to-Peer Applications on a Peer-to-Peer DCOM Architecture. In *Parallel and Distributed Computing and Systems*, Anaheim, California, USA, August 2001.
11. John P. Morrisson. *Condensed Graphs: Unifying Availability-Driven, Coercion-Driven and Control-Driven Computing*. PhD thesis, Technische Universiteit Eindhoven, October 1996.
12. Keith Power. *Compeer: A Scalable, Self-Organizing, Peer-to-Peer Metacomputer*. PhD thesis, National University Ireland Cork, December 2003.
13. Sylvia Ratnasamy, Paul Francis, Mark Handley, Richard Karp, and Scott Shenker. A Scalable Content Addressable Network. In *Proceedings of ACM SIGCOMM 2001*, 2001.
14. Antony Rowstron and Peter Druschel. Pastry: Scalable, Decentralized Object Location, and Routing for Large-Scale Peer-to-Peer Systems. *Lecture Notes in Computer Science*, 2218:329–??, 2001.
15. Ion Stoica, Robert Morris, David Karger, M. Francs Kaashoek, and Hari Balakrishnan. Chord: A scalable peer-to-peer lookup service for internet applications. In *Proceedings of the 2001 conference on applications, technologies, architectures, and protocols for computer communications*, pages 149–160. ACM Press, 2001.
16. Andrew S. Tanenbaum and Maarten van Steen. *Distributed Systems: Principles and Paradigms*. Prentice Hall, New York, 2002.
17. Jerome Verbeke, Neelakanth Nadgir, Greg Ruetsch, and Ilya Sharapov. Framework for Peer-to-Peer Distributed Computing in a Heterogeneous, Decentralized Environment. In *Proceedings Third International Workshop Grid Computing - GRID 2002*, Baltimore, MD, USA, November 2002.

A Distributed Location Identification Algorithm for Ad hoc Networks Using Computational Geometric Methods

Koushik Sinha and Atish DattaChowdhury

Honeywell Technology Solutions Lab,
Bangalore 560076, India
sinha_kou@yahoo.com, atish.chowdhury@honeywell.com

Abstract. We present here a novel approach where we identify a *region* within which a node is *guaranteed* to be found, in contrast to the existing approaches where no such confining region for a node can be guaranteed, but only the location could be estimated either with no definitive error bound or only with some probabilistic error. The location identification algorithm presented here minimizes the size of this region, using computational geometric methods. The proposed technique iteratively improves the *region of residence* of all the nodes in the network through the exchange of region information among neighbors in $O(nD)$ time, where n and D are the number of nodes and diameter of the network respectively. Simulation results also show encouraging results with this approach.

1 Introduction

Most location estimation systems in mobile and ad hoc networks utilize the fundamental method of using *trilateration* and *triangulation* of received signals to obtain an estimate of the receiver's position [8]. An LOS path or direct path, is the straight line connecting the transmitter and the receiver. NLOS signals occur due to multi-path conditions in which the received signals come from either reflected, diffracted or scattered paths, thus introducing excess path lengths in the actual euclidian distance between the transmitter and the receiver. The NLOS error is defined to be the excess distance traversed compared to the direct path and is always positive. The corruption of LOS signals by NLOS signals and Gaussian measurement noise are the major sources of error in all location estimation systems, the former being the dominant factor [12].

The Global Positioning System (GPS) [9,15] is perhaps the most widely publicized location-sensing system. Unfortunately, GPS does not scale well in dense urban areas or in indoor locations. Modeling of the radio propagation environment helps in providing a more accurate location estimate by mitigating the effect of NLOS errors. While reasonably accurate radio propagation models exist for outdoor conditions[7,14,16], unfortunately there are no such unanimously accepted models for indoor environments. Several authors have, however, attempted for mitigating the effect of NLOS errors [2,3,4,5,6,10,11]. In the absence of a suitable model for predicting the location of a mobile terminal, it is possible that the node may be far away from the estimated point.

In this paper, rather than doing a location prediction as mentioned above, we consider the location discovery problem in terms of finding the *region* where a node is

D.A. Bader et al. (Eds.): HiPC 2005, LNCS 3769, pp. 62–72, 2005.

guaranteed to be found. The objective of the location identification process then becomes minimizing the size of such a region. We assume that a small percentage of the terminals (nodes) in the network know their locations with a high degree of accuracy - possibly through GPS access or by some other means. We term such nodes as reference nodes. We propose a distributed algorithm using computational geometric techniques to compute the smallest *region of residence* where a node is guaranteed to be found, for all non-reference nodes in the network. A unique feature of our algorithm is that the location regions of the nodes in the network are improved through the exchange of location information between the neighbors in $O(nD)$ time, where n and D are the number of nodes and diameter of the network respectively. Simulation results also demonstrate that our algorithm succeeds in finding reasonably small stable regions of residence for all non-reference nodes.

2   System Model

We model an ad hoc network scenario as a graph $G = (V, E)$, consisting of n nodes. V is the set of all nodes, $|V| = n$ and E is the set of edges in the graph G. The nodes may be either stationary or mobile. All communication links are assumed to be bi-directional. We say node v is a *neighbor* of u, if they are within each other's hearing zone. The neighborhood, $N(i)$, of a node i consists of all nodes that within its transmission range. A small percentage of the nodes are assumed to know their individual locations with high precision, either through GPS or some other means. These nodes serve as the *reference nodes* (RN) in the network. Initially, nodes other than the RNs do not possess any knowledge of their location. The reference nodes are assumed to possess point locations (zero area regions) while the non-reference nodes are initially assumed to reside in a region of infinite size. However, in practice, the reference nodes can have any arbitrary shaped location region. $Ref_n = \{u : u \in V, u \text{ is a reference node}\}$ denotes the set of reference nodes.

Considering thermal noise at the receiver, NLOS errors and channel characteristics, the measured range between two nodes u and v can be expressed as

$$r_{uv} = d_{uv} + \eta_{uv} + c\tau_{uv} \tag{1}$$

where d_{uv} represents the unknown Euclidian distance between u and v. η_{uv} completely models the combined additive effects of thermal receiver noise, signal bandwidth and signal-to-noise ratio. η_{uv} has been shown to be a zero-mean normal random variable and hence can be either measured or pre-computed [5]. We assume η_{uv} to be always additive. $c\tau_{uv}$ represents the NLOS distance error and is the dominant error contributor [12], c being the speed of light in air. We define the set RR to be the set of all such measured ranges for all node-pairs in the network, i.e., $RR = \{r_{ij} : r_{ij} \in E, \forall i, j \in V\}$. Also, $RR_i = \{r_{ij}, j \in N(i)\}$.

3   Preliminaries

Our proposed algorithm is based on the triangulation technique to compute the region where a node is guaranteed to be found.

Definition 1. The *region of residence*, R_i of a node i is defined to be the region where i is guaranteed to be found.

The region of residence of a reference node is assumed to be a point location of zero area. All other nodes have a non-zero finite area region of residence. Our objective is to find the minimum region of residence of a node i.

Lemma 1. *The range measurements obtained for a node u, from a neighbor $v \in N(u)$ will always be greater than or equal to the Euclidian distance between u and v.*

Proof: Follows directly from equation 1.   □

Given two nodes u and v and a range measurement r_{uv} from v to u, the region of residence of u in the view of node v is the region formed by extending v's region residence R_v in every direction by the measured range value, r_{uv}. We denote this operation by the operator $\oplus$, whose left operand is a region of residence and the right operand is a range value. Thus the region of residence of u in the view of node v is $R_{uv} = R_v \oplus r_{uv}$. We call R_{uv} as the *viewed region of residence* of node u.

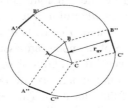

Fig. 1. Construction of viewed region of residence of a node

Example 1. Suppose node v has a triangular region of residence $\triangle ABC$, and the range of u measured by v is r_{uv} as shown in Figure 1. We draw a line $A'B'$ parallel to AB at a distance r_{uv} from AB and on the opposite side of the node C such that $AA'B'B$ forms a rectangle. Similarly, we draw a line $B''C'$ parallel to BC on the side opposite to that of A and distant r_{uv} from BC so as to form a rectangle $BB''C'C$, and also a line $C''A''$ parallel to CA on the side opposite to that of B and distant r_{uv} from CA so that $CC''A''A$ is a rectangle. Now from the point A, draw a circular arc of radius r_{uv} so as to cut the lines $A'B'$ and $C''A''$ at A' and A'' respectively. Similarly, draw two other circular arcs of radius r_{uv} : i) from B to cut the lines $A'B'$ and $B''C'$ at B' and B'' respectively, and ii) from C to cut the lines $C'B'$ and $C''A''$ at C' and C'' respectively. The closed convex region $A'B'B''C'C''A''$ is the resulting R_{uv}.

It may be mentioned here that the region R_{uv} can also be viewed as the Minkowski's sum [18] of the region of residence R of the node v and a circle of radius r_{uv} centered at origin. We assume that the initial regions of residence of all nodes are bounded either by straight line segments or by circular arcs. Hence, the region R_{uv} will also be bounded by straight line segments and/or circular arcs only. We state the following result without proof.

Lemma 2. *Node u is guaranteed to be found at some location inside R_{uv}.*

Theorem 1. *The current minimum region of residence, $\mathcal{R}_u$ of a node u, based on the information from its neighbors is the region formed by the intersection of the viewed regions of residence R_{ui}'s, $i \in N(u)$, i.e., $\mathcal{R}_u = \bigcap_{i \in N(u)} R_{ui}$.*

Proof: The proof follows from lemma 2 as the common intersection region is the smallest region that satisfies lemma 2 for all neighbors $i \in N(u)$.   □

Note that this current minimum region of residence may subsequently get refined (contracted in size) by improved viewed regions of residences from its neighbors.

Theorem 2. *The minimum region of residence of a node u, $\mathcal{R}_u$ (based on the information from its neighbors) can not subsequently be made larger by an altered viewed region of residence, R_{ui} from any neighbor i.*

Proof: The proof follows directly from theorem 1.   □

To find the minimum region of residence, our algorithm proceeds in two steps :

- In the first step, every node u in the network determines its current region of residence by ranging with each of its neighbors.
- Once u has determined its current minimum region of residence, it attempts to improve the regions of residence of each neighbor, using its own region of residence and the range measurements that it obtained from the respective neighbors.

Example 2 below illustrates the working of our algorithm :

Example 2. Consider a node u with three neighbors i, j and k. Figure 2 demonstrates a probable situation where the triangles $\triangle ABC$ and $\triangle DEF$ define the region of residence of nodes i and j respectively. $PQRS$ is the region of residence of node k. For simplicity, we assume the regions of residence as polygonal. Let r_{1u}, r_{2u} and r_{3u} be the range measurements that u obtains by ranging with i, j and k respectively. According to the view of node i, u lies in the region R_{iu}, dictated by the shape $A'B'B''C'C''A''$ as demonstrated in example 1. Similarly, $D'E'E''F'F''D''$ and $P'Q'Q''R'R''S'S''P''$ define the region of residence of u in the views of the nodes j and k respectively. Following theorem 1, the shaded region LMN defines the minimum region of residence of node u, where u is guaranteed to be found.

Once the minimum region of residence of node u is found, node u then tries to refine the minimum region of residence of a neighbor v, $\forall v \in N(u)$, using $\mathcal{R}_u$ and the corresponding measured range r_{uv} from v. The new minimum region of residence of node v, $\mathcal{R}_v'$ is defined as the intersection of the viewed region of residence of v by u, R_{vu} and the current minimum region of residence of v, $\mathcal{R}_v$. In Figure 3, region LMN defines the minimum region of residence of node u. r_{3u} is the measured range from k to u. The region defined by the dotted lines, $L'M'N'$ defines the viewed region of residence, R_{ku} of k by u. The region $UN'VQ$ is the intersection region of the current minimum region of residence of k, $PQRS$ and $L'M'N'$. Following theorem 1, the new minimum region of residence of node k is the region $UN'VQ$. Node u tries to similarly improve the regions of nodes i and j using the measured ranges r_{1u} and r_{2u} and its minimum region of residence, $\mathcal{R}_u$.

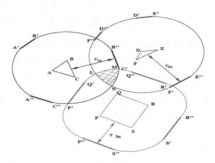

Fig. 2. Computation of minimum region of residence

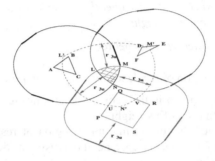

Fig. 3. Improving the minimum regions of residence of neighbors

A careful scrutiny of figure 3 in the above example reveals that the part of the boundary of the region of residence of k which causes a computation (improvement) in the region of residence for u, and the part of the boundary of the region of residence for k which is refined (improved) due to this computed part of the region of residence for u, are mutually disjoint. This observation holds even if the node k would have an initial region of residence of a different shape.

Lemma 3. *Given a minimum region of residence of a node u, $\mathcal{R}_u$ and a measured range r_{uv} from a neighbor v, the improved minimum region of residence of v, $\mathcal{R}'_v$ is given by the intersection of the viewed region of residence of v by u, R_{vu} and the current minimum region of residence of v, $\mathcal{R}_v$.*

$$\mathcal{R}'_v = \mathcal{R}_v \cap R_{vu} \qquad (2)$$

Proof: The proof follows directly from lemma 2 and theorem 1. □

From lemma 3, we get, $R'_v \subseteq R_v$. Note that R'_v is generated by introducing some extra arc and/or straight line segments on R_v due to the region computation initiated by the node u. Let us denote these set of new arcs and straight line segments by E^v_u. Elements of E^v_u are either parallel to some boundary edge of R_u or a circular arc of a circle with radius r_{vu}, centered at some vertex on the boundary of R_u.

Suppose there is a path in the network starting from a node u_0 to some node u_k, given by $u_0 \ u_1 \ u_2 \ \cdots u_k$. If u_0 initiates its region computation by its neighbors and gets its region computed by these neighbors as R_{u_0}, then R_{u_0} may cause an improvement in the region of node u_1. This, in turn, may cause an improvement in the region of the node u_2, and so on, so that the process of region refinements may successively follow through the nodes $u_1, u_2, \cdots, u_k$. In particular, if we now assume that $u_k = u_0$, i.e., $u_0 \ u_1 \ u_2 \ \cdots u_k$ is a cycle, then we claim that this process of successive region refinements will not be able to further refine the region R_{u_0} of u_0 after a finite number of steps. To establish this claim, we proceed as follows.

Let $E_{u_0}^{u_1}$ denote the set of newly introduced lines and/or arcs on the boundary of the minimum region of residence of u_1 due to the region computation of u_0 caused by all the immediate neighbors of u_0. The changed region R'_{u_1} of u_1 due to $E_{u_0}^{u_1}$ may cause a change in R_{u_2} by introducing some new lines and/or arcs which we denote by the set $E_{u_0, \, u_1}^{u_2}$. In general, we denote the set of newly added lines and/or arcs in the region of u_j, $1 \le j \le k$, by $E_{u_0, \, u_1 \, \ldots u_{j-1}}^{u_j}$. Because of the properties of Minkowski's sum of $R'_{u_{j-1}}$ and a circle of radius r_{u_{j-1}, u_j} (range value between nodes u_{j-1} and u_j) with center at the origin, we see that for any j, $1 \le j \le k$, two possible cases may arise :

Case 1: A line segment (arc) in $E_{u_0, \, u_1 \, \ldots u_{j-1}}^{u_j}$ is parallel to some line segment (arc) in $E_{u_0, \, u_1 \, \ldots u_{j-2}}^{u_{j-1}}$ (for $j > 1$) or in R_{u_0} (for $j = 1$).

Case 2: An arc in $E_{u_0, \, u_1 \, \ldots u_{j-1}}^{u_j}$ is

i) not parallel to any arc in $E_{u_0, \, u_1 \, \ldots u_{j-2}}^{u_{j-1}}$ (for $j > 1$) or in R_{u_0} (for $j = 1$),
ii) but is an arc of a circle with radius r_{u_{j-1}, u_j}, having center at one point on the region $R'_{u_{j-1}}$ which is the point of intersection of two different arcs or two different line segments or an arc and a line segment, at least one of which must be in $E_{u_0, \, u_1 \, \ldots u_{j-2}}^{u_{j-1}}$ (for $j > 1$) or in R_{u_0} (for $j = 1$).

This fact is illustrated in Figure 4 where the arcs α and β on $R'_{u_{j-1}}$ and R'_{u_j}, respectively are parallel to each other, while the arc γ on R'_{u_j} is derived from the point Q on $R'_{u_{j-1}}$ (with Q as center and a radius equal to r_{u_{j-1}, u_j}. We also see from Figure 4 that for every point on R'_{u_j}, there exists a unique point on $R'_{u_{j-1}}$ from which this point was derived. Thus, for the point T on R'_{u_j}, the corresponding point on $R'_{u_{j-1}}$ is Q, which is transitively derived from a point P on R_{u_0}.

Lemma 4. *Let T be any point on $E_{u_0, \, u_1 \, \ldots u_{j-1}}^{u_j}$, and P be the corresponding point on R_{u_0} from which T was derived. The Euclidean distance PT is always greater than or equal to the maximum of all $(r_{u_{j-1}, u_j}, \forall j, 1 \le j \le k)$.*

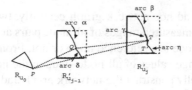

Fig. 4. Refinements of regions of successive neighbor nodes

Proof: We prove this by induction on j. Our claim is trivially true for $j = 1$. Suppose the claim is true for $j = 1, 2, \cdots, j - 1$. For $j \geq 1$, referring to Figure 4, the Euclidean distance PQ is then greater than or equal to the maximum of $(r_{u_0,u_1}, r_{u_1,u_2}, \cdots, r_{u_{j-2},u_{j-1}})$. Now, if the point T is on an arc or line segment in $E^{u_j}_{u_0, \, u_1 \ldots u_{j-1}}$ parallel to an arc or line segment in $E^{u_{j-1}}_{u_0, \, u_1 \ldots u_{j-2}}$, then the Euclidean distance $PT = PQ + QT$, from which the result follows as $QT = r_{u_{j-1},u_j}$. If, however, the point T is on an arc/line segment in $E^{u_j}_{u_0, \, u_1 \ldots u_{j-1}}$, not parallel to any arc/line segment in $E^{u_{j-1}}_{u_0, \, u_1 \ldots u_{j-2}}$, then the corresponding point Q on $R'_{u_{j-1}}$ from which T is derived, must be the point of intersection of two different arcs and/or line segments, as explained above. Without loss of generality, let Q be the point of intersection of two arcs α and δ in $E^{u_{j-1}}_{u_0, \, u_1 \ldots u_{j-2}}$, as shown in Figure 4. The arc δ is mapped to the parallel arc η in $E^{u_j}_{u_0, \, u_1 \ldots u_{j-1}}$. Let the arcs γ and η intersect at the point T'. Hence, the line QT' is normal to the tangent to the arc δ, and it follows that ΔPQT is an obtuse-angled triangle with the obtuse angle at point Q. This implies that the distance PT is greater than either of PQ and QT, and hence the lemma follows. $\qquad\square$

Thus, we see that the part of the region boundary of a neighbor u_j of u_0, $\forall j, 1 \leq j \leq k$ which gets modified (refined) due to the region computation initiated by u_0, is always at a distance greater than or equal to r_{u_0,u_1} from the corresponding part of R_{u_0} which caused this refinement of R_{u_j}. Hence we get the following important result which guarantees the termination of the successive refinement scheme.

Theorem 3. *If a node u initiates its region computation with the help of range readings from all of its neighbors, the computed region R_u of u may cause refinements of the successive neighbors through the whole network, but it will never be able to further refine R_u of u itself.* $\qquad\square$

Definition 2. The stable region of residence of a node u is the minimum region of residence of u which can not be further improved upon using the current global set of range readings for all node pairs in the network. We denote such a region by $\mathcal{S}_u$.

Theorem 4. *A node u can compute its stable region of residence once it gets the range readings of all possible directly communicating nodes in the network along with the initial region information of all nodes.*

Proof: Omitted due to brevity. $\qquad\square$

Theorem 5. *The computation of the stable regions of residence of all nodes in the network is functionally equivalent to an all-to-all broadcast of the range information of all node-pairs in the network (the set RR) and the set Ref_n.*

Proof: To reconstruct the ad hoc network graph centrally, two pieces of information would be required : (i) the measured ranges of all node pairs and, (ii) the information as to whether an individual node is a reference node or not. From theorem 4, we see that if a node possessed the range values of all node-pairs in the network (the set RR) and the set Ref_n, it could locally construct the network graph and then compute the stable regions of residence of all nodes. Since the possession of the set RR and the set Ref_n by a node in the network effectively implies a broadcast of these two sets, if every node were to locally compute the stable regions of residence, the problem maps to that of an all-to-all broadcast of the set RR_i and status (whether its a reference node) of each node i in the system. Each node on receiving this information from a neighbor would attach its own RR_i set and its status and broadcast the message again. $\square$

4   Proposed Location Identification Algorithm

Every node in the network maintains a local variable $status$, which is set to 1 if the node is a reference node, zero otherwise. Initially, the minimum regions of residence of all non-reference nodes are assumed to be infinity. Each node i does a ranging with its neighbors to obtain a set of measured ranges, RR_i. i then computes the viewed region of residence for every node $j \in N(i)$. Node i then exchanges three pieces of information with each neighbor $j \in N(i)$ - i) value of the status variable, $status_i$, ii) viewed region of residence of j, R_{ji}, iii) area of the current minimum region of residence of i, A_i.

Let $RR_i = \{r_{ij} : j \in N(i)\}$, $\mathcal{T}$ = set of viewed regions of residence, R_{ij}. Once node i has its viewed region of residence, R_{ij} from its every neighbor j, it computes its current minimum region of residence using the following algorithm :

Function compute_region : **Boolean**
var A_{old}, A_i : **Real**;
begin
    for each $R_{ij} \in \mathcal{T}$ such that $A_j \neq \infty$ **do**
        /* Compute the minimum region of residence of node i from the viewed regions */
        $\mathcal{R}_i \leftarrow \mathcal{R}_i \cap R_{ij}$;
    endfor;
    $A_i \leftarrow$ Area of $\mathcal{R}_i$;
    if $A_i < A_{old}$ **then return** true;   /* $\mathcal{R}_i$ improved */
    else return false;   /* No improvement in $\mathcal{R}_i$ */
end.

Once node i computes its minimum region of residence, it tries to improve the minimum region of residence of each of its neighbors as follows:

Procedure improve_region
begin
    for each $j \in N(i)$ such that $status_j \neq 1$ **do**
        /* Construct R_{ji}, the viewed region of j */
        $R_{ji} = \mathcal{R}_i \oplus r_{ij}$;
        Transmit R_{ji} to node j;
    endfor;
end.

The following location identification algorithm is executed by each node i until the node attains its stable region of residence, $\mathcal{S}_i$.

Algorithm location_region_identify
var $region\_change\_flag$: **Boolean**;
begin

```
    while (true)
        Get neighbor set N(i);
        Generate RR_i : measure range with every neighbor j ∈ N(i);
        region_change_flag = false;
        repeat
            Get T : viewed regions of residenc R_ij from every neighbor j;
            if status_i = 0 then region_change_flag = compute_region(i, T);
            improve_region(i, N(i), RR_i);
        until region_change_flag = false;    /* iterate until R_i = S_i */
    endwhile;
end.
```

4.1 Complexity Analysis

The operations *compute_region* and *improve_region* both need $O(\Delta)$ time in the worst case, where Δ is the maximum node degree in the network. Also, the operation *Get T* needs $O(\Delta)$ time. If we assign a distinct time slot to each node depending on its unique *id* number [13] to avoid collision, then each iteration of the *repeat* loop would need $O(n)$ time slots. From theorem 5, the algorithm terminates when an all-to-all broadcast of the sets RR and Ref_n is achieved. Assuming that the effect of the viewed range information of all node pairs can be transmitted in $O(1)$ time slots, the whole communication process will be completed in $O(D)$ rounds, D being the diameter of the network (each round consists of $O(n)$ time slots for message communication from all nodes in a layer to the nodes in the next layer [13]). Hence, the algorithm *location_region_identify* will need $O(nD)$ time.

5 Simulation Results

We experimented with various graph topologies by randomly generating static ad hoc graphs. All communications were assumed to be symmetrical. We assumed a transmission range of 30 units of distance for all nodes in the system, and an environment with a mix of LOS and Obstructed LOS (OLOS) signals. The LOS range errors are drawn from a gaussian distribution of mean μ and standard deviation σ, while the OLOS range errors are drawn from an exponential distribution with parameter λ [2]. The parameters μ, σ and λ characterize the channel characteristics and hence the amount of ranging errors. We evaluated the performance of our algorithm against varying channel characteristics, percentage of LOS/OLOS signals and percentage of reference nodes in the system. For each set of parameters, the experiments were repeated for 100 randomly generated graphs and the area of the stable region of residence of all nodes in the system were measured. Figures 5 and 6 demonstrate the median area of the stable regions of residence for random graphs of 100 nodes. The plots of mean area of the stable regions of residence look similar to their median counterpart, with the difference that the mean areas are slightly larger than the median areas. We chose to measure the median area as it is less affected by outlier cases. The mean area, on the other hand, can get significantly affected by situations such as a non-reference node, i having only one neighbor (say j), in which case $R_i = R_{ij}$ and the area of R_{ij} can be either very small or

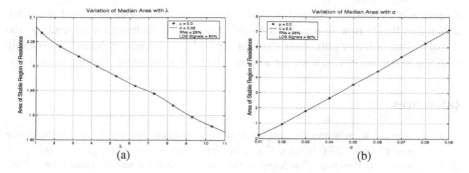

Fig. 5. Variation of size of stable region of residence with λ and σ

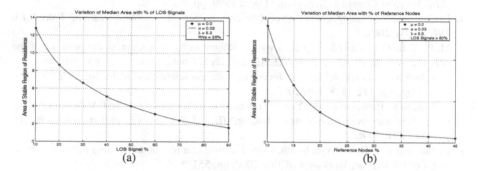

Fig. 6. Variation of size of stable region of residence with % of LOS signals and RNs

Table 1. Variation of median area of stable regions of residence with node density

Node Density (nodes/sq. units)	% of RNs	Median area of stable region (sq. units)
0.01	25	1.980
0.0075	25	2.656
0.005	25	5.249
0.003	25	10.275

quite large, depending on the ranged distance between the two nodes. The plots for the mean area of the stable regions of residence have been omitted due to space constraints.

Table 1 shows the variation of median area of stable regions of residence with node density, measured in number of nodes per square unit. The areas depicted are for $\lambda = 5.3$, $\mu = 0.0$, $\sigma = 0.03$, 80% LOS signals and 25% of the nodes designated as RNs.

6   Conclusion

We have presented a novel approach to the problem of location discovery in an ad hoc network using computational geometric methods. In contrast to the existing approaches of estimating a point location, algorithm *location_region_identify* computes the region of residence for a node, where it is *guaranteed* to be found. The proposed algorithm takes only $O(nD)$ time to identify the stable regions of residence for all nodes in the network. Simulation results show that our algorithm succeeds in finding reasonably small stable regions of residence for all non-reference nodes under varying conditions.

References

1. M. Kanaan and K. Pahlavan, "A comparison of wireless geolocation algorithms in the indoor environment," *Proc. IEEE Wireless Commun. Networking Conf.*, 2004, pp. 177 - 182.
2. B. Alavi and K. Pahlavan, "Modeling of the distance error for indoor geolocation," *Proc. IEEE Wireless Commun. Networking Conf.*, 2003, vol. 1, pp. 668-772.
3. P-C. Chen, "A non-line-of-sight error mitigation algorithm in location estimation," *Proc. IEEE Wireless Commun. Networking Conf.*, 1999, pp. 316 - 320.
4. L. Cong and W. Zhuang, "Non-line-of-sight error mitigation in mobile location," *Proc. IEEE INFOCOM*, 2004.
5. M. P. Wylie-Green and S. S. (Peter) Wang, "Robust range estimation in the presence of the non-line-of-sight error," *Proc. 54th IEEE Vehi. Tech. Conf.*, 2001, vol. 1, pp. 101-105.
6. K. Pahlavan, X. Li and J. Mäkelä, "Indoor geolocation science and technology," *IEEE Commun. Mag.*, pp. 112 - 118, Feb. 2002.
7. A. Falsafi, K. Pahlavan and G. Yang, "Tranmission techniques for radio LAN's - a comparative performance evaluation using ray tracing," *IEEE J. Sel. Areas Commun.*, vol. 14, no. 33, pp. 477 - 491, 1996.
8. K. Sinha and N. Das, "Exact location identification in a mobile computing network," *Proc. Intl. Conf. Par. Proc. Workshop* (ICPP), 2000, pp. 551 558.
9. J. Hightower and G. Borriello, "Location systems for ubiquitous computing," *IEEE Computer*, vol. 34, No. 8, pp. 57 - 65, 2001.
10. P. Bahl and V. Padmanabhan, "RADAR : an in-building RF-based user location and tracking system," *Proc. IEEE Infocom*, 2000, pp. 775 - 784.
11. S-S Woo, H. R. You and J. S. Koh, "The NLOS mitigation technique for position location using IS-95 CDMA networks," *Proc. IEEE Vehi. Tech. Conf.*, 2000, vol. 4, pp. 2556 - 2560.
12. J. J. Caffery and G. L. Stüber, "Subscriber location in CDMA cellular networks," *IEEE Trans. Vehi. Tech.*, vol 47, pp. 406 - 416, 1998.
13. S. Basagni, D. Bruschi and I. Chlamtac, "A mobility transparent deterministic broadcast mechanism for ad hoc networks", IEEE Trans. Network., Vol. 7, pp. 799-807, Dec. 1999.
14. L. J. Greenstein et. al., "A new path-gain/delay-spread propagation model for digital cellular channels," *IEEE Trans. on Vehi. Tech.*, vol. 46, no. 2, pp. 477 - 485, May 1997.
15. B. Parkinson et. al, "Global positioning system: theory and application," *Progress in Astronautics and Aeronautics*, vol. 163, 1996.
16. S. Fischer et. al, "Time of arrival estimation of narrowband TDMA signals for mobile positioning," *IEEE Intl. Sym. Pers., Indoor and Mobile Radio Comm.*, 1998, vol. 1, pp. 451-455.
17. S. Capkun, M. Hamdi and J-P. Hubaux, "GPS-free positioning in mobile ad hoc networks," *Proc. Hawaii Intl. Conf. Sys. Sci.*, Jan. 2001.
18. M. de Berg, M. van Kreveld, M. Overmars and O. Schwarzkopf, *Computational Geometry: Algorithms and Applications.* Springer-Verlag, Heidelberg, 1997.

A Symmetric Localization Algorithm
for MANETs Based on Collapsing
Coordinate Systems

Srinath Srinivasa and Sanket Patil

International Institute of Information Technology,
26/C, Electronics City, Bangalore 560100, India
{sri, sanket.patil}@iiitb.ac.in

Abstract. Localization in mobile ad hoc networks (MANETs) is the
process of fixing the position of a node according to some real or virtual
coordinate system. In many cases, solutions like Global Positioning System (GPS) are not feasible. As a result, several algorithms have been developed for localization based purely on local communication. However,
many of these suffer from one of the following: flooding of the network,
requirement for global knowledge, or the requirement of "beacon" nodes,
which know their absolute position according to GPS. At the very least,
localization algorithms require parts of the system to be either static
or relatively stable. In this paper, we propose a symmetric localization
algorithm that performs fairly accurate localization. No special elements
like beacons and other static elements are required; however, they are
not excluded.

1 Introduction

Mobile ad hoc networks comprise of computational nodes that can communicate
with one another within a given wireless range. There may be no fixed elements
in the network and the network itself may span areas much larger than the
wireless range of a single node. Depending on the amount of mobility in the
system, the network topology may be subject to frequent changes.

This paper addresses the issue of *localization* in mobile ad hoc networks. Localization is the process by which a node positions itself with respect to other
nodes in the network. This positioning may be based on a set of "real" coordinates like the latitude and longitude obtained from GPS, or based on "virtual"
coordinates that applies to the network alone.

GPS based localization has been addressed in [11, 12]. However, GPS based
localization is not always possible or desirable. This is because GPS based systems are expensive and they are susceptible to signal attenuation under thick
foliage, basements, etc.

Localization algorithms using virtual coordinates have been developed that
are based on measuring physical characteristics like signal strength. Some examples are [3, 7, 8, 9, 13]. Localization based on physical characteristics has its own
issues of signal attenuation by obstacles and multipath.

D.A. Bader et al. (Eds.): HiPC 2005, LNCS 3769, pp. 73–82, 2005.
© Springer-Verlag Berlin Heidelberg 2005

An alternative method of localization is purely algorithmic. Here, nodes are only concerned about the other nodes in their vicinity. They have no means for measuring signal strength, delay or movement. Some example algorithmic approaches include [1, 2, 4, 5, 10]. However, existing algorithmic approaches suffer from one or more of the following shortcomings: they require flooding of the network and/or require the use of beacons, which are special nodes that have knowledge of their actual coordinates, and/or knowledge of the perimeter of the network. At the very least, they assume that there is a set of relatively stable nodes that play a central role in the localization process [2].

The algorithm presented in this paper, which we call *Adorn*, is symmetric and requires none of the above. Each node in the network independently computes its coordinates by communication with its neighbours periodically. There is no special requirement on any nodes for the localization process. Beacon nodes are not necessary, although they are not excluded from the model. Beacon nodes can be used to aid in faster convergence of the algorithm.

2   Related Literature

Nagpal et al. [5] and Niculescu and Nath [6] propose variants over GPS based triangulation for localization. Both their algorithms require the presence of beacon nodes (or "landmarks" as they are called in [6]). Other nodes in the network calculate their positions based on their hop distances to at least three beacons.

Priyantha et al. [15] propose an approach that is GPS and beacon free. In the first phase, with the help of a few carefully chosen reference nodes, a fold-free graph embedding is produced that provides an estimate of the structure of the network. The other nodes use the hop counts from the reference nodes to approximate their polar coordinates.

While the above algorithms work well when the network is reasonably static and predictable, they are not suitable when the network's main feature is the unpredictable, highly mobile topology.

Perhaps, our algorithm is closest to the one proposed by Capkun et al. [2], which takes care of node mobility. The algorithm of Capkun et al. proceeds in two phases. In the first phase, every node builds its *local coordinate system*. In the second phase, each node reorients its coordinate system, with respect to a set of nodes that have lower mobility than the rest of the nodes, to achieve a common global orientation.

Iyengar and Sikdar [4] propose an algorithm that claims to be better than [2] by being more frugal in terms of communication.

However, both the above algorithms suffer from the requirement for a set of reasonably static subset of nodes, which form and maintain a global coordinate system. Localization accuracy and convergence depend on the stability of this set of nodes.

The *Adorn* approach do not have the above limitations. Our algorithm does not require any special nodes. Hence, the localization process is symmetric. In highly mobile networks, convergence is chaotic and it is not possible to predict the orientation of the ensuing coordinate system. Convergence can be aided with the help of beacon nodes which know their own coordinates apriori.

3   Adorn Localization

The Adorn model is based on the following assumptions: (a). In an ad hoc network, there need not be any fixed elements; (b) Ad hoc nodes may not be cognizant of their movement; (c). All nodes in the ad hoc network i have the same wireless range r; and (d). Nodes can communicate with other nodes which are in their wireless range; however they do not know the relative orientation of the other node with respect to themselves.

In the Adorn system, nodes maintain a set of 2-dimensional virtual coordinates. Isolated nodes have coordinates $(0, 0)$. When two or more nodes come within the range of one another, they evolve virtual coordinate axes where they place themselves on some points along each axis. The correctness of this placement is a measure of how well the nodes preserve the relative *ordering* among themselves.

Virtual coordinates are tagged with a "network-id" which specifies the coordinate system where its coordinates hold. Whenever two or more nodes with different network-ids come within range of one another, one of the network-ids dominates. Isolated nodes belong to a network whose id is the same as their node id.

Given a set of ad hoc nodes, formation of the virtual coordinate system begins independently at different locations in the network. Each coordinate system has its own network id. Eventually one or a small number of network-ids prevail over the entire network.

Nodes update their coordinates periodically based on the coordinates of their neighbours and the neighbours' neighbours. Nodes are not cognizant of their own movement. Even if they are, they may not be aware of the direction of their movement relative to the axes of the virtual coordinate system. Hence the updation interval does not depend on the movement of nodes.

The formation and maintenance of virtual coordinates is explained from two scenarios: *steady-state* behaviour and *formative* behaviour.

Steady-State Behaviour: Steady-state behaviour is when all nodes have been assigned coordinates and are updating their coordinates periodically.

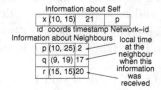

Fig. 1. Response to a RFP message

Whenever a node needs to update its coordinates (or a new node enters into a network that is in steady state), it sends out a RFP (request for positioning) message. Each node that is directly in range receives this message and returns the following information (schematically shown in Figure 1):

- Its network-id
- Its node id and virtual coordinates, tagged with its current timestamp
- Node ids and virtual coordinates of its neighbouring nodes, tagged with their timestamps

The timestamps that are stored along with the coordinates help a node to determine the latest information, especially about the neighbours of neighbours (or the 2-hop neighbours).

Network-ids specify the coordinate system nodes belong to. Whenever a node receives responses having different network ids, the node may decide to change its network id. This decision is based on one of the two strategies for *collapsing coordinate systems*, which are explained in section 4.

After the network-id and positions of the 2-hop neighbours are resolved, the node estimates its own position based on the following rationale: the node has to lie in the region that is the intersection of all the ranges of its 1-hop neighbours. However, it should not lie in any part of this region which lies within range of any of its 2-hop neighbours (schematically shown in Figure 2).

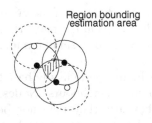

● One hop neighbours
○ Two hop neighbours

Fig. 2. Bounds for the area where the node originating the RFP is likely to be. It is the intersection of ranges of all the 1-hop neighbours(shown in solid lines) minus the common areas of the 2-hop neighbours(shown in dashed lines).

Let $hop1$ be the set of all 1-hop neighbours, and $hop2$ be the set of all 2-hop neighbours. Let $n = |hop1|$ and $m = |hop2|$. The node first places itself at the centroid of all the 1-hop neighbours. That is, $x = \frac{\sum_{i=1}^{n} x_i}{n}$ and $y = \frac{\sum_{i=1}^{n} y_i}{n}$. Here x_i is the x coordinate of the ith neighbour and y_i is the y coordinate of the ith neighbour.

At any estimated point (x, y), there is said to be a "repulsive" force from all 2-hop neighbours, and an "attractive" force from all 1-hop neighbours. The repulsive forces are computed along x and y dimensions as follows.

$$f_2^x = absmax_i(w_2 e^{-d} cos\theta), \ 1 \leq i \leq m$$
$$f_2^y = absmax_i(w_2 e^{-d} sin\theta), \ 1 \leq i \leq m.$$

Here $d = \sqrt{(x - x_i)^2 + (y - y_i)^2}$ is the distance between the currently estimated point and a given neighbouring node. For 2-hop neighbours, the angle between them (θ) is computed as $\theta = tan^{-1}\frac{y-y_i}{x-x_i}$. When $x > x_i$, the x component of the computed force would be positive. The same is true of the y component, and the net effect of the force would be directed towards widening the gap between x_i and x.

The magnitude of the repulsive force decreases exponentially as the distance between the estimated point and the 2-hop neighbour increases. The function *absmax* chooses the corresponding force component whose absolute value is the maximum. If the estimated node lies beyond the nearest 2-hop neighbour, it is sufficient. The term w_2 is a multiplicative factor for the repulsive forces from the 2-hop neighbours. It determines the number of steps by which coordinates are changed during refinement. In our experiments w_2 was kept at a value of 1.

The attractive forces from the 1-hop neighbours are given as follows.

$$f_1^x = absmax_i(w_1(1 - e^{-d})cos\theta), \ 1 \leq i \leq n$$
$$f_1^y = absmax_i(w_1(1 - e^{-d})sin\theta), \ 1 \leq i \leq n.$$

The magnitude of the attractive force increases with distance. The largest absolute value of the attractive forces is chosen to represent all the forces from 1-hop neighbours. The term w_1 is the multiplicative factor for the attractive forces.

The direction of the attractive forces is opposite to the direction of the repulsive forces. This is modeled by calculating the angle θ as follows: $\theta = tan^{-1}\frac{y_i-y}{x_i-x}$. This results in a negative force when $x > x_i$ and serves to diminish the distance between the neighbouring node and the estimated position.

Based on these force vectors improvements are computed separately along x and y dimensions. $x(k + 1) = x(k) + f_1^x + f_2^x$ and $y(k + 1) = y(k) + f_1^y + f_2^y$.

The improvement algorithm is said to converge if $f_1^x + f_2^x + f_1^y + f_2^y \leq \epsilon$, $\epsilon \to 0$. Computation proceeds until it converges or once it reaches a maximum limit on the number of iterations.

Formative Behaviour: In response to an RFP, if a node receives responses where every neighbour is in a different network, each network would have a cardinality of 1. Such a situation arises predominantly in the formative stages when all nodes have a coordinate of $(0, 0)$ and lie in a network by themselves.

In such a case, the node computing its RFP arbitrarily chooses one of the neighbouring nodes positioned at (x_i, y_i) and sets its own coordinates as $(x_i + \frac{r}{2}, y_i + \frac{r}{2})$. Hence, a coordinate system is "pulled out" from one of the origins.

4   Strategies for Collapsing Coordinate Systems

Locally Most Popular (LMP):   Whenever a node receives responses from its neighbours, the node determines the network-id that most of the neighbours (including itself) belong to. The node then changes its network-id to this "locally most popular" network-id.

Network Size Estimate (NSE): Here, we assume that every node maintains *nse*, an *estimate* of the size of the network it belongs to. To start with, each node has nse equal to 1 (signifying only one node, which is itself). Whenever a node receives responses from its neighbours, it finds the network with the highest nse. The node then changes its network-id to the network-id of this network. Once a node changes its network-id to that of the network with the highest nse, nse of the network is incremented by 1. Similarly if a node finds a higher value for the nse of its network from one of its neighbours, it increments its own nse value to the new value. The nse is thus a monotonically increasing function.

Evaluation results have shown a faster convergence rate for the second (NSE) strategy as compared to the first (LMP).

Beacons: A third strategy for collapsing coordinates is the use of special nodes called "beacons" which know their absolute coordinates. All beacons belong to a single special network id (usually a negative number). Whenever a node finds this special id in its neighbourhood, it simply changes to this special network.

Beacons with special ids are known to result in much faster convergence than the above two strategies. However, they do make the algorithm asymmetric, bringing in the need for special nodes.

5 Analysis

Resolution: The resolution of the coordinate system is the wireless range r. When all nodes in the network are in range with all other nodes (i.e. when the network is a clique), two or more nodes could be indistinguishable with respect to their virtual coordinates even though their real coordinates are not the same.

Minimum Density: It follows from the above that the best setting for computing virtual coordinates is when both *hop*1 and *hop*2 neighbours of a node are non-empty.

Hence for any node, there has to be at least 2 neighbours within its range such that the neighbours are not in range among themselves. This ensures the availability of non-empty *hop*1 and *hop*2 sets.

Let n be the total number of nodes and N the 2-dimensional surface area in which the network is situated. Given this, the *density* of the network is $\rho = \frac{n}{N}$. Given any geographic area of radius r within the network $\rho \pi r^2$ gives the number of nodes in that area.

In order for the system of coordinates to be reliable, each node should have at least one *hop*1 neighbour and at least one *hop*2 neighbour. This condition is both necessary and sufficient. The algorithm relies upon only *hop*1 and *hop*2 neighbours for computing its own coordinates. With only *hop*1, the node places itself in the centroid of the neighbours, which may not be accurate.

Given a node's location, a *hop1* neighbour can be found around a radius r with a probability of ρ. A *hop2* neighbour is one which is in range with the *hop1* neighbour but not in range with the original node.

Hence, we have a random variable Φ which defines the following relation for any three nodes x_i, x_j and x_k in the network: $d(x_i, x_j) \leq r \wedge d(x_j, x_k) \leq r \wedge d(x_i, x_k) > r$. Here $d()$ computes the Euclidian distance between two nodes.

Given any three nodes x_i, x_j and x_k the probability that x_k lies within range of x_j and not within range x_i is given by the ratio of the area of the shaded region in Figure 3 to the area of a node's range.

Fig. 3. Probability of finding a *hop1* and a *hop2* neighbour

This probability is the ratio of the shaded region to the area of the circle spanning the range of x_j. The area of the shaded region is given by [14]:

$$S(u) = \pi r^2 - 2r^2 cos^{-1}(\frac{u}{2r}) + \frac{u}{2}\sqrt{(4r^2 - u^2)}$$

Here, $u = d(x_i, x_j)$ is the displacement between the nodes x_i and x_j. The random variable Φ is now defined as:

$$\Phi = \int_0^r \frac{\rho S(u)}{\pi r^2} du$$

In order to calculate the lower bound on node density, we set $\Phi = 1$ and solve the above equation. This results in a value of ρ as $(1.14865r - 0.8372)^{-1}$. For a given value of r (say 5 meters), the minimum required density is given by $\rho = 0.2039$ nodes per square meter. The number of nodes in this region is given by $\rho \pi r^2 \simeq 16$. The resolution of the distance between nodes is given by the units of r (in this case 1 meter, i.e. 16 uniformly distributed nodes are sufficient in the neighbourhood when wireless range is 5 meters and node distances are measured with a resolution of 1 meter).

6   Simulation Results

Most of the evaluation experiments were based on a simulated environment depicting a node space size of 100 by 100 units and node radius of 5 units. Time is in terms of logical time ticks. Nodes move after random periods of time whose limit is controlled by a parameter called the "mobility factor" mf. Performance evaluation tests shown here address the localization algorithm itself and not necessarily its applicability to a wireless ad hoc network. This is because, other physical factors affecting a wireless ad hoc network like signal attenuation, multipath, etc. are ignored.

The variables that were observed from the simulation runs were the following: *error* and *numnets*. The first refers to the average error in positioning. This is calculated for each node as follows:

Error Calculation: At random intervals, nodes generate data packets. When data packets are generated, they are tagged with the current virtual coordinates (vx, vy) of the node. The *real* coordinates (x, y) of the location are also recorded. A timer is then started that runs for a time period of at least mf, allowing the node to move away. When the timer reaches 0, virtual coordinates (vx', vy') are calculated at the location (x, y) where the packet was generated. The error in positioning is the Eucledian distance between (vx, vy) and (vx', vy').

The rationale for the above method of error calculation is the absence of any absolute coordinate system against which errors can be measured. A measure of the goodness of the algorithm is to calibrate how stable the coordinate system is.

The second variable that is observed is the number of networks that prevail in the system for each tick of the simulation clock.

Figure 4 plots the cumulative average error across time ticks for different node densities. It is evident that error decreases as node density increases and is almost close to 0 for node densities greater than 0.2 nodes per square unit when wireless range is 5 units. This is consistent with the analytical result computed in the previous section. In these tests mf was set to a moderate value of 50 ticks between moves.

Figure 5 compares the two strategies for *collapsing coordinates*. It is clear that *NSE* has a much higher rater of convergence than *LMP*. The figure plots the number of networks in the system across clock ticks for both the strategies, for high and moderate mobility values, which can be controlled by the parameter mf.

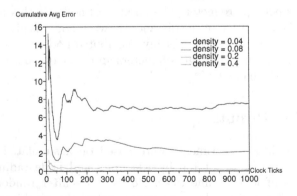

Fig. 4. Cumulative average error versus node density

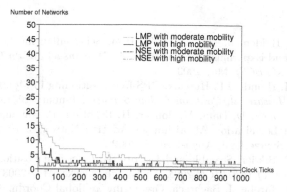

Fig. 5. Impact of *LMP* and *NSE* on convergence

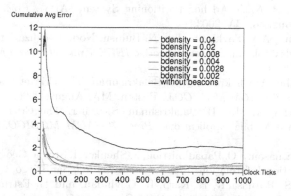

Fig. 6. Impact of beacons on cumulative average error

Figure 6 plots cumulative average error with and without beacons. The difference is significant and the availability of beacons result in much smaller positioning errors. The density of beacons does not seem to affect the overall positioning error. Convergence of the network onto a single coordinate system is also faster with the presence of beacons.

7   Conclusions

It is possible to generate a system of virtual coordinates without the need for flooding, perimeter nodes or even beacons as shown by the *Adorn* model. The model performs fairly accurate localization even in the face of high mobility. While the model does not require beacons, their availability greatly enhances localization accuracy. A limitation of this model is that the nodes can only determine their relative coordinates in a virtual coordinate system. This is not a major limitation since many applications like *data centric* routing do not require absolute positions.

References

1. N. Bulusu, J. Heidemann, D. Estrin, T. Tran. Self-configuring Localization Systems: Design and Experimental Evaluation. *ACM Transactions on Embedded Computing Systems(TECS)*, May 2003.
2. S. Capkun, M. Hamdi, J.P. Hubaux. GPS-free Positioning in Mobile Ad-Hoc Networks. *Proc. Hawaii Int. Conf. on System Sciences*, January 2001.
3. L. Evers, W. Bach, D. Dam, M. Jonker, H. Scholten, P. Havinga. An Iterative Quality-based Localization Algorithm for Ad Hoc Networks. *Proc. of Pervasive 2002*, Zurich, Switzerland, August 26-28, 2002.
4. R. Iyengar, B. Sikdar. Scalable and Distributed GPS free Positioning for Sensor Networks. *Proc. of IEEE ICC 2003*, Anchorage, Alaska, May 2003.
5. R. Nagpal, H. Shrobe, J. Bachrach. Organizing a Global Coordinate System from Local Information on an Ad Hoc Sensor Network. *Proc. of the Second International Workshop on Information Processing in Sensor Networks*, Palo Alto, April 2003.
6. D. Niculescu, B. Nath. Ad hoc Positioning System (APS). *Proc. of INFOCOM 2003*, San Francisco, CA, 2003.
7. P.N. Pathirana, A.V. Savkin, S. Jha, N. Bulusu. Node Localization Using Mobile Robots in Delay-tolerant Sensor Networks. *IEEE Transactions on Mobile Computing (TMC)*, 2004.
8. N.B. Priyantha, A. Chakraborty, H. Balakrishnan, The Cricket Location-Support System. *Proc. 6th ACM MOBICOM*, Boston, MA, August 2000.
9. N.B. Priyantha, A. Miu, H. Balakrishnan, S. Teller, The Cricket Compass for Context-Aware Mobile Applications, *Proc. 7th ACM MOBICOM*, Rome, Italy, July 2001.
10. A. Rao, S. Ratnasamy, C. Papadimitriou, S. Shenker, I. Stoica. Geographic Routing without Location Information. *Proc. of MobiCom'03*, San Diego, California, 2003.
11. S. Shenker, S. Ratnasamy, B. Karp, R. Govindan, and D. Estrin, "Datacentric storage in sensornets," ACM SIGCOMM Computer Communication Review, vol. 33, no. 1, pp. 137-142, 2003.
12. S. Ratnasamy, B. Karp, Y. Li, F. Yu, D. Estrin, R. Govindan, S. Shenker. GHT: A Geographic Hash Table for Data-Centric Storage. *Proc. of WSNA '02*, Atlanta, Georgia, USA, September 2002.
13. A. Savvides, C.-C. Han, M.B. Srivastava. Dynamic Fine-grained Localization in Ad-Hoc Networks of Sensors. *Proc. of MOBICOM 2001*, Rome, Italy, 2001.
14. E.W. Weisstein. Circle-Circle Intersection. From MathWorld–A Wolfram Web Resource. *http://mathworld.wolfram.com/Circle-CircleIntersection.html*
15. N. Priyantha, H. Balakrishnan, E. Demaine, and S. Teller. Anchor-free distributed localization in sensor networks. Technical Report TR-892, MIT LCS, Apr.2003.

Performance Study of LU Decomposition on the Programmable GPU*

Fumihiko Ino, Manabu Matsui, Keigo Goda, and Kenichi Hagihara

Graduate School of Information Science and Technology, Osaka University,
1-3 Machikaneyama, Toyonaka, Osaka 560-8531, Japan
ino@ist.osaka-u.ac.jp

Abstract. With the increasing programmability of graphics processing units (GPUs), these units are emerging as an attractive computing platform not only for traditional graphics computation but also for general-purpose computation. In this paper, to study the performance of programmable GPUs, we describe the design and implementation of LU decomposition as an example of numerical computation. To achieve this, we have developed and evaluated some methods with different implementation approaches in terms of (a) loop processing, (b) branch processing, and (c) vector processing. The experimental results give four important points: (1) dependent loops must be implemented through the use of a render texture in order to avoid copies in the video random access memory (VRAM); (2) in most cases, branch processing can be efficiently handled by the CPU rather than the GPU; (3) as Fatahalian et al. state for matrix multiplication, we find that GPUs require higher VRAM cache bandwidth in order to provide full performance for LU decomposition; and (4) decomposition results obtained by GPUs usually differ from those by CPUs, mainly due to the floating-point division error that increases the numerical error with the progress of decomposition.

1 Introduction

The GPU [1] is a single-chip processor, which is designed to accelerate rendering tasks for interactive visualization. Recently, GPUs on commodity PC graphics cards are emerging as a novel high performance computing (HPC) platform with providing faster floating-point operations than CPUs [2]. Newly added functionalities such as programmability and branch capability make them an attractive HPC platform not only for visualization purposes but also for general purposes.

Such new functionalities also activate the use of modern GPUs for solving numerical problems. Thompson et al. [3] implement matrix multiplication on a GPU, achieving three times higher performance compared with a simple CPU implementation. Larsen et al. [4] compare their GPU implementation with ATLAS [5], a cache-optimized CPU implementation. They present two requirements for making their GPU implementation competitive against ATLAS: one is a significant increase of VRAM access speed and the other is that of graphics chip core clock. To approach these requirements from

---

* This work was partly supported by JSPS Grant-in-Aid for Scientific Research on Priority Areas (16016254).

D.A. Bader et al. (Eds.): HiPC 2005, LNCS 3769, pp. 83–94, 2005.

the software side, Hall et al. [6] propose a VRAM cache and bandwidth aware algorithm with its theoretical evaluation. Their algorithm is evaluated on real graphics cards by Fatahalian et al. [2]. This experimental evaluation shows that higher VRAM cache bandwidth is yet essential for GPUs to outperform ATLAS.

In addition to the problem of matrix multiplication, there are a wide variety of numerical applications running on the GPU: the conjugate gradient method [7, 8, 9], the Gauss-Seidel method [7], the projected Jacobi method [9], and the fast Fourier transform [10]. Thus, many researchers try to accelerate numerical computations using the GPU. However, it is still not clear what kinds of design guidelines will yield higher performance on GPUs, mainly due to the rapid advances in GPU architectures. Furthermore, most vendors rarely disclose the details of their GPU architectures.

The goal of our work is to analyze the performance behavior of the GPU, aiming at making clear the design guidelines for GPU accelerated numerical computations. To achieve this, we focus on the problem of LU decomposition, which is used for ranking top 500 supercomputers. We present its design with different implementation approaches in terms of (a) loop processing, (b) branch processing, and (c) vector processing. We also show some performance studies using commodity PC graphics cards.

To the best of our knowledge, the key contributions of the paper are (1) the design guidelines for the above implementation issues (a)–(c) and (2) the first GPU implementation for LU decomposition.

The paper is organized as follows. Section 2 presents a brief overview of the GPU architecture and summarizes prior strategies for solving numerical problems by the GPU. Section 3 describes the implementation approaches that compose our methods, and then Section 4 shows the performance studies obtained on modern graphics cards. Finally, Section 5 concludes the paper.

2   Graphics Processing Unit (GPU)

2.1   Overview of Architecture

The rendering task, which GPUs originally accelerate, is to compute pixels on the 2-D image by projecting polygonal objects (triangles) located in the 3-D space. In order to accelerate this compute-intensive task, modern GPUs [11, 12] employ a pipeline architecture as shown in Fig. 1. Due to limited space, we introduce only the programmable part of this pipeline, namely vertex processors (VPs) and fragment processors (FPs).

VPs and FPs are vector processors with 4-length vector registers. These processors have the following characteristics.

VP: VPs are capable of fast geometric transformation in order to accelerate the projection of vertices of polygons onto the 2-D space. They are based on a MIMD structure [13] that allows applying different operations simultaneously to multiple vertices. Here, the polygonal data must be transferred from the main memory to the VRAM by using graphics APIs such as OpenGL [14] and DirectX [15].

FP: FPs are capable of rapid mapping of textures onto the 2-D image, aiming at producing more realistic images. To perform this, they obtain projected pixels (called

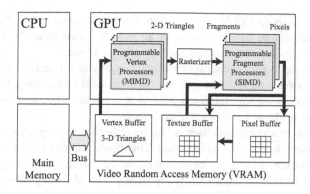

Fig. 1. GPU pipeline architecture

fragments) from the rasterizer, and then execute some mathematical operations between the pixels and textures. Here, the textures are read from the VRAM, and the mapping results are written to any buffer on the VRAM. FPs are based on a SIMD structure [13] that allows applying the same operation simultaneously to multiple fragments. There are two characteristics that must be mentioned. (C1) FPs support 4-length vector operations, because they deal with 4-channel (RGBA) data representing red, green, blue colors and opacity. (C2) Because textures are generally 2-D images, FPs can be regarded as vector processors that execute independent operations on each element on a matrix.

As we mentioned in Section 1, recent advances have removed many limitations that earlier GPUs have. For example, earlier GPUs do not have any control flow mechanism. Furthermore, only short programs were executable due to the limitation on instruction count. In contrast, modern GPUs allow more instructions with branch capability and some GPUs follow a 32-bit single floating-point number representation based on the IEEE standard [16]. Furthermore, by using the graphics APIs mentioned above, the rendered results can be transferred (readback) from the VRAM to the main memory.

2.2 Prior Strategies for Accelerating Numerical Computations on the GPU

Recent work [2, 6, 7, 8, 9, 10] uses only FPs for numerical computation while earlier work [3] uses VPs. This is due to lower performance of current VPs, which are not competitive against CPUs [6]. For example, as we show later, VPs in nVIDIA's GeForce cards provide 338 million vertices/s (namely, vectors/s), whereas FPs provide 3.6 billion texels/s (vectors/s). Due to this performance characteristic, we also use only FPs for LU decomposition.

In order to maximize the efficiency on FPs, prior work focuses on characteristics C1 and C2. For example, in a case of matrix multiplication $\mathbf{XY} = \mathbf{Z}$, each of elements Z_{ij} can be computed independently. Therefore, FPs are allowed to simply render the result matrix $\mathbf{Z}$ into a VRAM buffer by referring two textures each containing matrix data $\mathbf{X}$ and $\mathbf{Y}$, respectively. Thus, a doubly-nested loop without any dependencies between loop iterations can be efficiently processed by a single pass of the data through the pipeline. Furthermore, to enable vectorization, some researchers pack the matrix data into the 4-channel texture format. They store an $N \times N$ matrix in an $N/4 \times N$ texture, multiplying the elements on four rows and one column at once.

Although the single-pass rendering approach mentioned above is effective for independent nested loops, it must not be applied to a dependent nested loop, because such a loop cannot be processed simultaneously due to dependencies between loop iterations. This is one of the problems addressed in the paper.

The single-pass rendering approach is also inapplicable to programs whose size exceeds the limitation on the instruction count. This limitation can be resolved by multipass rendering, which aims at emulating the entire execution by dividing the program into small parts. In this method, data is repeatedly passed through the pipeline with varying the program parts at each pass.

In summary, prior work presents the following four guidelines for yielding higher performance on GPUs:

- Apply single-pass rendering to independent doubly-nested loops that contain a large amount of computation;
- Pack matrix data into the 4-channel texture format to enable vectorization and to reduce the usage of VRAM;
- Reduce the amount and number of data transfer between the GPU and the CPU;
- Reduce the number of VRAM accesses to save the VRAM bandwidth.

3   LU Decomposition on the GPU

This section presents the design of LU decomposition on the GPU. Table 1 summarizes our methods with their theoretical performance.

3.1   LU Decomposition

LU decomposition is a method for solving a linear system $\mathbf{Ax} = \mathbf{b}$. It factorizes a matrix $\mathbf{A}$ into two triangular matrices: a lower matrix $\mathbf{L}$ and an upper matrix $\mathbf{U}$. Then, the solution $\mathbf{x}$ is computed by forward and backward substitution. There are three algorithms for this method: right-looking (see Fig. 2), left-looking, and Crout algorithms

Table 1. Theoretical performance of proposed methods M1, M2, M3, and M4

	Vectorization	Branch	Loop	Rendering pass Number	Weight	VRAM copy Number	Amount (B)
M1	No	CPU	Copying	2N	1	2N	$32N(N^2 - 1)/3$
			Switching			0	0
M2	No	GPU	Copying	N	1	N	$16N(N^2 - 1)/3$
			Switching			0	0
M3	Yes	CPU	Copying	8N	1/4	2N	$2N(4N^2 + 3N - 4)/3$
			Switching			0	0
M4	Yes	GPU	Copying	4N	1/4	2N	$2N(4N^2 + 3N - 4)/3$
			Switching			0	0

```
1: Algorithm RightLookingLUDecomposition {
2:    for (i = 0; i < N; i++) {
3:       for (j = i + 1; j < N; j++) {
4:          A_ji = A_ji/A_ii; /* update L */
5:          for (k = i + 1; k < N; k++)
6:             A_jk- = A_ik * A_ji; /* update U */
7:       }
8:    }
9: }
```

Fig. 2. Right-looking LU decomposition algorithm

[17]. Given an $N \times N$ matrix, these algorithms require the same $O(N^3)$ time but differ in parallelism and locality of data access.

Among these algorithms, we select the right-looking version for GPUs, which have much smaller caches than CPUs [12]. The reason why we select this version is that its reference area at each decomposition step is smaller than the others. Thus, we think that current GPUs are not suited to algorithms that require larger caches. Note here that we currently do not consider pivoting because our main focus is the performance study of GPUs.

3.2 Design Policy

To realize LU decomposition on the GPU, the following three issues must be resolved.

(a) Loop processing: There are dependencies between the outer i loop iterations in Fig. 2. Due to these dependencies, we cannot simply apply single-pass rendering to LU decomposition. A naive solution is to use a multi-pass rendering approach.

(b) Branch processing: While matrix multiplication applies the same operation to all matrix elements, LU decomposition uses two different operations, each for computing matrices **L** and **U** (lines 4 and 6 in Fig. 2). Therefore, we have to select the correct operation depending on the location of matrix elements. Thus, branch processing is required for this selection because FPs are SIMD processors.

(c) Vector processing: As same as for matrix multiplication, we should pack matrix data into the 4-channel format to enable vectorization.

In the following we describe our implementation approaches that address the issues mentioned above. We present two approaches for each issue. As shown in Table 1, our proposed methods are combinations of these approaches. The design policies of these methods are as follows:

- Method M1 eliminates branch operations from the GPU program, though it requires more passes;
- In contrast, method M2 achieves less passes, though it includes branch operations in the GPU program;
- The remaining methods M3 and M4 exploit vectorization on the basis of methods M1 and M2, respectively.

3.3 Loop Processing

To consider the issue of loop processing, we first characterize the dependencies between loop iterations. The dependencies in the right-looking algorithm are as follows:

1. The outer i loop iterations cannot independently be processed;
2. The inner k loop iterations can independently be processed;
3. The inner k loop iterations must be processed after completing the assignment (line 4) in the middle j loop.

Due to the first dependency mentioned above, multi-pass rendering is necessary for LU decomposition. The key idea here is that, according to characteristic C2, single-pass rendering can be applied to the inner jk loops if these loops are restructured into an independent loop. The following reconstruction methods can be considered.

– Loop decomposition (method M1, Fig. 3(a)): In this method, a nested loop for updating $\mathbf{L}$ and $\mathbf{U}$ (lines 3–7 in Fig. 2) is decomposed into two loops (lines 3–4 and lines 5–7 in Fig. 3). These decomposed loops cannot be processed at once. However, each loop can be processed in parallel. Therefore, this method requires two passes for rendering $\mathbf{L}$ and $\mathbf{U}$, as illustrated in Fig. 4(a) and (b).
– Loop fusion (method M2, Fig. 3(b)): In this method, the assignment for updating $\mathbf{L}$ (line 4 in Fig. 2) is moved into the inner k loop (line 5–6 in Fig. 3(b)). This eliminates the dependencies so that enables parallel processing of the inner jk loops. Thus, this method requires only a single pass for updating $\mathbf{L}$ and $\mathbf{U}$, as illustrated in Fig. 4(c). However, it increases the time complexity because the assignment is moved into the inner loop (line 6 in Fig. 3(b)).

The next issue to be addressed is how data can be iteratively passed through the pipeline. There are two strategies for this issue.

– Copying strategy using a pixel buffer: FPs write their computation results into a pixel buffer. After this, the buffer context is copied to a texture for the next succeeding pass. This strategy requires the copy overhead.
– Switching strategy using a render texture: This strategy uses two textures, each for input and output of FPs. At every pass, FPs switch these textures to prevent VRAM copies. This strategy requires the switch overhead instead of the copy overhead.

```
1: Algorithm TwoPassLUDecomposition {
2:   for (i = 0; i < N; i + +) {
3:     for (j = i + 1; j < N; j + +) /* rendering */
4:       A_ji = A_ji/A_ii; /* update L */
5:     for (j = i + 1; j < N; j + +) /* rendering */
6:       for (k = i + 1; k < N; k + +)
7:         A_jk− = A_ik * A_ji; /* update U */
8:   }
9: }
```

(a)

```
1: Algorithm OnePassLUDecomposition {
2:   for (i = 0; i < N; i + +) {
3:     for (j = i + 1; j < N; j + +) { /* rendering */
4:       for (k = i; k < N; k + +) {
5:         if (i == k) A_ji = A_ji/A_ii; /* update L */
6:         else A_jk− = A_ik * A_ji/A_ii; /* update U */
7:       }
8:     }
9:   }
10: }
```

(b)

Fig. 3. Proposed methods. (a) Two-pass method M1 and (b) single-pass method M2.

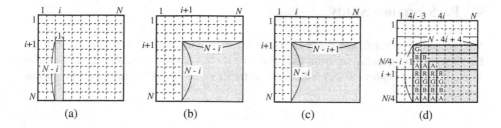

Fig. 4. Matrix data rendered at the i-th pass, where $1 \leq i \leq N$. (a,b) Method M1 renders matrices **L** and **U** in two passes. (c) Method M2 renders them at once. (d) Method M4 integrates vectorization with M2 by packing matrix data into the 4-channel (RGBA) format.

3.4   Branch Processing

We now describe how methods M1 and M2 resolve the issue of branch processing. Branches in method M1 are handled by the CPU. In this method, the entire loop is mapped to two rendering passes, as we mentioned in Section 3.3. Therefore, the CPU takes the responsibility for loading the appropriate GPU program with its rendering area. Thus, branches are naturally handled by the CPU. As a result, the GPU is allowed to concentrate on executing the given program without any control flow. On the other hand, method M2 requires the GPU to process branches. This can be easily implemented by comparing i and k, the location of matrix elements as shown in Fig. 3(b).

In summary, although CPU implementations do not include branch operations, their GPU versions may include them due to the SIMD architecture. If branch conditions are expressed by the location in matrix data, such branches can be eliminated from the GPU program but with more rendering passes.

3.5   Vector Processing

As same as for matrix multiplication [2, 6], we also apply vectorization to our methods M1 and M2 in order to reduce the execution time. Fig. 4(d) illustrates how the matrix data are mapped to the texture format. In this method, an $N/4 \times N$ texture represents an $N \times N$ matrix, enabling applying vector operations to the data on four rows and one column. Note here that we cannot apply them in other directions, for example, to the data on one row and four columns, because there are dependencies between different columns (the outer i loop iterations).

Applying vectorization then raises another issue to be addressed. The issue is that, as shown in Fig. 4(d), the appropriate channel must be selected in order to perform correct rendering for each column. For example, at the top-left corner of rendering area, we can see that all of the RGBA channels must be rendered for $i = 4, 8, \ldots$, whereas only the GBA channels must be rendered for $i = 1, 5, \ldots$.

This issue is the same branch issue addressed in Section 3.4, because its branch condition can be expressed by the location in matrix data. Therefore, we solve it in the same manner. That is, we implement four GPU programs, each renders the GBA, BA, A, and RGBA channels, respectively, and then switch them in a cyclic manner.

4   Performance Study

We now show performance studies in order to analyze the performance behavior of the GPU. We study its behavior from the following three viewpoints: design guidelines for implementation issues (a)–(c); efficiency in terms of cache bandwidth; and numerical error.

Table 2. Specification of experimental environments

GPU	nVIDIA GeForceFX 5900Ultra	nVIDIA QuadroFX 3400
Core clock	450MHz	350MHz
Texture fill-rate	3.6Gpixels/s	5.6Gpixels/s
VRAM capacity	128MB	256MB
VRAM bandwidth	27.2GB/s	28.8GB/s
Texture cache capacity	Undisclosed	Undisclosed
Texture cache bandwidth	11.4GB/s	15.6GB/s
Graphics bus	AGP8X	PCI Express
CPU	Pentium 4 2.6GHz	Pentium 4 2.8GHz
OS	Red Hat Linux 9	Windows XP

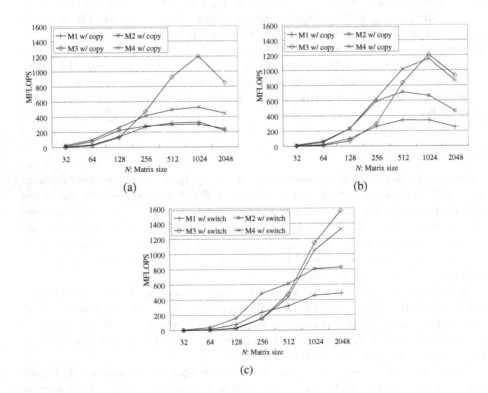

Fig. 5. Measured performance for different matrix sizes. (a) GeForce card with the copying strategy. Quadro card (b) with the copying strategy and (c) with the switching strategy.

Table 2 shows the specification of our two machines used for the study. We have implemented the four methods using the C++ language, the OpenGL library, and the Cg language [18]. Note here that render-to-texture functionality is not yet available on Linux systems. Therefore, we used only the copying strategy for the Linux system.

Fig. 5 shows the measured performance for different matrix sizes N. We can see that the Quadro card yields the best performance of 1.6 GFLOPS for $N = 1024$ by using method M3 with the switching strategy. On the other hand, the GeForce card reaches 1.2 GFLOPS for $N = 1024$ by using method M3 with the copying strategy.

In this figure, we can also see that methods M2 and M4 on Quadro provide relatively higher performance than those on GeForce. This indicates that the newer generation of Quadro reduces the branch overhead on FPs, as compared to GeForce, because these methods differ from methods M1 and M3 in the use of branch operations.

We next investigate the breakdown of execution time to present design guidelines for implementation issues (a)–(c). Table 3 shows the breakdown measured on Quadro.

- (a) On loop processing. The switching strategy prevents copies in the VRAM, so that spends no time T for the VRAM copy, as shown in Table 3. However, instead of this overhead, the switch overhead is observed in the CPU time C. For example, method M2 increases time C from 95 ms to 128 ms. However, this switch overhead is small enough to the entire time A. This is also true for methods M3 and M4, which require more switches due to vectorization. Therefore, in most cases, the switching strategy seems better than the copying strategy. One concern is portability because Linux systems currently support only the copying strategy.
- (b) On branch processing. As compared to method M1, method M2 increases the GPU time G as matrix size N increases. We can see this increase also in its vectorization version M4. This increase is due to the branch overhead occurred on FPs. Thus, for larger matrices, branch operations should be processed by the CPU in order to obtain higher performance. In contrast, for smaller matrices, method M1

Table 3. Breakdown of measured time on Quadro. A, G, C, and T represent the entire time, the GPU calculation time, the CPU calculation time, and the VRAM copy time, respectively.

N	M1 w/ copying (ms) A	G	C	T	M2 w/ copying (ms) A	G	C	T	M3 w/ copying (ms) A	G	C	T	M4 w/ copying (ms) A	G	C	T
32	9	6	2	1	7	6	1	1	39	12	26	1	51	1	48	1
64	13	8	3	2	10	7	2	1	55	15	38	3	64	3	56	5
128	26	14	6	6	19	12	4	3	86	20	60	6	94	6	82	6
256	79	41	11	26	55	36	6	13	160	36	108	15	160	17	126	16
512	438	250	24	164	335	240	13	82	365	102	201	62	360	84	214	63
1024	3291	2022	64	1205	2691	2050	37	604	1306	566	391	350	1334	592	391	351
2048	34942	21629	108	13206	30545	23752	95	6698	10079	5875	781	3422	10489	6307	761	3421

N	M1 w/ switching (ms) A	G	C	T	M2 w/ switching (ms) A	G	C	T	M3 w/ switching (ms) A	G	C	T	M4 w/ switching (ms) A	G	C	T
32	8	5	3	—	6	5	1	—	50	28	21	—	63	43	21	—
64	15	7	8	—	9	6	3	—	69	38	31	—	77	46	32	—
128	26	13	13	—	17	10	7	—	103	42	61	—	114	57	57	—
256	67	38	29	—	49	36	13	—	208	68	140	—	206	69	136	—
512	318	243	75	—	306	264	42	—	418	149	269	—	409	164	245	—
1024	1603	1470	133	—	1756	1690	66	—	1096	596	500	—	1135	650	485	—
2048	11564	11249	315	—	13309	13181	128	—	4477	3483	994	—	5048	4124	924	—

provides higher performance than method M2. This opposite result is due to the switch overhead mentioned above. That is, though method M1 eliminates branch operations from the GPU program, it requires an additional overhead for switching GPU programs. This overhead results in longer time C, increasing the entire time A especially for smaller N. Thus, there is a tradeoff relation between the GPU branch and the CPU branch. The tradeoff point is determined by the computational amount associated with a single pass of rendering. In this experiment, the tradeoff point is $N = 512$.

– (c) On vector processing. The timing benefits of vectorization can be observed in time G. For example, applying vectorization to M1 reduces time G from 11249 ms to 3483 ms. Thus, the vectorization effect is almost the same value as the vector length. In addition to this obvious result, vectorization also contributes to the reduction of the VRAM copy time T when using the copying strategy. This reduction comes from the data packing required for vectorization. Actually, time T in M3 is 3422 ms, which is almost 1/4 of time T in M1. Thus, vectorization is essential to reduce both the GPU time and the VRAM copy time.

We next investigate the efficiency from the viewpoint of cache bandwidth. As Fatahalian et al. [2] did, we also modified GPU programs such that the programs only access the matrix data without performing mathematical operations. Then, by measuring the GPU time, namely the access time, and using the theoretical amount of data accesses in Table 1, we obtain the throughput of our methods. The best throughput on GeForce is 8.6GB/s when using method M3 with the copying strategy for $N = 1024$ and that of Quadro is 11.4GB/s when using M3 with the switching strategy for $N = 2048$. According to these results and theoretical bandwidth in Table 2, the efficiency of cache bandwidth reaches 75% on GeForce and 73% on Quadro. These values are similar to those of matrix multiplication [2]. Thus, we find that GPUs require higher VRAM cache bandwidth in order to provide full performance for LU decomposition.

In terms of FLOPS, the efficiency of our methods is estimated as at most 30%. This efficiency is not competitive against that of CPU implementations. For example, some CPU implementations [19, 20, 21] optimize locality to achieve higher cache utilization, so that achieve an efficiency of more than 80%. Therefore, more efficient methods are required to make GPUs a competitive HPC platform.

Finally, we investigate computation results in terms of numerical errors. In most cases, there are differences between the CPU and GPU results. These differences are due to the floating-point error, as presented in Table 4. As Hillesland et al. [22] observed, our GPUs also do not establish error bounds compatible with the IEEE standard, though they have the same floating-point representation. In particular, division has larger error than the other operations, because it is implemented by a combination of reciprocal and multiplication. Unfortunately, this division error is critical for LU decomposition, because it increases and propagates the entire error at each decomposition step.

Furthermore, though recent GPUs deal with single-precision floating-point numbers, they do not support double-precision numbers. Thus, errors caused by this limited precision are not essentially addressed yet.

Table 4. Floating-point errors in unit in last place. Errors on Quadro are measured by Paranoia [22]. In the IEEE standard [16], the result is rounded to the nearest representable number.

Operation	IEEE754	Quadro
Multiplication	$[-0.5, 0.5]$	$[-0.78125, 0.625]$
Division	$[-0.5, 0.5]$	$[-1.19902, 1.37442]$
Subtraction	$[-0.5, 0.5]$	$[-0.75, 0.75]$
Addition	$[-0.5, 0.5]$	$[-1, 0]$

5   Conclusions

We have presented the design and implementation of LU decomposition on the programmable GPU. To study the performance behavior of modern GPUs, we have developed and evaluated some implementation approaches in terms of (a) loop processing, (b) branch processing, and (c) vector processing.

The experimental results give four important points: (1) for dependent loops, the switching strategy using a render texture avoids copies in the VRAM, reducing execution time by 50%; (2) there is a tradeoff relation between the CPU branch and the GPU branch, and the CPU branch provides higher performance for the decomposition of matrices larger than 512×512; (3) the efficiency of floating-point operations is at most 30%, and as Fatahalian et al. state for matrix multiplication, GPUs also require higher cache bandwidth in order to provide full performance also for LU decomposition; and (4) GPUs usually provide different decomposition results from those obtained using a CPU, mainly due to the floating-point division error that increases the numerical error with the progress of decomposition.

Thus, as same as for matrix multiplication, we find that current GPUs are not so suited well for LU decomposition. However, as Moreland et al. [10] pointed out, GPUs are rapidly increasing their performance beyond the Moore's law [23]. Therefore, we believe that this architecture will emerge as an attractive HPC platform, at least for applications where the error is not a critical problem.

References

1. Fernando, R., ed.: GPU Gems: Programming Techniques, Tips and Tricks for Real-Time Graphics. Addison-Wesley, Reading, MA (2004)
2. Fatahalian, K., Sugerman, J., Hanrahan, P.: Understanding the efficiency of GPU algorithms for matrix-matrix multiplication. In: Proc. SIGGRAPH/EUROGRAPHICS Workshop Graphics Hardware (GH'04). (2004) 133–137
3. Thompson, C.J., Hahn, S., Oskin, M.: Using modern graphics architectures for general-purpose computing: A framework and analysis. In: Proc. 35th IEEE/ACM Int'l Symp. Microarchitecture (MICRO'02). (2002) 306–317
4. Larsen, E.S., McAllister, D.: Fast matrix multiplies using graphics hardware. In: Proc. High Performance Networking and Computing Conf. (SC2001). (2001)
5. Whaley, R.C., Petitet, A., Dongarra, J.J.: Automated empirical optimizations of software and the ATLAS project. Parallel Computing **27** (2001) 3–35
6. Hall, J.D., Carr, N.A., Hart, J.C.: Cache and bandwidth aware matrix multiplication on the GPU. Technical Report UIUCDCS-R-2003-2328, University of Illinois (2003)

7. Krüger, J., Westermann, R.: Linear algebra operators for GPU implementation of numerical algorithms. ACM Trans. Graphics **22** (2003) 908–916
8. Bolz, J., Farmer, I., Grinspun, E., Schröder, P.: Sparse matrix solvers on the GPU: Conjugate gradients and multigrid. ACM Trans. Graphics **22** (2003) 917–924
9. Moravánszky, A.:   Dense Matrix Algebra on the GPU.   (2003) http://www.shaderx2.com/shaderx.PDF.
10. Moreland, K., Angel, E.: The FFT on a GPU. In: Proc. SIGGRAPH/EUROGRAPHICS Workshop Graphics Hardware (GH'03). (2003) 112–119
11. Fernando, R., Harris, M., Wloka, M., Zeller, C.: Programming graphics hardware. In: EUROGRAPHICS 2004 Tutorial Note. (2004) http://download.nvidia.com/developer/presentations/2004/Eurographics/EG_04_TutorialNotes.pdf.
12. Pharr, M., Fernando, R., eds.: GPU Gems 2: Programming Techniques for High-Performance Graphics and General-Purpose Computation. Addison-Wesley, Reading, MA (2005)
13. Grama, A., Gupta, A., Karypis, G., Kumar, V.: Introduction to Parallel Computing. second edn. Addison-Wesley, Reading, MA (2003)
14. Shreiner, D., Woo, M., Neider, J., Davis, T., eds.: OpenGL Programming Guide. fourth edn. Addison-Wesley, Reading, MA (2003)
15. Microsoft Corporation: DirectX (2005) http://www.microsoft.com/directx/.
16. Stevenson, D.: A proposed standard for binary floating-point arithmetic. IEEE Computer **14** (1981) 51–62
17. Dongarra, J.J., Duff, I.S., Sorensen, D.C., Vorst, H.V.D., eds.: Solving Linear Systems on Vector and Shared Memory Computers. SIAM, Philadelphia, PA (1991)
18. Mark, W.R., Glanville, R.S., Akeley, K., Kilgard, M.J.: Cg: A system for programming graphics hardware in a C-like language. ACM Trans. Graphics **22** (2003) 896–897
19. Naruse, A., Sumimoto, S., Kumon, K.: Optimization and evaluation of linpack benchmark for Xeon processor. IPSJ Trans. Advanced Computing Systems **45** (2004) 62–70 (In Japanese).
20. Goto, K., van de Geijn, R.: On reducing TLB misses in matrix multiplication. Technical Report CS-TR-02-55, The University of Texas at Austin (2002)
21. Dongarra, J.J., Luszczek, P., Petitet, A.: The LINPACK benchmark: past, present and future. Concurrency and Computation: Practice and Experience **15** (2003) 803–820
22. Hillesland, K.E., Lastra, A.:   GPU floating point paranoia.   In: Proc. 1st ACM Workshop General-Purpose Computing on Graphics Processors (GP²'04). (2004) C–8 http://www.cs.unc.edu/~ibr/projects/paranoia/.
23. Moore, G.E.: Cramming more components onto integrated circuits. Electronics **38** (1965) 114–117

PENCAPS: A Parallel Application
for Electrode Encased Grounding Systems Project

Marco Aurélio S. Birchal[1], Maria Helena M. Vale[2], and Silvério Visacro[2]

[1] PUC Minas – Pontifícia Universidade Católica de Minas Gerais,
GSDC – Computing and Digital Systems Group,
Av. Dom José Gaspar, Belo Horizonte, 30535-610, Brazil
birchal@pucminas.br
http://www.pucminas.br
[2] UFMG – Universidade Federal de Minas Gerais, LRC – Lightning Research Center,
Av. Antônio Carlos 6627, Belo Horizonte, 31270-910, Brazil
{mhelena, visacro}@cpdee.ufmg.br
http://www.ufmg.br

Abstract. The design of concrete encased electrode grounding systems by conventional computation procedures is a time-consuming task. It happens once the electromagnetic representation of the physical system requires the calculation of large full matrices. Recently, the possibility of paralleling the procedures involved in such calculations led the authors to implement a C language parallel application, based on MPI (Message Passing Interface). This article presents the engineering problem associated to this development and the fundamental aspects regarding this application, including the evaluation of its efficiency for solution of large grounding systems.

1 Introduction

Grounding systems are relevant elements for assuring safety conditions in electrical systems and to reduce damages during different occurrences, such as short-circuits and faults. They also play an important role in practices intended to prevent damage to sensitive electric and electronic equipments.

Therefore, in electrical engineering, the design of efficient grounding systems is always required to avoid risk of serious injury and economical losses. Computational models were developed to improve the ability to perform the design process. Sometimes, environmental conditions (very high soil resistivity) and restrictions in the available area for system installation make it difficult to develop such efficient designs, even with the available models.

Very frequently the metallic structure of the building foundations is used to complement the grounding system, in order to achieve the desired efficiency [1]. In this case, concrete reinforcing steel bars are placed inside concrete structures to work as grounding electrodes. This type of grounding element is called *concrete encased electrodes*. Nevertheless, the inclusion of the concrete encased electrodes in computational models consists in a complex task. This motivated the researchers of LRC - Lightning Research Center (UFMG) to develop an approach to contemplate this aspect, based on the application of the boundary element technique to the electromag-

D.A. Bader et al. (Eds.): HiPC 2005, LNCS 3769, pp. 95–105, 2005.

netic problem corresponding to the concrete encased electrode [2]. Earlier works [3] implemented this approach on sequential program solutions, with no satisfactory time-based results.

This paper presents a new application software, called PENCAPS, that takes advantage of the parallel processing paradigm to solve the concrete-encased grounding electrode problem, with a significant performance gain. The application is able to calculate those quantities of real interest for this engineering design problem that are basically translated by means of the grounding resistance and the potential distribution over the soil.

Following, section 2 presents the mathematical model that represents the concrete encased electrode grounding system. The parallel algorithm implementation is commented in section 3. The results obtained by simulations are shown in section 4. Finally, some remarkable aspects of this work are emphasized in section 5.

2   The Mathematical Method

In order to evaluate the grounding resistance and its potential distribution, the proposed solution applies the electric charge surface method. It takes advantage of an electromagnetic property that allows an analogy between surface current density of a body and its charge surface, once the conditions imposed to the electric potential and to the electric field in the proximities of such a body are the same.

Fig. 1 represents a concrete block in which an encased electrode is inserted [4]. The mathematical model divides the block surface into parts, represented by Si (varies from a to f).

If a surface S is charged with a charge density η, the potential over this surface can be computed by equation 1. In this way the potential over a point r, near a second point r´, both over the surface S, can be calculated [5].

$$V(r) = \oint_S \frac{\eta(r')}{4\pi\varepsilon|r - r'|} dS \ . \tag{1}$$

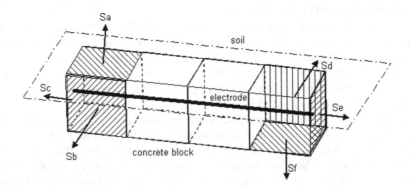

Fig. 1. A concrete encased electrode. The charge surfaces are depicted as Sa through Sf with its unitary vectors.

Applying boundary conditions to the potential, the electric potential over the electrode can be expressed using the relation of equation 2, where S_e is the electrode surface.

$$V_{elet} = \int_S \frac{\eta(r')dS'}{4\pi\varepsilon|r-r'|}, r \in S_e .$$ (2)

Once V_{elet} is a known value, which is the electric potential that the electrode has been submitted to, during the occurrence into consideration, the corresponding charge density $\eta(r')$ can be found.

Equation 3 represents the final relation that express the main solution to the problem of the concrete encased electrode. In this equation, ρ_s and ρ_c represent the soil and the concrete resistivities, respectively.

$$\eta(r) = 2\varepsilon_0 \frac{(\rho_s - \rho_c)}{(\rho_s + \rho_c)} \int_S \frac{\eta(r')n.(r-r')}{4\pi\varepsilon_0|r-r'|^3} dS .$$ (3)

The solution consists in finding all the charge densities, represented by $\eta(r)$ and $\eta(r')$. The total current that flows over the electrode can be calculated by equation 4.

$$I_t = \int_{S_e} \frac{\eta(r)}{\varepsilon_0 \rho_c} dS .$$ (4)

And, the grounding resistance, by the Ohm's law, as shown by equation 5.

$$R_t = \frac{V_{elet}}{I_t} .$$ (5)

2.1 Numeric Solution

The solution of the problem consists in finding a set of $\eta_h(r)$ approximated charge density that, given a surface divided into n parts, can be represented by the equation 6 [3]. N_i are functions that worth 1 to all points over S_i and zero to all the others.

$$\eta_h(r) = \sum_{i=1}^{n} \eta_i N_i(r) .$$ (6)

Equation 7 represents the system of linear equations that can be solved to find η vector.

$$
\begin{bmatrix} V_1 \\ V_2 \\ ... \\ V_n \\ 0 \\ 0 \\ ... \\ 0 \end{bmatrix}
=
\left[
\begin{array}{ccccc|c}
A_{11} & A_{12} & ... & & ... & A_{1n} \\
A_{21} & A_{22} & ... & & ... & A_{2n} \\
... & ... & ... & & & \\
\hline
& & & & & \\
& & & & & ... \\
... & ... & ... & & & \\
A_{n1} & A_{n2} & ... & & ... & (A_{nn}-1)
\end{array}
\right]
\cdot
\begin{bmatrix} \eta_1 \\ \eta_2 \\ ... \\ \eta_n \end{bmatrix} .
$$ (7)

Matrix A elements are calculated by equations 8 and 9, where r_i is the central point over each one of the n S_i surfaces; r_j are all the n-1 central points over the rest of the S_j surfaces.

$$A_{ij} = 2\varepsilon_0 \frac{(\rho_s - \rho_c)}{(\rho_s + \rho_c)} \int_{S_j} \frac{N_j(r_j)n_i.(r_i - r_j)}{4\pi\varepsilon_o |r_i - r_j|^3} dS, \ i \neq j \ . \tag{8}$$

$$A_{ii} = 2\varepsilon_0 \frac{(\rho_s - \rho_c)}{(\rho_s + \rho_c)} \int_{S_i} \frac{N_i(r_i)n_i.(r_i - r_i)}{4\pi\varepsilon_o |r_i - r_i|^3} dS, \ i = j \ . \tag{9}$$

3   The Parallel Implementation

The main problem to be solved in the design of a grounding system is to calculate the resulting resistance. This calculation involves the construction of matrix A (equation 7), whose elements are computed by numerical integrals, as described in equations 8 and 9. In addition, the matrix that represents the system geometry is a full matrix. These considerations emphasize the large amount of computational effort related to grounding problem solution.

In a recent past, the authors had developed a first computational model dedicated to the solution of concrete encased electrodes, implementing a sequential program to calculate the resistance of earthing systems. This was the first version of PENCAPS [3], using conventional processing logic. Although this tool was successfully implemented and had confirmed its potentiality to solve concrete-encased grounding systems problem, it could not work on large-scale problems. The main difficulty was concerned to deal with large dimension matrices and the consequent long time spent to find the results.

This difficulty became a strong motivation for investigating the application of parallel processing in grounding project. Therefore, authors decide to implement a parallel version of PENCAPS.

The parallel approach presented in this work was developed as a set of tools called "front-end" and "back-end".

A front-end tool, developed as a Microsoft Windows application, constitutes the user interface with PENCAPS. It was designed as a graphical program with grounding project facilities. Geometry visualization is provided and 3D potential profile graphs construction is also allowed.

The front-end is, indeed, a totally functional grounding system project tool, capable of doing all the steps needed to prepare the project. Nevertheless, this front-end is a sequential application that can not take advantage of the parallel computing paradigm. All sequential steps can be performed by this tool. This brings to the user a reliable way of controlling the project progress. Fig. 2 shows the front-end interface.

The back-end is the second PENCAPS tool. It is a text based application, written in C standard code with MPI (Message Passing Interface) [6]. This is a performance-tuned application built upon code portable concern. It was designed to run over any cluster-based architecture or MPP (Massively Parallel Processor) capable of compiling and running C/MPI code.

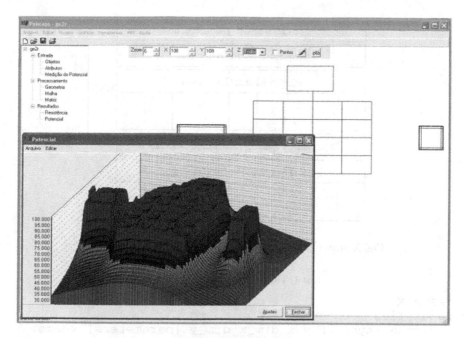

Fig. 2. The Front-End

In a computational ambient that has some kind of sharing file resource like a NFS (Network File System) with SMB (Server Message Block) protocol, the user can dispatch back-end to run, directly from the front-end. This is done by a built-in telnet client that logs on the remote host and passes shell commands to be executed on the computer that runs the MPI daemon. When using the back-end over an UNIX compatible system, it is necessary to have telnet access to the system. In this way, the front-end will be able to run the back-end automatically.

3.1 Steps of a Concrete Encased Grounding Project

The concrete encased grounding project consists of a sequence of steps that starts from geometry determination and finalizes with grounding resistance calculation. Fig. 3 summarizes these steps and shows where parallel approaches are indicated.

The early steps that comprise geometric aspects do not need to be parallelized once they don't demand great computational effort. The first, the geometry construction, is prepared by specifying the electrical system components.

The user creates two input files to describe the physical system. The first one is the "file of objects", which contains each element (grid, ring, wire and block), described by its position in the space, its length and name. List 1 shows a sample geometry file. The second is the "file of attributes". This file contains the electrical description of each component, as its resistivity and electrical potential in the presence of a fault.

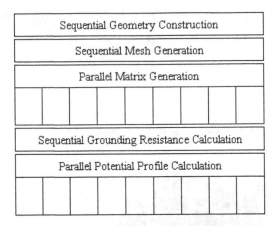

Fig. 3. Steps of a concrete encased grounding system project

List 1. The file of objects.Table 1

```
*REMARK
input file - objects
<nº obj> <type> x y z dim_x dim_y [parameters] <name>
*OBJECT
1   GRID -25.0 -15.0 0.5 50.00 30.00 5 5 16 16 GRID
2   BLCK 46.0 -4.0 0.5 8.0 1.0 16 8 BLOCK X
3   RING 45.2 -4.8 0.3 9.6 9.6 16 16 RING1
```

The object file, just prepared, provides the necessary information for PENCAPS to create the "geometry file". This contains a long list of coordinates, in the three-dimensional Cartesian plan, that describes either the end point of an element (as a single wire) or an intersection among different elements (two or more wires, grids or faces of a block). It was implemented a sequential routine that performs this task.

Next step in grounding project procedure is the mesh generation. The boundary element method works dividing the structures into small parts or frames. These frames are allocated in a mesh, as the framing result. To do this, informations from the new created geometry file and those from attributes file are crossed. A specific sequential routine does this work and generates another output, the "mesh file".

The third project step is the coefficient matrix generation. This is the most time-consuming task of all procedure. It also demands great processor resources. It is necessary that PENCAPS crosses the mesh data with the electrical components characteristics (attribute file) and calculates, for each one of the mesh's rows, the applied numeric integral (as depicted in equations 8 and 9). The results are then stored within the coefficient matrix, for further calculation of grounding resistance.

To calculate matrix A elements, it was implemented a parallel routine, in the gather fashion way, that breaks the mesh into small pieces. Each one of the parallel processes works on a part of the mesh. They compute the values corresponding to their own part and send the partial results to the root process. The root, that also works in its matrix slice, gathers the partial results and builds the final full matrix.

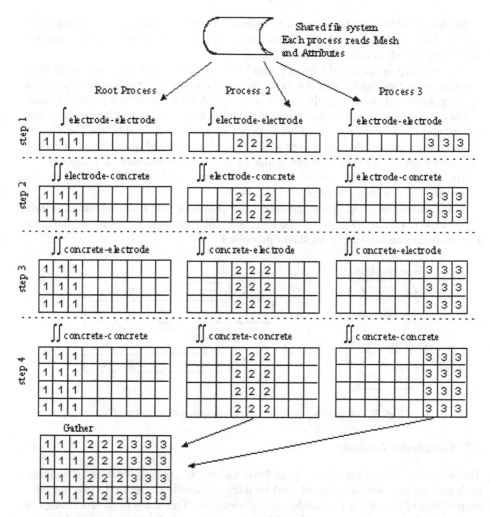

Fig. 4. Parallel building of the matrix of the problem by three processes

The central problem in this approach is that A is not a homogeneous matrix. Different parts of this matrix must be calculated according to different numeric integral equations.

The process has to be made in four parts, changing the computing equation from part to part. In each part, or step, the influence among the two different type of elements (concrete and electrodes) are considered. First, the influences of electrodes over themselves are computed. The second step computes the influence of electrodes over concrete surfaces. The third, calculates the influence of concrete block over electrodes and the forth and last, evaluates the influence of concrete blocks over themselves. Fig. 4 illustrates this procedure.

All calculations are functions of pre-computed data stored in mesh file. So, computations can be done in parallel, for different slices of the same matrix. A range of elements is attributed to each process that can work independent of the others. All processes can work at the same time.

By the end of these calculations, all processes, including root, have one slice of the matrix. They send their results to the root and, after the sending step, a new data file, containing the full matrix, is written in the disk.

The next step is the calculation of grounding resistance which consists on the main result of the grounding system. It is done by a sequential routine that solves the linear equation system, described by the matrix just created, for a given fault potential value.

Once grounding resistance had been calculated, it is possible to determine the potential distribution over the soil, as a post processing task. This is done by the last routine. An input file indicates the points over which it will be computed the electrical potential. The routine applies the electrical charge and the resistance to evaluate the potential.

This is another parallel routine that works in a master-slave style, sending the coordinates to be calculated to the slave processes.

At the end of the computation, an output file containing the resultant values is written, and the potential 3D graph can be plotted by the front-end. Fig. 5 shows the interactions among master (root) and slaves processes.

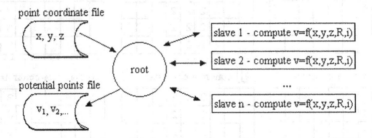

Fig. 5. Master-slave processes to compute potential soil profile

3.2 Complexity Analysis

The main parallel routine is the matrix building one. It has four separate subroutines, each one dealing with a different kind of integral equation. The first equation has a loop of size n, where n is the number of line elements. This leads to an $O(n)$ complexity in this phase of the computation. The second has an $O(n^2)$ complexity, due to an internal loop for solving a double integral equation. The third subroutine solves an $O(m.n)$ equation, where m is the number of the four-sided polygons created by the mesh structure. The fourth, the heaviest subroutine, is $O(m^2)$.

In all cases, the parallel implementation causes the complexity to a $1/p$ multiply factor, leading the four subroutines complexity to $O(n/p)$, $O(n^2/p)$, $O(m.n/p)$ and $O(m^2/p)$, respectively.

4 Results

A case study was proposed to test the program [7]. The grounding system is represented by the geometry showed in Fig. 6. It is a connected system with two concrete encased electrodes, one grid and one direct bared ring. The right encased electrode is

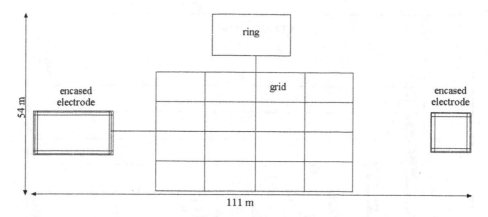

Fig. 6. Geometry of the grounding system

supposed to have an aerial connection to the grid. The soil resistivity is considered 2,500 Ω.m and the concrete resistivity is 400 Ω.m. A 100 kV potential fault was applied in the entire system. The mesh file for this simulation has 4,002 nodes and 3,528 elements.

Two testbeds were used in the case study. The first one is a cluster of 10 Pentium II / 300 Mhz at the LRC Center, and the second is a high performance cluster of 10 SMP Pentium III / 1 GHz with a Giga bit Ethernet network at CENAPAD MG/CO (High Performance Processing National Center - Minas Gerais and Midwest Region) in Belo Horizonte, Brazil. The execution time for one processor was taken as speedup 1. Fig. 7 shows the speedup [8] as a function of the number of processors, as it was measured in the LRC cluster.

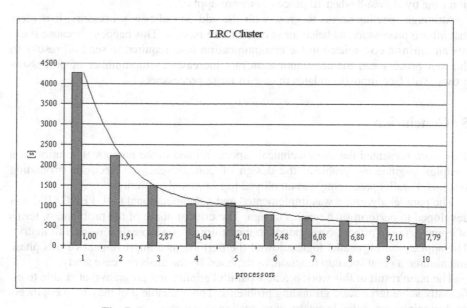

Fig. 7. Execution time and speedup for the LRC cluster

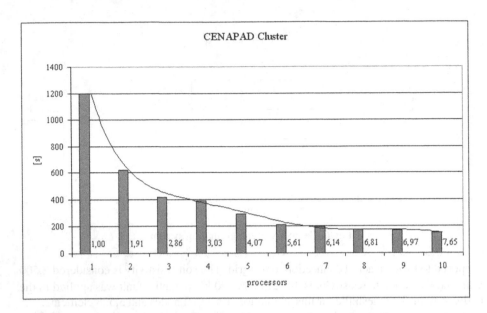

Fig. 8. Execution time and speedup for the CENAPAD cluster

The program was run in the CENAPAD cluster and the same measurements were taken, as depicted in Fig. 8.

One can see that the program showed a great speedup as it could find more processors to spread across. In both cases, it reached a 7 to 8 speedup, reducing the execution time by almost 8 when 10 processors were applied.

Although speedup seems to grow with the addition of new processors, it is clear that after 6 processors, its behavior is much more modest. This happens because it exists an intrinsic cost related to the communication time required to send all results to the root process and the communication rate increases as the number of processors grows. This fact almost annulates the use of more processors.

5   Conclusion

This work presented the main technical aspects related to the numerical solution of a complex engineering problem: the design of concrete-encased-electrode grounding systems. Parallel processing was employed to improve the solution.

The parallel algorithm was implemented and a computational tool, PENCAPS, was developed to perform such type of design. The critical phase of the problem, in terms of computational efforts, corresponds to the building of a large dimension full matrix. The application of parallel logic allowed the performance improvement of this phase and assured a great speedup function, as depicted by the analyzed case study.

The main result of this work is a new parallel application program that is able to efficiently solve large scale grounding problems. This valuable tool may help engineers on designing more reliable electrical protection systems.

References

1. Ufer, H. G.: Investigation and Testing of Footing-Type Grounding Electrodes for Electrical Installations. IEEE Trans. Power Apparatus and Systems, vol 83 (1964) 1042–1048
2. Visacro, S. F., Ribeiro, H. A.: Some Evaluations Concerning the Performance of Concrete-Encased Electrodes: an Approach by the Boundary Elements. In: Proceedings of the International Conference on Grounding and Earthing - GROUND′98, Belo Horizonte, Brazil (1998) 63-67
3. Ribeiro, H. A.: Development of a Computational Tool for the Performance Evaluation of the Concrete Encased Grounding Systems Over Low Frequency Phenomena. M.SC Thesis – (in Portuguese) UFMG, Belo Horizonte, Brazil (2000)
4. Kostic, M.B., Popovic, B. D., Jovanonic, M. S.: Numerical Analysis of a Class of Foundation Grounding Systems. IEE Proceedings – C, vol. 137 n° 2 (1990) 123-128
5. Visacro, S. F., Ribeiro, H. A., Palmeira, P. F. M.: Evaluation of Potential Distribution at Vicinities of Grounding Configurations Comprising Both Concrete Encased Electrodes and Conventional Meshes. In: Proceedings of the International Conference on Grounding and Earthing - GROUND′2000, Belo Horizonte, Brazil (2000) 123-126
6. Snir, M., Otto, S., Lederman, S. H., Walker, D., Dongarra, J.: MPI: The Complete Reference. The MIT Press, London (1996)
7. Birchal, M. A. S., Vale M. H. M., Visacro, S. F.: Analysis of Risk Conditions on Interconnected Grounding Systems: Concrete Encased Electrodes and Grid. In: Proceedings of the International Conference on Grounding and Earthing – GROUND′2004, Belo Horizonte, Brazil (2004) 297-301
8. Quinn, M. J.: Parallel Programming in C with MPI and OpenMP. McGraw-Hill, New York (2004)

Application of Reduce Order Modeling to Time Parallelization

Ashok Srinivasan[1], Yanan Yu[2], and Namas Chandra[3]

[1] Computer Science, Florida State University,
Tallahassee FL 32306, USA
asriniva@cs.fsu.edu
[2] Computer Science, Florida State University,
Tallahassee FL 32306, USA
yu@cs.fsu.edu
[3] Mechanical Engineering, Florida State University,
Tallahassee FL 32310, USA
chandra@eng.fsu.edu

Abstract. We recently proposed a new approach to parallelization, by decomposing the time domain, instead of the conventional space domain. This improves latency tolerance, and we demonstrated its effectiveness in a practical application, where it scaled to much larger numbers of processors than conventional parallelization. This approach is fundamentally based on dynamically predicting the state of a system from data of related simulations. In earlier work, we used knowledge of the science of the problem to perform the prediction. In complicated simulations, this is not feasible. In this work, we show how reduced order modeling can be used for prediction, without requiring much knowledge of the science. We demonstrate its effectiveness in an important nano-materials application. The significance of this work lies in proposing a novel approach, based on established mathematical theory, that permits effective parallelization of time. This has important applications in multi-scale simulations, especially in dealing with long time-scales.

1   Introduction

Many problems in science are formulated as initial value problems. The initial state of a physical system at some time is given, along with, possibly, some boundary conditions. A differential equation describes how the state changes with time, and possibly space. The problem is solved by iteratively computing the states at successive points in time, using a differential equation solver. We shall refer to each iteration as a *time step*. A large computational effort can be involved when the state of the system is large, or when the number of time steps is large. In order to reduce the computation time, parallelization is often used, especially with large physical systems.

Even when the state space is not large, the computational effort can be large if we need to compute for a large number of time steps. This has been identified

D.A. Bader et al. (Eds.): HiPC 2005, LNCS 3769, pp. 106–117, 2005.

as one of the important challenges in nanoscale simulations and computational materials science [3]. Conventional parallelization is not effective in such problems, because the granularity becomes fine, limiting scalability.

We recently proposed [5,6] a time parallelization approach, to improve scalability. Here, results from related simulations are used to parallelize along the time domain. The basic idea is to have each processor simulate a different interval of time. The difficulty here is that each processor needs the state of the system at the beginning of the time interval it simulates, since it solves an initial value problem. We use the fact that typically the results of prior, related, simulations are available, to predict, in parallel, the states of the current simulation at desired points in time. The prediction mechanism also 'learns", thereby attempting to predict better as the simulation proceeds. The predicted states are verified in parallel to ensure accuracy of the results.

An important limitation of the earlier approach was that the prediction mechanism required some detailed knowledge of the the science of the problem. This is not easy to obtain in complicated simulation conditions. The main focus of this work is to show how reduced order modeling can be used to perform predictions for time parallelization.

The significance of this work lies in presenting an approach to time parallelization that is based on stronger mathematical foundations. This makes time parallelization of more complex problems too feasible. Time parallelization, especially for Molecular Dynamics (MD) simulations, has important applications to long time simulation simulations in computational Chemistry, Physics, Biology, Materials, and Engineering, making this work important.

The outline of the rest of the paper is as follows. In § 2, we describe a nanomaterials application that will be used to demonstrate the effectiveness of our technique. We then summarize the time-parallelization approach in § 3. In § 4, we describe prior and related work. We outline a particular reduced order modeling method, called *Proper Orthogonal Decomposition* (POD), and describe its use in time parallelization, in § 5. In § 6, we show the details of the steps involved in using POD to time-parallelize our application. We present conclusions and future work in § 7.

2   Carbon Nanotube Application

Tensile Test on a Carbon Nanotube. The physical system we consider is a Carbon Nanotube (CNT) [4]. One important application of CNTs is in nano-composites, where CNTs are embedded in a polymer matrix. In such applications, it is important to determine the mechanical properties of the CNT. Some important data for this is obtained from the "tensile test", in which one end of the CNT is held fixed, while the other end is pulled at a constant velocity. The response of the material is characterized by the *stress* (force required to pull the tube, divided by it cross-sectional area) experienced at a given *strain* (the elongation of the nanotube, relative to its original length). A stress-strain curve, as shown in Fig. 3 later, describes the response of the material when it is pulled at the specified velocity (more formally, strain-rate). Another important property is the strain at which the CNT starts to break.

Molecular Dynamics Simulation of a CNT. The state S_t of the CNT at any time t is defined by the position and velocity, at time t, of the atoms in the CNT. For a CNT with N atoms, there are $6N$ quantities (three position coordinates and the three velocity coordinates per atom) that define the state. The mechanical properties of the CNT at time t can be determined from S_t. Given S_t, we compute the state $S_{t+\Delta t}$ at the next time step $t + \Delta t$, by first computing forces on each atom due to other atoms [4], and then using Newton's laws of motion. A numerical time integration scheme (fourth-order Nordsiek in our implementation) is used in the latter. In MD computations, the time step size Δt is typically required to be less than a femto second (10^{-15}s), to ensure stability and accuracy.

This small Δt is an impediment to effective MD computation, because it makes a large number of time steps necessary. Furthermore, this computation will not parallelize efficiently using conventional parallelization, for physical systems of realistic sizes. As an alternative, researchers simulate by pulling the CNT at a faster rate than is realistic. The faster-rate simulation requires less time than that for a realistic rate, because the same strain is reached in less time with the former. It is assumed that the stress-strain relationship determined at the higher strain rate is the same as that at a lower strain rate. However, it is known that such an assumption is not accurate when the strain rates vary by several orders of magnitude [7]. On the other hand, if we could parallelize the computation efficiently on a large number of processors, then we could reach the desired time scale with more realistic strain-rates too.

3   Time Parallelization Through Guided Simulations

We recently [5,6] introduced the idea of guided simulations to parallelize along the time domain. We outline the approach below in Algorithm 1. Some details are deferred to § 6.

Let us call a few, say 1000, time steps as a *time interval*. Divide the total number of time steps needed into a number of time intervals. Ideally, the number of intervals should be much greater than the number of processors. Let t_i denote the beginning of the i th interval. Each processor $i \in \{1..P\}$, somehow predicts the states at times t_{i-1} and t_i, with the state at time t_0 being a known initial state S_0. Then it performs accurate computations (MD in our application), starting from the predicted state at time t_{i-1} up to time t_i. It then *verifies* if the prediction for t_i is close to the computed result. Both prediction and verification are done in parallel. If the predicted result is close to the computed one, then

TimeParallelize(Initial State S_0, Number of processors P, Number of time intervals m)

- $i=0$; $\hat{S}_0 = S_0$
- WHILE $i < m$
  - FOR each processor $j \in \{1.. \min(P, m - i - 1)\}$
    * $T_{i+j-1} = $ PredictStateAt (time = $i+j$-1)
    * $T_{i+j} = $ PredictStateAt (time = $i+j$)
    * $\hat{S}_{i+j} = $ AccurateComputation(StartState=T_{i+j-1}, StartTime=$i+j$-1, EndTime=$i+j$)
    * UpdatePredictionParameters(CurrentParameters, $\hat{S}_{i+j}$, T_{i+j})
    * IF IsDifferenceTooLarge($\hat{S}_{i+j}$, T_{i+j})
      · THEN $Next_j = j$
      · ELSE $Next_j = $ P
  - k = AllReduce(Next, min)
  - IF $j=k$
    * THEN Broadcast(Prediction Parameters)
  - SendReceive($\hat{S}_{i+k}$, From Processor k,To Processor 0)
  - FOR each processor $j \in \{1..P\}$
    * $i = i + k$

Fig. 1. The time parallelization algorithm

the initial state for processor $i + 1$ was accurate, and so the computed result for processor $i + 1$ too is accurate, provided the predicted state for time t_{i-1} was accurate. If the results differ significantly, then the predicted state for t_i was inaccurate, and we say that processor i *erred*. Computations for subsequent points in time too have to be discarded, since they might have started from an incorrect start state. The next phase starts from computed state for the latest time that is known to be accurate. The errors observed in the previous verification step are used to improve the predictor by better determining the relationship between the current simulation and old ones.

Note the following: (i) Processor 0 always starts from a state known to be accurate. (ii) The algorithm *always progresses* at least one time interval, since the accurate computations on processor 0 lead to accurate results on that processor. (iii) The prediction mechanism has two components – one that uses only prior simulation data, and another that "learns" the relationship between prior data and the current simulation based on the difference between the predicted and the computed states. The learning is represented using some prediction parameters. These parameters need to be identical on all processors. Otherwise, verification of prediction at time t_i at processor i does not imply that the prediction for initial state at time t_i on processor $i + 1$ was correct. So these prediction parameters are broadcast in Algorithm 1. (iv) The answers are *always accurate*, if we correctly define "sufficient closeness" of the predicted and computed states; a good predictor enables greater speedup, while a poor one leads to it becoming a sequential computation. (v) If the time interval consists of a large number of time steps, then the communication cost can be made negligible, leading to a very *latency tolerant* algorithm.

4   Related Work

Prior Work. In Algorithm 1, we need implementations of the following functions: (i) PredictStateAt, (ii) UpdatePredictionParameters, and (iii) IsDifferenceTooLarge. We used knowledge of the science to implement those functions in [6]. A large amount of data, totaling around 500 MBytes, were required for accurate prediction. Determining the permissible difference between predicted and computed states is application dependent. We defined two states to be equivalent, in our application, if the differences in positions, potential energy (also a function of positions), and kinetic energy (a function of velocities) were less than inherent fluctuations, as described in [5,6]. Using data from several time points of a single simulation that pulled a 1000-atom CNT at 10m/s, we predicted the behavior of a CNT pulled at 1m/s. The resulting time parallel code ran on 990 processors of the Xeon cluster at NCSA with efficiency greater than 97% [6].

Other Approaches. Due to its importance, there have been several works on the spatial parallelization of MD, including CNT computations specifically [4]. Time parallelization using the Parareal approach [1] is another promising alternative to conventional parallelization. We described its limitations in detail in [5]. The speedup and efficiency obtained have been limited. Speedups on simulated experiments (ignoring communication costs – the experiments were not on actual parallel machines) ranged between 8 to 130, with efficiency between 25% and 1.3% respectively on some model problems.

5   Application of Reduced Order Modeling to Time Parallelization

The basic idea behind reduced order modeling is that, while the state space involved in a simulation might be high dimensional, the states lie close to a lower dimensional subspace of the larger space. For example, the state of the CNT with N atoms is defined by $6N$ quantities, which can be represented by a vector $x \in \Re^{6N}$. If the CNT is pulled in the z direction, then most of the interesting changes take place in the z coordinates of the atoms. So we expect to find a smaller dimensional subspace of $\Re^{6N}$, close to which the states lie.

We use Proper Orthogonal Decomposition (POD) for reduced order modeling. The same idea is known by other names, such as Principal Component Analysis or Karhunen-Loève analysis. We next outline this method, and the intuition behind it. Further details can be found in [2].

Let $\hat{x}(t; v) \in \Re^m$ denote the state of a system at time t, simulated with parameters v. For example, in our application, v is a single parameter – the velocity at which the CNT is pulled. We assume that the states of the simulations lie close to a smaller dimensional affine subspace (a shifted linear subspace) of $\Re^m$. The shift is given by some vector μ, and the linear subspace S is given by the span of some vectors, which we represent as the columns of a matrix U. POD attempts to find a μ and a U that define this affine subspace $\mu + \text{span}\{U\}$.

We assume that a database of simulation results exists, with the states of simulations conducted under different parameters stored, for various values of time t. We choose n of these states to find a suitable basis U. Proper choice of these states is an important issue, which we shall not discuss here. Let us call the chosen states $\hat{x}_i$, $1 \leq i \leq n$. We consider the case where $n < m$. We define $\mu = 1/n \sum_{i=1}^{n} \hat{x}_i$. Let $x_i = \hat{x}_i - \mu$. We then construct the *snapshot matrix* $A = [x_1 x_2 \cdots x_n] \in \Re^{m \times n}$.

Let $A = \tilde{U} \tilde{\Sigma} \tilde{V}^T$ be the singular value decomposition of A. Here, $\tilde{U} = [u_1 \cdots u_m] \in \Re^{m \times m}$ and $\tilde{V} \in \Re^{n \times n}$ are orthogonal matrices, and the rectangular diagonal matrix $\tilde{\Sigma} \in \Re^{m \times n}$ contains the singular values of A (which are non-negative) in descending order. The columns of $\tilde{U}$ form a basis for $\Re^m$. Column i of $\tilde{\Sigma} \tilde{V}^T$ gives the coefficients of x_i in the basis consisting of columns of $\tilde{U}$. The coefficients of columns u_i, $i \geq n$, are zero. If the x_i's lie close to a d dimensional linear subspace of $\Re^m$, then only the first d singular values are large. So we define $U \in \Re^{m \times d}$ to consist of the first d columns of $\tilde{U}$, $\Sigma \in \Re^{d \times d}$ the corresponding submatrix of $\tilde{\Sigma}$, and $V \in \Re^{n \times d}$ the first d columns of $\tilde{V}$. Then $A \approx U \Sigma V^T$, and $\sigma_i v_{ji}$ is the component of x_j along the direction of u_i. That is, $x_j \approx \Sigma_{i=1}^{d} c_{ji} u_i$, where $c_{ji} = \sigma_i v_{ji}$.

The linear subspace S spanned by the columns of U is optimal in the sense that among all linear subspaces S' of dimension d, it has a minimum value of $\Sigma_{i=1}^{n} D(S', x_i)$, where $D(S', x_i)$ is the square of the distance between x_i and S'. Note that if the x_i's lie close to S, then the original data $\hat{x}_i$'s lie close to the affine subspace $S + \mu$.

If we just wished to represent the database with less storage, then we can store μ, U, and the coefficients of U for each state in the database. Since the columns of U are orthogonal, the coefficients for any vector $\hat{x}$ are easily obtained as $U^T(\hat{x} - \mu)$, involving just a vector subtraction and d dot products. If there are M states in the database, then the total storage is $Md + (d+1)m$, with $d \ll M, m$, in contrast to Mm storage required for the original data. In simulating dynamical systems, the state is expressed as $\hat{x} \approx \mu + \Sigma_{i=1}^{d} c_i u_i$, and this is substituted in the equations defining the evolution of the state, to come up with equations that define the evolution of c_i's.

6   Experimental Results

We first show various aspects of the application of POD to prediction, and then present speedup results. The physical system considered is a 1000-atom CNT at $300K$. The database consisted of tensile test simulation results at velocities given below, from zero strain to until the CNT starts to break. The time parallel simulations were performed for velocities of 2m/s and 0.1m/s.

Basis Vectors for the CNT. The states at the following points in time were used to construct the snapshot matrix A: (i) Velocity = 10m/s: after time steps $-$ 50,000, 100,000, 200,000, 300,000, and 350,000. (ii) Velocity = 5m/s: after time steps $-$ 100,000, 200,000, 400,000, 600,000, and 700,000. (iii) Velocity = 1m/s:

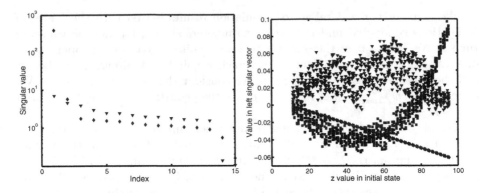

Fig. 2. Left: Plot of singular values against its index on a semi-log scale. The diamonds show the values for z and the triangles for y. Right: Plot of basis vectors against the z coordinate value of the corresponding atom in the initial state. The diamonds show the values for z's u_1, the squares for z's u_2, and the triangles for x's u_1.

after time steps – 500,000, 1,000,000, 2,000,000, 3,000,000, and 3,500,000. (iv) The initial state, which is identical for all velocities. The above times indicate that we chose a set of five different strain values, and noted the state at these strains for each of the three parameters. We generate separate bases for the x, y, and z coordinates, and so we created three different snapshot matrices, one for each set of coordinates. Each snapshot matrix is of dimension 800×16, with each row corresponding to the position coordinate of an atom moved using MD (the temperature is held constant, and so we did not include the velocities in the state).

We then determined μ, $\tilde{U}$, $\tilde{\Sigma}$, and $\tilde{V}$ as described in § 5. For each coordinate, *columns of $\hat{U}$ that corresponded to large singular values were usually placed in U*. Apart from the singular value, we also checked to see if the vector represented a pattern, or just corresponded to random "noise". Fig. 2 shows that the z coordinate has one very high singular value, followed by a moderately large one, followed by several smaller ones. We look further into each u_i corresponding to the larger singular values. Fig. 2 shows the values of the components of u_1 and u_2 for z are non-random. On the other hand, u_1 of x appears random.

Based on many such observations of the singular values and randomness, we used u_1 and u_2 as basis for z's lower dimensional subspace, and none for x and y. Consequently, the predicted values for the x and y coordinates are always μ_x and μ_y respectively, while for the z coordinate, we use a two dimensional subspace $\mu_z + \text{span}\{u_1, u_2\}$.

Relationship Between Velocity and Time. The relationship between velocity and time is important for the following reason. We expect time parallelization to be effective if we can predict the behavior of a simulation with parameter v, from those of prior simulations performed under different parameters. We are, in effect, assuming similarity of behavior under different parameters. However, the

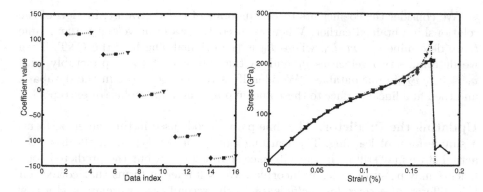

Fig. 3. Left: Plot of z's u_1 coefficient against the the index of the data point (that is, its column number in the A snapshot matrix). The diamonds show the values for velocity 1m/s, the squares for 5m/s, and the triangles for 10m/s. The dotted lines connect points at the same values of strain. Right: Plot of stress versus strain at 0.1m/s. The solid line represents the exact sequential MD results. The squares represents the time parallel code on 400 processors. The dash-dotted line with triangles represents direct prediction.

behavior at time t with parameter v may be similar to the behavior at a different time t_1 when simulated with parameter v_1. For example, in our prior work, we expected long time behavior at a low strain rate to be related to short time behavior at a large strain rate. This was based on knowledge of the science of the problem, and is not often easy to predict. Instead, *we identify similar behavior as having similar values of coefficients for basis vectors with large singular values.* In our application, only z's u_1 has a very large singular value. Fig. 3 shows the coefficients for z's u_1 are similar for similar strains. This provides a more formal justification for the assumption used in the previous work, which we use in this paper too.

Direct Prediction Using Interpolation. We can predict the state $\hat{x}(t; v)$ for a time-parameter combination not in the database by first predicting the vector of coefficients, c. This is done by *interpolating or extrapolating coefficients of similar states from the database.* The parameter-time relation obtained from the previous step tells us which states can be expected to be similar. Then the actual state is easily obtained as $\hat{x}(t; v) = Uc + \mu$. We call this prediction as a *direct prediction.* We later explain how the learning mechanism modifies this prediction.

We now give some details of the implementation. We precomputed coefficients for 40 states at different times (or, equivalently, strains) for each of the velocities: 10m/s, 5m/s, and 1m/s. We also computed coefficients for the initial state, which was identical for all velocities. These coefficients are easily obtained as $U^T(\hat{x} - \mu)$ for a known state $\hat{x}$. The amount of data is small, and is stored in an array in our code. Note that even the number of prior results required (121) is fairly small, requiring less than 10 MBytes of disk space.

We consider the coefficients to be functions of v and strain ϵ, in view of the relationship obtained earlier. When we need the state for velocity v and time t, we determine $\epsilon = vt/L_0$, where L_0 is the original length of the CNT. Then we determine two velocities v_1 and v_2 that are closest to v (preferably with $v_1 \leq v \leq v_2$) in our database. We identify strains closest to ϵ in the database, and then fit a linear surface to these known points and interpolate or extrapolate.

Updating the Predictor. The time parallel code used in this paper performs a simple form of learning. The learning can be of two types. In the first, the actual state may still be in the subspace spanned by U, but the predicted coefficients may be systematically incorrect. We will refer to terms that correct for this difference as *corrector coefficients*. In the second case, we may need a new basis vector, which corresponds to a new physical process. Singular value decomposition of the residuals in the verification step, orthogonal to the subspace spanned by U, can yield such a basis. We have incorporated the former type of learning in our implementation.

The corrector coefficients are initially 0. Let processor i start from a predicted state at time t_i and perform accurate MD computations up to time t_{i+1}. Let the state reached be $\hat{S}$ and the predicted state at time t_{i+1} be T. Processor i then computes coefficients $c = U^T(\hat{S} - T)$ and adds it to the current corrector coefficients. The lowest indexed processor that erred, or processor P if none erred, broadcasts its corrector coefficients to all processors. In the next phase of computations, let T be the state predicted from the interpolation. Then the actual prediction for that time is taken to be $T + Uc$. If the systematic error in coefficients varies slowly with time, then this correction accounts for it. It does not account for errors that arise from the current simulation leaving the low dimensional subspace.

Direct prediction can be performed for any point in time. Time parallelization can be accurate even when direct parallelization is not, for the following reasons. (i) The corrector coefficients perform a simple form of learning, and make predictions better. (ii) The verification step can detect errors, and so the results are accurate, even though the computation slows down. (iii) The computed state can depart from the low-dimensional subspace, and come closer to the correct state. The computation may slow down in this case too, since differences between predicted and computed states are treated as errors in the verification step.

Validation. Fig. 3 shows the stress-strain relationship from the exact sequential simulation, direct prediction, and the time parallel code. This relationship is the primary material property of interest. Away from the point when the CNT starts to break (stress reduces with increase in strain then), direct prediction is quite accurate. However, close to the point of breakage, it errs. This error can be traced to its higher errors in potential energy. However, it predicts the time of start of breakage correctly as at around 17% strain. Note that the stress-strain relationship obtained from the time parallel code is accurate until the point of breakage. From a practical point of view, the behavior of the CNT after its starts breaking is not relevant. We performed similar validation against the exact results for 2m/s simulations, and with different numbers of processors. We also compared other quantities, such as positions and potential energy.

The above observations suggest that direct prediction may be acceptable when interpolating in a parameter-time range between existing data. However, extrapolation can lead to errors. For example, the points close to the CNT breaking involve extrapolation in strain and in velocity. The time parallel code performs accurately until the point of breakage. After this, since the code does not correct for the new phenomenon of breaking, the predictions fail. Of course, this is detected during verification, and so the computation progresses slowly.

When extrapolating data over a wide range, such as performing a very low strain rate calculation, where new phenomena are likely to occur, direct prediction may not be accurate. Time parallelization, on the other hand, can be effective there, since it does not give erroneous results. The savings in time using direct prediction are enormous, when it can be applied. Determining the stress-strain relationship at a velocity of 0.1m/s, for example, requires about a week of sequential computing time when we simulate until the CNT starts to break. This can be done in 10^{-5}s per time point using direct prediction. Time parallelization yields accurate results up to the point of breakage, and we have simulated it on hundreds of processors with nearly ideal speedup. So the above simulation can be completed in less than an hour, accurately, in parallel.

Speedup Results. Speedup results are reported on the *Tungsten* Xeon cluster at NCSA. This cluster consists of Dell PowerEdge 1750 servers, with each node containing two Intel Xeon 3.2 GHz processors, 3 GB ECC DDR SDRAM memory, 512 KB L2 cache, 1 MB L3 cache, running Red Hat Linux. Myrinet 2000 and Gigabit Ethernet interconnects are available. We used the Myrinet interconnect. The ChaMPIon/Pro MPI implementation was used with gcc/g77 compilers for our mixed C/Fortran code, compiled with '-O3' optimization flags set. The MPI calls were purely in the C code. The timing results are based on wall clock time when run in non-dedicated mode.

Fig. 4 shows the speedup results, for a simulation at 0.1m/s. Similar results were obtained with 2m/s simulations, which was run on up to 250 processors. We can see that speedup is good on up to 400 processors on the Xeon cluster. The efficiency is greater than 95% for all cases except for 400 processors, where the efficiency is around 90%. The processors never err in the course of the simulation (up to the point where the CNT starts to break), and so loss in speed is only due to the overheads of prediction and communication. These overheads are small, compared with the computation time. For example, the prediction related computations take less than 10^{-4}s, the AllReduce $\approx 10^{-4} - 10^{-3}$s for 50-1000 processors, Broadcast $\approx 10^{-4}$s for 50-1000 processors, and the Send/Recv about 10^{-4}s. Load imbalance is not an issue, since each processor performs, essentially, the same amount of computation. All the overheads are insignificant, relative to the computation time (≈ 13s) for simulating a single time interval. We expect the loss in efficiency, especially for the 400 processor run, to be due to running in a non-dedicated mode. For example, we got over 97% efficiency on up to 1000 processors with our previous predictor, when run in dedicated mode. However,

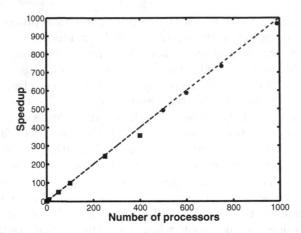

Fig. 4. Speedup curve. The dashed line shows the ideal speedup, and squares show the speedup on the NCSA Xeon cluster for a velocity of 0.1m/s, run in *non-dedicated* mode, with the new prediction scheme. The circles show speedup on the same machine with the earlier predictor, for a velocity of 1m/s, run in *dedicated* mode.

the actual prediction and communication overheads of the previous predictor are slightly larger than for the new one. In particular, the earlier prediction related computations took $\approx 10^{-3}$s, and Broadcast $\approx 10^{-1} - 10^{-2}$s for 50-1000 processors. In any case, the efficiency is very good with both predictors and they scale to much larger numbers of processors than conventional parallelization.

7   Conclusions

We have shown that reduced order modeling can provide a systematic procedure for choosing a basis for modeling the data, without much apriori information on the physical processes the system is undergoing. Such modeling enables us to predict the states, for different time and parameter values. Furthermore, the amount of prior data needed is less. Parallelization of time, using our approach, scales to at least two orders of magnitude larger numbers of processors than conventional parallelization. Our results are, therefore, of much significance, since they suggest a general technique for more complicated problems, were less knowledge of the physical processes is available.

Future work will consist of simulating multiple parameter systems, such as strain rate, temperature, and CNT diameter. We will also include, in the implementation, the ability to learn about new phenomena the CNT undergoes, orthogonal to the selected subspace. We plan to use other reduced order modeling techniques too, such as Centroidal Voronoi Tesselations [2]. We will also perform time parallel simulations under more experimental conditions, apart than tensile tests, and at lower strain rates, and include the material responses in multi-scale FEM simulations.

Acknowledgments

This work was funded by NSF grant # CMS-0403746. We thank ORNL (CNMS /NTI grant #CNMS2004-028) and NCSA (proposal #ASC050004) for providing computer time. We also thank Xin Yuan at Florida State University for permitting use of his Linux cluster, where our codes are first tested, and Max Gunzburger, for drawing our attention to reduced order modeling literature. Most of all, A.S. thanks Sri S.S. Baba, whose inspiration and help were crucial to the success of this work.

References

1. L. Baffico, S. Bernard, Y. Maday, G. Turinici, and G. Zerah. Parallel-in-time molecular-dynamics simulations. *Physical Review E (Statistical, Nonlinear, and Soft Matter Physics)*, 66:57701–57704, 2002.
2. J. Burkardt, Q. Du, M. Gunzburger, and H. Lee. Reduced order modeling of complex systems. In D. F. Griffiths and G. A. Watson, editors, *Proceedings of the 20 th biennial conference on Numerical Analysis*, pages 29–38, Dundee, Scotland, U.K., 2003. University of Dundee.
3. Theory and modeling in nanoscience, May 2002. Report of the May 10-11, 2002 Workshop conducted by the basic energy sciences and advanced scientific computing advisory committees to the Office of Science, Department of Energy.
4. J. Kolhe, U. Chandra, S. Namilae, A. Srinivasan, and N. Chandra. Parallel simulation of Carbon nanotube based composites. In L. Bougé and V. K. Prasanna, editors, *Proceedings of the 11 th International Conference on High Performance Computing (HiPC), Lecture Notes in Computer Science – 3296*, pages 211–221. Springer-Verlag, 2004.
5. A. Srinivasan and N. Chandra. Latency tolerance through parallelization of time in scientific applications. *Parallel Computing*, 31:777–796, 2005.
6. A. Srinivasan, Y. Yu, and N. Chandra. Scalable parallelization of molecular dynamics simulations in nano mechanics, through time parallelization. Technical Report TR-050426, Department of Computer Science, Florida State University, 2005.
7. B. I. Yakobson, M. P. Campbell MP, and C. J. Brabec. High strain rate fracture and C-chain unraveling in Carbon nanotubes. *Computational Materials Science*, 8:341–348, 1997.

Orthogonal Decision Trees for Resource-Constrained Physiological Data Stream Monitoring Using Mobile Devices

Haimonti Dutta, Hillol Kargupta*, and Anupam Joshi

Department of Computer Science and Electrical Engineering,
University of Maryland Baltimore County,
1000 Hilltop Circle Baltimore, MD 21250, USA
{hdutta1, hillol, joshi}@cs.umbc.edu

Abstract. This paper considers the problem of monitoring physiological data streams obtained from resource-constrained wearable sensing devices for pervasive health-care management. It considers Orthogonal decision trees (ODTs) that offer an effective way to construct a redundancy-free, accurate, and meaningful representation of large decision-tree-ensembles often created by popular techniques such as Bagging, Boosting, Random Forests and many distributed and data stream mining algorithms. ODTs are functionally orthogonal to each other and they correspond to the principal components of the underlying function space. This paper offers experimental results to document the performance of ODTs on grounds of accuracy, model complexity, and resource consumption.

1 Introduction

Monitoring time-critical data streams using mobile and wearable devices in a ubiquitous manner is important in many applications. Online classification of data streams in such resource-constrained environments is a challenging task that requires light-weight classifiers that are accurate but compact in representation. This paper considers the classification and monitoring of physiological data streams using decision tree (e.g., CART[1] and C4.5 [2]) ensembles. Ensemble learning techniques are often used where a single decision tree does not provide sufficient accuracy. Boosting [3,4], Bagging[5], Stacking [6], and random forests [7] are some of the well-known ensemble-learning techniques. Many of these techniques often produce large ensembles that combine the outputs of a large number of trees for producing the overall output. In many time-critical applications such as monitoring data streams [8], particularly for resource constrained environments [9], using a large ensemble for continuous monitoring is computationally challenging. A redundancy free and meaningful compact representation of large ensembles is therefore needed.

---

* Also affiliated with AGNIK, LLC, USA.

D.A. Bader et al. (Eds.): HiPC 2005, LNCS 3769, pp. 118–127, 2005.
© Springer-Verlag Berlin Heidelberg 2005

This paper reports an application of redundancy-free decision trees-ensembles by constructing Orthogonal Decision Trees (ODTs) [10]. The technique first constructs an algebraic representation of trees using multivariate discrete Fourier bases. The new representation is then used for eigen-analysis of the covariance matrix generated by the decision trees in Fourier representation. The proposed approach converts the corresponding principal components to decision trees using a technique reported elsewhere [9]. These trees are functionally orthogonal to each other and they span the underlying function space. These orthogonal trees are in turn used for accurate (in many cases with improved accuracy) and redundancy-free (in the sense of orthogonal basis set) compact representation of large ensembles. We use this compact ODT ensemble to implement a system for monitoring physiological health data streams that can run on resource constrained PDA/wearable devices.

The rest of the paper is organized as follows. Section 2 discusses the importance of monitoring physiological data streams using wearable devices and arm-bands available in the market. Section 3 presents the underlying theory for representation of decision trees using their Fourier spectra. Section 4 describes the process of removing redundancy from decision tree ensembles and Section 5 explains the method of constructing ODTs. Sections 6 presents experimental results for ODTs and compares it with a well known ensemble learning technique. Finally, Section 7 concludes this paper.

2 Physiological Data Stream Monitoring

Consider a real time environment to monitor the health effects of environmental toxins or disease pathogens on humans. There are significant advances being made today in biochemical engineering to create low cost sensors for various toxins[11] that could constantly monitor the environment and generate data streams over wireless networks. It is not unreasonable to assume that similar sensors could be developed to detect disease causing pathogens. In addition, most state health/environmental agencies and the federal government entities such as CDC and EPA have mobile labs and response units that can test for the presence of pathogens or dangerous chemicals. The mobile units will have handheld devices with wireless connections on which to send the data and/or their analysis. In addition, each hospital today generates reports on admissions and discharges, and often reports that to various monitoring agencies. Given these disparate data streams, one could analyze them to see if correlates can be found, alerting experts to potential cause-effect relations (Pfiesteria found in Chesapeake Bay and hospitals report many people with upset stomach who had seafood recently), potential epidemiological events (field units report dead infected birds and elderly patients check in with viral fever symptoms, indicating tests needed for west Nile virus and preventive spraying), and more pertinent in present times, low grade chemical and biological attacks (sensors detect particular toxins, mobile units find contaminated sites, hospitals show people who work at or near the sites being admitted with unexplained symptoms). At present, much of this analysis is done "post facto" – experts hypothesize on possible causes of ailments, then gather the data from disparate sources to confirm their hypotheses. Clearly, a more proactive environment which could mine these diverse data streams to detect emergent patters would be extremely useful. This scenario, of course, has some futuristic elements.

On a more present day note, there are now several wearable sensors on the market such as SenseWear armband from BodyMedia[1] , Wearable West[2], and LifeShirt Garment from Vivometrics[3] that can be used to monitor vital signs for a person such as temperature, heart-rate, heat flux, SpO_2 etc. The sensors in the SenseWear armband can measure the following: (1) Heat flux (the amount of heat dissipated by the body), (2) acceleration (3) galvanic skin response (electrical conductivity between two points on the wearer's arm, (4) skin temperature, (5) near-body temperature (air temperature immediately around the wearer's armband).

This body monitoring device can be worn continuously, and it can store up to five days of physiological data before it had to be retrieved. The LifeShirt Garment is yet another example of an easy to wear shirt, that allows measurement of pulmonary functions via sensors woven into the shirt. Analyzing these vital signs in real time using small form factor wearable computers has several valuable near term applications. For instance, one could monitor senior citizens living in assisted or independent housing, to alert physicians and support personnel if the signs point to distress. Similarly, one could monitor athletes during games or practice. Other potential applications include battlefield monitoring of soldiers, or monitoring first responders such as firefighters.

This paper offers a method for on-line monitoring of physiological data using ODT ensembles running on wearable or handheld devices. The following section presents the foundation material needed to understand ODTs.

3   Fourier Spectrum of Decision Trees

This section briefly discusses the background material [9] necessary for the development of the proposed technique to construct orthogonal decision trees. The proposed approach makes use of linear algebraic representation of the trees. In order to do that that we first need to convert the trees into a numeric tree just in case the attributes are symbolic. This can be done by simply using a codebook that replaces the symbols with numeric values in a consistent manner. Since the proposed approach of constructing orthogonal trees uses this representation as an intermediate stage and eventually the physical tree is converted back, the exact scheme for replacing the symbols (if any) does not matter as long as it is consistent.

Once the tree is converted to a discrete numeric function, we can also apply any appropriate analytical transformation if necessary. Fourier transformation is

---

[1] http://www.bodymedia.com/index.jsp
[2] http://www.smartextiles.info
[3] http://www.vivometrics.com

one such interesting possibility. Fourier bases are orthogonal functions that can be used to represent any discrete function. Consider the set of all ℓ-dimensional feature vectors where the i-th feature can take λ_i different categorical values. The Fourier basis set that spans this space is comprised of $\Pi_{i=0}^{\ell}\lambda_i$ basis functions. Each Fourier basis function is defined as, $\psi_{\mathbf{j}}^{\overline{\lambda}}(\mathbf{x}) = \frac{1}{\sqrt{\Pi_{i=1}^{l}\lambda_i}}\Pi_{m=1}^{l}\exp^{\frac{2\pi i}{\lambda_m}x_m j_m}$, where $\mathbf{j}$ and $\mathbf{x}$ are strings of length ℓ; x_m and j_m are m-th attribute-value in $\mathbf{x}$ and $\mathbf{j}$, respectively; $x_m, j_m \in \{0, 1, \cdots \lambda_i\}$ and $\overline{\lambda}$ represents the feature-cardinality vector, $\lambda_0, \cdots \lambda_\ell$; $\psi_{\mathbf{j}}^{\overline{\lambda}}(\mathbf{x})$ is called the $\mathbf{j}$-th basis function. The vector $\mathbf{j}$ is called a *partition*, and the *order* of a partition $\mathbf{j}$ is the number of non-zero feature values it contains. A Fourier basis function depends on some x_i only when the corresponding $j_i \neq 0$. If a partition $\mathbf{j}$ has exactly α number of non-zeros values, then we say the partition is of order α since the corresponding Fourier basis function depends only on those α number of variables that take non-zero values in the partition $\mathbf{j}$.

A function $f : \mathbf{X}^{\ell} \rightarrow \Re$, that maps an ℓ-dimensional discrete domain to a real-valued range, can be represented using the Fourier basis functions: $f(\mathbf{x}) = \sum_{\mathbf{j}} w_{\mathbf{j}}\overline{\psi}_{\mathbf{j}}^{\overline{\lambda}}(\mathbf{x})$. where $w_{\mathbf{j}}$ is the Fourier Coefficient (FC) corresponding to the partition $\mathbf{j}$ and $\overline{\psi}_{\mathbf{j}}^{\overline{\lambda}}(\mathbf{x})$ is the complex conjugate of $\psi_{\mathbf{j}}^{\overline{\lambda}}(\mathbf{x})$; $w_{\mathbf{j}} = \sum_{\mathbf{x}} \psi_{\mathbf{j}}^{\overline{\lambda}}(\mathbf{x})f(\mathbf{x})$. The *order* of a Fourier coefficient is nothing but the order of the corresponding partition. We shall often use terms like *high order* or *low order* coefficients to refer to a set of Fourier coefficients whose orders are relatively large or small respectively. Energy of a spectrum is defined by the summation $\sum_{\mathbf{j}} w_{\mathbf{j}}^2$. Let us also define the inner product between two spectra $\mathbf{w}_{(1)}$ and $\mathbf{w}_{(2)}$ where $\mathbf{w}_{(i)} = [w_{(i),1}w_{(i),2},\cdots w_{(i),|J|}]^T$ is the column matrix of all Fourier coefficients in an arbitrary but fixed order. Superscript T denotes the transpose operation and $|J|$ denotes the total number of coefficients in the spectrum. The inner product, $< \mathbf{w}_{(1)}, \mathbf{w}_{(2)} >= \sum_{\mathbf{j}} w_{(1),\mathbf{j}}w_{(2),\mathbf{j}}$. We will also use the definition of the inner product between a pair of real-valued functions defined over some domain Ω. This is defined as $< f_1(\mathbf{x}), f_2(\mathbf{x}) >= \sum_{\mathbf{x} \in \Omega} f_1(\mathbf{x})f_2(\mathbf{x})$.

Fourier transformations of bounded-depth decision trees have several properties that makes it an efficient one. Some of the relevant ones are listed below:

1. Energy of the Fourier coefficients decay exponentially with respect to $o(\mathbf{j})$ where $o(\mathbf{j})$ denotes the order of the partition $\mathbf{j}$.
2. the Fourier spectrum of a decision tree can be efficiently computed [9] and
3. the Fourier spectrum can be directly used for constructing the tree [12].
4. Fourier transformation of decision trees also preserves inner product.

More details can be found elsewhere [13,12].

4 Removing Redundancies from Ensembles

Existing ensemble-learning techniques work by combining (usually a linear combination) the output of the base classifiers. They do not structurally combine the classifiers themselves. As a result they often share a lot of redundancies. The Fourier representation offers a unique way to fundamentally aggregate the trees and perform further analysis for constructing an efficient representation.

Let $f_e(\mathbf{x})$ be the underlying function representing the ensemble of m different decision trees where the output is a weighted linear combination of the outputs of the base classifiers. Then we can write,

$$f_e(\mathbf{x}) = \sum_i^m \alpha_i \tau_{(i)}(\mathbf{x}) = \sum_i^m \alpha_i \sum_{j \in \mathcal{J}_1} w_{(i),\mathbf{j}} \overline{\psi}_\mathbf{j}^\lambda(\mathbf{x}) \tag{1}$$

Where α_i is the weight of the ith decision tree and Z_i is the set of all partitions with non-zero Fourier coefficients in its spectrum. Therefore, $f_e(\mathbf{x}) = \sum_{j \in \mathcal{J}} w_{(e),\mathbf{j}} \overline{\psi}_\mathbf{j}^\lambda(\mathbf{x})$, where $w_{(e),\mathbf{j}} = \sum_{i=1}^m \alpha_i w_{(i),\mathbf{j}}$ and $\mathcal{J} = \cup_{i=1}^m \mathcal{J}_i$. Therefore, the Fourier spectrum of $f_e(\mathbf{x})$ (a linear ensemble classifier) is simply the weighted sum of the spectra of the member trees.

Consider the matrix D where $D_{i,j} = \tau_{(j)}(\mathbf{x}_i)$, where $\tau_{(j)}(\mathbf{x}_i)$ is the output of the tree $\tau_{(j)}$ for input $\mathbf{x}_i \in \Omega$. D is an $|\Omega| \times m$ matrix where $|\Omega|$ is the size of the input domain and m is the total number of trees in the ensemble.

An ensemble classifier that combines the outputs of the base classifiers can be viewed as a function defined over the set of all rows in D. If $D_{*,j}$ denotes the j-th column matrix of D then the ensemble classifier can be viewed as a function of $D_{*,1}, D_{*,2}, \cdots D_{*,m}$. When the ensemble classifier is a linear combination of the outputs of the base classifiers we have $F = \alpha_1 D_{*,1} + \alpha_2 D_{*,2} + \cdots \alpha_m D_{*,m}$, where F is the column matrix of the overall ensemble-output. Since the base classifiers may have redundancy, we would like to construct a compact low-dimensional representation of the matrix D. However, explicit construction and manipulation of the matrix D is difficult, since most practical applications deal with a very large domain. In the following we demonstrate a novel way to efficiently perform a PCA of the matrix D, defined over the entire domain. The approach uses the Fourier spectra of the trees and works without explicitly generating the matrix D. It is important to note that existing PCA-based regression schemes [14] offer a way to find the weights for the members of the ensemble. They do not offer any way to aggregate the tree structures and construct a new representation of the ensemble which the current approach does.

The following analysis will assume that the columns of the matrix D are mean-zero. This restriction can be easily removed with a simple extension of the analysis. Note that the covariance of the matrix D is $D^T D$. Let us denote this covariance matrix by C. The (i,j)-th entry of the matrix,

$$C_{i,j} = <D(*,i), D(*,j)> = <\tau_{(i)}(\mathbf{x}), \tau_{(j)}(\mathbf{x})> = <\mathbf{w}_{(i)}, \mathbf{w}_{(j)}> \tag{2}$$

Now let us consider the matrix W where $W_{i,j} = w_{(j),(i)}$, i.e. the coefficient corresponding to the i-th member of the partition set $\mathcal{J}$ from the spectrum of the tree $\tau_{(j)}$. Equation 2 implies that the covariance matrices of D and W are identical. Note that W is an $|\mathcal{J}| \times m$ dimensional matrix. For most practical applications $|\mathcal{J}| << |\Omega|$. Therefore analyzing W using techniques like PCA is significantly easier. The following discourse outlines a PCA-based approach.

PCA of the matrix W produces a set of eigenvectors which in turn defines a set of Principal Components, $V_1, V_2, \cdots V_k$. Let $\gamma_{(j),q}$ be the j-th component of the q-th eigenvector of the matrix $W^T W$.

$$V_q = \sum_{j=1}^{n} \gamma_{(j),q} D(*,j) = \left[\sum_{j=1}^{n} \gamma_{(j),q} \tau_{(j)}(\mathbf{x}) \right]_{\mathbf{x} \in \Omega} = \left[\sum_{i} a_{i,q} \overline{\psi_i}^{\lambda}(\mathbf{x}) \right]_{\mathbf{x} \in \Omega}.$$

Where $a_{i,q} = \sum_{j=1}^{n} \gamma_{(j),q} w_{(j),i}$. The eigenvalue decomposition constructs a new representation of the underlying domain where the feature corresponding to column vector V_q is $v_q = \sum_i a_{i,q} \overline{\psi_i}^{\lambda}(\mathbf{x})$ i.e., $V_q = [v_q]_{\mathbf{x} \in \Omega}$. Note that v_q is a linear combination of a set of Fourier spectra and therefore it is also a Fourier spectrum. Also note that V_q-s are orthogonal [10].

Therefore, we conclude that the spectra corresponding to the orthonormal basis vectors V_q and V_r are themselves orthonormal. Let f_q and f_r be the functions corresponding to the spectra $\mathbf{a}_q$ and $\mathbf{a}_r$. In other words, $f_q(\mathbf{x}) = \sum_i a_{i,q} \psi_i(\mathbf{x})$ and $f_r(\mathbf{x}) = \sum_i a_{i,r} \psi_i(\mathbf{x})$. Then we can conclude that, $< V_q, V_r > = < \mathbf{a}_q, \mathbf{a}_r > = < f_q(\mathbf{x}), f_r(\mathbf{x}) >$. This implies that the inner product between the output vectors of the corresponding functions are also orthonormal to each other. The following section defines ODTs that makes use of these principal components.

5   Orthogonal Decision Trees

The analysis presented in the previous sections offers a way to construct the Fourier spectra of a set of functions that are orthogonal to each other and therefore redundancy-free. These functions also define a basis and can be used to represent any given decision tree in the ensemble in the form of a linear combination. Orthogonal decision trees (ODTs) can be defined as an immediate extension of this framework.

A pair of decision trees $f_1(\mathbf{x})$ and $f_2(\mathbf{x})$ are orthogonal to each other if and only if $< f_a(\mathbf{x}), f_b(\mathbf{x}) > = 0$ when $a \neq b$ and $< f_a(\mathbf{x}), f_b(\mathbf{x}) > = 1$ otherwise. The second condition is actually a slightly special case of orthogonal functions—orthonormal condition. A set of trees are pairwise orthogonal if every possible pair of members of this set satisfy the orthogonality condition.

The principal components $V_1, V_2, \cdots V_k$ computed using the eigenvectors of the covariance matrix C are orthogonal to each other themselves. Since each of these principal components is a Fourier spectrum in itself we can always construct a decision tree from this spectrum [10]. Although the tree looks physically different from the Fourier spectrum, they are functionally identical. Therefore, the trees constructed from the principal components $V_1, V_2, \cdots V_k$ also maintain the orthogonality condition. These orthogonal trees now can be used to represent the entire ensemble in a very compact and efficient manner. The orthogonality condition guarantees that the representation is not redundant. These ODTs form a basis set that spans the entire function space of the ensemble. The overall output of the ensemble is computed from the output of these orthogonal trees. The following section reports some experimental results.

6   Experimental Results

This section documents the performance of ODTs on a physiological data set obtained from the Physiological Data Modeling Contest[4] held as part of the International Conference on Machine Learning, 2004. It is comprised of several months of data from more than a dozen subjects collected using BodyMedia wearable body monitors.

In our experiments, the training set consisted of 50,000 instances and 11 continuous and discrete-valued attributes[5]. The test set had 32,673 instances. The continuous-valued attributes were discretized using the WEKA software. The final training and test data sets had all discrete valued attributes. A binary classification problem was formulated, which monitored whether an individual was engaged in a particular activity(class label=1) or not(class label=0) depending on the physiological sensor readings.

C4.5 decision trees were built on data blocks of size 150 instances; the classification accuracy and tree complexity (number of nodes in the tree) were noted. These were then used to compute their Fourier spectra and the matrix of the Fourier coefficients was subjected to principle component analysis. ODTs corresponding to the significant components were constructed and combined using an uniform aggregation scheme. The accuracy and size of the ODTs are noted and compared with the corresponding characteristics of a Bagging ensemble with the same number of decision trees in the ensemble.

Figure 1 illustrates the distribution of tree complexity and error in classification for the original C4.5 trees used to construct an ODT ensemble. The total number of nodes in the original C4.5 trees varied between three and thirteen. The trees had an error of less than 25(%). In comparison, the average complexity of the ODTs was found to be 3 for all the different ensemble sizes. In fact, for this particular dataset, the sensor reading corresponding to transverse accelerometer attribute was found to be the most interesting. All the ODTs used this attribute as the root node for building the trees. The Figure 2(Left) illustrates the distribution of classification-error for an ODT ensemble of 75 trees.

We compared the accuracy obtained from an aggregated ODT to that obtained from a bagging ensemble (using the same number of trees in each case). Figure 2(Right) plots the error in classification of the aggregated ODT and bagging versus the number of decision trees in the ensemble. We found that the classification from an aggregated ODT was better than Bagging when the num-

---

[4] http://www.cs.utexas.edu/users/sherstov/pdmc/

[5] The attributes used for the classification experiments were gender, galvanic skin temperature, heat flux, near body temperature, pedometer, skin temperature, readings from the longitudinal and transverse accelerometer and time for recording an activity called session time.

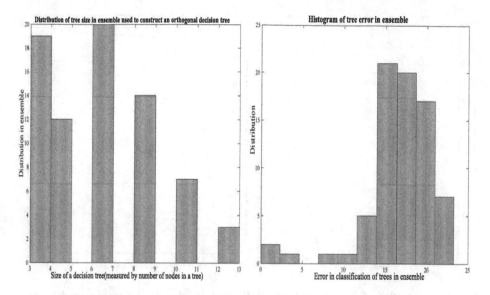

Fig. 1. (Left) Histogram of tree complexity and (Right) error in classification for the original C4.5 trees

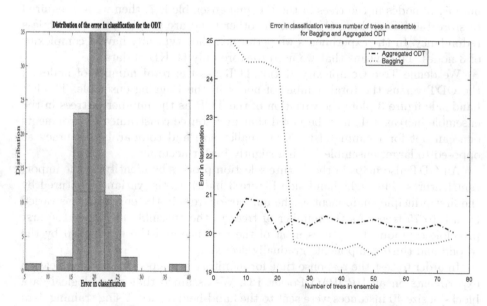

Fig. 2. (Left) Classification error for the ODT ensemble. (Right) Comparison of error for trees in the ensemble for aggregated ODT versus Bagging.

ber of trees in the ensemble was smaller. With increase in number of trees in the ensemble Bagging provided a slightly better accuracy.

In the current implementation storing a node data structure in a tree requires approximately 1 KB of memory. Consider an ensemble of 20 trees. If the average

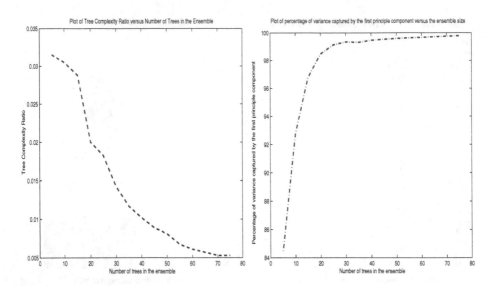

Fig. 3. (Left) Plot of Tree-Complexity-Ratio and (Right) Variance captured by the first principle component versus number of trees in the ensemble

number of nodes in the trees in the Bagging ensemble is 7, then we are required to store 140 KB of data. ODTs on the other hand are smaller in size, with less redundancy. In the experiments we performed they typically have a complexity of 3 nodes. This means that we need to store only 60 KB of data.

We define Tree Complexity Ratio (TCR) as the total number of nodes in the ODT versus the total number of nodes in the Bagging ensemble. The left hand side figure 3 plots the variation of the TCR as the number of trees in the ensemble increases. It may be noted that in resource constrained environments one can opt for meaningful trees of smaller size and comparable accuracy as opposed to larger ensembles with a slightly better accuracy.

An ODT also helps in the feature selection process by identifying the important features. The right hand side Figure 3 indicates the variance captured by the first principle component as the number of trees in the ensemble was varied from 5 to 75 trees. As the number of trees in the ensemble increases, the first principle component captures most of the variance and those occupied by the second and third components gradually decreases.

In order to test the response time for monitoring, we performed classification experiments on an HP iPAQ Pocket PC. We assumed that physiological data blocks of size 40 instances were sent to the hand-held device. Using training data obtained previously, we pre-computed C4.5 decision trees. The Fourier spectra of the trees were evaluated(preserving approximately 99(%) of the total energy) and the coefficient matrix was projected onto the most significant principal components. We also measured the given time required to produce an accuracy rate from all the instances available by the specified classification scheme. The equivalent ODT consistently outperformed a bagging ensemble. We could not report detailed experimental results here because of limited space.

7   Conclusions

Orthogonal decision trees offer an effective way to construct redundancy-free ensembles that are easier to understand and apply. They are particularly useful in monitoring data streams using resource constrained platforms where storage and CPU computing power are limited but fast response is important. ODTs are constructed from the Fourier spectra of the decision trees in the ensemble. Redundancy is removed from the ensemble by performing a PCA of these Fourier spectra. This paper described an application of ODT ensembles for monitoring physiological data streams in time-critical resource-constrained environments. We plan to explore additional applications of ODTs in other domains. We are also working on developing techniques that makes use of the spectral representation of an ensembles for identifying its various functional and structural properties (e.g. stability).

Acknowledgments

This work is supported by NSF CAREER award IIS-0093353 and IIS-0203958.

References

1. Breiman, L., Freidman, J.H., Olshen, R.A., Stone, C.J.: Classification and Regression Trees. Wadsworth, Belmont, CA (1984)
2. Quinlan, J.R.: C4.5: Programs for Machine Learning. Morgan Kauffman (1993)
3. Freund, Y.: Boosting a weak learning algorithm by majority. Information and Computation **121** (1995) 256–285
4. Drucker, H., Cortes, C.: Boosting decision trees. Advances in Neural Information Processing Systems **8** (1996) 479–485
5. Breiman, L.: Bagging predictors. Machine Learning **24** (1996) 123–140
6. Wolpert, D.: Stacked generalization. Neural Networks **5** (1992) 241–259
7. Breiman, L.: Random forests. Machine Learning **45** (2001) 5–32
8. Street, W.N., Kim, Y.: A streaming ensemble algorithm (sea) for large-scale classificaiton. In: Seventh ACM SIGKDD International Conference on Knowledge Discovery and Data Mining, San Francisco, CA (2001)
9. Kargupta, H., Park, B.: A Fourier spectrum-based approach to represent decision trees for mining data streams in mobile environments. IEEE Transactions on Knowledge and Data Engineering **16** (2002) 216–229
10. Kargupta, H., Dutta, H.: Orthogonal Decision Trees. In: Fourth IEEE International Conference on Data Mining (ICDM). (2004) 427–430
11. Kostov, Y., Rao, G.: Low-cost optical instrumentation for biomedical measurements. Review of Scientific Instruments **71** (2000) 4361–4373
12. Park, B.H., Kargupta, H.: Constructing simpler decision trees from ensemble models using Fourier analysis. In: Proceedings of the 7th Workshop on Research Issues in Data Mining and Knowledge Discovery, ACM SIGMOD. (2002) 18–23
13. Linial, N., Mansour, Y., Nisan, N.: Constant depth circuits, fourier transform, and learnability. Journal of the ACM **40** (1993) 607–620
14. Merz, C.J., Pazzani, M.J.: A principal components approach to combining regression estimates. Machine Learning **36** (1999) 9–32

Throughput Computing
with Chip MultiThreading and Clusters

Mukund Buddhikot and Sanjay Goil

Sun Microsystems,
Santa Clara, CA-95054, USA
{mukund.buddhikot, sanjay.goil}@sun.com

Abstract. Chip MultiThreading (CMT) based systems are being introduced in the market by several computer platform vendors. At the same time, cluster computing platforms are becoming prevalent in market segments which tend to be highly price/performance driven. This paper analyzes the architectural space of these two prominent computing paradigms in technical computing markets. This analysis is carried out in terms of the application turnaround time, throughput, and scalability across multiple threads of execution. Additionally, we introduce various subscription models to optimize application throughput and turnaround time.

1 Introduction

Over the past few years, major computer companies have introduced processors incorporating Chip MultiThreading (CMT)[1]. As compute densities become more important in data centers, new paradigms in system design are to integrate systems functionalities as close to the silicon as possible. Power, floor space, and cost are of primary importance in such design considerations. On the performance side, the thread level parallelism (TLP) inherent in the enterprise and high-performance computing applications provides a better opportunity to optimize across several threads, either of the same multi-threaded workload, or of different workloads. In this paper we explore the ideas of throughput computing from the aspect of optimizing workloads for a collection of users, reflecting real use of commercial systems, instead of the traditional quest for performance for the single threaded job of one single user. Thus, we deal with the interplay between the application throughput and the application turnaround time for the CMT model for various subscription modes. Compute nodes based on the AMD™ Opteron processor and Intel™ Xeon processor have proven very successful for grid computing as cluster nodes connected with an appropriate interconnect. We contrast throughput computing and cluster computing to set the stage for results with different subscription models and report major throughput gains on several applications. Future CMT implementations will further exploit concurrent placement of user threads on shared CPU resources, significantly reducing the overheads [7] seen today. We present results with CMT with several important applications in the energy and life sciences industry and over subscription work model results for SPEC OMPM2001 [8].

D.A. Bader et al. (Eds.): HiPC 2005, LNCS 3769, pp. 128–136, 2005.

Section 2 deals with the Chip MultiThreading technology. Section 3 explores the cluster platforms. Section 4 outlines the subscription work models. Conclusions and future work are covered in Section 5.

2 Chip MultiThreading

Most server applications these days are multithreaded or multiprocess applications with a large amount of Thread Level Parallelism (TLP). These applications handle thousands-to-millions of transactions daily, thus necessitating high throughput as the primary metric for the computing systems for such applications. These server applications tend to have limited Instruction Level Parallelism (ILP) and a substantial operating system activity. They also have limited data and code locality leading to high data and instruction cache misses. Similar behavior is also seen on several parallel technical computing workloads.

Traditional methods for enhancing performance have been to increase single thread performance by emphasizing higher clock rate, out-of-order execution, and increasing numbers of functional units resulting in a complex computing core. With increasing gap between processor and memory performance, it becomes important to design new methods to mask latencies to memory and also the latencies associated with the long running I/O operations. These methods need to exploit the design advantages of the 65 nm process technologies and the needs of the throughput oriented applications.

Simultaneous MultiThreading (SMT)[5][6] approach involves multiple threads (contexts) executing concurrently and utilizing processor resources relying on the instruction dependencies within the two or more contexts. A major advantage of SMT is that it requires only 5-10% extra transistors for an extra context resulting in almost the same manufacturing efficiency as the non-SMT. One of the drawback of the SMT approach is the verification complexity. Performance of the SMT core is limited by the context switching overhead and the available memory bandwidth.

Chip MultiProcessing (CMP) provides multiple cores in a single chip. The cores are identical. Sun's US-IV is a CMP processor. These first-generation CMP processors are derived from earlier uniprocessor designs and the two cores generally may not share any resources, except for the off-chip data paths.

CMT (Chip MultiThreading) combines SMT and CMP by providing multiple cores per chip and multiple threads per core. CMT threads can be categorized into two types, namely, light weight threads and heavy weight threads. Light weight threads have modest single thread performance, since they tend to have less complexity (one/two-issue, in-order pipeline, little speculation). Light weight threads enable large number of strands on the chip and the result in less power-consumption. Heavy weight strands offer good to great single thread performance and may not result in power-savings. Chip MultiThreading thus focuses on systems integration and addresses the issues of power, floorspace and cost. On the performance side, throughput computing aspect focuses on performance metrics for multiple users and multiple jobs, by exploiting thread level parallelism (TLP) more aggressively than it exploits instruction level parallelism (ILP).

Sun has developed its second generation CMT processor, US-IV+ [3], in which the on-chip L2 and off-chip L3 caches are shared between the two cores. Sun also recently announced a 32-way CMT SPARC processor, code-named Niagara [2]. Niagara has eight cores and each core is a four-way SMT with its own private L1 caches. All eight cores share a 3MB, 12-way L2-cache. Niagara represents a third-generation CMT processor, where the entire design, including the cores, is optimized for a CMT design point. Niagara is an example of light weight thread CMT design.

Early results with the Niagara-based platforms illustrate the benefits of the CMT implementation. These results and benefits with the Niagara-based platforms for the E-commerce applications and bioinformatics codes have been very encouraging. Some results on this platform will be shared at the conference presentation.

3 Performance Space

Compute clusters have become appealing with high speed interconnect offerings such as the Myrinet [12] and InfiniBand interconnect from SilverStorm [13]. These high bandwidth and low latency interconnects result in cluster scalability comparable to that achievable with small and medium scale symmetric multiprocessors; due to which the compute clusters have become platforms of choice in the HPTC (High Performance Technical Computing) markets such as oil & gas, bioinformatics and computational fluid dynamics in the manufacturing industry. Efforts are going on to employ clusters for on line transaction processing as well.

3.1 Performance Metrics

Scalability and computing efficiency are primary performance metrics for the cluster platforms. Scalability of a cluster can be defined as the speed up (obtained on a given problem size) as a function of the number of cluster nodes. Better the speed up with the addition of the number of cluster nodes, better is the scalability. Major factors deciding the cluster scalability are the bandwidth and latency characteristics of the cluster interconnect and the amount of inter-node communication during the application execution. Computing efficiency of a cluster is the ratio of compute power delivered by the cluster and the theoretical peak compute power of the cluster.

3.2 Benchmarks

Compute power in terms of number of floating point operations delivered by the cluster is traditionally measured by the HPL (High Performance Linpack)[10], which is a parallel implementation of the linpack benchmark. HPL is employed for the cluster rankings by the Top500 organization. HPL performance is not representative of typical application performance due to its heavy reuse of memory and extensive optimization of the benchmark software. HPC Challenge (HPCC) benchmarks [9] were created to provide a complete performance metric for high performance cluster computing platform. HPCC benchmark consists of components to measure the floating point capability of a cluster for Matrix Multiplication, FFT (Fast Fourier Transform) computation, the HPL and modules to measure the inter-node communication capabilities. HPCC results are available at [9]. For more relevance to end-users, we analyze the cluster performance space with prominent applications in the energy and life sciences sector as examples below.

3.3 ISV Applications

A market leading reservoir simulation application and a major seismic processing application in the oil and gas industry are chosen for these investigations. The reservoir simulation application employed computes flows in a reservoir by solving mass conservation equations for multi-component multiphase fluid flows in porous media. (For details of the applications, please contact the authors).

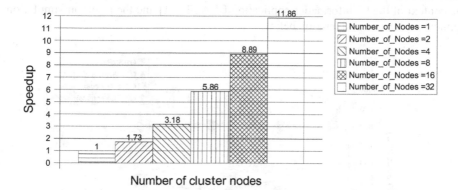

Fig. 1. Scalability of Reservoir Simulation Application on a Cluster

Cluster configuration C under test is as follows.

AMD™ Opteron 2.0 GHz, SuSE Linux SLES8 64bit, 2GB/node, mpich v1.2.5.10, Myrinet GM v2.0.9.

Figure 1 shows reservoir simulation application scalability. Scalability of the application is limited by the inter node communication after each iteration during the solution process, which results in dismal 37% efficiency on 32 nodes.

3.4 Seismic Simulation

Seismic processing application chosen delineates hydrocarbon reserves in the area of interest. Figure 2 shows scalability of seismic processing application on Cluster C.

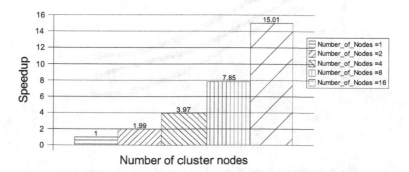

Fig. 2. Scalability of Seismic Processing Application on a Cluster

Load put by the seismic application on the cluster interconnect is little; which is reflected in near perfect scalability of the application as the number of cluster nodes is increased. The application under consideration achieves 94% efficiency on 16 nodes.

3.5 Scalability

Scalability of the applications under cluster, Symmetric MultiProcessing (SMP) and CMT models as the number of processors in the models is increased is presented in the context of the bioinformatics application BLAST [11] and the reservoir simulation application under consideration below.

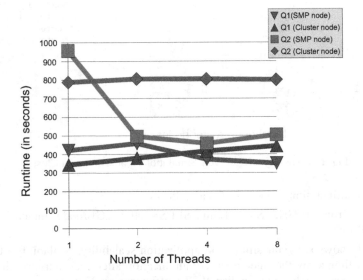

Fig. 3. Scaling of BLAST in CMT and Cluster Environment

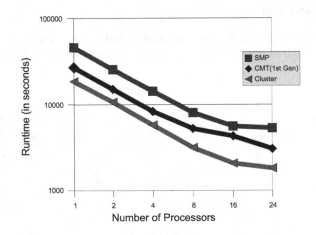

Fig. 4. Reservoir Simulation in CMT and Cluster Environment

Figure 3 below shows the performance of BLAST query1 and query2 in the cluster and SMP environments. Cluster environment under consideration is the cluster C described in Section 3.3. SMP environment discussed is the Sun Fire 6800 (Solaris 9, 32 GB, 24 US-III processors). For query1, it is noticed that the query execution performance, as measured by the query runtime, is better in the cluster environment for one node in the cluster after which the SMP environment shows better performance; this being due to the fact that the SMP memory latency being better than the cluster interconnect latency as the number of participating nodes in the computation increases. Query 2 shows better performance in the SMP environment than the cluster environment as the number of computing threads reaches 4. This example illustrates the power parallelization on the SMP model for these queries. These differences could be due to the threading model in use apart from the latency and interconnect differences between the clusters and the SMP.

Figure 4 shows the performance of the reservoir simulation application in cluster, SMP and (first generation) CMT environments. Cluster environment is the cluster C and the SMP environment is the configuration described above. CMT environment is the 1st generation CMT-based Sun Fire 6900 (Solaris 9, 32 GB, 24 US-IV chips, 48 cores or threads). Cluster performance is better due to the better clock frequency of the constituent processor nodes. It is also noticed that the first generation CMT delivers good scalability and has better performance than the traditional SMP environment due to double the number of threads of computation in a chip.

4 Subscription Work Model

With the availability of a large number of threads for CMT and nodes in the cluster environment, it is interesting to analyze the impact of subscription models on the application turnaround time and throughput. Subscription models to be considered include fully subscribed model, under subscribed model and overly subscribed model. A fully subscribed model indicates that the number of application threads spawned equals the number of available threads. An overly subscribed model indicates that the number of the application threads spawned is greater than the number of available threads, where as the under subscribed model consists of the number of application threads less than the number of available threads/nodes.

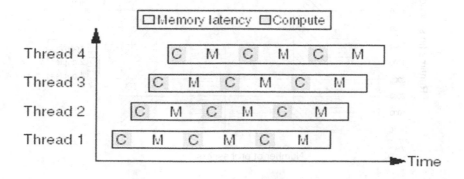

Fig. 5. CMT Threads in Time and Space

Figure 5 shows the spatio-temporal arrangement for the CMT threads. It shows the compute cycles of the threads [2-4] during the memory access by the thread 1, thereby showing 100% processor utilization. In actual practice, the processor utilization will be less than the 100% due to the memory contention and delays such as the lock acquisition latencies. An analytical model for the processor utilization in a time quantum $T = T_c + T_d$ can be developed as follows.

$$P = pT_c/(T_c + T_d) \qquad (1)$$

where p is the number of active threads/cores

$\quad$ T_c is the computation cycle time of a thread.

$\quad$ T_d is the average delay associated with the thread.

From equation (1), it can be seen that if the application involves delays due to long disk accesses, contention at the system bus leading to less than optimal utilization of the processor pipeline, such an environment is suited for oversubscribing.

Turnaround (response) time Vs Throughput is an important metric for throughput computing. Our major focus needs to be on finding out the achievable throughput while maintaining the turnaround (response) time within acceptable levels. Trade off between turnaround time and throughput is driven by the choice of the computing (CMT or cluster) model, platform characteristics (number of nodes, CPU clock frequency, interconnect latency, interconnect bandwidth and memory subsystem performance), subscription model, operating system capabilities and application characteristics. The trade off between turnaround time and throughput and interplay of resource utilization and the subscription models is explained below.

Figure 6 shows the impact of running multiple reservoir simulation application instances on cluster C. Running two instances of the application results in double the throughput but only 20% increase in the turnaround time, where as running four instances of the application results in up to 40% increase in turnaround time for a throughput increase by a factor of 4.

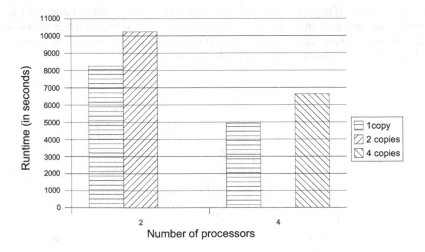

Fig. 6. Subscription Models in a Cluster for Reservoir Simulation

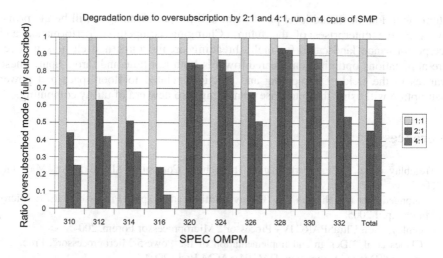

Fig. 7. Oversubscribed mode studies on CMT with SPEC OMPM2001

Another result we share in this version is our experiments with SPEC OpenMP Medium benchmarks (SPEC OMPM2001) [8] in over subscribed mode. In the over-subscribed mode several benchmarks are performing close to a single threaded execution, meaning that we can execute 2 or 4 threads with a minor overhead of <10%. Examples of these types of codes are the 328 and 330 benchmarks in the Figure below. On the other hand there are several benchmarks that are resource limited, like 310, 312, 314, 316 that degrade severely in oversubscribed modes. Memory bandwidth limits and scheduling overheads are mainly responsible for such degradation. It is noticed that for all the SPEC OMPM2001 codes, except the 328 and 330 benchmarks, the fully subscribed mode is the best at present.

Figure 6 illustrates an example of an application (reservoir simulation) in the cluster environment where over subscription does not affect the turnaround time substantially, where as Figure 7 shows an example of a benchmark (SPEC OMPM2001) where over subscription affects the turnaround time to a major extent. Behaviors in Figure 6 and Figure 7 can be explained by the Equation (1). Figure 6 indicates an application environment where the T_d is high. T_d is high due to the synchronization delays at each iteration. The availability of system resources (due to wait from threads in the first application instance) leads to better system utilization with scheduling and execution of the threads from the second instance of the application, leading to better throughput. Execution environment of Figure 7 is an example consisting of threads competing for system bus and memory leading to contention for the resources in the oversubscribed mode leading to higher throughput at the cost of longer response times.

5 Conclusions and Future Work

Chip MultiThreading based systems with multiple threads of computation on a single die are now becoming available from Sun Microsystems and other vendors. These systems with lots of thread level parallelism and power advantages will be the prime movers of the enterprises of the future. Changing competitive performance and price/performance landscape, for multi-threading architectures and clusters will require applications optimizations geared towards both run time and throughput. These advances in the field of computing and practices related to floor space and power consumption will usher the enterprise computing into new era of utility computing.

References

1. Tremblay et. al, ``High Performance Throughput Computing``,IEEE Micro, May-June 2005.
2. P. Kongetira et al,``Niagara : A 32-way Multithreaded Sparc Processor", IEEE Micro, March-April 2005
3. Greenley,``Sun UltraSPARC IV+ Processor", Microprocessor Forum, 2004.
4. J. Clabes et al, ``Design and Implementation of the Power5 Microprocessor", Proc. 41st Ann. Conf. Design Automation (DAC 04), ACM Press 2004.
5. ``Hyper Threading Technology Architecture and Microarchitecture", Intel Technology Journal, Vol 6.
6. Tulllsen,``Simultaneous Multithreading:Maximizing On-Chip Parallelism", Proceedings of the 22nd Annual International Symposium on Computer Architecture, June 1995.
7. M.Buddhikot, "Distributed Lock Protocol for a Clustered Multiprocessor", European Simulation Symposium, Delft, Netherlands, 1993.
8. S.Goil et. al., The SPEC OMPM2001 Benchmarks on the Sun UltraSPARC Multiprocessors, WOMPAT 2002.
9. HPC Challenge Benchmarks, University of Tennessee, Knoxville.
10. A Portable Implementation of the High-Performance Linpack Benchmark for Distributed-Memory Computers, A. Petitet, R. C. Whaley, J. Dongarra, A. Cleary, Innovative Computing Laboratories, University of Tennessee, Knoxville.
11. BLAST http://www.ncbi.nlm.nih.gov/blast
12. Myrinet Interconnects http://www.myricom.com/
13. SilverStorm Interconnects http://www.silverstorm.com

Supporting MPI-2 One Sided Communication on Multi-rail InfiniBand Clusters: Design Challenges and Performance Benefits*

Abhinav Vishnu, Gopal Santhanaraman, Wei Huang,
Hyun-Wook Jin, and Dhabaleswar K. Panda

Department of Computer Science and Engineering,
The Ohio State University, Columbus, OH 43210
{vishnu, santhana, huanwei, jinhy, panda}@cse.ohio-state.edu

Abstract. In cluster computing, InfiniBand has emerged as a popular high performance interconnect with MPI as the *de facto* programming model. However, even with InfiniBand, bandwidth can become a bottleneck for clusters executing communication intensive applications. Multi-rail cluster configurations with MPI-1 are being proposed to alleviate this problem. Recently, MPI-2 with support for one-sided communication is gaining significance. In this paper, we take the challenge of designing high performance MPI-2 *one-sided communication* on multi-rail InfiniBand clusters. We propose a unified MPI-2 design for different configurations of multi-rail networks (*multiple ports, multiple HCAs* and *combinations*). We present various issues associated with one-sided communication such as *multiple synchronization messages, scheduling of RDMA (Read, Write) operations, ordering relaxation* and discuss their implications on our design. Our performance results show that multi-rail networks can significantly improve MPI-2 one-sided communication performance. Using PCI-Express with two-ports, we can achieve a peak *MPI_Put* bidirectional bandwidth of 2620 Million Bytes/s, compared to 1910 MB/s for single-rail implementation. For PCI-X with two HCAs, we can almost double the throughput and reduce the latency to half for large messages.

1 Introduction

High computational power of commodity PCs combined with the emergence of low latency and high bandwidth interconnects has led to the trend of *cluster computing*. In this area, Message Passing Interface (MPI) [6] has become the *de facto* standard for writing parallel applications. MPI-2 has been introduced as a successor of MPI-1 with *one-sided communication* as one of its main additional

---

* This research is supported in part by Department of Energy's grant #DE-FC02-01ER25506; National Science Foundation's grants #CNS-0403342 and #CCR-0311542; grants from Intel and Mellanox; and equipment donations from Intel, Mellanox, AMD, Apple and Sun Microsystems.

D.A. Bader et al. (Eds.): HiPC 2005, LNCS 3769, pp. 137–147, 2005.

features. Recently, InfiniBand Architecture [8] has been proposed as the next generation interconnect for inter-process communication and I/O. Due to its open standard and high performance, InfiniBand is becoming increasingly popular for cluster computing. However, even with InfiniBand, network bandwidth can become the performance bottleneck for communication intensive applications. This is especially the case for clusters built with SMP (2-16 way symmetric multiprocessor systems) machines, in which multiple processes may run on a single node and must share the node bandwidth. Multi-rail [11](*multiple ports, multiple HCAs and combinations*) cluster configurations with MPI-1 are being proposed to alleviate this problem. Compared to MPI-1, MPI-2 is the next generation MPI standard with one-sided operations (such as MPI_Put and MPI_Get). This leads to the following challenges:

1. *How to design support for one-sided operations on multi-rail InfiniBand clusters?*
2. *How much benefits can be achieved compared to the single-rail implementation?*

In this paper, we take on these challenges. We propose a unified MPI-2 design with different configurations of multi-rail networks (*multiple ports, multiple HCAs* and *combinations*) for one-sided communication. We present various issues associated with one-sided communication (*multiple synchronization messages, scheduling of RDMA (Read, Write) operations, scheduling policies, ordering relaxation*) and discuss their implications on our design.

We implement our design on MVAPICH2[1] and evaluate it with micro-benchmarks on different multi-rail configurations. Our performance results show that multi-rail networks can significantly improve MPI-2 one-sided communication performance. Using two-ports on EM64T cluster with PCI-Express, we can achieve an *MPI_Put* bandwidth of 1500 Million Bytes/s (MB/s), and a bidirectional bandwidth of 2620 MB/s. Using two-HCAs on IA32 cluster with independent PCI-X buses, we can achieve a *MPI_Put* bandwidth of 1750 MB/s, and a bidirectional bandwidth of 1810 MB/s.

The rest of the paper is organized as follows: In section 2, we provide background information for InfiniBand, MVAPICH2 and multi-rail configurations. In section 3, we describe the multi-rail MPI-2 design for one-sided communication and discuss the design issues. In section 4, we present performance results of our multi-rail MPI-2 implementation. In section 5, we present the related work. In section 6, we conclude and discuss our future directions.

2   Background

In this section, we provide background information for our work. First, we provide a brief introduction of InfiniBand. Then, we discuss some of the internals

---

[1] MVAPICH/MVAPICH2 [13] are high performance MPI-1 and MPI-2 implementations from The Ohio State University, currently being used by more than 250 organizations across 28 countries.

of MPI-2 one-sided communication and their implementations over InfiniBand. We also present a brief overview of multi-rail InfiniBand clusters.

2.1 Overview of InfiniBand

The InfiniBand Architecture (IBA) [8] defines a switched network fabric for interconnecting processing nodes and I/O nodes. It provides a communication and management infrastructure for inter-processor communication and I/O. In an InfiniBand network, processing nodes and I/O nodes are connected to the fabric by *Host Channel Adapters* (HCA). HCAs sit on processing nodes. InfiniBand Architecture supports both channel and memory semantics for Reliable Connection service. In channel semantics, send/receive operations are used for communication. In memory semantics, InfiniBand supports *Remote Direct Memory Access (RDMA)* operations, including RDMA write and RDMA read. RDMA operations are one-sided and do not incur software overhead at the remote side. In these operations, the sender can directly access remote memory by posting RDMA descriptors.

2.2 MPI-2 One-Sided Communication

In MPI-2 one-sided communication, the sender can access the remote address space directly. Such one-sided communication is also referred to as *Remote Memory Access* or *RMA* communication. In this model, the *origin process* (the process that issues the RMA operation) provides necessary parameters needed for communication. The area of memory on the *target process* accessible by the origin process is called a *Window*. MPI-2 specification defines various communication operations:

1. *MPI_Put* operation transfers the data to a window in the target process
2. *MPI_Get* operation transfers the data from a window in the target process
3. *MPI_Accumulate* operation combines the data movement to target with a reduce operation

As per the semantics of one-sided communication, the return of the one-sided operation call does not guarantee the completion of the operation. In order to guarantee the completion of one-sided operation, explicit synchronization operations must be used. We mainly focus on active synchronization in this paper.

2.3 Multi-rail InfiniBand Configurations

Multi-rail networks can be built by using *multiple HCAs* on a single node, or by using *multiple ports* in a single HCA. In an MPI application, any pair of processes can communicate with each other. This is implemented in MPICH2 designs by an abstraction called *virtual channel*. A virtual channel can be regarded as an abstract communication channel between two processes. In [11], we have proposed enhanced virtual abstraction to provide a unified solution to support multiple HCAs, multiple ports, and multiple paths in a single port. In our proposed design, a virtual channel can consist of multiple *virtual subchannels* (referred as subchannels from here on-wards). Each subchannel refers to a path of communication between end nodes.

2.4 MVAPICH2

MVAPICH2[13] is our high performance implementation of MPI-2 over Infini-Band. The implementation is based on MPICH2. As a successor of MPICH[6], MPICH2[1] supports MPI-1 as well as MPI-2 extensions including one-sided communication. One sided communication can be implemented using a variety of approaches. One approach is to use the point to point implementation provided by MPICH2 for one-sided communication. This approach involves the remote host for communication and synchronization operations. In the second approach, the one-sided operations are implemented at the CH3 level by extending the CH3 interface [10, 9]. This approach shows benefits with respect to latency and bandwidth for regular communication patterns. It also provides better overlap between computation and communication along with scalability. We refer to the first approach as *Point to Point Based* and second approach as *Direct One Sided*. Fig. 1 shows the path taken by these approaches. In this paper, we design the *Direct One Sided* over multi-rail InfiniBand clusters implementation along with *active* mode of synchronization.

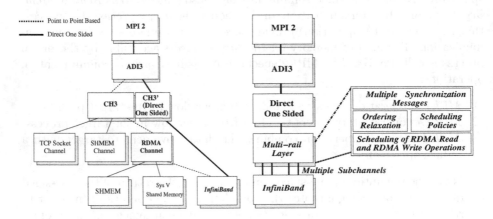

Fig. 1. Implementations of one-sided communication in MVAPICH2

Fig. 2. Basic Architecture

3   Multi-rail Layer Design for MPI-2 One Sided Communication

In this section, we present the design issues involved with MPI-2 one-sided communication on multi-rail InfiniBand clusters.

3.1   Basic Architecture

The basic architecture of our design to support multi-rail networks for MPI-2 one-sided communication is shown in Figure 2. In the figure, we can see that besides the MPI-2, Direct One Sided layer and InfiniBand layer, our design consists of an intermediate layer, *Multi-rail Layer*.

This layer takes the responsibility of scheduling messages on the available subchannels. Besides this, it takes care of the correctness issues like *Multiple Synchronization Messages* and efficiency issues like *Scheduling Policies, Ordering Relaxation* and *Scheduling of RDMA Read and RDMA Write Operations*.

In this section, we discuss the design challenges involved for multi-rail MPI-2 design associated at the Multi-rail Layer.

Multiple Synchronization Messages: In order to initiate the one-sided communication, the origin process calls *win_start* to open a window. The target process *posts* the buffers for the window. Once the one-sided communication is done, a synchronization message needs to be sent to the target process. The receipt of synchronization message guarantees the data transfer of previously issued RMA operations. However, when multiple subchannels are used, data transfer on one subchannel might not have finished even though other subchannels would have received the synchronization message. Hence, we need to issue synchronization messages on each subchannel. It is to be noted, that when the load on subchannels is balanced, the transfer of synchronization messages along multiple subchannels takes place in parallel, incurring very small overhead.

Scheduling of RDMA Read and RDMA Write Operations: In MPI-1, usually the two sided communication uses either RDMA Write or RDMA Read for data transfer in InfiniBand. For many MPI-2 applications, in one-sided communication, the *MPI_Put* and *MPI_Get* operations are implemented using RDMA Write and RDMA Read, respectively. Since RDMA Read and RDMA Write utilize bandwidth in different directions, it is important to schedule them independently with respect to each other's load on different subchannels.

In order to achieve this, we propose a load based fragmentation policy discussed in the next section, which maintains independent queues of *MPI_Put* and *MPI_Get* operations issued in an *epoch*. Trivally, this policy would fragment the messages equally on all subchannels in the presence of only one kind of one-sided operation. In presence of a combination of one-sided operations, each having the same size, this policy would fall back to equal fragmentation.

Scheduling Policies Classification Based on Message Size: In this paper, we classify the policies used for scheduling based at different layers. As proposed in [7], we use *reordering* and *no reordering* policies at the CH3' (Direct One Sided) Layer. At the multi-rail layer, we do a classification of the policies based on the message size. We employ the following policies:

– *Round Robin*
– *Load Balanced Fragmentation*

For small messages, we employ round robin policy. In this policy, the complete message is sent using one of the available subchannels in a round robin fashion. Fragmentation incurs overhead of posting descriptors on multiple subchannels, which is significant for small messages. Hence, we employ a *switchover threshold*, messages of size less than this threshold are scheduled in a round robin fashion. For large messages, we primarily use Load Balanced Fragmentation policy. In this policy, we divide the message in chunks and schedule them, so that the load on all subchannels is balanced. This policy leads to optimal utilization of all subchannels for medium to large messages.

Ordering Relaxation: Two-sided communication requires messages to be processed in order at the receiver side. One sided communication imposes no ordering requirements for messages within an *epoch*, by the definition from the semantics. As a result, the one-sided approach does not need to maintain ordering at the receiver side. We simplify our design by incorporating this fact, reducing the overhead of bookkeeping at the receiver side.

4 Performance Evaluation

In this section, we evaluate the performance of our multi-rail MPI-2 design over InfiniBand. We show the performance benefit which can be achieved with multi-rail design compared to the single-rail implementation.

4.1 Experimental Testbed

We evaluated our implementation with multiple HCAs on IA32 systems comprising of independent PCI-X buses, and on EM64T systems comprising of PCI-Express bus and multiple ports per adapter. Our experimental testbed comprises of two clusters.

IA32 Cluster with Multiple HCAs: This cluster consists of two SuperMicro SUPER X5DL8-GG nodes with ServerWorks GC LE chipsets. Each node has dual Intel Xeon 3.0 GHz processors, 512 KB L2 cache, and PCI-X 64-bit 133 MHz bus. We have used InfiniHost MT23108 Dual-Port 4x HCAs from Mellanox. The ServerWorks GC LE chipsets have two separate I/O bridges and three PCI-X 64-bit 133 MHz bus slots. To reduce the impact of I/O bus contention, the two HCAs are connected to separate PCI-X buses connected to different I/O bridges.

EM64T Cluster with Multiple Ports: This cluster consists of two EM64T nodes having 8X PCI Express slots. Each node has two Intel Xeon CPUs running at 3.4 GHz processors, 512 KB L2 cache and 1 GB of main memory. This cluster uses III Generation MT25208 4X Dual Port HCAs from Mellanox. A combined unidirectional bandwidth of 8X can be used, when both ports are used for communication.

4.2   One Sided Communication Micro-benchmarks

In this section, we introduce the micro-benchmarks used to evaluate the MPI-2 one-sided operation performance. We use such as uni- and bi-directional bandwidth, as well as micro-benchmarks with other communication patterns.

Two processes are involved in uni-directional bandwidth test. The origin process starts a window access epoch, issues a *window* of RMA operations (MPI_Put for MPI_Put bandwidth test, MPI_Get for MPI_Get bandwidth test), and ends the access epoch. The target process just starts and ends a window exposure epoch. This step is repeated for multiple iterations. For bidirectional bandwidth test, both processes starts and end a window exposure epoch.

In the Interleaving test, the origin process issues a *window* of *MPI_Put* operations followed by a *window* of *MPI_Get* operations. Due to the impact of reordering at CH3' layer, interleaving of these operations provides almost bidirectional bandwidth throughput in comparison to unidirectional throughput.

4.3   Performance Benefits of Multi-rail Design

To evaluate the performance benefits of our multi-rail MPI-2 design, we compare it with our original MVAPICH2 design, which can only use only one-port of a nic. In the multi-rail design, we use *load balanced fragmentation* for large messages and *round robin* scheme for small messages. We present performance comparisons using latency for *MPI_Get* operation and bandwidth and bidirectional bandwidth for *MPI_Put* operations.

Microbenchmark Evaluation for Basic One Sided Operations: In Figures 3 and 5 we present the results for *MPI_Put* bandwidth and bidirectional bandwidth respectively for the IA32 cluster with multiple HCAs. We show the results for EM64T with two-ports on PCI-Express in Figures 4 and 6.

In Figure 3, we observe that for small messages (less than or equal to 1KBytes), both multi-rail design and the original implementation perform comparably. For large messages, multi-rail design outperforms the original implementation considerably. With multi-rail design, we can achieve a maximum peak unidirectional *MPI_Put* bandwidth of 1750 MB/s in comparison to 880 MB/s for our original implementation. We also notice, that due to the absence of rendezvous protocol, medium size messages (2KB - 16KB) can take advantage of load balanced fragmentation policy for multi-rail design.

We observe a similar trend for dual-port on EM64T in Figure 4. For messages of size greater than 8KBytes, we use fragmentation policy. We can achieve a peak bandwidth of 1500 MB/s using multi-rail design, in comparison to 971 MB/s for the original implementation capable of using only one-port of a nic.

In figures 5 and 6, we compare the performance of *MPI_Put* bidirectional bandwidth for IA32 cluster and EM64T cluster, respectively. For IA32 cluster, due to the bottleneck of PCI-X, we can achieve only 941 MB/s for original implementation. However, using multi-rail design we can achieve a peak bidirectional

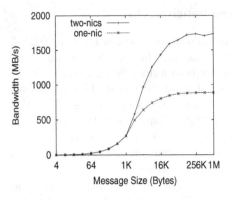

Fig. 3. MPI_Put Bandwidth on the IA32 Cluster

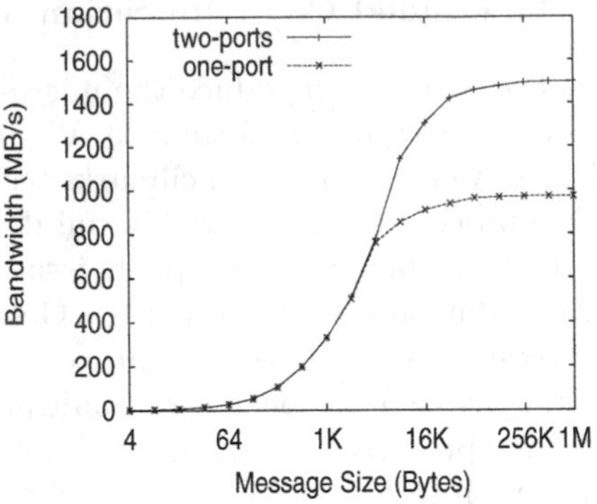

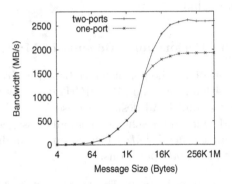

Fig. 4. MPI_Put Bandwidth on the EM64T Cluster

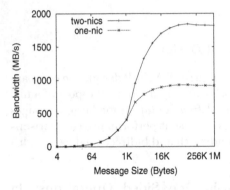

Fig. 5. MPI_Put Bidirectional Bandwidth on the IA32 Cluster

Fig. 6. MPI_Put Bidirectional Bandwidth on the EM64T Cluster

bandwidth of 1810 MB/s. For EM64T cluster, we can achieve a peak bidirectional bandwidth of 2620 MB/s with two-ports in comparison to 1910 MB/s using the original implementation.

In figures 7 and 8, we present the results for *MPI_Get* latency for IA32 and EM64T cluster, respectively. We observe that we perform almost similar with the original implementation for small messages. For large messages, we can improve the latency by 45% for IA32 cluster and 33% for EM64T cluster by using multi-rail design.

Impact of Reordering on One Sided Communication: Figure 9 shows the performance achieved by a combination of policies at the CH3' layer and Multi-rail layer. At the multi-rail layer we use load balanced fragmentation policy. At the CH3' layer, we compare impact of reordering with no reordering, when combined with the multi-rail policy specified above.

For IA32 cluster using two-nics, we can achieve almost 1703 MB/s without reordering, which is close to the multi-rail peak unidirectional bandwidth.

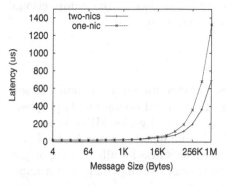

Fig. 7. MPI_Get Latency on the IA32 Cluster

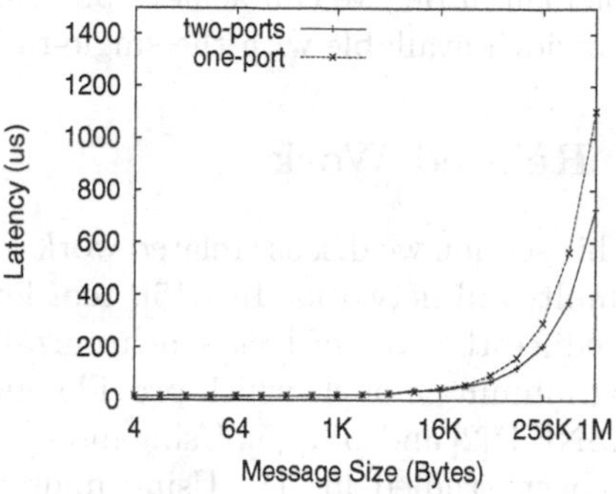

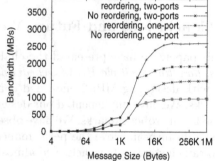

Fig. 8. MPI_Get Latency on the EM64T Cluster

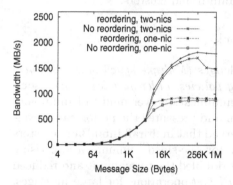

Fig. 9. Interleaved throughput on the IA32 Cluster

Fig. 10. Interleaved throughput on the EM64T Cluster

With single-rail implementation, we can achieve a peak bandwidth of 880 MB/s without reordering. We notice that with reordering for two-nics, we can almost achieve 1800 MB/s, almost the peak bidirectional bandwidth with two-nics. With single-rail implementation, due to the limitation of PCI-X, we can achieve only 907 MB/s.

In figure 10, we evaluate the performance of CH3' layer reordering, compared to the no reordering policy for the EM64T cluster. We use the load balanced fragmentation at the multi-rail layer. Using two-ports and reordering, we can achieve 2604 MB/s, which is almost the peak bidirectional bandwidth available with two-ports. It is interesting to notice, that reordering with single-rail implementation outperforms the combination of no reordering with multi-rail implementation. We attribute it to the fact that, PCI-Express can achieve 8X bidirectional bandwidth with one-port. However, due to the contention at the NIC, we cannot achieve a combined 8X unidirectional bandwidth using two-ports. Using reordering with single-rail implementation, we can achieve 1900 MB/s. However we can only achieve a peak bandwidth of 1474 MB/s using multi-rail implementation with no reordering. With no reordering for single-rail implementation, we can achieve 962 MB/s, which is close to the unidirectional bandwidth available with the single-rail implementation.

5   Related Work

In this section we discuss related work on one-sided communication model as well as multi rail networks. In [15], reordering of one sided operations is proposed to reduce the cost of lock synchronization operation. Besides MPI, some other programming models which provide one-sided communication are ARMCI [14], GASNET [2] and BSP [5]. Using interconnection networks for different topologies has been studied in [4]. Using multirail networks to build high performance clusters is proposed in [3].

However, none of the above works have focussed on design of MPI-2 one-sided communication operations with multirail InfiniBand clusters.

6   Conclusions and Future Work

In this paper, we have presented the challenges *(Multiple synchronization messages, handling multiple HCAs, scheduling policies, ordering relaxation)* associated with desigining MPI-2 one-sided communication over multirail Infiniband networks. We have implemented our design and presented the performance evaluation for microbenchmarks. We have observed that multirail InfiniBand clusters can significantly improve the performance for one-sided communication. Using a two rail cluster, we have achieved almost doubled the throughput and reduced the latency to half with *MPI_Put* and *MPI_Get* operations for large messages. We have also observed that reordering policy can significantly improve the performance for communication patterns with a mix of one-sided operations.

In future, we plan to evaluate our implementation on large scale clusters for applications with one-sided communication. We also plan to evaluate the scheduling policies in depth, to take care of different communication patterns for one-sided communication.

7   Software Distribution

As indicated earlier, the open-source MVAPICH2 [12] software is currently being used by more than 250 organizations world-wide. The latest release is 0.6.5. The proposed MPI-2 multirail one-sided communication solution will be available in the 0.7.0 release.

References

1. Argonne National Laboratory. MPICH2. http://www-unix.mcs.anl.gov/mpi/mpich2/.
2. D. Bonachea. GASNet Specification, v1.1. Technical Report UCB/CSD-02-1207, Computer Science Division, University of California at Berkeley, October 2002.

3. S. Coll, E. Frachtenberg, F. Petrini, A. Hoisie, and L. Gurvits. Using multirail networks in high-performance clusters. In *CLUSTER '01: Proceedings of the 3rd IEEE International Conference on Cluster Computing*, page 15, Washington, DC, USA, 2001. IEEE Computer Society.

4. J. Duato, S. Yalamanchili, and L. Ni. *Interconnection Networks: An Engineering Approach*. The IEEE Computer Society Press, 1997.

5. M. Goudreau, K. Lang, S. B. Rao, T. Suel, and T. Tsantilas. Portable and Effcient Parallel Computing Using the BSP Model. *IEEE Transactions on Computers*, pages 670–689, 1999.

6. W. Gropp, E. Lusk, N. Doss, and A. Skjellum. A High-Performance, Portable Implementation of the MPI Message Passing Interface Standard. *Parallel Computing*, 22(6):789–828, 1996.

7. W. Huang, G. Santhanaraman, H.-W. Jin, and D. K. Panda. Scheduling of MPI-2 One Sided Operations On InfiniBand. In *Int'l Parallel and Distributed Processing Symposium (IPDPS '05)*.

8. InfiniBand Trade Association. InfiniBand Architecture Specification, Release 1.0, October 24 2000.

9. W. Jiang, J.Liu, H. W. Jin, D. K. Panda, D. Buntinas, R.Thakur, and W.Gropp. Efficient Implementation of MPI-2 Passive One-Sided Communication on Infini-Band Clusters. EuroPVM/MPI, September 2004.

10. W. Jiang, J. Liu, H.-W. Jin, D. K. Panda, W. Gropp, and R. Thakur. High Performance MPI-2 One-Sided Communication over InfiniBand. International Symposium on Cluster Computing and the Grid (CCGrid 04), April 2004.

11. J. Liu, A. Vishnu, and D. K. Panda. Building multirail infiniband clusters: Mpi-level design and performance evaluation. In *SC '04: Proceedings of the 2004 ACM/IEEE conference on Supercomputing*, page 33, Washington, DC, USA, 2004. IEEE Computer Society.

12. J. Liu, J. Wu, S. P. Kini, D. Buntinas, W. Yu, B. Chandrasekaran, R. Noronha, P. Wyckoff, and D. K. Panda. MPI over InfiniBand: Early Experiences. Technical Report, OSU-CISRC-10/02-TR25, Computer and Information Science, the Ohio State University, January 2003.

13. Network-Based Computing Laboratory. MVAPICH: MPI for InfiniBand on VAPI Layer. http://nowlab.cse.ohio-state.edu/projects/mpi-iba/index.html, January 2003.

14. J. Nieplocha and B. Carpenter. ARMCI: A Portable Remote Memory Copy Library for Distributed Array Libraries and Compiler Run-Time Systems. *Lecture Notes in Computer Science*, 1586, 1999.

15. R. Thakur, W. Gropp, and B. Toonen. Minimizing Synchronization Overhead in the Implementation of MPI One-Sided Communication. In *EuroPVM/MPI*, September 2004.

High Performance RDMA Based All-to-All Broadcast for InfiniBand Clusters*

S. Sur, U.K.R. Bondhugula, A. Mamidala, H.-W. Jin, and D.K. Panda

Department of Computer Science and Engineering,
The Ohio State University, Columbus, Ohio 43210
{surs, bondhugu, mamidala, jinhy, panda}@cse.ohio-state.edu

Abstract. The All-to-all broadcast collective operation is essential for many parallel scientific applications. This collective operation is called MPI_Allgather in the context of MPI. Contemporary MPI software stacks implement this collective on top of MPI point-to-point calls leading to several performance overheads. In this paper, we propose a design of All-to-All broadcast using the Remote Direct Memory Access (RDMA) feature offered by InfiniBand, an emerging high performance interconnect. Our RDMA based design eliminates the overheads associated with existing designs. Our results indicate that latency of the All-to-all Broadcast operation can be reduced by 30% for 32 processes and a message size of 32 KB. In addition, our design can improve the latency by a factor of 4.75 under no buffer reuse conditions for the same process count and message size. Further, our design can improve performance of a parallel matrix multiplication algorithm by 37% on eight processes, while multiplying a 256x256 matrix.

1 Introduction

The Message Passing Interface (MPI) [1] has become the *de-facto* standard in writing parallel scientific applications which run on High Performance Clusters. MPI provides *point-to-point* and *collective* communication semantics. Many scientific applications use collective communication to synchronize or exchange data [2]. The All-to-all broadcast (MPI_Allgather) is an important collective operation used in many applications such as matrix multiplication, lower and upper triangle factorization, solving differential equations, and basic linear algebra operations.

InfiniBand [6] is emerging as a high performance interconnect for interprocess communication and I/O. It provides powerful features such as Remote DMA (RDMA) which enables a process to directly access memory on a remote node. To exploit the benefits of this feature, we design collective operations directly on top of RDMA. In this paper we describe our design of the All-to-all Broadcast operation over RDMA which allows us to eliminate messaging overheads like extra message copies, protocol handshake and extra buffer registrations. Our designs utilize the basic choice of algorithms [17] and extend that for a high performance design over InfiniBand.

* This research is supported in part by Department of Energy's grant #DE-FC02-01ER25506, National Science Foundation's grants #CNS-0403342, #CNS-0509452 and #CCR-0311542; grants from Intel and Mellanox; and equipment donations from Intel, Mellanox, AMD, Apple and Sun Microsystems.

D.A. Bader et al. (Eds.): HiPC 2005, LNCS 3769, pp. 148–157, 2005.

We have implemented and incorporated our designs into MVAPICH [14], a popular implementation of MPI over InfiniBand used by more than 250 organizations world wide. MVAPICH is an implementation of the Abstract Device Interface (ADI) for MPICH [4]. MVAPICH is based on MVICH [10]. Our performance evaluation reveals that our designs improve the latency of MPI_Allgather on 32 processes by 30% for 32 KB message size. Additionally, our RDMA design can improve the performance of MPI_Allgather by a factor of 4.75 on 32 processes for 32 KB message size, under no buffer reuse conditions. Further, our design can improve the performance of a parallel matrix multiplication algorithm by 37% on eight processes, while multiplying a 256x256 matrix.

The rest of this paper is organized as follows: in Section 2, we provide a background on the topic. Our motivation is described in Section 3. In Section 4, we describe our RDMA based design in detail. Our experimental evaluation is described in Section 5. Various related works are mentioned in Section 6. Finally, this paper concludes in Section 7.

2 Background

2.1 Overview of InfiniBand Architecture

The InfiniBand Architecture [6] defines a switched network fabric for interconnecting processing and I/O nodes. In an InfiniBand network, hosts are connected to the fabric by Host Channel Adapters (HCAs). InfiniBand utilities and features are exposed to applications running on these hosts through a *Verbs* layer. InfiniBand Architecture supports both channel semantics and memory semantics. In channel semantics, send/receive operations are used for communication. In memory semantics, InfiniBand provides Remote Direct Memory Access (RDMA) operations, including RDMA Write and RDMA Read. RDMA operations are one-sided and do not incur software overhead at the remote side. Regardless of channel or memory semantics, InfiniBand requires that all communication buffers to be "registered". This buffer registration is done in two stages. In the first stage, the buffer pages are pinned in memory (i.e. marked unswappable). In the second stage, the HCA memory access tables are updated with the physical addresses of the pages of the communication buffer.

2.2 MPI_Allgather Overview

MPI_Allgather is an All-to-all broadcast collective operation defined by the MPI standard [12]. It is used to gather contiguous data from every process in a communicator and distribute the data from the jth process to the jth receive buffer of each process. MPI_Allgather is a blocking operation (i.e. control does not return to the application until the receive buffers are ready with data from all processes).

2.3 Related Algorithms and Their Cost Models

Several algorithms can be used to implement MPI_Allgather. Depending on system parameters and message size, some algorithms may outperform the others. Currently,

MPICH [4] 1.2.6 uses the Recursive Doubling algorithm for power-of-two process numbers and up to medium message sizes. For non-power of two processes, it uses the Bruck's algorithm [3] for small messages. Finally, the Ring algorithm is used for large messages [17]. In this section, we provide a brief overview of the Recursive Doubling and Ring algorithms. We will use these algorithms in our RDMA based design.

Recursive Doubling: In this algorithm, pairs of processes exchange their buffer contents. But in every iteration, the contents collected during all previous iterations are also included in the exchange. Thus, the collected information *recursively doubles*. Naturally, the number of steps needed for this algorithm to complete is $log(p)$, where p is the number of processes. The communication pattern is very dense, and involves one half of the processes exchanging messages with the other half. On a cluster which does not have constant bisection bandwidth, this pattern will cause contention. The total communication time of this algorithm is:

$$T_{rd} = t_s * log(p) + (p-1) * m * t_w \qquad (1)$$

Where, t_s = Message transmission startup time, t_w = Time to transfer one byte, m = Message size in bytes and p = Number of processes.

Ring Algorithm: In this algorithm, the processes exchange messages in a ring-like manner. At each step, a process passes on a message to its neighbor in the ring. The number of steps needed to complete the operation is $(p-1)$ where p is the number of processes. At each step, the size of the message sent to the neighbor is same as the MPI_Allgather message size, m. The total communication time of this algorithm is:

$$T_{ring} = (p-1) * (t_s + m * t_w) \qquad (2)$$

3   Can RDMA Benefit Collective Operations?

Using RDMA, a process can directly access the memory locations of some other process, with no active participation of the remote process. While it is intuitive that this approach can speed up point-to-point communication, it is not clear how *collective* communications can benefit from it. In this section, we present the answer to this question and present the motivation of using RDMA for collective operations.

3.1   Bypass Intermediate Software Layers

Most MPI implementations [4] implement MPI collective operations on top of MPI point-to-point operations. The MPI point-to-point implementation in turn is based on another layer called the ADI (Abstract Device Interface). This layer provides abstraction and can be ported to several different interconnects. The communication calls pass through several software layers before the actual communication takes place adding unnecessary overhead. On the other hand, if collectives are directly implemented on top of the InfiniBand RDMA interface, all these intermediate software layers can be bypassed.

3.2 Reduce Number of Copies

High-performance MPI implementations, MVAPICH [14], MPICH-GM [13] and MPICH-QsNet [15] often implement an eager protocol for transferring short and medium-sized messages. In this eager protocol, the message to be sent is copied into internal MPI buffers and is directly sent to an internal MPI buffer of the receiver. This causes two copies for each message transfer. For a collective operation, there are either $2 * log(p)$ or $2 * (p - 1)$ sends and receives (every send has a matching receive). It is clear that as the number of processes in a collective grows, there are increasingly more and more message copies. Instead, with RDMA based design, messages can be directly transferred without undergoing several copies, as described in section 4.

3.3 Reduce Rendezvous Handshaking Overhead

For transferring large messages, high-performance MPI implementations often implement the Rendezvous Protocol. In this protocol, the sender sends a RNDZ_START message. Upon its receipt, the receiver replies with RNDZ_REPLY containing the memory address of the destination buffer. Finally, the sending process sends the DATA message directly to the destination memory buffer and issues a FIN completion message. By using this protocol, zero-copy message transfer can be achieved.

This protocol imposes bottlenecks for MPI collectives based on point-to-point design. The processes participating in the collective need to continuously exchange addresses. However, these address exchanges are redundant. Once the base address of the collective communication buffer is known, the source process can compute the destination memory address for each iteration. This computation can be done locally by the sending process by calculating the array index for the particular algorithm and iteration number. Thus, for each iteration, RDMA can be directly used without any need for address exchange [16].

3.4 Reduce Cost of Multiple Registrations

InfiniBand [5], like most other RDMA capable interconnects, requires that all communication buffers be registered with the InfiniBand HCA. This "registration" actually involves locking of pages into physical memory and updating HCA memory access tables. After registration, the application receives a "memory handle" with keys which can be used by a remote process to directly access the memory. Thus, for performing each send or receive, the memory area needs to be registered.

Collective operations implemented on top of point-to-point calls would need to issue several MPI sends or receives to different processes (with different array offsets). This will cause multiple registration calls. For current generation InfiniBand software/hardware stacks, each registration has high setup overhead of around 90 μs (section 5). Thus, point-to-point implementation of collectives requires multiple registration calls with significant overhead. However, the RDMA based design would need only *one* registration call. The entire buffer passed to the collective call can be registered in one go. Thus, this will eliminate unnecessary registration calls.

4   Proposed RDMA Based All-to-All Broadcast Design

In this section, we describe our RDMA based design in detail. However, before we use RDMA, we have to deal with several design choices that are posed by the RDMA semantics.

4.1   RDMA Design Choices: Copy Based or Zero Copy

As stated in section 2.1, InfiniBand (like other modern RDMA capable interconnects) requires communication buffers to be registered. Thus, for transferring a message, either (1) the message is copied to a pre-registered buffer, or (2) the message buffer itself is registered at both sender and receiver ends. Approach (2) allows us to achieve zero-copy. Since copy cost is small for smaller messages, approach (1) is used for small messages. As the copy cost is prohibitive for larger messages, approach (2) is used for messages exceeding a certain threshold.

4.2   RDMA-Based Design for Recursive Doubling

We propose a RDMA based design for Recursive Doubling (RD) algorithm. In RD, the size of the message exchanged by pairs of nodes doubles each iteration along with the distance between the nodes. If m is the message size contributed by each process, the amount of data exchanged between two processes increases from m in the first iteration to $\frac{mp}{2}$ in the $\log(p)^{th}$ iteration. As we observed in section 4.1, the optimal method to transfer short messages is copy based and for longer messages, we need to use zero copy. However, since in the RD algorithm, the actual message size in each iteration changes, we also have to dynamically switch between copy based and zero copy protocols to achieve an optimal design.

Hence, we switch between the two design alternatives at an iteration k ($1 \leq k \leq \log(p)$) such that the message size being exchanged, $2^{k-1}m$, crosses a fixed threshold M_T. The threshold M_T is determined empirically. Hence, message exchanges in the first k dimensions use a copy-based approach, and those in higher dimensions from $k + 1$ through $\log(p)$ use a zero copy approach.

For performing the copy based approach, we need to maintain a pre-registered buffer. We call it "Collective Buffer". The design issues relating to maintaining this buffer and buffering schemes are described as follows:

Collective Buffer: This buffer is registered at communicator initialization time. Processes exchange addresses of their collective buffers also during that time. Some pre-defined space in the collective buffer is reserved to store the peer addresses and completion flags required for zero-copy data transfers. Data sent in any iteration comprises data received in all previous iterations along with the process' own message.

Buffering Scheme: In RD, data is always sent from and received to contiguous locations in either the collective buffer or the user's receive buffer. Since the amount of data written to a collective buffer cannot exceed M_T, the collective buffer never needs to be more than $2M_T$ which is 8 KB (ignoring space for peer addresses and completion flags) for a single Allgather call.

4.3 RDMA Ring for Large Messages

We implement the Ring algorithm for MPI_Allgather over RDMA only for large messages and large clusters. As observed in [17], large clusters may have better near-neighbor bandwidth. Under such scenarios, it is beneficial for MPI_Allgather to mainly communicate between neighbors. The Ring algorithm is ideal for such cases. Since we implement this algorithm for only large messages, we use a complete zero copy approach here. The design in this case is much simpler. The benefit of the RDMA-based scheme comes from the fact that we have a single buffer registration and a single address exchange performed by each node instead of p registrations, and $(p-1)$ address exchanges in the point-to-point based design. We use this Ring algorithm for messages larger than 1 MB and process numbers greater than 32.

5 Experimental Evaluation

In this section, we evaluate the performance of our designs. We use three cluster configurations for our tests:

1. *Cluster A:* 32 Dual Intel Xeon 2.66 GHz nodes with 512 KB L2 cache and 2 GB of main memory. The nodes are connected to Mellanox MT23108 HCA using PCI-X 133 MHz I/O bus. The nodes are connected to Mellanox 144-port switch (MTS 14400).
2. *Cluster B:* 16 Dual Intel Xeon 3.6 GHz nodes (EM64T) with 1MB L2 cache and 4 GB of main memory. The nodes are connected to Mellanox MHES18-XT HCA using PCI-Express (x8) I/O bus.
3. *Cluster C:* 8 Dual Intel Xeon 3.0 GHz nodes with 512 KB L2 cache and 2 GB of main memory. The nodes are connected to the same InfiniBand network as Cluster A.

We have integrated our RDMA based design in the MVAPICH [14] stack. We refer to the new design as "MVAPICH-RDMA". The current implementation of MPI_Allgather over point-to-point is referred to as "MVAPICH-P2P". Our experiments are classified into three types. First, we demonstrate the latency of our new RDMA design. Secondly, we investigate performance of the new design under low buffer re-use conditions. Finally, we evaluate the impact of our design on a Matrix Multiplication application kernel which uses All-to-all broadcast.

5.1 Latency Benchmark for MPI_Allgather

In this experiment, we measure the basic latency of our MPI_Allgather implementation. All the processes are synchronized with a barrier and then MPI_Allgather is repeated 1000 times, using the same communication buffer. The results are shown in Figures 1 and 2 for Cluster A and in Figures 3 for Cluster B. The results from both Clusters A and B follow the same trends. The results are explained as follows:

Small Messages: As described in section 3, the RDMA based design can avoid the various copy and layering overheads in different layers of the MPI point-to-point implementation. The results indicate that latency can be reduced by 17%, 13% and 15%

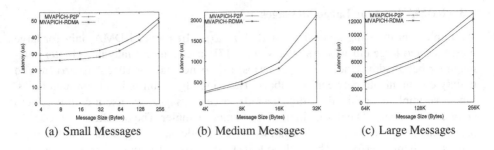

Fig. 1. MPI_Allgather Performance on 16 Processes (Cluster A)

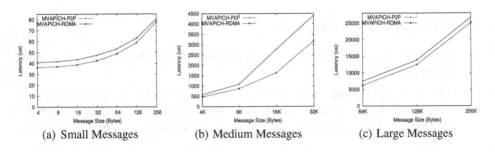

Fig. 2. MPI_Allgather Performance on 32 Processes (Cluster A)

Fig. 3. MPI_Allgather Performance on 16 Processes (Cluster B)

for 16 processes on Cluster A (Fig 1(a)), 32 processes on Cluster A (Fig 2(a)) and 16 processes on Cluster B (Fig 3(a)) for 4 byte message size, respectively.

Medium Messages: For medium sized messages, the point-to-point based design required rendezvous address exchange for transferring messages at every step of the algorithm. However, for the RDMA based MPI_Allgather, no such exchange is required (section 4). We note from section 2.3, the number of steps increases as the number of processes, and so does the cumulative cost of address exchange. Our RDMA based design is able to successfully avoid this increasing cost.

The results indicate that latency can be reduced by 23%, 30% and 37% for 16 processes on Cluster A (Fig 1(b)), 32 processes on Cluster A (Fig 2(b)) and 16 processes on Cluster B (Fig: 3(b)) for 32 KB message size, respectively.

Large Messages: Large messages are also transferred using the same zero copy technique used for medium sized messages. Hence, the same address exchange cost can be saved (as described in the previous case). However, since the message sizes are large, the address exchange forms a lesser portion of the overall cost of MPI_Allgather. The results for large messages indicate that latency can be reduced by 7%, 6% and 21% for 16 processes on Cluster A (Fig 1(c)), 32 processes on Cluster A (Fig: 2(c)) and 16 processes on Cluster B (Fig 3(c)) for 256 KB message size, respectively.

Scalability: We plot the MPI_Allgather latency numbers with varying process counts, for a fixed message size to see the impact of RDMA design on scalability. Figure 5.1 shows the results for 32 KB message size. We observe that as the number of processes increase, the gap between the point-to-point implementation and RDMA design increases. This is due to the fact that the RDMA design eliminates the need for address exchange (which increases as the number of processes).

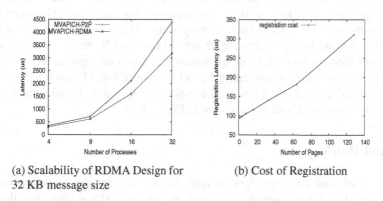

(a) Scalability of RDMA Design for 32 KB message size (b) Cost of Registration

Fig. 4. Scalability and Registration Cost on Cluster A

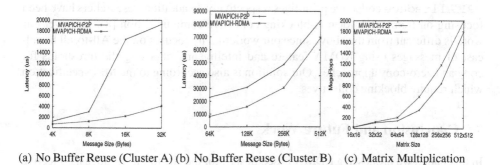

(a) No Buffer Reuse (Cluster A) (b) No Buffer Reuse (Cluster B) (c) Matrix Multiplication

Fig. 5. Impact of Buffer Registration and Performance of Matrix Multiplication

5.2 `MPI_Allgather` Latency with No Buffer Reuse

In the above experiment, we measured the latency of `MPI_Allgather` when utilizing the same communication buffers for a large number of iterations. The cost of
registration was thus amortized over all the iterations by the registration cache maintained by MVAPICH. However, it is not necessary that all MPI applications will always
reuse their buffers. In the case where applications use `MPI_Allgather` with different
buffers, the point-to-point based design will be forced to register the buffers separately,
thus incurring high cost. The cost of just memory registration is shown in Figure 5.1.
We observe that memory registration is in fact quite costly.

In the following experiment, we conduct the same latency test (as mentioned in
previous section), but the buffers used for each iteration are different. Figures 5(a) and
5(b) show the results for Clusters A and B, respectively.

The RDMA based `MPI_Allgather` performs 4.75 and 3 times better for Cluster
A and B for 32 KB message size, respectively.

5.3 Matrix Multiplication Application Kernel

In the previous sections, we have seen how the RDMA design impacts the basic latency
of `MPI_Allgather`. In order to evaluate the impact of this performance boost on end-
MPI applications, we build a distributed-memory Matrix Multiplication routine over the
optimized BLAS provided by the Intel Math Kernel Library [7]. We use a simple row-
block decomposition for the data. In each iteration, the matrix multiplication is repeated
with a fresh set of buffers. This application kernel is run on Cluster C using 8 processes.
We observe that using our RDMA design, the application kernel is able to perform 37%
better for an array size of 256x256, as shown in Figure 5(c).

6   Related Work

Recently, a lot of work has been done to improve the performance of collective operations in MPI. In [17] the authors have implemented several well known collective
algorithms over the MPI point-to-point primitives. In [9], [16] and [11] the authors
have shown the benefits of using RDMA for `MPI_Barrier`, `MPI_Alltoall` and
`MPI_Allreduce` collective primitives respectively. In addition, researchers have been
focusing on framework for non-blocking collective communication [8]. However our
work is different from the above since our work mainly focuses on the All-to-all broadcast of messages using RDMA feature and intelligently choosing the thresholds for
copy and zero-copy approaches. Our solution is also according to the MPI specification
which require blocking collectives.

7   Conclusion and Future Work

In this paper, we proposed a RDMA based design for the All-to-all Broadcast collective
operation. Our design reduces software overhead, copy costs, protocol handshake – all
required by the implementation of collectives over MPI point-to-point. Performance

evaluation of our designs reveals that the latency of MPI_Allgather can be reduced by 30% for 32 processes and a message size of 32 KB. Additionally, the latency can be improved by a factor of 4.75 under no buffer reuse conditions for the same process count and message size. Further, our design can speed up a parallel matrix multiplication algorithm by 37% on 8 processes, while multiplying a 256x256 matrix.

In the future. We will investigate impact on real world applications in much larger clusters. We will also consider utilizing the RDMA based All-to-all broadcast for designing other collectives like RDMA based All-to-all personalized exchange [16].

References

1. MPI: A Message-Passing Interface Standard.   http://www.mpi-forum.org/docs/mpi-11-html/mpi-report.html.
2. D. H. Bailey, E. Barszcz, J. T. Barton, D. S. Browning, R. L. Carter, D. Dagum, R. A. Fatoohi, P. O. Frederickson, T. A. Lasinski, R. S. Schreiber, H. D. Simon, V. Venkatakrishnan, and S. K. Weeratunga. The NAS parallel benchmarks. volume 5, pages 63–73, Fall 1991.
3. J. Bruck, C.-T. Ho, S. Kipnis, E. Upfal, and D. Weathersby. Efficient Algorithms for All-to-All Communications in Multiport Message-Passing Systems. *IEEE Transactions in Parallel and Distributed Systems*, 8(11):1143–1156, November 1997.
4. W. Gropp, E. Lusk, N. Doss, and A. Skjellum. A High-Performance, Portable Implementation of the MPI, Message Passing Interface Standard. Technical report, Argonne National Laboratory and Mississippi State University, 1996.
5. InfiniBand Trade Association. InfiniBand Architecture Specification, Volume 1, Release 1.2. http://www.infinibandta.com.
6. InfiniBand Trade Association. InfiniBand Trade Association. http://www.infinibandta.com.
7. Intel Corporation.   The Intel Math Kernel Library.   http://www.intel.com/cd/software/products/asmo-na/eng/perflib/mkl/index.htm.
8. L. V. Kale, S. Kumar, and K. Vardarajan. A Framework for Collective Personalized Communication. In *International Parallel and Distributed Processing Symposium*, 2003.
9. S. P. Kini, J. Liu, J. Wu, P. Wyckoff, and D. K. Panda. Fast and Scalable Barrier using RDMA and Multicast Mechanisms for InfiniBand-Based Clusters. In *Euro PVM/MPI*, 2003.
10. Lawrence Berkeley National Laboratory. MVICH: MPI for Virtual Interface Architecture. http://www.nersc.gov/research/FTG/mvich/index.html, August 2001.
11. A. Mamidala, J. Liu, and D. K. Panda. Efficient Barrier and Allreduce on IBA clusters using hardware multicast and adaptive algorithms. In *IEEE Cluster Computing*, 2004.
12. Message Passing Interface Forum. *MPI-2: Extensions to the Message-Passing Interface*, Jul 1997.
13. Myricom Inc. Portable MPI Model Implementation over GM, March 2004.
14. Network-Based Computing Laboratory. MPI over InfiniBand Project. http://nowlab.cse.ohio-state.edu/projects/mpi-iba/.
15. Quadrics. MPICH-QsNet. http://www.quadrics.com.
16. S. Sur, H.-W. Jin, and D.K. Panda. Efficient and Scalable All-to-All Exchange for InfiniBand-based Clusters. In *International Conference on Parallel Processing (ICPP)*, 2004.
17. Rajeev Thakur and William Gropp. Improving the Performance of Collective Operations in MPICH. In *Euro PVM/MPI 2003*, 2003.

Providing Full QoS Support in Clusters Using Only Two VCs at the Switches*

A. Martínez[1], F.J. Alfaro[1], J.L. Sánchez[1], and J. Duato[2]

[1] Departamento de Informática, Escuela Politécnica Superior,
Universidad de Castilla-La Mancha, 02071 - Albacete, Spain
{alejandro, falfaro, jsanchez}@info-ab.uclm.es
[2] Dept. de Informática de Sistemas y Computadores, Facultad de Informática,
Universidad Politécnica de Valencia, 46071 - Valencia, Spain
jduato@disca.upv.es

Abstract. Current interconnect standards providing hardware support for quality of service (QoS) consider up to 16 virtual channels (VCs) for this purpose. However, most implementations do not offer so many VCs because they increase the complexity of the switch and the scheduling delays. In this paper, we show that this number of VCs can be significantly reduced. Some of the scheduling decisions made at network interfaces can be easily reused at switches without significantly altering the global behavior. Specifically, we show that it is enough to use two VCs for QoS purposes at each switch port, thereby simplifying the design and reducing its cost.

1 Introduction

The last decade has witnessed a vast increase in the variety of computing devices as well as in the number of users of those devices. In addition to the traditional desktop and laptop computers, new handheld devices like pocket PCs, PDAs, and multimedia cellular phones have now become household names.

The main reasons for the widespread use of computing devices are the availability of cheaper and more powerful devices and, even more importantly, the huge amount of information and services available through the Internet. These services rely on applications executed in many servers all around the world. Clusters of PCs have emerged as a cost-effective platform to implement these services and run the required Internet applications. These clusters provide service to thousands or tens of thousands of concurrent users. Many of these applications are multimedia applications, which usually present bandwidth and/or latency requirements [1]. These are known as QoS requirements.

In the next section, we will be looking at some of the proposals to provide QoS in clusters and system area networks. Most of them incorporate 16 or even

---

* This work was partly supported by the Spanish CICYT under grant TIC2003-08154-C06, Junta de Comunidades de Castilla-La Mancha under grant PBC-05-005-1, and by the Spanish State Secretariat of Education and Universities under FPU grant.

D.A. Bader et al. (Eds.): HiPC 2005, LNCS 3769, pp. 158–169, 2005.

more VCs, devoting a different VC to each traffic class. This increases the switch complexity and also prevents the use of these VCs for other purposes (for instance, to provide adaptive routing or fault tolerance). Moreover, it seems that, when the technology enables it, the trend is to increase the number of ports instead of increasing the number of VCs per port [2].

In the recent switch designs, the buffers at the ports are implemented with a memory space organized in logical queues. These queues consist in linked lists of packets, with pointers to manage them. Our experience with communications' hardware manufacturers [3] has taught us that the complexity of the switch and the scheduling delays heavily depend on the number of queues at the ports. VCs, which can be used for many purposes, are implemented as queues of this kind. Then, a reduction of the number of VCs necessary to support QoS can be very helpful in the switch design.

In this paper, we show that it is enough to use two VCs at each switch port for the provision of QoS. One of these VCs is used for QoS traffic and the other one for best-effort traffic. Although this is not a new idea, the novelty of our proposal lies in the fact that the global behavior of the network is very similar as if it had much more VCs. This can be achieved by reusing in the switches some of the scheduling decisions made at network interfaces. Simulation results show that applications achieve an adequate QoS performance, but with a reduced processing delay and using fewer VCs, which results in less chip area.

The remainder of this paper is structured as follows. In the following section the related work is presented. In Section 3 we present our strategy to reduce the number of VCs required for QoS support. Details on the experimental platform and the performance evaluation are presented in Section 4. Finally, Section 5 summarizes the results of this study and identifies directions for future research.

2 Related Work

During the last decade several switch designs with QoS support have been proposed. All of them incorporate VCs in order to provide QoS support. In these proposals, different scheduling algorithms are used to arbitrate between the different existing traffic flows, providing each one with QoS according to its requirements.

The Multimedia Router (MMR) [4] is a hybrid router. It uses pipelined circuit switching for multimedia traffic and virtual cut-through for best-effort traffic. Pipelined circuit switching is connection-oriented and needs one VC per connection. This is the main drawback of the proposal because the number of VCs per physical link is limited by the available buffer size and there may not be enough VCs for all the possible existing connections. Therefore, the number of multimedia flows allowed is limited by the number of VCs. Moreover, the scheduling among hundreds of VCs is a complex task.

MediaWorm [5] was proposed to provide QoS in a wormhole router. It uses a refined version of the Virtual Clock algorithm [6] to schedule the existing VCs. These VCs are divided into two groups: One for best-effort traffic and the other for real-time traffic. Several flows can share a VC, but 16 VCs are still needed to provide QoS. Besides, it is well known that wormhole is more likely to produce congestion than virtual cut-through. In [7], the authors propose a preemption mechanism to enhance MediaWorm performance, but in our view that is a rather complex solution.

InfiniBand was proposed in 1999 by the most important IT companies to provide present and future server systems with the required levels of reliability, availability, performance, scalability and QoS [8]. Specifically, the InfiniBand Architecture (IBA) proposes three main mechanisms to provide the applications with QoS. These are traffic segregation with service levels, the use of VCs (IBA ports can have up to 16 VCs) and the arbitration at output ports according to an arbitration table. Although IBA does not specify how these mechanisms should be used, some proposals have been made to provide applications with QoS in InfiniBand networks [9].

Finally, PCI Express Advanced Switching (AS) architecture is the natural evolution of the traditional PCI bus [10]. It defines a switch fabric architecture that supports high availability, performance, reliability and QoS. AS ports incorporate up to 20 VCs that are scheduled according to some QoS criteria. In the AS specifications, three possible arbiters are proposed, one of them being table-based.

All the proposals studied use a significant number of VCs to provide QoS support. If a great number of VCs is implemented, it would require a significant fraction of chip area and would make packet processing a more time-consuming task. Moreover, in all the cases, the VCs are used to segregate the different traffic classes. Therefore, it is not possible to use the available VCs to provide other functionalities like adaptive routing or fault tolerance when all the VCs are used to provide QoS support.

On the other hand, there have been proposals which use only two VCs. For instance, the Avici TSR [11] is a well-known example of this. It is able to segregate premium traffic from regular traffic. However, it is limited to this classification and cannot differentiate among more categories. In the recent IEEE standards, it is recommended to consider seven traffic classes [12]. So, although being able to differentiate two categories is a big improvement, it could be insufficient.

In contrast, the novelty of our proposal lies in that although we use only two VCs in the switches, the global behavior of the network is very similar as if the switches were using much more VCs. This is because we are reusing at the switch ports the scheduling decisions performed at the network interfaces, which have as many VCs as traffic classes. In the end, the network provides a differentiated service to all the traffic classes considered.

To the best of our knowledge, only Katevenis and his group [13] have proposed something similar before. The basic idea of their architecture is to map the multiple priority levels onto the two existing queues, for each crosspoint buffer. The operation of the system is analogous to a two-lane highway, where cars drive in one lane and overtake using the other. However, this proposal is complex because it needs specific hardware and signaling. Furthermore, it is more limited in its scope than ours, because it is aimed to a single-stage router based on a single buffered crossbar. This crossbar has small buffers at the crosspoints that the authors split into two VCs. In contrast, our proposal is a simpler and more general technique, as we will see in the next section.

3   Reducing the Number of VCs for QoS Support

In this section, we present our proposal for QoS provision with reduced resources. The basic idea consists in using only two VCs at the switch ports. One of these VCs would be used for QoS packets and the other for best-effort packets. We reuse at switches the scheduling decisions performed at network interfaces. This allows us to achieve a performance similar to that obtained by systems with much more VCs.

Figure 1 shows an example of a network interface that is connected to a switch. Note that at both the network interface and the switch input port, there are several VCs. When a packet arrives at the switch, the header is analyzed and the packet is then usually stored in a VC according to the flow or class to which it belongs to. However, packets arriving at the switch have been previously sorted by the network interface according to some criteria. If we separate the packets in different VCs, we are losing this order, which may contain enough information to simplify the scheduling at the switch. However, it is not enough to put all the packets in the same VC to reuse the scheduling decisions. It is also necessary that the arbiter implementing this technique considers the global priority level of the packets, as opposed to the traditional 2 VC design, which would only consider two categories: premium and regular.

In order for our proposal to be effective we need two assumptions. The first assumption we make is that a static priority criterion exists to order packets. In this way, every packet would be stamped with a priority level. This is necessary because we will maintain the incoming order along the whole network. This is not a big deal because queuing delays for QoS traffic will be short, and therefore, the packet ordering established at network interfaces does not need to be changed at any switch in the path.

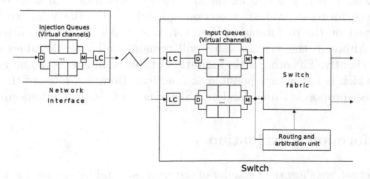

Fig. 1. QoS support at the network interface and the switch

The second assumption is that there must be a connection admission control (CAC) for the traffic with QoS requirements so that no link is oversubscribed by QoS traffic. This requirement is needed to provide bandwidth guarantees and avoid starvation of the QoS traffic. The CAC is also necessary to assure that this kind of traffic will flow with short delays.

Note that although we assume that QoS traffic does not oversubscribe any link, no assumption is made about best-effort traffic. Thus, if we did not separate QoS traffic and best-effort traffic, the total bandwidth demand for a given output link could exceed the available bandwidth. For this reason, we cannot use just one VC, and therefore we propose to use two VCs at the switches.

It is important to note that the order of the different best-effort traffic classes is also kept with this design. Although we use only one VC for best-effort traffic, we also consider the different priorities of packets belonging to this group. This means that the switch will also differentiate among this kind of packets, as we will see in the next section.

Now, we will proceed to describe in depth how our proposal works. Let us suppose that several packets arrive at a switch from a network interface. Taking into account that the interface implements a priority-based arbiter, the first packet should be the one with the highest priority. So, instead of separating the packets among several VCs according to their traffic classes, we put them all in the same queue in the arrival order. Later, when the switch must decide which packets should be transmitted, it will seek in the input queues. It is only necessary to look at the first packet in each queue, because its position at the front of the queue indicates that it had a higher priority when it left the network interface.

Obviously, the network interface can only arbitrate among the packets it holds at a given moment. Therefore, when no more high-priority packets are available, a low-priority QoS packet can be transmitted. If this packet has to wait at a switch input queue, and other packets with higher priority are transmitted from the network interface, they would be stored in the same VC as the low-priority packet, and be placed after it in the queue. Thus, the arbiter would penalize the high-priority packets, because they would have to wait until the low-priority packet is transmitted. But this situation, which we call *order error*, has a small impact on performance because there is bandwidth reservation for QoS packets. This means that all the QoS packets will flow with short delay.

In the performance evaluation section, we will see that the order errors have a low impact on the performance. However, they make latency and jitter more variable. Although the average value will remain similar, peak values will be increased slightly. The other main limitation of this technique is the necessity of a CAC, which is not always possible or practical. However, most of the recent interconnect proposals (InfiniBand or AS) include a CAC in their specification.

4   Performance Evaluation

In this section, we show the behavior of our proposal and we compare it with the performance of traditional switches. First, we will explain the simulated network architecture. Next, we will give details on the parameters of the network and the load used for the evaluation. Finally, we present and comment the results.

4.1 Simulated Architecture

Our objective is to evaluate the performance of our technique under fair conditions. In order to achieve this, we will define a complete network architecture, including all the elements necessary to work. However, many of them are not directly related with our work. We do not aim at achieving the best performance or proposing a whole new switch design, but, instead, we aim at defining a fair scenario in which compare several switch architectures. For that reason, we have not used state-of-the-art routing techniques, congestion control mechanisms, etc. We have used the most popular and well-known solutions for these issues.

The network used to test the proposals is a perfect-shuffle multi-stage interconnection network (MIN) with 64 end-points. We have chosen a MIN because it is a usual topology for clusters. However, our proposal is valid for any network topology, including both direct networks and MINs. The switches use a combined input and output buffer architecture, with a crossbar to connect the buffers. No packets are dropped because we use credit-based flow control. The parameters of the network elements used in this performance study are given in Table 1. Note that we are assuming some internal speed-up ($\times 1.5$), as it is usually the case in most commercial switches.

Virtual output queuing (VOQ) is implemented to solve the head-of-line blocking problem at the switch level. However, this does not increase the necessary buffer memory, only the crossbar scheduling time. Note that, nowadays, the queues are implemented logically over a shared space. That means that adding more queues implies more pointers and more complex arbiters, but not more buffer space. Furthermore, most of current commercial switches include VOQ as the technique to minimize the effects of head-of-line blocking.

In Table 1 we also show the amount of memory at each port. It is the minimum necessary space to achieve the peak throughput. Taking into account that there is credit-based flow control and we are using Virtual Cut-Through switching, it would be necessary a space at each VC of one maximum packet size plus one round-trip-time. However, since it is common to use maximum packet size, we round the number up, resulting in two whole packets, which results in 4 Kbytes for each VC. Note that switches using our proposal save memory (and thus chip area) at the ports by a factor of 4 compared with the 8 VC case.

Table 1. Simulation parameters

Switch ports	8	Port buffer size	32 Kbytes
Packet size	64 to 2048 bytes	Channel bandwidth	2 Gb/s
Packet header size	8 bytes	Crossbar bandwidth	3 Gb/s
Control message size	8 bytes	Network interfaces	64
Crossbar scheduling time	20/10 ns	Output scheduling time	10/5 ns

The scheduling delays have been chosen to reflect the actual delays in real implementations and the use of fewer VCs with our proposal. We have considered the research in the area, like Peh's work [14], which indicates that the arbitration time is logarithmic in the number of VCs. In Table 1, the number before the slash is the delay for 8 VC switches and the number after the slash is the delay for 2 VC switches (both traditional and using our technique). Therefore, we obtain a speed-up of 2.0 using our proposal.

The CAC we have implemented is a simple one, based on average bandwidth. Each connection is assigned a path where enough resources are assured. We also use a load balancing mechanism, which consists in assigning the least occupied route among those possible. There exist more complex and powerful mechanisms, but this is enough to test our proposal.

4.2　Traffic Model

In Table 2 the characteristics of the traffic injected in the network are included. We have considered the traffic classes (TCs) defined by the IEEE standard 802.1D-2004 [12] at the Annex G, which are particularly appropriate for this study. However, we have added an eighth TC, *Preferential Best-effort*, with a priority between *Excellent-effort* and *Best-effort*. In this way, the workload is composed of 8 different TCs. Each TC has decreasing priority, such that TC 7 has the highest priority and TC 0 has the lowest.

The proportion of each category has been chosen to provide meaningful results. Our intention is to put the network in a situation where the different TCs have to compete for limited resources. We also want to have diversity between the sources, combining different packet sizes and different traffic distributions, that is, constant bit rate (CBR) flows combined with variable bit rate (VBR) flows. It is possible that this mix of traffic is not actually present in a real-life cluster, but it serves perfectly to evaluate the different architectures we are testing.

The destination pattern is based on Zipf's law [15], as recommended in [16], with $k = 1$. In this way, the traffic is not uniformly distributed, but, instead, for each TC and input port it is established a ranking among all the possible

Table 2. Traffic injected per host

TC	Name	% BW	Packet size	Notes
7	Network Control	1	64 bytes	self-similar
6	Audio	15.625	128 bytes	CBR 64 Kb/s connections
5	Video	15.625	$\leq$ 2 Kbytes	750 Kb/s MPEG-4 traces
4	Controlled Load	15.625	2 Kbytes	CBR 1 Mb/s connections
3	Excellent-effort	13.031	2 Kbytes	self-similar
2	Preferential Best-effort	13.031	2 Kbytes	self-similar
1	Best-effort	13.031	2 Kbytes	self-similar
0	Background	13.031	2 Kbytes	self-similar

destinations. Therefore, there will be destinations with a higher chance of being elected by a group of flows, where this probability is obtained with the aforementioned Zipf's law. The global effect is a potential full utilization of the network, but with a reduced performance compared with a uniform distribution.

The packets are generated according to different distributions, as can be seen in Table 2. *Audio, Video,* and *Controlled Load* traffic are composed of point-to-point connections of the given bandwidth. The self-similar traffic is bursty traffic generated with on/off sources, governed by Pareto distributions, as recommended in [17].

4.3 Simulation Results

In this section, the performance of our proposal is shown. We have considered three traditional QoS indices for this performance evaluation: Throughput, latency, and jitter. Note that packet loss is not considered because no packets are dropped due to the use of credit-based flow control. Maximum jitter determines the receiver's user space for audio and video. Inappropriate results of latency or jitter may lead to dropped packets at the application level. For that reason, we also show the cumulative distribution function (CDF) of latency and jitter, which represents the probability of a packet achieving a latency or jitter equal to or lower than a certain value.

We have performed the tests considering three cases. First, we have tested the performance of our proposal, which uses 2 VCs at each switch port. It is referred in the figures as *New 2 VCs*. Note that, with our proposal, the network interfaces still use 8 VCs. Second, we have decided to perform the test with traditional switches using 8 VCs because this number matches the number of TCs. In this case, it is referred in the figures as *Traditional 8 VCs*. Third, we have also tested a traditional approach with 2 VCs, noted in the figures as *Traditional 2 VCs*. In this case, the network interfaces also use 2 VCs. Therefore, we have two references to compare our proposal, one being the lower bound (*Traditional 2 VCs*) and the other the upper bound (*Traditional 8 VCs*).

Figure 2 shows the latency results for TC 7, which corresponds to *Network Control* traffic. The throughput of this TC, which is not shown, is the optimum in the three cases. We can see that the three cases succeed in getting a reasonable average latency. However, the *Traditional 2 VCs* case achieves the worst

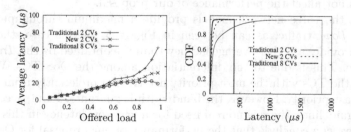

Fig. 2. Latency results for *Network Control* traffic

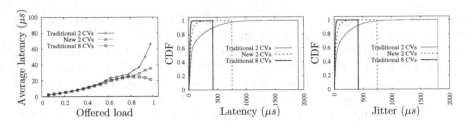

Fig. 3. Performance of *Audio* traffic

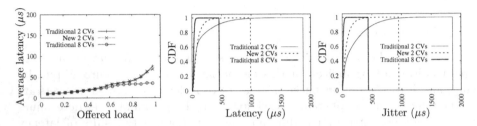

Fig. 4. Performance of *Video* traffic

performance when the input load is high. In Figure 2, we can also see the CDF of latency at a network load of 1.0. For the *Traditional 8 VCs* and the *New 2 VCs* cases, the results are quite good. However, for the *Traditional 2 VCs* case the results are not so suitable, because many packets have a latency far above the average. In this figure, maximum values are represented by vertical lines.

In Figure 3, we show the performance of *Audio* traffic. According to the IEEE guidelines, this TC should achieve latency and jitter values lower than 10 ms. This is achieved in the three cases. However, the *Traditional 2 VCs* case yields an inadequate performance, in terms of latency and jitter, at high load. Note that our proposal increases slightly the maximum latency and jitter. However, the results are still acceptable and a significant improvement over the *Traditional 2 VCs* approach.

The *Video* traffic results (Figure 4) are very similar to the audio traffic. Again, the performance of the network using our technique is quite close to the results obtained with the *Traditional 8 VCs* approach. Note that the use of VBR traffic does not affect the performance of our proposal.

Finally, the three network models provide a maximum throughput to the *Controlled Load* traffic, as can be seen in Figure 5. In this case, the *New 2 VCs* case provides a higher average latency than the obtained by the *Traditional 2 VCs* case. This is due to the differentiation among the QoS TCs. With our technique, the TCs with the most priority are treated preferently from the TCs with the least priority. However, the bandwidth is guaranteed and, according to the IEEE guidelines, there is no real need for a very low latency in this case. At this point we can conclude that the performance of our proposal for QoS traffic is very similar to that obtained with a switch design with more VCs.

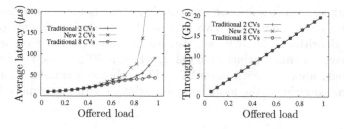

Fig. 5. Average latency and throughput for *Controlled Load* traffic

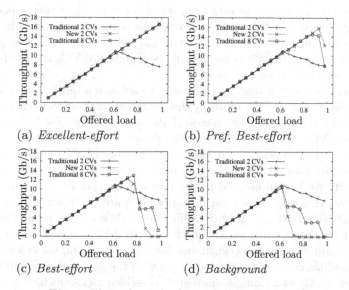

Fig. 6. Throughput for best-effort traffic classes

In Figure 6 we can see the throughput for the best-effort TCs. In these cases, the *Traditional 2 VCs* approach produces the same performance for all the TCs, which is an inadequate behavior, because *Excellent-effort* and *Preferential Best-effort* traffic should have better performance. The reason for this inadequate behavior is that in the *Traditional 2 VCs* model, all the best-effort classes look the same for the schedulers at both the network interfaces and the switches.

On the other hand, the arbiters using our technique take into account the priority of the packets, even if they share the same VC. For that reason, our proposal, which devotes a single VC in the switches for all the best-effort TCs, can provide a behavior similar to that of the *Traditional 8 VCs* approach, which uses 4 VCs for the best-effort TCs.

The *Best-effort* and *Background* TCs obtain a slightly worse performance with our technique if we compare it with the performance of the *Traditional 8 VCs* case. This is due to a lower global throughput of the network using our technique, but it only affects the TCs with the least priority, which is alright.

According to these results, we can conclude that our proposal can provide an adequate QoS. We only need two VCs at the switches, which simplifies the arbitration algorithm. The switches also incorporate 1/4 the memory at the ports and have a reduced arbitration delay, which is 50% of the necessary time in a traditional approach with more VCs. This scheme is simpler than today's trends but as powerful as the more complex arbiters with many more VCs.

5   Conclusions

The proposal of this paper consists in making the network elements cooperate, building together ordered flows of packets. Consequently, the switches try to respect the order in which packets arrive at the switch ports, which is probably correct. This allows a drastic reduction in the number of VCs required for QoS purposes at each switch port.

This study has shown that it is possible to achieve a more than acceptable QoS with only two VCs. We reuse in the switches some of the scheduling decisions made at the network interfaces. This opens up the possibility of using the remaining VCs for other concerns, like adaptive routing or fault tolerance. Furthermore, it is also possible to reduce the number of VCs supported at the switches, thereby simplifying the design, or increasing the number of ports.

The results we have presented in the previous section have shown that our proposal provides a very similar performance compared with a traditional architecture with 8 VCs, both for the QoS traffic and the best-effort traffic. Comparing our technique with a traditional architecture with 2 VCs, our proposal provides a significant improvement in performance for the QoS traffic, while for the best-effort, the traditional model is unable to provide the slightest differentiation.

We are currently examining a number of possible extensions to the work here presented. First, we are preparing a study on more complex switch models that can benefit from our proposal, such as switches using EDF arbitration. Second, we intend to use this technique in other environments, like Internet routers or networks on chip. Finally, we are considering to code the switches and network interfaces with a hardware description language, which can then be implemented in FPGAs to examine the actual reductions of delays and area.

References

1. Miras, D.: A survey on network QoS needs of advanced internet applications. Technical report, Internet2 - QoS Working Group (2002)
2. Minkenberg, C., Abel, F., Gusat, M., Luijten, R.P., Denzel, W.: Current issues in packet switch design. In: ACM SIGCOMM Computer Communication Review. (2003)
3. Duato, J., Johnson, I., Flich, J., Naven, F., García, P., Nachiondo, T.: A new scalable and cost-effective congestion management strategy for lossless multistage interconnection networks. In: Proceedings of the 11th International Symposium on High-Performance Computer Architecture. (2005)

4. Duato, J., Yalamanchili, S., Caminero, M.B., Love, D., Quiles, F.: MMR: A high-performance multimedia router. Architecture and design trade-offs. In: Proceedings of the 5th Symposium on High Performance Computer Architecture. (1999)

5. Yum, K.H., Kim, E.J., Das, C.R., Vaidya, A.S.: MediaWorm: A QoS capable router architecture for clusters. IEEE Trans. Parallel Distrib. Syst. **13** (2002) 1261–1274

6. Zhang, L.: VirtualClock: A new traffic control algorithm for packet switched networks. ACM Transaction on Computer Systems **9, 2** (1991) 101–124

7. Yum, K., Kim, E., Das, C.: QoS provisioning in clusters: An investigation of router and NIC design. In: Proceedings of the 28th Annual International Symposium on Computer Architecture, IEEE Computer Society (2001)

8. InfiniBand Trade Association: InfiniBand architecture specification volume 1. Release 1.0. (2000)

9. Alfaro, F.J., Sánchez, J.L., Duato, J.: QoS in InfiniBand subnetworks. IEEE Trans. Parallel Distrib. Syst. **15** (2004) 810–823

10. Advanced switching core architecture specification. Technical report, (available at http://www.asi-sig.org/specifications for ASI SIG members)

11. Dally, W., Carvey, P., Dennison, L.: Architecture of the Avici terabit switch/router. In: Proceedings of the 6th Symposium on Hot Interconnects. (1998)

12. IEEE: 802.1D-2004: Standard for local and metropolitan area networks. http://grouper.ieee.org/groups/802/1/ (2004)

13. Chrysos, N., Katevenis, M.: Multiple priorities in a two-lane buffered crossbar. In: Proceedings of the IEEE Globecom 2004 Conference. (2004)

14. Peh, L., Dally, W.: A delay model and speculative architecture for pipelined routers. In: Proceedings of the 7th International Symposium on High-Performance Computer Architecture. (2001)

15. Zipf, G.K.: The Psycho-biology of Languages. Houghton-Miffin, MIT (1965)

16. Elhanany, I., Chiou, D., Tabatabaee, V., Noro, R., Poursepanj, A.: The network processing forum switch fabric benchmark specifications: An overview. IEEE Network (2005)

17. Jain, R.: The art of computer system performance analysis: techniques for experimental design, measurement, simulation and modeling. John Wiley and Sons, Inc. (1991)

Offloading Bloom Filter Operations to Network Processor for Parallel Query Processing in Cluster of Workstations

V. Santhosh Kumar[1], M.J. Thazhuthaveetil[1,2], and R. Govindarajan[1,2]

[1] Supercomputer Education and Research Centre
[2] Department of Computer Science and Automation,
Indian Institute of Science, Bangalore 560 012, India
gvsk@hpc.serc.iisc.ernet.in
{mjt, govind}@{csa, serc}.iisc.ernet.in

Abstract. Workstation clusters have high performance interconnects with programmable network processors, which facilitate interesting opportunities to offload certain application specific computation on them and hence enhance the performance of the parallel application. Our earlier work in this direction achieves enhanced performance and balanced utilization of resources by exploiting the programmable features of the network interface in parallel database query execution. In this paper, we extend our earlier work for studying parallel query execution with Bloom filters. We propose and evaluate a scheme to offload the Bloom filter operations to the network processor. Further we explore offloading certain tuple processing activities on to the network processor by adopting a network interface attached disk scheme. The above schemes yield a speedup of up to 1.13 over the base scheme with Bloom filter where all processing is done by the host processor and achieve balanced utilization of resources. In the presence of a disk buffer cache, which reduces both the disk and I/O traffic, offloading schemes improve the speedup to 1.24.

1 Introduction

Cluster computer systems, assembled from commodity off-the-shelf components, have emerged as a viable alternative to high-end custom parallel computer systems for applications demanding high performance [1]. An important component in such cluster computer systems is their high performance interconnect with programmable network interfaces such as Myrinet, Quadrics, and Infiniband [6]. The network processors available in such programmable interfaces often have the capability to perform application related processing, thus facilitating interesting opportunities to enhance application performance in cluster systems.

Our earlier work [10] demonstrates offloading the tuple splitting computation (to determine on which cluster node a tuple is to be processed), and tuple processing computation to the network processor results in significant performance improvement. Further, this work [10] also explored the benefits of attaching the disks to network interface.

In this paper, we first study the performance of a base case where all the application related tuple processing is performed by the host processor (HP). Using a Bloom filter [7] reduces the host processor workload as it eliminates the processing of tuples that

D.A. Bader et al. (Eds.): HiPC 2005, LNCS 3769, pp. 170–179, 2005.
© Springer-Verlag Berlin Heidelberg 2005

are not selected (whose join key attribute do not match) by the join operation. Further, the Bloom filter reduces the data transferred over the network. Our simulation results indicate that although query execution time reduces significantly, HP utilization still remains high. We therefore propose offloading the Bloom filter operations from the host processor (HP) to the network processor (NP). Performance evaluation indicates a reduction in execution time by about 8%. Further offloading of tuple processing activities is possible by attaching the disk directly to NI as in [10]. This results in a performance improvement of upto 13%. With this offloading the disk becomes the bottleneck. To overcome this we study the impact of caching of tuples of frequently used relations or intermediate relations. With a cache hit ratio of 0.5, our schemes result in an execution time speedup of 1.24 over the Base scheme with an identical cache hit ratio. Further the utilization of key resources *viz*, HP, NP, and Disk are well balanced.

In the following section we provide the necessary background. In Section 3 we describe the Petri net model developed in [10] for the *Base Scheme* and extend the same for the Base scheme with Bloom filter. Section 4 discusses the *NP Bloom* scheme where Bloom filter operations are offloaded to NP. In Section 5 we describe the *Network Interface with attached Disk and Bloom Filter* scheme and evaluate its performance. Concluding remarks are provided in Section 7.

2   Background

Clusters: A cluster of workstations is a distributed memory machine where each node is a stand-alone system, with CPU and memory connected by the memory bus, and peripherals like disk, and network interface attached to the I/O bus. Typical high performance cluster interconnects like the Myrinet network interface (NI) [13] have NIs with a programmable network processor (NP), on board memory (SRAM), a host DMA engine (HDMA), an EBUS (External Bus) Interface (64-bit), a send DMA engine (SDMA) and a receive DMA engine (RDMA). The HDMA is used to transfer data across the I/O bus to the node memory. SDMA and RDMA are used to transfer data from the NI SRAM to the communication network (switch), and vice-versa. It is also equipped with a memory-to-memory copy engine (MCE). The switch employs wormhole routing to transfer packets between network interfaces. Research [14] in communication layers for high performance scientific applications has led to the development of user-level communication techniques which have reduced the involvement of HP in communication to deliver better application performance. We, therefore, assume such a user-level communication layer in our system.

Parallel Query Processing: In a relational database system queries composed of relational operators like select and join are used to manipulate data. The join operator, which combines tuples from two relations based on a common attribute, is the most crucial and expensive operator [8]. Hash-based join algorithms are more efficient than other join algorithms, such as sort-merge or nested-loop, in systems with large main memories [8]. So, in this work, join operators in queries are executed using hash-join algorithms. Several parallel query processing techniques have been devised and employed in parallel database machines to improve query execution time [4]. A cluster of workstations is essentially a shared-nothing architecture, in which intra-operator parallelism is better exploited [4,11] with horizontally partitioned relations. We therefore consider only exploiting intra-operator parallelism and horizontally partitioned relations.

When a query involves multiple joins, a query tree or a query execution plan is used to represent the scheduling sequence of the constituent operations. Query trees are characterized as left-deep, right-deep or bushy trees. Right-deep and bushy trees have multiple operators simultaneously active, and are suitable for pipelined implementations in multiprocessor systems [12,3]. Left-deep trees plan allows only one operator is active on all the nodes and is well suited to shared-nothing architectures. Hence, we adopt a left-deep query tree plan.

In the hash join, there are two phases, namely (i) the Build Phase, where the inner relation is hashed on the join attribute and the hash table is built and (ii) the Probe Phase, where the outer relation is hashed on the join attribute using the same hash function used to probe the hash table, and result tuples are generated on successful matches. In the parallel version, first a separate hash function is used to determine the cluster node where the tuples will be processed. We refer to this activity as tuple splitting. Tuples are routed to their cluster nodes and then the join operation (build or probe) takes places on that node. Thus both phases of parallel join involve communication between cluster nodes. In case select or project operation exists along with join, select or project is first applied on the tuples before the join operation.

Bloom Filter: Bloom filters [2] are used in distributed databases to improve the join performance by reducing the amount of tuple processing being performed by the host processor and the amount of data transferred over the network [7]. Bloom filter is a bit vector representation of the set of keys which can be queried to check if a key is present. This is used in the join process as follows: Initially the bit vector is initialized to all zeros. Each tuple of the inner relation is hashed on the join attributes to the corresponding bit in the bit vector. Then the same hash function on the join attribute of the second relation is used to generate an index. If the corresponding bit of the bit vector is zero, then the actual hash table probe operation can be avoided. On the other hand a successful check in the bloom filter does not necessarily indicate that the key is present in the hash table, and so the probe operation has to be done. Thus, the Bloom filter can give rise to false positives.

3   Base Scheme

In our Base Scheme all activities related to query processing are performed by the host processor (HP). The network processor (NP) is involved only in message sends and receives among the cluster nodes. We evaluated the performance of the Base Scheme through simulation of a Petri Net (PN) model.

3.1   Base Scheme Description

Figure 1 shows our Petri net model of a single node in the cluster performing the parallel query execution on a cluster. Since our modifications are centered around the hash join, for clarity, the PN model for the join operations are described. We use the name of a timed transition *e.g., T_Build*, to represent the duration of the timed transition.

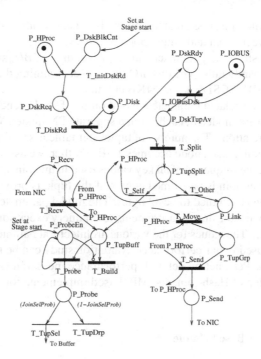

Fig. 1. Petri net model for Disk and Join operations

The number of tuples to be processed is modeled by available tokens in place P_TupAv. T_Split transition models the duration to compute the cluster $node\_id$ to which tuples are to be routed. T_Split fires depending on the availability of HP (modeled by place P_HProc). We assume that the tuples are uniformly distributed across nodes and hence T_Self fires with probability $1/N$ and and T_Other with a probability $(N-1)/N$, where N is the number of nodes in the cluster. T_Build represents the duration of per tuple build operation, which fires with the availability of HP, during the build phase. T_Probe represents the HP duration for per-tuple probe operation which is enabled during the probe phase. Based on the join selectivity (denoted by $JoinSelProb$), certain tuples qualify in the join process (modeled by T_TupSel) while others are dropped (modeled by T_TupDrp). Tuples which fire T_Other are grouped into messages (T_Move) and enqueued by HP to be transmitted by NI. T_Send and T_Recv model the software overhead incurred by HP for initiating a send/recv operation.

The architectural parameters of our model are set to represent contemporary high performance computing nodes. We measured the time taken for various tasks on a 2.4GHz Pentium 4 processor based system and scaled them to 3.6GHz, so as not to overestimate the benefits of offloading. The parameters for host communication overheads were obtained from measurements on Myrinet user-level messaging software running on Myrinet LANai 9.2 processor [6]. Network Interface Parameters values are estimated assuming a NI SRAM bandwidth of 2.664GB/s (333MHz, 64 bit bus). The network link bandwidth is assumed to be 4GB/s based on the Myrinet specifications [13]. This value is also used to fix the switch delay for a packet; the bisection bandwidth of the switch must be greater than the link bandwidth to support simultaneous connection between input and output ports of the switch.

The I/O bus transfer time parameters are set assuming 2GB/s, similar to the PCI-Express-1X bus [9] bandwidth. The disk I/O time is set assuming the ability to deliver 1.28GB/s (4 * 320MB/s SCSI disk), in 64KB chunks.

We set the database related parameters based on the TPC-H benchmark [16]. TPC-H queries 3,5,7-10 were modeled, each using a separate PN model. We assume a tuple size of 128B for all relations. The number of tuples per table is set so that the horizontal partition of the relation in each node occupies 1GB. Further, we assume that there is no skew in the data, *i.e.*, the frequency of all key values used in the join attributes occur with equal frequency. When join is performed, a part of the tuple is projected out. The size of the projected tuple is assumed to be 32 bytes. The model parameter values for selectivity ratios were obtained using measurements from query execution on a single node running PostgreSQL. The values for the various architectural parameters and those for the TPC-H queries used in our performance evaluation study can be found in [10].

We validated our Petri net model with performance measurements from a uniprocessor implementation of Hash-join, and MPI based implementations on 2 nodes and 4 nodes [10].

3.2   Performance of Base Scheme

We simulated our Petri net models using CNET, an event-driven petri net simulator [17]. The simulator reports the total simulation time for the Petri net model, as well as the total firing times for each timed transition. All reported results are for 8 node cluster[1] averaged across 3 independent runs for each query.

We use relative speedups of query execution times for performance comparison and the utilization of resources like host processor (HP), Disk, I/O Disk, I/O NIC, Switch (SW) and network processor (NP) to identify the bottleneck resources. In discussing the results, we report the average of the execution time of all the queries. Relative speedup of a scheme, is the ratio of the average query execution time in the Base scheme (with BS-1X parameters) to that in the proposed model.

Table 1 shows that query execution time is dominated by the tuple processing cost of HP, which has the maximum resource utilization of 97.4%. We found that doubling tuple processing cost of HP for Base Scheme (BS-2X) power yielded a speedup of 1.78 with respect to BS-1X configuration, showing that tuple processing activities done by HP are a significant factor in query execution time. This motivated us to study the effect of Bloom filter operations on the performance of query execution time.

3.3   Base Scheme with Bloom Filter

In the context of a parallel join operation, the Bloom filter can be incorporated in two ways depending on whether the filter operations are performed before the tuples are transmitted to their destination node (referred to as Global Bloom Filter), or after they

---

[1] We have studied the performance for 4 node, 8 node and 16 node clusters in our earlier study [10], and found that the performance study exhibits similar results.

Table 1. Comparison of Resource Utilization of Base Schemes with Bloom Filter

Model	Disk	HP	I/O Disk	I/O NIC	SW	NP	Exec Time(s)	Relative Speedup
BS-1X	35.7	97.4	22.5	7.9	16.4	0.1	2.05	1.00
BS-2X	63.5	86.6	39.9	14.0	29.2	0.1	1.15	1.78
GBF	60.5	73.6	38.1	5.4	11.2	0.1	1.21	1.69
LBF	58.9	85.2	37.1	13.0	27.1	0.1	1.24	1.65

arrive at the destination node (referred to as Local Bloom Filter). The advantage with Global Bloom Filter (GBF) is that, it does not transmit unwanted tuples to remote nodes by performing the filter operation first. This reduces I/O and network traffic. However it becomes necessary to exchange the Bit filters after the build phase. In contrast, in the Local Bloom Filter (LBF), the filter is built for tuples that arrive at the destination node.

To estimate Bloom filter model parameters we used an implementation of Jenkins's hash function[5] and measured the execution time for 6M keys, requiring 1 bit for 1 key [2]. We found that per key time for filter operations is $0.051\mu s$, with the computation component of filter operation (hash function computation) is $0.0195\mu s$ and memory access time is $0.0315\mu s$[3].

Table 1 shows the relative speedup and resource utilizations for Global and Local Bloom filter implementation referred to as GBF and LBF schemes, compared to the Base scheme (BS-1X). We find that both schemes give a significant speedup in execution time (1.69 for GBF and 1.65 LBF). Also, HP utilization reduces from 97.4% for the Base scheme to 73.6% and 85.2% for GBF and LBF respectively due to the early dropping of tuples and the reduction in probe operations during the Bloom filter operation.

Further, we observe that utilization of HP for GBF scheme (73.6%) is lower than that of LBF (85.2%) scheme. This is because GBF reduces the communication overhead incurred by HP. This results in an improved speedup in execution time for GBF scheme (1.69) as compared to LBF scheme (1.65).

While Bloom filter achieves a significant speedup in query execution, we find that HP has a high utilization and NP a low utilization. Hence we ran additional experiments for the GBF by doubling the host processing power and found an increase in speedup from 1.69 to 2.25, implying that the host processing power is the dominant factor with BS-1X configuration for the GBF scheme. We next study the impact of offloading Bloom filter operations to NP.

4 Network Processor Running Bloom Filter

Offloading the Bloom filter operations to the NP results in two schemes, NP-LBF and NP-GBF, depending on whether its a local or global filter scheme. In the NP-GBF,

---

[2] The Bloom filter gives a false positive rate of 22.1%.

[3] We found that the time for BuildFilter and CheckFilter operations are approximately the same. This is reasonable due to the high memory access cost, which is required for both operations.

Table 2. Comparison of Resource Utilization and Speedup: GBF vs. NP-LBF

Scheme	HP:NP	Disk	HP	Utilization I/O Disk	I/O NIC	SW	NP	MCE	Relative Speedup
GBF		60.5	73.6	38.1	5.4	11.2	0.1	0.0	1.00
	1:1/8	56.9	63.3	35.8	2.3	26.2	59.8	3.10	0.97
NP-LBF	1:1/6	63.2	69.2	39.8	2.5	28.9	50.6	3.40	1.04
	1:1/4	65.3	72.7	41.1	2.6	30.0	36.5	3.50	1.08
	1:1/2	65.3	72.7	41.1	2.6	30.0	21.0	3.50	1.08
	1:1	65.3	72.7	41.1	2.6	30.0	13.2	3.50	1.08

since the filter operations are to be performed before the transmitting the tuples to the destination, all the tuples would have be transferred to the NI. Earlier work [10] has shown that this could cause the I/O bus to be a bottleneck, restricting the performance enhancements possibilities. So we consider only the NP-LBF scheme where the filter operations are performed by the NP after the tuples arrive at the destination node.

Table 2 shows the resource utilizations and speedup for GBF and NP-LBF. We use GBF for comparison purposes as it was better of the schemes compared in Table 1. When $HP : NP$ ratio, which models the relative processing power of HP and NP, is 1:1/8, NP-LBF does not perform as well as GBF, having a relative speedup of 0.97. As the processing power of NP is increased, and at $HP : NP$ ratio of 1:1/4, the speedup improves to 1.08. Further, we see a reduction in NP utilization (from 59.8% to 36.5%), implying that it is no longer a bottleneck and its processing power is sufficient for Bloom filter operations to be performed by NP. The HP utilization increases to 72.7% suggesting that tuple processing costs are a dominant factor in query execution as NP processing power increases. Disk utilization is also comparable to that of HP, indicating that it is also a bottleneck resource.

Since HP is the resource with highest utilization, we also simulated the NP-LBF scheme for $HP : NP$ ratio of 1:1/4, with double the HP and NP processing powers. We observed that the relative speedup increases from 1.08 (NP-LBF-1X) to 1.42 (NP-LBF-2X). This motivated us to next consider further options to offload some tuple processing work from HP to NP.

5   Network Interface with Attached Disk and Bloom Filter

Here we consider the idea of attaching the disk to the network interface directly instead of to the system bus, proposed in our earlier work [10], which gives opportunity for additional tuple processing to be offloaded to the NP. The operations of $node\_id$ computation and communication of tuples can then be performed by NP, along with the Bloom filter activities.

There are two ways to organize the Bloom filter in the NID model as in the Base Scheme a) Local Bloom filter with NID (NID-LBF) b) Global Bloom filter with NID (NID-GBF). In both NID-GBF and NID-LBF schemes, the NP does the job of reading the tuples directly from the disk, routing the tuples to the destination node and the filter operations, and the HP performs the hash build and hash probe operations. They differ

Table 3. Comparison of Resource Utilization and Speedup for NID Schemes

Scheme	HP:NP	Disk	HP	Utilization I/O Disk	I/O NIC	SW	NP	MCE	Relative Speedup
GBF		60.5	73.6	38.1	5.4	11.2	0.1	0.0	1.00
	1:1/8	55.4	41.3	0.0	3.5	25.5	59.3	29.9	0.92
	1:1/6	61.1	45.5	0.0	3.8	28.1	50.4	33.0	1.01
NID-LBF	1:1/4	64.8	48.4	0.0	4.1	29.8	37.6	35.0	1.07
	1:1/2	67.5	50.3	0.0	4.2	31.0	23.2	36.4	1.12
	1:1	68.7	51.3	0.0	4.3	31.6	15.5	37.1	1.13
	1:1/8	57.3	42.9	0.0	3.6	9.4	61.2	6.4	0.95
	1:1/6	63.4	47.5	0.0	4.0	10.4	52.1	7.1	1.05
NID-GBF	1:1/4	68.2	51.0	0.0	4.3	11.2	39.3	7.6	1.13
	1:1/2	68.2	51.0	0.0	4.3	11.2	23.0	7.7	1.13
	1:1	68.2	51.1	0.0	4.3	11.2	14.9	7.6	1.13

only in when the filter operations are performed, i.e., before sending the tuples to the remote node or on receiving the tuples at the remote node.

Table 3 compares the resource utilizations and speedup for NID-LBF, NID-GBF, and GBF schemes. When for $HP : NP$ ratio is 1:1/8 both NID-LBF and NID-GBF schemes have a speedup less than 1. Further NP is the resource with highest utilization (59.3% for NID-LBF, 61.2% for NID-GBF), suggesting that NP's is the bottleneck resource. With an increase in the processing power of NP from 1:1/8 to 1:1/4, NID-LBF and NID-GBF attain speedups of 1.07 and 1.13 respectively. Also, NP utilization decreases to 37.6% and 39.3% for NID-LBF and NID-GBF respectively, indicating that NP is no longer the bottleck resource. Further increase in NP's processing power does not yield any benefit, as the disk unit becomes the resource with higher utilization (greater than 64.8%).

6   Performance with Disk Caching

The discussions in the previous sections indicate that when the processing power of NP is greater than 1/4th the processing power of HP, the bottleneck shifts to disk. Using large memory caches for tuples of often used relations can help by reducing the load on the disk. Such caches can be modeled by means of a Buffer Hit Probability (BHP). or intermediate relations (in a sequence of database operations) improve performance. In this section we explore the possibility of using such a cache in GBF, NP-LBF, NID-LBF and NID-GBF schemes. For the later two schemes the cache has to be maintained in the NIC.

We observe from Table 4 that the speedup is less than 1 when NP's processing power lower than 1/4th of HP. At lower processing power, NP is the bottleneck resource, as seen from the high utilization of NP $> 56.7\%$. As NP's processing power increases to 1:1, we see a relative speedup of 1.10, 1.09, and 1.24 for NP-LBF, NID-LBF and NID-GBF respectively. The additional improvement in performance is due to the buffer cache. We also find the the key resources HP, NP, Disk achieve balanced resource utilizations.

Table 4. Comparison of Resource Utilizations and Speedup with Memory Cache for BHP=0.5

Scheme	HP:NP	Disk	HP	I/O Disk	I/O NIC	SW	NP	MCE	Relative Speedup
GBF	38.1	90.8	24.0	6.7	13.9	0.1	0.0	0.0	1.00
	1:1/8	34.1	68.4	21.5	2.4	29.6	68.8	3.6	0.89
	1:1/6	37.1	78.5	23.4	2.8	33.0	57.7	3.9	0.97
NP-LBF	1:1/4	40.9	88.2	25.7	3.2	36.6	44.5	4.3	1.07
	1:1/2	41.7	91.1	26.2	3.3	37.6	26.3	4.4	1.09
	1:1	41.7	91.1	26.2	3.3	37.6	16.5	4.4	1.10
	1:1/8	31.3	45.8	0.0	3.9	28.3	65.7	33.2	0.82
	1:1/6	34.9	51.4	0.0	4.3	31.7	56.8	37.1	0.92
NID-LBF	1:1/4	35.4	52.0	0.0	4.4	32.0	40.5	37.5	0.93
	1:1/2	40.6	59.5	0.0	5.0	36.7	27.4	43.1	1.07
	1:1	41.8	61.0	0.0	5.1	37.6	18.4	44.0	1.09
	1:1/8	32.5	48.1	0.0	4.1	10.5	68.6	7.2	0.86
	1:1/6	36.8	54.1	0.0	4.6	11.8	59.4	8.1	0.97
NID-GBF	1:1/4	41.6	61.7	0.0	5.2	13.5	47.5	9.2	1.10
	1:1/2	47.2	69.3	0.0	5.8	15.2	31.2	10.4	1.24
	1:1	47.4	69.6	0.0	5.9	15.2	20.3	10.4	1.24

7 Conclusions

Optimizing the performance of these parallel query processing clusters is commercially important. In our earlier work we had evaluated both software and hardware modifications, exploiting the programmable features of NP to a achieve higher performance with balanced utilization of system resources. Using a Bloom filter reduces the host processor workload as it eliminates the processing on tuples that are not selected (whose join key attribute do not match) by the join operation. Further the Bloom filter also reduces the data transferred over network. In this work we offload the Bloom filter activities to the network processor and evaluated its benefits. We evaluate the performance of the proposed modifications using timed Petri net models. We find that offloading Bloom filter processing from the host processor to the network processor results in execution time speedup of upto 1.24, and achieves a balance resource utilization. We suggest that if future network interfaces are equipped with programmable processor of high power, applications should be able to exploit them in improving system performance.

References

1. T. Anderson, D. Culler, and D. Patterson. A Case for NOW (Networks of Workstations). *IEEE Micro,* 16(1):54-64, Feb 1995.
2. B. Bloom. Space/time trade-offs in hash coding with allowable errors, *Communications of the ACM* 13(7):422-426, July 1970
3. M.-S. Chen, M.-L. Lo, P. S. Yu, and H. C. Young. Using Segmented Right-Deep Trees for the Execution of Pipelined Hash Joins. *In Proc. of 18th Very Large Data Bases,* pages 15-26, Aug 1992.
4. D. DeWitt and J. Gray. Parallel Database Systems: The future of High Performance Database Systems. *Communications of the ACM* 35(6):85-98, Jun 1992.

5. B. Jenkins. http://burtleburtle.net/bob/c/lookup2.c 1996
6. J. Liu, B. Chandrasekaran, W Yu *et al* . Microbenckmark Performance Comparison of High-Speed Cluster Interconnects. *IEEE Micro,* 24(1):42-51, Jan-Feb 2004.
7. L. F. Mackert, and G. M. Lohman. R* Optimizer Validation and Performance Evaluation for Distributed Queries. In *Proc. of 12th Intl. Conf. on Very Large Data Bases*, 149-159, Aug 1986.
8. P. Mishra and M. H. Eich. Join Processing in relational databases. *ACM Computing Surveys*, 24(1):63-113, Mar 1992.
9. PCI-SIG Home, http://www.pcisig.com/, 2004.
10. V. Santhosh Kumar, M. J. Thazhuthaveetil, R. Govindarajan. Exploiting Programmable Network Interfaces for Parallel Query Execution in Workstation Clusters. TR-HPC-10/2005, LHPC, SERC, IISc, 2005 (http://hpc.serc.iisc.ernet.in/Publications/gvsk2005.ps).
11. D. Schneider and D. DeWitt. A performance evaluation of four parallel join algorithms in a shared-nothing multiprocessor environment. In *Proc. of ACM SIGMOD Conference*, Jun 1989.
12. D. Schneider and D. DeWitt. Tradeoffs in processing complex queries via hashing in multiprocessor database machines. In *Proc. of 16th Intl. Conf. on Very Large Data Bases*, Aug 1990.
13. C. L. Seitz. Myrinet Technology Roadmap. *Myrinet Users Group Conf.* May 2002.
14. V. Karamcheti, and A. Chien. Software Overhead in Messaging Layers: Where Does the Time Go? In *Proc. of 6th Intl. Conf. on Architectural Support for Programming Languages and Operating Systems*, Oct 1994.
15. T. Tamura, M. Oguchi, and M. Kitsuregawa. Parallel Database Processing on a 100 Node PC Cluster: Cases for Decision Support Query Processing and Data Mining. In *Proc. of Supercomputing*, Nov 1997.
16. TPC Benchmark^{TM} H (Decision Support) Standard Specification Revision 1.3.0. Transaction Processing Performance Council(TPC), 1999.
17. W. M. Zuberek. Modeling using Timed Petri Nets - event-driven simulation, Technical Report No. 9602, Dept. of Computer Science, Memorial Univ. of Newfoundland, St. John's, Canada, 1996 (ftp://ftp.cs.mun.ca/pub/techreports/tr-9602.ps.Z).

A High-Speed VLSI Array Architecture for Euclidean Metric-Based Hausdorff Distance Measures Between Images

N. Sudha and E.P. Vivek

Department of Computer Science and Engineering,
Indian Institute of Technology, Madras
sudha@cs.iitm.ernet.in, vivekep@cse.iitm.ernet.in

Abstract. A new parallel algorithm to compute Euclidean metric-based Hausdorff distance measures between binary images (typically edge maps) is proposed in this paper. The algorithm has a running time of $O(n)$ for images of size $n \times n$. Further, the algorithm has the following features: (i) simple arithmetic (ii) identical computations at each pixel and (iii) computations using a small neighborhood for each pixel. An efficient cellular architecture for implementing the proposed algorithm is presented. Results of implementation using field-programmable gate arrays show that the measures can be computed for approximately 88000 image pairs of size 128×128 in a second. This result is valuable for real-time applications like object tracking and video surveillance.

1 Introduction

Model-based recognition is an important problem in computer vision. This is the problem of locating an object, of which the computer has a model, in an image. A powerful method for model-based recognition is based on the *Hausdorff distance* [1, 2, 3, 4, 5, 6] between two images. Some real-time applications of Hausdorff distance are object tracking [4], video sequence matching [5] and face recognition-based video surveillance [6]. The distance measures the degree of mismatch between the images. Tolerance to the presence of outliers and missing feature pixels due to occulusion can be accomplished with suitable modifications [1].

A direct method of computation of Hausdorff distance $H(A, B)$ between a pair of binary images A and B, each of size $n \times n$, is based on finding the nearest foreground pixel in B to every foregound pixel in A and vice versa. The computation takes $O(n_A n_B)$ time where n_A and n_B are the number of foreground pixels of A and B (and in the worst case, $n_A = n_B = O(n^2)$). The computations of Hausdorff distance-related measures which this paper deals with have the same time complexity by direct method. Efficient sequential algorithms whose running time is linear in the size of the image have been developed [7, 8]. For large image sizes, further speedup is desirable.

D.A. Bader et al. (Eds.): HiPC 2005, LNCS 3769, pp. 180–189, 2005.
© Springer-Verlag Berlin Heidelberg 2005

Some parallel algorithms for Hausdorff distance computation are available in the literature [9, 10, 11, 12]. Implementations of these algorithms are on transputers [9], distributed systems [10], massively parallel computers such as MasPar [11] and SYMPHONIE [12]. The arrangements involve numerous general-purpose processors. The recent work on Hausdorff distance has been on its applications. To the best of our knowledge, a cost-effective application-specific architectural solution for its computation in real-time applications is not available in literature. This paper reports a new hardware algorithm to compute two Euclidean metric-based measures related to $H(A, B)$: (i) $H^2(A, B)$ and (ii) an integer approximation to $H(A, B)$. Both of these quantities take integer values (unlike $H(A, B)$) and they can be computed readily in hardware. The new algorithm can be easily mapped on to a cellular array of processors. Results of FPGA-based implementation of the architecture show that the hardware can process images much faster than the video rate and confirm the suitability for real-time image matching. Further, a speed-up by a factor in excess of 8000 over a C-based PC implementation is shown.

2   Definitions and Notations

Consider two binary images A and B (typically edge maps) each of size $n \times n$ and consisting of foreground (black) and background (white) pixels. The *Hausdorff distance* (based on Euclidean distance metric) between the images with foregrounds F_A and F_B is $H(A, B) = \max\{h(A, B), h(B, A)\}$ where $h(A, B) = \max_{p_A \in F_A} d(p_A, F_B) = \max_{p_A \in F_A} \min_{p_B \in F_B} d(p_A, p_B)$ $h(A, B)$ and $h(B, A)$ are the *directed Hausdorff distances* from A to B and B to A respectively, and $d(p_A, F_B)$ is the Euclidean distance between pixel p_A of A and the foreground F_B of B.

While $H(A, B)$ is a real number, $H^2(A, B)$ is an integer (since pixel coordinates are integers). The latter may be used for digital hardware based image matching. $H_i(A, B)$ may be expressed as $\max[h_i(A, B), h_i(B, A)]$ where $h_i(A, B)$ and $h_i(B, A)$ are integer approximations to the directed Hausdorff distances $h(A, B)$ and $h(B, A)$ respectively. The relationship between h_i and h is given by $(h_i(A, B) - 0.5) < h(A, B) \leq (h_i(A, B) + 0.5)$. Similar relationship exists between H_i and H. Since H_i differs from H only by ± 0.5 pixel unit, H_i is adequate for image matching applications. Other quantities used in this paper are as follows:

$b_A(p_0)$: binary value of pixel p_0 in image A

$d(p_0)$, $sqd(p_0)$: Euclidean distance and its square in pixel units between p_0 and its nearest foreground pixel

$(\Delta r(p_0), \Delta c(p_0))$: displacement vector pointing to the nearest object pixel

$nw(p_0)$: flag (set to 1 when p_0 is newly added to dilated image)

The neighbourhood of p_0, $N(p_0)$, consists of eight immediate neighbours surrounding p_0. If p_0 is a corner pixel in the image, then its neighbours inside the image grid are only of interest.

3   Main Ideas in Computation

We sketch the novelty in the computation of two measures here. The computation of these quantities is based on two subalgorithms to compute (i) $h_i(A, B)$ and $h^2(A, B)$, and (ii) $h_i(B, A)$ and $h^2(B, A)$. The procedure to compute $h_i(A, B)$ and $h^2(A, B)$ is described below. An analogous procedure applies to $h_i(B, A)$ and $h^2(B, A)$.

$h_i(A, B)$ is related to $h(A, B)$ and the following result on $h(A, B)$ is important for the development of our algorithm.

$h_i(A, B)$ is computed by dilating uniformly the foreground F_B of image B (For $h_i(B, A)$, dilation of F_A is performed). $h_i(A, B)$ is initialized to 0 and dilation is performed iteratively. At each iteration, the foreground is dilated by one pixel unit and $h_i(A, B)$ is incremented by 1. The procedure terminates when the dilated F_B fully covers F_A upon placing image B over image A. The dilation based on city-block and chessboard metrics can be done recursively by considering a small neighbourhood for each pixel. However, Euclidean metric involves nonlinear operations that are hard to decompose into local neighbourhood operations for recursive dilation. Here, we propose an algorithm for Euclidean metric-based dilation involving only local neighbourhood operations on each pixel.

A number of quantities related to $h_i(A, B)$ need to be initialized and updated during the iterative process. The definition of $h_i(A, B)$ suggests use of $d(p_0)$.

The algorithm for dilation sets $(\Delta r, \Delta c)$ of pixels in F_B to $(0, 0)$ and then iteratively computes those of others. At each iteration k, there are certain background pixels p_0 near the boundary of the 'current' dilated foreground whose d lies between $k - 0.5$ and $k + 0.5$. These pixels get added and form the new boundary of the dilated foreground. This can be stated in terms of the integer $sqd(p_0)$ as the set of background pixels that satisfy $(k^2 - k) < sqd(p_0) \leq (k^2 + k)$. $sqd(p_0)$ is computed using $(\Delta r, \Delta c)$ of neighbours $p_i \in N(p_0)$ belonging to dilated F_B. Each $(\Delta r(p_i), \Delta c(p_i))$ gives rise to an estimate $(\Delta r_i, \Delta c_i)$ for $(\Delta r(p_0), \Delta c(p_0))$. $\Delta r_i = \Delta r(p_i) + 1$ if p_0 and p_i lie in different rows and $\Delta r_i = \Delta r(p_i)$ otherwise. Similarly, $\Delta c_i = \Delta c(p_i) + 1$ if they lie in different columns and $\Delta c_i = \Delta c(p_i)$ otherwise. $(\Delta r(p_0), \Delta c(p_0))$ takes the value of $(\Delta r_i, \Delta c_i)$ that yield the minimum in Equation (1).

$$sqd(p_0) = \min[\Delta r_i{}^2 + \Delta c_i{}^2] \tag{1}$$

Since we are simply comparing sqd with $(k^2 + k)$ to determine 'new dilated foreground' pixels, we can define a quantity called α as $(k^2 + k) - sqd(p_0)$. It is worth noting that $\alpha(p_0)$ can be expressed in terms of $\alpha(p_i)$, $\Delta r(p_i)$ and $\Delta c(p_i)$ without explicitly computing $sqd(p_0)$ or k^2. Expanding Δr_i^2 and Δc_i^2 in terms of $\Delta r(p_i)$ and $\Delta c(p_i)$ in Equation (1), $sqd(p_0)$ can be expressed as $sqd(p_i) + \beta r_i + \beta c_i$ where $\beta r_i = 2\Delta r(p_i) + 1$ if p_0 and p_i lie in different rows and $\beta r_i = 0$ otherwise. Similarly, $\beta c_i = 2\Delta c(p_i) + 1$ if they lie in different columns and $\beta c_i = 0$ otherwise. Substituting this expression for $sqd(p_0)$ in $\alpha(p_0)$, we have

$$\alpha(p_0) = \max_{p_i \in N(p_0)} [\alpha_i] \tag{2}$$

where $\alpha_i = \alpha(p_i) - \beta r_i - \beta c_i$.

A central feature of the proposed algorithm is that each α is only $O(\log n)$ and further, $\alpha(p_0)$ (for background pixels $p_0 \in B$) is obtained at a given iteration incrementally using Δr, Δc and α of neighbours p_i (obtained in previous iteration) belonging to dilated F_B. If $\alpha(p_0) \geq 0$, p_0 will get added to form the new dilated F_B and

$$(\Delta r(p_0), \Delta c(p_0)) = (\Delta r_i, \Delta c_i) \tag{3}$$

where i corresponds to neighbour p_i that satisfies Equation (2).

Once $(\Delta r, \Delta c)$ of a pixel is known, its α needs to be updated for two successive iterations as it depends on k. $\alpha^{(k)}(p_0)$ at iteration k is expressed in terms of $\alpha^{(k-1)}(p_0)$ at iteration $k - 1$ as follows:

$$\alpha^{(k)}(p_0) = 2k + \alpha^{(k-1)}(p_0) \tag{4}$$

can be obtained using $\alpha^{(k)}(p_0) = k^2 + k - sqd(p_0)$ and $\alpha^{(k-1)}(p_0) = (k-1)^2 + (k-1) - sqd(p_0)$. $h_i(A, B)$ is the final value of k while $h^2(A, B)$ is now computed using $h_i(A, B)$ and α values of $p_0 \in F_A$ that lie on the boundary of the final dilated F_B. The farthest pixel from F_B is one among these pixels. The minimum α corresponds to the maximum sqd and hence $h^2(A, B)$ is given by

$$h^2(A, B) = h_i^2(A, B) + h_i(A, B) - \min_{p_0 \in F_A} [\alpha(p_0)] \tag{5}$$

4   Proposed Algorithm

The algorithm initializes Δr, Δc and α to 0 for all pixels. Three quantities, namely b_A, b_B and nw, are introduced to facilitate hardware implementation of the algorithm. $b_A(p_0)$ and $b_B(p_0)$ represent the binary values of p_0 in images A and B respectively. $b_A(p_0)$ is initialized to 1 for $p_0 \in F_A$ and to 0 for others. $b_B(p_0)$ is initialized similarly. nw is a flag assigned to each pixel for keeping track of the boundary of dilated F_B. nw is initialized to 0 for all pixels.

ALGORITHM HD_AtoB
Inputs: Images A and B.
Outputs: $h_i(A, B)$ and $h^2(A, B)$.
Step 1: Computation of $h_i(A, B)$

   repeat iteration {dilation process}
     for all pixels p_0 **do in parallel**
    switch $(b_B(p_0))$
    1: { pixel p_0 is in dilated F_B }
       Update $\alpha(p_0)$ as in Equation (4)
       $b_A(p_0)$=0 { $b_A(p_0)$ is set to 0 since $p_0 \in F_B$ }
       $nw(p_0)$=0 {p_0 is not new in dilated F_B}
       break { switch $b_B(p_0) = 1$ }

```
0: { check whether p_0 belongs to dilated boundary of F_B }
   if there exists p_I ∈ N(p_0) such that b_B(p_I) = 1 do
      Compute α(p_0) as in Equation (2)
      if (α(p_0) ≥ 0) do
         Compute (Δr(p_0), Δc(p_0)) as in Equation (3)
         b_B(p_0) = 1
         b_A(p_0) = 0
         nw(p_0) = 1 {p_0 is newly added to dilated F_B}
      end if
   end if
   break { switch b_B(p_0) = 0 }
   end for
   k = k + 1
until b_A(p_0) = 0 for all pixels
h_i(A, B) = k
```

Step 2: Computation of $h^2(A, B)$

```
{Filter valid α values}
for all pixels p_0 do in parallel
   if not[p_0 ∈ F_A and nw(p_0) = 1] do
      α(p_0) = MAX {maximum value assigned to invalid α. MAX can be n^2 }
   end if
end for
{Let α values of pixels be represented by α(1), α(2), ..., α(n²). Assume n² to be power of 2.}
for l = 1 to log_2 n² do {l: a level in the tree }
   for i = 1 to n²/2^l do in parallel {i: a node in a level}
      α(i) = min[α(2i), α(2i − 1)]
   end for
end for
h²(A, B) = h_i²(A, B) + h_i(A, B) − α(1) {α(1) has the minimum }
```

The computation of $h^2(A, B)$ relies on $h_i(A, B)$ obtained in Step 1. A procedure similar to the one given will output $h_i(B, A)$ and $h^2(B, A)$. $H_i(A, B)$ is the maximum of $h_i(A, B)$ and $h_i(B, A)$ while $H^2(A, B)$ is the maximum of $h^2(A, B)$ and $h^2(B, A)$.

4.1   Complexity Analysis

The time complexity purely depends on the number of iterations to compute $h_i(A, B)$ and the height of the tree in the computation of $h^2(A, B)$. Since the dilation for the computation of $h_i(A, B)$ is based on Euclidean distance, the dilation distance cannot be greater than the length of the diagonal of the image grid. For an $n \times n$ image, this length is $\sqrt{2}n$. The number of iterations is directly proportional to the dilation distance and hence $O(n)$ iterations are required for dilation. The height of tree is of $O(\log n)$. The time complexity of the entire algorithm is therefore $O(n)$. The space complexity of the algorithm is $O(n^2 \log n)$ since at most $O(\log n)$ space is required for storing $\Delta r, \Delta c$ and α for a pixel (Entities such as nw, b_A and b_B require constant space).

5   VLSI Architecture

The proposed algorithm is amenable for mapping on to a hardware consisting of (i) a cellular architecture that implements Step 1 of algorithm and (ii) an additional logic (h^2AdlLogic) for $h^2(A, B)$ that implements Step 2.

5.1   Cellular Architecture

The architecture has $n \times n$ array of identical cells. Each cell is connected to the eight neighbouring cells surrounding it. A cell consists of storage elements Δr, Δc, α, b_A and b_B, and logic circuits for computing them. The iteration number k is generated by an external counter and the cells are updated synchronously with respect to an external clock. The speed of operation of the cellular architecture depends primarily on the delay due to a cell.

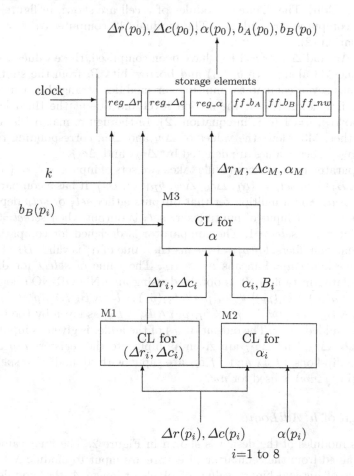

Fig. 1. Different modules of a cell. CL denotes 'combinational logic'.

Table 1. Value of *select* for different cases of inputs to comparator-multiplexer module. "-" indicates "don't care".

$b_B(p_j)$	$b_B(p_k)$	B_j	B_k	cmp	$select$	comments
0	0	-	-	-	-	both α_j & α_k are invalid
1	0	-	-	-	0	α_j is only valid
0	1	-	-	-	1	α_k is only valid
1	1	0	0	1	0	both are valid and positive. $\alpha_j > \alpha_k$
1	1	0	0	0	1	both are valid and positive. $\alpha_k > \alpha_j$
1	1	1	1	1	0	both are valid and negative. $\alpha_j > \alpha_k$
1	1	1	1	0	1	both are valid and negative. $\alpha_k > \alpha_j$
1	1	0	1	-	0	α_j is positive. α_k is negative
1	1	1	0	-	1	α_j is negative. α_k is positive

Design of a Cell. The different modules of a cell are shown in figure 1. The module M1 computes Δr_i and Δc_i. The module M2 computes α_i, $i = 1$ to 8, given in equation (2).

Once α_i, Δr_i and Δc_i for $i=1$ to 8 have been computed, these values are given to the module M3 along with $b_B(p_i)$ and borrow bits B_i from the subtracters of M2. M3 has seven comparator-multiplexer modules arranged in three levels (four at the first level, two at the second level and one at the third level) to compute $\max[\alpha_i]$, $i = 1$ to 8, in equation (2). In the figure, $\max[\alpha_i]$ is denoted by α_M. Further, M3 allows the values of Δr_i and Δc_i, corresponding to the i that yields α_M. These values are denoted by Δr_M and Δc_M.

The comparator-multiplexer module takes two sets of inputs, $set_j = \{\alpha_j, \Delta r_j, \Delta c_j, b_B(p_j), B_j\}$ and $set_k = \{\alpha_k, \Delta r_k, \Delta c_k, b_B(p_k), B_k\}$. It has a comparator to compare $\alpha_j > \alpha_k$ and a multiplexer that outputs either set_j or set_k depending on the value of $select$ input of multiplexer. set_j is output when $select=0$ while set_k is output when $select=1$. The comparator is designed for comparing two unsigned binary numbers. $b_B(p_j)=1$ means the value of α_j is valid. $B_j=1$ means α_j is negative and $cmp=1$ means $\alpha_j > \alpha_k$. The value of $select$ for different cases is tabulated in table 1. It is obtained using an AND-OR-NOT logic that implements $[\overline{b_B(p_j)} \wedge \overline{b_B(p_k)}] \vee [b_B(p_j) \wedge b_B(p_k) \wedge \overline{B_j} \wedge \overline{B_k} \wedge \overline{cmp}] \vee [b_B(p_j) \wedge b_B(p_k) \wedge B_j \wedge B_k \wedge \overline{cmp}] \vee [b_B(p_j) \wedge b_B(p_k) \wedge B_j \wedge \overline{B_k}]$ as given by the table.

α_M is then added to $2k$. The output $\alpha(p_0)$ of the adder is given as input to the register reg_α. Δr_M and Δc_M are given as inputs to the registers $reg\_\Delta r$ and $reg\_\Delta c$. The flip-flops ff_b_A and ff_b_B are input with '0' and '1' respectively. A T flip-flop (ff_nw) is used for nw.

5.2   Design of $h^2 AdlLogic$

The various modules of the design is shown in Figure 2. The final values of α and nw obtained from the cellular architecture are input to module M1. Based on the values of nw and binary values of pixels of image A, the module filters valid α values and assigns a maximum value (say, MAX) to each invalid α.

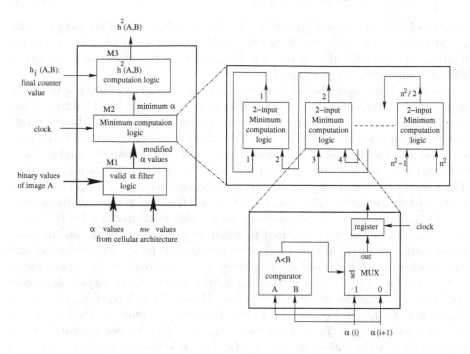

Fig. 2. Block diagram of the design for $h^2(A, B)$

MAX is the maximum possible value that reg_α in the cell can store. For each α corresponding to pixel p_0, a 2-input MUX whose inputs are MAX and α realizes the filter logic. The *select* input of MUX is generated with an AND-NOT logic whose inputs are $nw(p_0)$ and binary value of p_0 in A.

The modified values of α are input to module M2 where the minimum α is found. The binary-tree structured minimum finding is implemented with 2-input minimum computation logic blocks as shown in the figure. A sequential logic with $n^2/2$ such blocks whose outputs are fed back to the inputs of the first $n^2/4$ blocks realizes the module. Once the inputs are initialized with the modified values of α, the module finds the minimum when clocked. The output of the first block gives the minimum after $\log_2 n^2$ (height of tree) clock pulses.

The minimum α along with the final value ($h_i(A, B)$) of external counter (that generates iteration number k for the cellular architecture) are given to module M3 which computes $h^2(A, B)$. The module is realized with a multiplier, an adder and a subtracter.

5.3   Implementation and Performance

The design has been coded in VHDL (a hardware description language) and its functional behavior has been tested in ModelSim. The design has then been implemented in Xilinx FPGA. It has been mapped onto one of the target devices of Xilinx. An appropriate device has been chosen taking into consideration the number of logic blocks and pins available. The specifications of the target device are as follows - family:Virtex-E; device:xcv3200e; package:fg1156; speed grade:-6. After mapping the design on to a Xilinx device, placement and routing of FPGA components are performed.

Since a cellular architecture consists of locally connected identical cells, the delay due to interconnections is negligible. Hence, the time (T_{hi}) taken to compute $h_i(A, B)$ can be obtained from the maximum clock rate (f_c) obtained from the implementation of a cell. The number of clock pulses (n_{hi}) needed for computation equals the number of iterations taken which is less than or equal to d_{max}, the maximum possible distance value. That is, $n_{hi} \leq d_{max}$, where $d_{max} = \sqrt{2}n$. The period (T_c) of a clock pulse is $1/f_c$ seconds and hence $T_{hi} \leq d_{max} \times T_c$. The time to compute both $h_i(A, B)$ and $h_i(B, A)$ is $2T_{hi}$. The number of image pairs (N_{hi}) per second that can be processed on a cellular architecture is $\lceil 1/2T_{hi} \rceil$.

The additional time T_a required to compute $h^2(A, B)$ depends on the maximum path delays $(T_{M1}$ and $T_{M3})$ of combinational logic of modules M1 and M3 as also on maximum clock frequency f_{M2} required for the operation of M2. The time (T_{M2}) taken for finding minimum in M2 is $\log_2 n^2/f_{M2}$ since the operation requires $\log_2 n^2$ clock pulses. Hence, $T_a = T_{M1} + T_{M2} + T_{M3} = T_{M1} + \log_2 n^2/f_{M2} + T_{M3}$. The net time T_h taken to compute $h^2(A, B)$ is less than or equal to $T_{hi} + T_a$. The number of image pairs N_h per second that can be processed by the entire design is $1/2T_h$. The percentage increase in time $(t_\%)$ for the computation of $h^2(A, B)$ due to $h^2 AdlLogic$ is $(T_a/T_{hi}) \times 100$.

Let G_c, $G_{\alpha M1}$, G_{bM2} and G_{M3} be respectively the equivalent gate count of Xilinx components consumed by the following designs: a cell, the logic that processes a single α in M1 of $h^2 AdlLogic$, a block in M2, and M3. The equivalent gate counts G_{hi} and G_a for $n \times n$ cellular architecture and $h^2 AdlLogic$ are estimated to be $n^2 G_c$ and $n^2 G_{\alpha M1} + (n^2/2)G_{bM2} + G_{M3}$. The percentage increase in gate count $(g_\%)$ for the computation of $h^2(A, B)$ is $(G_a/G_{hi}) \times 100$.

Some results for an image of size 128×128 are as follows. f_c, T_{M1}, f_{M2}, T_{M3}, G_c, $G_{\alpha M1}$, G_{bM2} and G_{M3} obtained from the implementation are 33 MHz, 8.6 ns, 159 MHz, 26.5 ns, 196, 66, 153 and 6800 respectively. $T_{hi} \leq 5.5$ μs and $N_{hi} \geq 91150$ while $T_h \leq 5.62$ μs and $N_h \geq 88920$. N_{hi} and N_h are much greater than the video rate, which is about 30 frames per second and hence the computation of Hausdorff distance on a cellular architecture-based hardware is well-suited for real-time applications. The equivalent gate counts G_{hi} and G_a are computed to be roughly 3.2 million and 2.3 million respectively. Each of the designs for $h_i(A, B)$ and $h^2 AdlLogic$ will therefore fit into one Virtex-E chip whose gate count is approximately 4.1 million. $t_\%$ is 2.2 % while $g_\%$ is 72 %. Comparing with the computation of $h_i(A, B)$, there is negligible increase in time but a moderate increase in gate count for the computation of $h^2(A, B)$.

6   Conclusions

In this paper, a new cellular architecture-directed parallel algorithm with a running time of $O(n)$ has been proposed to compute two measures, namely $H^2(A, B)$

and $H_i(A, B)$ between binary images A and B, each of size $n \times n$. $H_i(A, B)$ is adequate for matching applications as it differs from $H(A, B)$ by atmost ± 0.5 pixel unit only. $H(A, B)$ may be obtained from $H^2(A, B)$ in software when required. Features of the proposed algorithm are (i) simple arithmetic operations (ii) identical computations for each pixel and (iii) computations of a local nature at each iteration. Results of implementation show that the hardware-based Hausdorff distance computation can process images much faster than the video rate and is therefore appropriate for real-time image matching applications such as object detection and tracking in video and face detection for biometric surveillance system.

References

1. Huttenlocher, D., Klanderman, G., Rucklidge, W.: Comparing images using the Hausdorff distance. IEEE Transactions on Pattern Analysis and Machine Intelligence **15** (1993) 850–863
2. Sim, D.G., Kwon, O.K., Park, R.H.: Object matching algorithms using robust Hausdorff distance measures. IEEE Transactions on Image Processing **8** (1999) 425–429
3. Kwon, O.K., Sim, D.G., Park, R.H.: Robust Hausdorff distance matching algorithms using pyramidal structures. Pattern Recognition **34** (2001) 2005–2013
4. Ayala-Ramírez, V., Parra, C., Devy, M.: Active tracking based on hausdorff matching. In: ICPR. (2000) 4706–4709
5. Kim, S.H., Park, R.H.: An efficient algorithm for video sequence matching using the modified hausdorff distance and the directed divergence. IEEE Trans. Circuits Syst. Video Techn. **12** (2002) 592–596
6. Gao, Y.: Efficiently comparing face images using a modified hausdorff distance. IEE Proceedings - Vision, Image and Signal Processing **150** (2003) 346–350
7. Shonkwiler, R.: An image algorithm for computing the hausdorff distance efficiently in linear time. Information Processing Letters **30** (1989) 87–89
8. Shonkwiler, R.: Computing the hausdorff set distance in linear time for any l_p point distance. Information Processing Letters **38** (1991) 201–207
9. Bossomaier, T., Loeff, A.: Parallel computation of the Hausdorff distance between images. Parallel Computing **19** (1993) 1129–1140
10. You, J., Cohen, H., Zhu, W., Pissaloux, E.: A robust and real-time texture analysis system using a distributed workstation cluster. Proceedings of IEEE International Conference on Acoustics, Speech and Signal Processing **4** (1996) 2207–2210
11. Gualtieri, J., Moigne, J.L., Packer, C.: Distance between images. Proceedings of the fourth IEEE symposium on the frontiers of massively parallel computation (1992) 216–223
12. Azencott, R., Durbin, F., Paumard, J.: Robust recognition of buildings in compressed large aerial scenes. Proceedings of IEEE International Conference on Image Processing **2** (1996) 617–620

Sensor Selection Heuristic in Sensor Networks

Vaishali P. Sadaphal and Bijendra N. Jain

Department of Computer Science and Engineering,
Indian Institute of Technology Delhi,
Hauz Khas, New Delhi 110016, India
{vaishali, bnj}@cse.iitd.ernet.in

Abstract. We consider the problem of estimating the location of a moving target in a 2-D plane. In this paper, we focus attention on selecting an appropriate 3rd sensor, given two sensors, with a view to minimize the estimation error. Only the selected sensors need to measure distance to the target and communicate the same to the central "tracker". This minimizes bandwidth and energy consumed in measurement and communication while achieving near minimum estimation error. In this paper, we have proposed that the 3rd sensor be selected based on three measures viz. (a) collinearity, (b) deviation from the ideal direction in which the sensor should be selected, and (c) proximity of the sensor from the target. We assume that the measurements are subject to multiplicative error. Further, we use least square error estimation technique to estimate the target location. Simulation results show that using the proposed algorithm it is possible to achieve near minimum error in target location.

1 Introduction

We consider the problem of estimating the location of a moving target 'T' in a 2-D plane. The target is moving at varying speed and direction (see Figure 1). For the present, we assume that the target is not aware of its own location. Or, if it is aware of its location, then it does not share this information with any other device. In either case, we assume that it is possible for sensors, such as s^1 located at $[x_1, y_1]$, to "measure" the distance from/to the target, located at (x_0, y_0), and thereby estimate the location of target 'T'. Several methods for measuring distance between a sensor and the target are available. See [1], [2] for methods based on radio signal strength (RSSI) and [3], [4] for methods based on time difference of arrival (TDOA).

Irrespective of the method used to measure distance, all such measurements are subject to error. Two models have been studied, viz. (a) additive, and (b) multiplicative. In this paper, we confine ourselves to using the multiplicative model for errors. Let d_A^i and d_M^i be the actual and measured distance between target and the ith sensor. The multiplicative error in measurement by i^{th} sensor, $e_m^i = d_A^i - d_M^i = \gamma \frac{\epsilon_{meas}}{2} d_A^i$ where, ϵ_{meas} is a measure of the amount of measurement error, and $-1 \leq \gamma \leq 1$. That is, the error is uniformly distributed with endpoints $(1 - \frac{\epsilon_{meas}}{2})d_A^i$ and $(1 + \frac{\epsilon_{meas}}{2})d_A^i$, zero mean and standard deviation of $\sigma_m^i = \frac{1}{\sqrt{3}}(\frac{\epsilon_{meas}}{2})d_A^i$.

D.A. Bader et al. (Eds.): HiPC 2005, LNCS 3769, pp. 190–200, 2005.

Even though the distance between a given sensor and the target is known only with some error, it is possible to use distance measurements from 3 or more sensors to 'estimate' the location of the target assuming that (a) the location of the sensors is known to the central device responsible for estimating the location of the target, (b) the clocks of the sensors are synchronized so that the sensors "measure" distance at approximately the same time, and (c) the sensors are able to communicate their measurements to this central device, also referred to as the "tracker".

Since the target is moving and since a sensor must be within a certain distance from the target (before it can detect the presence of the target and measure distance), we assume that there are several sensors, $\{s^i\} = \Sigma$, spread across the 2-D plane. In fact, we assume that there are three or more sensors located in and around every point in the 2-D plane so that we can compute an estimate based on measurements from a subset of three sensors suitably selected to minimize estimation error. This approach also allows one to minimize communication overheads and conserve battery power available to sensors. Further, since the target is moving, the collection of sensors changes every time an estimate is required to be obtained. Specifically, we assume that as the target moves, if sensors $\{s^1, s^2, s^3\}$ have made measurements at time t_k, then at time t_{k+1}, we drop one of the sensors s^1, s^2, or s^3 and select a sensor s^4 suitably so as to minimize the error in estimated location of the target. Accordingly, this paper is about suitably selecting the 3rd sensor from a set of N_{k+1} sensors.

2   Mobile Target Tracking

We now outline the overall scheme for tracking the target as it moves in the 2-D plane.

Let $L_k = [x_k, y_k]$ be the actual location of the target at time t_k. Let $\hat{L}_{k-1} = [\hat{x}_{k-1}, \hat{y}_{k-1}]$ and $\hat{L}_k = [\hat{x}_k, \hat{y}_k]$ be the *estimated* location of the target at time t_{k-1} and t_k, respectively (see Figure 2). The latter estimate $\hat{L}_k$ is obtained using (a) an *a-priori* estimate of the target's location $\bar{L}_k = [\bar{x}_k, \bar{y}_k]$, and (b) measurements made at t_k by sensors σ_k^1, σ_k^2, and σ_k^3 located at λ_k^1, λ_k^2, and λ_k^3, respectively. The *a-priori* estimate $\bar{L}_k = [\bar{x}_k, \bar{y}_k]$ may be based on its estimated location at time instants t_{k-1} and t_{k-2}. (see Figure 2).

The new estimate of the location of the target at time t_{k+1} is obtained on the basis of new distance measurements d_{k+1}^1, d_{k+1}^2 from $\sigma_{k+1}^1, \sigma_{k+1}^2$ respectively, and d_{k+1}^3 from a newly selected sensor σ_{k+1}^3. Sensor σ_{k+1}^3 is selected assuming knowledge of *a-priori* estimate of the target location. The latter is an extrapolation of its location assuming that the average velocity during $[t_k, t_{k+1}]$ is the same as the average velocity during and $[t_k - 1, t_k]$. The new estimate of the location of the target at time t_{k+1} is obtained thus:

– *Step 1:* Given the estimated location $\hat{L}_k$ (based on measurements from σ_k^1, σ_k^2, and σ_k^3) and $\hat{L}_{k-1}$ at times t_k and t_{k-1}, respectively, compute an *a-priori* estimate

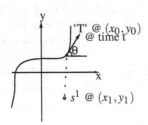

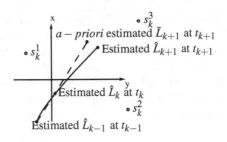

Fig. 1. Trajectory of the target 'T' in 2-D plane

Fig. 2. Estimated location of the target on the basis of *a-priori* estimate

$$\bar{L}_{k+1} = \alpha_{k+1}(\hat{L}_k - \hat{L}_{k-1})$$

where, $\alpha_{k+1} = \frac{t_{k+1}-t_k}{t_k-t_{k-1}}$. (See Figure 2.)

- *Step 2:* Given its approximate location $\bar{L}_{k+1}$, identify an appropriate subset of three sensors, viz. $\sigma_{k+1}^1, \sigma_{k+1}^2$, and σ_{k+1}^3 from a given subset of sensors $\Sigma_{k+1} \subset \Sigma$.
- *Step 3:* Sensors $\sigma_{k+1}^1, \sigma_{k+1}^2$, and σ_{k+1}^3 obtain distance measurements d_{k+1}^1, d_{k+1}^2, and d_{k+1}^3, respectively, and communicate the same to the central "tracker".
- *Step 4:* The "tracker" computes a least square error estimate, $\hat{L}_{k+1}$, such that $\sum_{i=1}^{i=3} e_i^2$ is minimized. Here, $e_i = \|\hat{L}_{k+1} - \lambda_{k+1}^i\|^{\frac{1}{2}} - d_{k+1}^i$, where $\|\hat{L}_{k+1} - \lambda_{k+1}^i\|$ is the Euclidean distance between $\hat{L}_{k+1}$ and $\lambda_{k+1}^i = [x_{k+1}^i, y_{k+1}^i]$, the location of sensor σ_{k+1}^i. This method is described in Section 4.

In Step 2, we stipulate that sensors σ_{k+1}^i are selected as follows: $\sigma_{k+1}^1 = \sigma_k^i, j = 1, 2,$ or 3; $\sigma_{k+1}^2 = \sigma_k^i, j = 1, 2,$ or 3; $\sigma_{k+1}^2 \neq \sigma_{k+1}^1$, $\sigma_{k+1}^3 \in \Sigma_{k+1}; \sigma_{k+1}^3 \neq \sigma_{k+1}^1, \sigma_{k+1}^3 \neq \sigma_{k+1}^2$. That is, one of the sensors used to estimate the location of the target at time k is replaced by another more suitable sensor from amongst the remaining set of sensors that are within the range of the target (see Sections 5 and 6).

Section 3 describes related work in the area of target tracking and sensor selection. Location estimation has been described in Section 4, while Section 5 and Section 6 describe the method and the algorithm for sensor selection, respectively. We present simulation results in Section 7. Section 8 concludes the paper.

3   Related Work

Research on estimating the location of a *fixed* target, given measurements from a subset of sensors has been reported in [1], [3], [5], [6], [7], [8], [9], [10], and [11]. These differ from each other on the basis of (a) the number of sensors required, (b) the nature of measurements, and (c) the technique for estimating the location.

- Priyantha et al [3] estimate target location using trilateration using distance measurements based on TDOA from 3 different sensors.
- Bahl et al [1] estimate target location using trilateration, but using RSSI measurements. Additionally, they build a radio map of the site and then locate the target based on radio signal strength measurements.
- Triangulation is used in robotics [11] to estimate the location of a robot. This requires 3 or more angle of arrival (AoA) measurements.

Tracking of mobile targets using sensor networks has been studied in [12], [13], [14].

- In [12] and [13] the target location is approximated by the location of a sensor when the target comes within its range. The resulting resolution is the same as that of the sensors. This may be, however, improved using multiple sensors.
- In [14] acoustic measurements from all sensors in the cluster around the target are used by a cluster head to minimize the error in estimating the location.

In this paper, and as described earlier in Section 2, a group of sensors in and around an *a-priori* estimated location of the target is selected. This group changes as and how the target moves within the 2-D plane.

Several papers (see [6], [7], [9], and [10]) have used different approaches to study selection of sensors.

- Zhao et al have proposed in [6] and [7] that sensors should be selected such that (a) communication overhead is minimized, and (b) the error in locating the target is minimized using a Bayesian maximum likelihood estimator. However, this assumes that an *a-priori* estimate of the location is available.
- Wang et al [9] also assume *a-priori* knowledge of the target location, expressed in the form of a Gaussian probability distribution function. The error in TDOA or AoA measurements is assumed to be Gaussian. While the estimate is based upon Bayesian filtering, the sensors selected are those that maximize the entropy difference between the *a-priori* and *posteriori* estimates of the target location.
- In Isler et al [10] the target location is estimated based on the region of intersection of 2-D cones resulting from uncertainty in AoA measurements from multiple sensors. Sensors which minimize the area of such intersection are the one selected. This scheme also requires an *a-priori* knowledge of the location of the target.

In this paper, we have proposed that a 3rd sensor be selected given two sensors based on three measures viz. (a) collinearity, (b) deviation from the ideal direction, and (c) proximity of the sensor from the target. We use least square error estimation technique to estimate the target location.

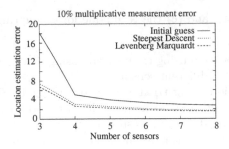

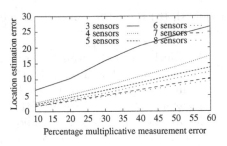

Fig. 3. Location estimation error in initial guess and in optimized result

Fig. 4. Measurement error vs. Location estimation error

4 Target Location Estimation

Before discussing selection of the 3rd sensor, we discuss the method to obtain a least square error estimate of the location at time t_k. For convenience, we drop the subscript k in t_k and instead work with time t. The estimation problem can be stated thus: given measurements d^1, d^2, ..., d^n from sensors s^1, s^2, ..., s^n, respectively located at $\lambda^1 = [x^1, y^1]$, $\lambda^2 = [x^2, y^2]$, ..., $\lambda^n = [x^n, y^n]$, compute an estimate $\hat{L} = [\hat{x}, \hat{y}]$ such that

$$\sum_{i=1}^{i=n}\{\sqrt{(\hat{x} - x^i)^2 + (\hat{y} - y^i)^2} - d^i\}^2$$

is minimum. The measurements are possibly subject to errors.

In order to compute the optimal $[\hat{x}, \hat{y}]$, we have experimented with two algorithms: (a) Steepest Descent algorithm [15], and with (b) Levenberg Marquardt algorithm [15] to compute the optimal $[\hat{x}, \hat{y}]$. In either case, the method requires an initial "guess". We have proposed that the initial estimate be obtained by solving $\binom{n}{2}$ linear equations[1].

Our experience (see also [16]) shows that (a) the above method for computing the initial guess is reasonably adequate in helping one to descend to the optimum (see Figure 3), (b) the error in location estimation increases (almost) linearly with error in measurement (see Figure 4), and (c) the number of iterations required in Steepest Descent method is generally less than 20, while the same is less than 6 in case of Levenberg Marquardt. In this paper, however, we use the Steepest Descent algorithm in all our simulations. (See Figure 3.)

5 Sensor Selection

The position of the 3rd sensor relative to that of the other two sensors and the target plays an important role in location estimation accuracy. In a simulation,

---

[1] These are obtained by subtracting equations of the type $\sqrt{(\hat{x} - x^i)^2 + (\hat{y} - y^i)^2} - d^i = 0$ from one another, thereby resulting in $\binom{n}{2}$ linear equations.

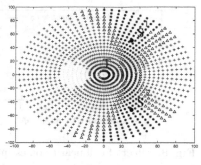
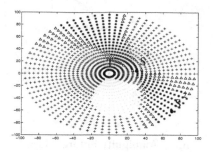

Fig. 5. $d^1 = d^2$, $\alpha > \frac{\pi}{2}$ **Fig. 6.** $d^1 \neq d^2$, $\alpha < \frac{\pi}{2}$

the 3rd sensor, s^3, was placed at various locations and the resulting error in location estimation was computed. The sensing range, r_0, was assumed to be 100m. The measurement error was assumed to be 30%, that is $\epsilon_{meas} = 0.3$. The target was assumed to be at $[0,0]$. The two sensor, s^1 and s^2, are placed at different positions such that (a) the visual angle made by them with the target, $\alpha > \frac{\pi}{2}$ or $\alpha < \frac{\pi}{2}$, and (b) $d^1 = d^2$ or $d^1 \neq d^2$.

Part of the results[2] are given in Figures 5 and 6. It can be seen that (a) when sensors s^1, s^2, and s^3 are almost collinear, the error in estimated location is large, (b) there is a preferred region in which to locate the 3rd sensor. This is the region where a dot, ".", is printed.

The sensor selection technique proposed in this paper is based on three factors, viz. (a) collinearity of sensors, (b) ideal direction in which 3rd sensor should be selected (for the given position of two sensors), and (c) proximity of selected sensor from the target. This selection is done before any new measurement is made, and assuming that the target is indeed located near the *a-priori* estimate of its location.

5.1 Collinearity

Consider the distribution of sensors s^1, s^2, and s^3 in a 2-D plane shown in Figure 7. If, for the moment, we assume that the error in distance measurements in near zero, then it can be concluded that the target is either located at position A, or at position B. Further, measurement from a third sensor does not add value

---

[2] A different symbol is printed for each range of estimation error. In particular, if μ is the average estimated error and ϵ_{loc} is the estimation error when the third sensor is placed in the specific position, then

- if $\epsilon_{loc} \geq \frac{3}{2}\mu$, then print "*", else
- if $\mu \leq \epsilon_{loc} < \frac{3}{2}\mu$, then print "$\triangle$", else
- if $\frac{1}{2}\mu \leq \epsilon_{loc} < \mu$, then print "+", else
- if $\epsilon_{loc} \leq \frac{1}{2}\mu$, then print "."

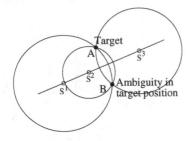
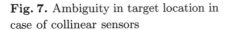
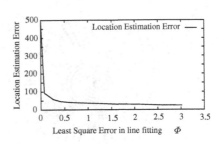

Fig. 7. Ambiguity in target location in case of collinear sensors

Fig. 8. Effect of collinearity on estimation error

since s^1, s^2 and s^3 are collinear. Distance measurement from a 3rd sensor which is not collinear is necessary to resolve whether the target is at A or B.

We define the collinearity coefficient Φ for an sensor s^i as the residual error resulting from a linear least square fit through the given two sensors and the sensor s^i:

$$\Phi(s^1, s^2, s^i) = min_{(m,c)}\{(y^1 - mx^1 - c)^2 + (y^2 - mx^2 - c)^2 + (y^i - mx^i - c)^2\} \quad (1)$$

Figure 8 shows a plot of estimation error vs. Φ, the residual. When Φ is small, the sensors are almost collinear and the estimation error is high. But if the collinearity coefficient is large, the location estimation error is likely to be small.

5.2   Ideal Direction

Below we argue that, given the positions of two sensors, there is a preferred direction in which to locate the 3rd sensor. Consider the region of intersection of the error annuli corresponding to the computed distance between the *a-priori* estimated location of the target and the two sensors (see Figure 9). Clearly, the region formed around the *a-priori* estimate of its location is the one in which the probability that the target is located is the maximum. For simplicity, this latter region is approximated by the parallelogram obtained by intersecting

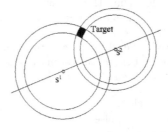

Fig. 9. Region of intersection of the error annuli

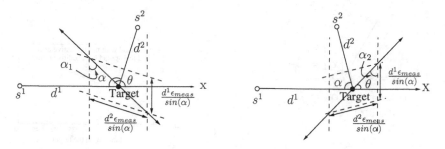

Fig. 10. Ideal direction in case $\alpha > \frac{\pi}{2}$ **Fig. 11.** Ideal direction in case $\alpha < \frac{\pi}{2}$

bands formed by tangents to the error annuli. This is shown in Figures 10 and 11 for the two cases viz. (a) $\alpha > \frac{\pi}{2}$ and (b) $\alpha < \frac{\pi}{2}$, respectively. Here α is the visual angle made by the two sensors with the target. Irrespective of whether $\alpha > \frac{\pi}{2}$ or $\alpha < \frac{\pi}{2}$, we propose that the 3rd sensor be placed along the larger diagonal of the parallelogram, since the resulting *maximum* uncertainty in the location of the target is minimized. This is discussed in detail below.

The ideal direction (see Figures 10 and 11), specified with respect to the axis passing from sensor s^1 to the target, is given by

$$\theta = \frac{3\pi}{2} + tan^{-1}\frac{d^2 sin(\alpha)}{d^1 + d^2 cos(\alpha)}, \text{ if } \alpha > \frac{\pi}{2}, \text{ and}$$

$$\theta = \frac{3\pi}{2} + \alpha - tan^{-1}\frac{d^1 sin(\alpha)}{d^2 - d^1 cos(\alpha)}, \text{ if } \alpha < \frac{\pi}{2}.$$

Selecting a sensor located approximately in the direction θ ensures that the resulting polygon of intersection after inclusion of measurement will have a smaller longest axis. A measure which captures the deviation from the ideal direction, and which may be used to select the 3rd sensor, is defined thus:

$$\Psi = |\theta_{s^i} - \theta| \qquad (2)$$

where θ_{s^i} is the direction of sensor s^i with respect to the axis from s^1 to the target.

5.3   Proximity of Sensors to Target

We now argue that the 3rd sensor is preferably placed as close as possible to the target. This is so only if measurement error is multiplicative.

From simulations (see Figure 13), it has been observed that the location estimation error increases almost linearly with the distance of the three sensors from the target. Since we have assumed a multiplicative error model, a measure of proximity Δ of sensor s^i from the target is defined as:

$$\Delta(s^i) = \|s^i - L\|, \qquad (3)$$

where $\|s^i - L\|$ is the Euclidean distance between s^i and L. Since the location of the target L is unknown, we use $\bar{L}$, the *a-priori* estimate of the location.

$$\Delta(s^i) = \|s^i - \bar{L}\|. \qquad (4)$$

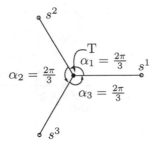

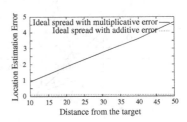

Fig. 12. Three sensors with equal visual angles α

Fig. 13. Effect of distance on location estimation error

6   Multi-objective Sensor Selection Algorithm

In this paper, we propose that for given sensors s^1 and s^2, a 3rd sensor s^3 is selected, from those sensors that are within the range of the target, have adequate battery power, and such that (a) the coefficient of collinearity is maximized, (b) the deviation of the direction of the sensor from the ideal direction is minimized, and (c) the distance of the sensor from the target is minimized. Since this is a multi-objective optimization problem, we propose the following algorithm:

- *Step 1:* Eliminate all sensors for which the collinearity coefficient $\Phi \leq \phi_0$.
- *Step 2:* Of the remaining sensors, consider only those for which deviation from the ideal direction $\Psi \leq \psi_0$.
- *Step 3:* Finally, of the remaining sensors, select the one for which the measure of proximity Δ is the minimum.

7   Simulation Results

Simulations were carried out with N sensors ($N = 10$, 20, or 30), each of which is randomly placed within 100m of the "actual" location of the target. The sensing range, r_0, is 100m. The target was assumed to be at $[0, 0]$, and the *a-priori* location of the target was also assumed to be $[0, 0]$. The measurement error is assumed to be 30%. Values of ψ_0 and ϕ_0 required in the above algorithm are specified thus: $\psi_0 = 30^\circ$, $\phi_0 = 625m^2$.

We have compared the results obtained using the proposed algorithm with results obtained using other criteria for selecting the 3rd sensor, viz. (a) proximity only, (b) collinearity only, and (c) deviation from ideal direction only (see Table 1). The error in estimated location based on our algorithm is also compared with the error resulting from selecting the "best" possible sensor. The latter is obtained by exhaustively computing the least square error estimate for each possible sensor.

It may be noted that using the proposed algorithm, we achieve near minimum error in estimated location. Further, as N increases, there is improvement in the

Table 1. Comparison of location estimation error with proposed algorithm and with other algorithms

N	Collinearity only	Proximity only	Ideal direction only	Best sensor	Proposed algorithm
10	11.7581	24.6160	15.2932	9.1745	9.4021
20	11.5150	19.4959	16.9380	8.2882	8.7331
30	10.2022	13.1000	17.4704	8.1151	8.5329

error in estimated location. However, the error in estimated location does not improve when (a) the 3rd sensor is selected based only on the ideal direction since the resulting three sensors may be collinear, in which case the estimation error may be large.

8   Conclusion

In this paper, we have proposed that the sensors be selected based on three measures viz. (a) collinearity, (b) ideal direction in which the sensor should be selected so that the error is minimized, and (c) proximity of the sensor from the target. We use measurements that are subject to multiplicative error. Further, we use least square error estimation technique to estimate the target location.

The sensor selection is done by the central "tracker" and only the selected sensors measure distance to the target and communicate to the central "tracker" for estimating the target location.

We propose that a 3rd sensor be selected in the ideal direction calculated on the basis of given two sensor positions. The knowledge of *a-priori* target position is assumed to be available. The results obtained with proposed algorithm are very very encouraging.

References

1. P. Bahl and V. N. Padmanabhan, "RADAR: An in-building RF-based user location and tracking system," in *In INFOCOM, TelAviv, Israel*, 2000.
2. D. Niculescu and B. Nath, "Ad Hoc Positioning System," in *In GLOBECOM, San Antonio*, 2001.
3. N.B.Priyantha, A.Chakraborty, and H.Balakrishnan, "The cricket location-support system," in *In ACM MOBICOM, Boston, MA*, 2000.
4. A.Savvides, C.C.Han, and M.Srivastava, "Dynamic fine-grained location-support system," in *In ACM MOBICOM*, 2001.
5. B. Parkinson and J. Spilker, "Global positioning system: Theory and application," in *American Institute of Astronautics and Aeronautics*, 1996.
6. F. Zhao, J. Liu, L. Guibas, and J. Reich, "Collaborative signal and information processing: An information directed approach," in *Proceedings of the IEEE, vol. 91, no. 8*, 2003.
7. J. Liu, J. Reich, and F. Zhao, "Collaborative in-network processing for target tracking," in *EURASIP JASP, vol. 4, no. 378-391*, 2003.

8. E. Ertin, J. W. Fisher, and L. C. Potter, "Maximum mutual information principle for dynamic sensor query problems," in *2nd International Workshop on Information Processing in Sensor Networks*, 2003.

9. H. Wang, K. Yao, G. Pottie, and D. Estrin, "Entropy-based sensor selection heuristic for target localization," in *IPSN 2004, Proceedings of the third international symposium on Information processing in sensor networks. ACM Press*, 2004.

10. V. Isler and R. Bajcsy, "The sensor selection problem for bounded uncertainty sensing models," in *Center for Information Technology Research in the Interest of Society*, 2003.

11. M.Betke and L.Gurvits, "Mobile robot localization using landmarks," in *Proceedings of the IEEE International Conference on Robotics and Automation, volume 2, pages 135-142*, 1994.

12. S.Chits, S.Sundresh, Y.Kwon, and G.Agha, "Cooperative tracking with binary-detection sensor networks," in *Technical Report UIUCDCS-R-2003-2379, Computer Science Dept., University of Illinois at Urbana-Champaign*, 2003.

13. J. Aslam, Z. Butler, V. Crepi, G. Cybenko, and D. Rus, "Tracking a moving object with a binary sensor network," in *ACM International Conference pn Embedded Networked Sensor Systems (SenSys)*, 2003.

14. Q.X.Wang, W.P.Chen, R.Zheng, K.Lee, and L.Sha, "Acoustic target tracking using tiny wireless sensor devices," in *International Workshop on Information Processing in Sensor Networks (ISPN)*, 2003.

15. W. H. Press, B. P. Flannery, S. A. Teukolsky, and W. T. Vetterling in *Numerical Recipes in C*, 2002.

16. V. P. Sadaphal and B. N. Jain, "Spread-based heuristics for sensor selection in sensor networks," Technical Report in preparation.

Mobile Pipelines: Parallelizing Left-Looking Algorithms Using Navigational Programming

Lei Pan[1,2], Ming Kin Lai[2], Michael B. Dillencourt[2], and Lubomir F. Bic[2,*]

[1] Jet Propulsion Laboratory, California Institute of Technology,
Pasadena, CA 91109-8099, USA
Lei.Pan@jpl.nasa.gov
[2] Donald Bren School of Information & Computer Sciences,
University of California, Irvine, CA 92697-3425, USA
{pan, mingl, dillenco, bic}@ics.uci.edu

Abstract. We consider the class of "left-looking" sequential matrix algorithms: consumer-driven algorithms that are characterized by "lazy" propagation of data. Left-looking algorithms are difficult to parallelize using the message-passing or distributed shared memory models because they only possess pipeline parallelism. We show that these algorithms can be directly parallelized using mobile pipelines provided by the Navigational Programming methodology. We present performance data demonstrating the effectiveness of our approach.

1 Introduction

In computational science, array-based algorithms (e.g., matrix factorization algorithms) are sometimes classified as "right-looking" or "left-looking" algorithms [1]. In both cases, the array is scanned from left to right. Right-looking algorithms are producer-driven: at each stage, the algorithm performs computations on the current element, and then immediately performs updates to the elements to the right of the current element. The fundamental data flow is eager propagation to the right, or scattering, as illustrated in Fig. 1(a). Left-looking algorithms, in contrast, are consumer-driven: at each stage, the algorithm updates the current element using previously computed values of elements to its left, after which the algorithm performs computations on the newly updated current element. Here the fundamental data flow is gathering previously computed data from the left, as illustrated in Fig. 1(b). Skeleton right-looking and left-looking algorithms are shown in Fig. 2(a) and Fig. 2(b), respectively. In a matrix application, each element x[i] would be a matrix column.

Right-looking algorithms are easy to parallelize directly. In each iteration of the outer loop in Fig. 2(a), the iterations of the inner loop are independent of one another and hence can be parallelized using data-parallel constructs such as *doall* or *forall*. In contrast, the inner loop in Fig. 2(b) carries dependence and hence admits only pipelined parallelism.

* The authors gratefully acknowledge the support of a U.S. Department of Education GAANN Fellowship.

D.A. Bader et al. (Eds.): HiPC 2005, LNCS 3769, pp. 201–212, 2005.
© Springer-Verlag Berlin Heidelberg 2005

Fig. 1. Data access patterns. (a) Right-looking (scattering); (b) Left-looking (gathering).

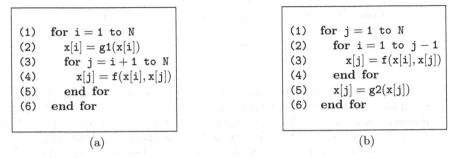

Fig. 2. (a) Right-looking code; (b) Left-looking code

A simple contrived left-looking algorithm is shown in Fig. 3(a). The jth iteration of the outer loop, which computes $a[j]$, requires the values of $a[i]$ computed by all the previous iterations. In this particular case, the algorithm can be parallelized by first transforming it by switching the order of loop nesting, in effect turning it to a right-looking algorithm. However, the general problem of converting a left-looking algorithm to an equivalent right-looking algorithm is not always completely straightforward. For example, even in the simple algorithm from Fig. 3(a), correctly reversing the loop order requires another modification, to the statement at line (5). A symbolic analysis technique for verifying the legality of program transformations is available, but there is no known transformation sequence to convert one to another [2].

Even if it were possible to automatically transform left-looking algorithms to their equivalent right-looking forms, this would not necessarily be a good approach to achieving parallelism. One reason is *algorithmic integrity* [3]: keeping the parallel implementation closer in structure to the original algorithm makes the parallel implementation easier to understand and hence to maintain or modify. Even a minor modification of a left-looking sequential algorithm may totally invalidate the corresponding parallel implementation if it relies on conversion to a right-looking algorithm [4]. A second reason is that it is sometimes useful to update only a portion of an array—for example, when refactorizing a portion of a matrix. In Fig. 2(a), this can be done simply by changing the bounds on the outer loop. In Fig. 2(b), the change is more subtle.

A third, very important, reason for not transforming a left-looking algorithm to a right-looking algorithm is performance: a sequential program may have been carefully crafted to make effective use of the cache or a particular data layout.

(1) **for** j = 2 **to** N	(1) **for** j = 2 **to** N
	(1.1) hop(node_map[j]); x ← a[1[j]]
(2) **for** i = 1 **to** j − 1	(2) **for** i = 1 **to** j − 1
	(2.1) hop(node_map[i])
(3) a[j] ← j * (a[j] + a[i])/(j + i)	(3) x ← j * (x + a[1[i]])/(j + i)
(4) **end for**	(4) **end for**
	(4.1) hop(node_map[j]); a[1[j]] ← x
(5) a[j] ← a[j]/j	(5) a[1[j]] ← a[1[j]]/j
(6) **end for**	(6) **end for**
(a)	(b)

Fig. 3. Pseudocode for a simple algorithm. (a) Sequential; (b) DSC using NavP.

The closer the parallel algorithm is to the sequential algorithm the more likely it is to preserve such performance enhancements that were in the original sequential code. Experimental evidence in [2] shows that converting between a left-looking and a right-looking algorithm can have a significant effect on performance.

In this paper we examine an alternative approach to parallelizing left-looking algorithms: rather than converting them to right-looking algorithms, we parallelize the original code directly, thus preserving the integrity of the original sequential algorithm. As we show, this can be done quite easily using the paradigm of Navigational Programming (NavP), in which multiple migrating threads carry out the computation. In this model, computations are programmed to migrate among the processors. They follow the locations of large-sized data, while carrying along small-sized data. The individual migrating computations generally follow each other, thus forming a **Mobile Pipeline**. Figure 4 illustrates the principle by comparing a conventional (stationary) pipeline with a mobile pipeline. In the figure, $C1$, $C2$, and $C3$ are computations, and a, b, c, d, and e are the data being computed. In a conventional pipeline, $C1$, $C2$, and $C3$ are stationary, whereas in a mobile pipeline they migrate. The essence of our NavP approach is to use Distributed Sequential Computing (DSC) [3] threads to construct mobile pipelines to exploit pipeline parallelism in the left-looking algorithms. The NavP view naturally describes efficient distributed algorithms, with regular or irregular communication patterns, using code that is structurally the same as the original sequential algorithm [5].

If we attempt to directly parallelize the code using either a Distributed Shared Memory (DSM) or Message Passing (MP) paradigm, we find that we either have to use considerably more memory—enough that the solution is no longer scalable—or asymptotically increase the communication cost. The reason why NavP is superior for this problem can be summarized as follows, in the context of Fig. 3(a). We can think of the computation of $a[j]$ as being a pipeline of $j − 1$ stages, with the ith stage being the incorporation of the value of $a[i]$. If each pipeline is stationary and remains on one processor, then each needed value of $a[i]$ must, at some point during the execution of the pipeline, be on that processor.

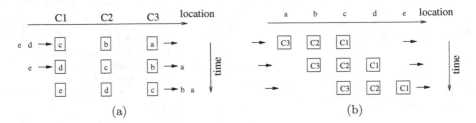

Fig. 4. The comparison of two pipelines. (a) Conventional; (b) Mobile.

If the values are stored there permanently, additional memory is required; if they are stored there temporarily, additional communication is required. The NavP solution, in contrast, avoids this problem by having a moving pipeline visit the necessary data, so that no element of the array needs to be replicated or re-communicated.

We describe our approach in more detail in Sect. 2. In Sect. 3, we discuss the results of applying the same pipelining technique to a numerical kernel, Crout factorization. We present performance data in Sect. 4, and we conclude by discussing some related work and some final remarks.

2   A Simple Example

In this section, we discuss and analyze the parallelization of the sequential algorithm introduced in Sect. 1. To make the discussion more concrete, assume that the parallel computation is being performed on P processors, each of which stores N/P array entries. In Sect. 2.1 we describe a NavP implementation that requires a communication cost of $O(N \cdot P)$ communications and $O(N/P)$ memory on each processor. In our full length technical report [4], we show that any direct parallel implementation of the sequential algorithm using either MP or DSM either requires $\Omega(N^2)$ communication cost or requires $\Omega(N)$ memory usage on at least one processor. The first case represents an asymptotic increase in communication cost, whereas the second essentially requires that the entire input array be stored on one processor, which is not a scalable solution if the number of processors is large.

2.1   NavP Solution

In NavP, we use multiple self-migrating threads to carry out computations for distributed parallel computing. We insert statements of the form $hop(dest\_node)$ into sequential codes to provide computation mobility. The threads carry data to remote nodes using thread-private variables, and they communicate with each other using shared node variables (stationary on a node, and shared by all threads that currently reside on that node). Concurrent self-migrating threads residing on the same node use events, with $signalEvent()$ and $waitEvent()$, to synchronize with each other. This is necessary because the daemon on each node uses multiple threads to handle communication and computation. NavP is essentially distributed concurrent shared variable programming. It provides a different view of parallel distributed computing [5] from the classical SPMD (Single Program Multiple Data) view.

Our NavP approach uses the MESSENGERS [6, 7, 8] migrating thread environment. This parallel execution environment is efficient because, as pointed out in our technical report [4], we do not need to move code, we keep the cost of book keeping small, and we use user level multithreading to efficiently schedule the migrating threads.

In the NavP approach, the parallelization of a given sequential algorithm proceeds in two steps. The first step is referred to as DSC (Distributed Sequential Computing) [3]. In this step, we start with a data distribution pattern, and insert *hop*() statements in the sequential code so that the computation follows the data it accesses through the network. The resulting DSC program is a distributed program, but with a single locus of computation.

Figure 3(b) lists the DSC code of the simple example. Three *hop*() and load/unload compound statements are inserted (at lines (1.1), (2.1), and (4.1)). Code structure is not changed. In the pseudocode, x is a thread-private variable that is available to the thread wherever it hops, and $a[.]$ is a distributed shared variable that is logically one big array but physically a distributed collection of sub-arrays. The auxiliary array *node_map*[.] provides the node ID of a given array entry, and $l[.]$ contains the local array index of an entry with a given global index. The DSC code works for arbitrary data distributions (e.g., block, cyclic, or block cyclic).

```
(1)   for j = 2 to N
```

```
(0.1)signalEvent(evt, 1)
(1)    for j = 2 to N
(1.2)   inject(entry_proc(j))
(6)   end for
```

```
(1.1)  hop(node_map[j]); x ← a[l[j]]
(2)     for i = 1 to j − 1
(2.1)    hop(node_map[i])

(3)       x ← j ∗ (x + a[l[i]])/(j + i)

(4)     end for
(4.1)  hop(node_map[j]); a[l[j]] ← x
(5)     a[l[j]] ← a[l[j]]/j
(6)   end for
```

```
(1.0)Thread entry_proc(j)
(1.1)   hop(node_map[j]); x ← a[l[j]]
(2)      for i = 1 to j − 1
(2.1)     hop(node_map[i])
(2.2)     if (i = 1) waitEvent(evt, j − 1)
(3)       x ← j ∗ (x + a[l[i]])/(j + i)
(3.1)     if (i = 1) signalEvent(evt, j)
(4)      end for
(4.1)   hop(node_map[j]); a[l[j]] ← x
(5)      a[l[j]] ← a[l[j]]/j
(5.1)end Thread
```

(a)   (b)

Fig. 5. The simple algorithm. (a) DSC using NavP; (b) Pipelining using NavP.

```
(1) for J = 1 to num_blocks
(1.1) hop(node_map[J]); load blk J to x[]
(2)    for I = 1 to J − 1
(2.1)    hop(node_map[I])
(3)      Call F(I, J, a, x)
(4)    end for
(4.1) hop(node_map[J])
(5)    Call F(J, J, x, x); unload x[] to blk J
(6) end for
```

```
(1) Procedure F(I, J, a, x)
(2) for j = start(J) to end(J)
(3)    for i = start(I) to min(end(I), j − 1)
(4)      x[1[j]] ← j ∗ (x[1[j]] + a[1[i]])/(j + i)
(5)    end for
(6)    if (I = J) a[1[j]] ← a[1[j]]/j
(7) end for
(8) end Procedure
```

(a)   (b)

Fig. 6. Block pseudocode for the simple algorithm. (a) DSC; (b) The function F().

The next step of NavP is called DPC (Distributed Parallel Computing). In this step, transformations are used to cut the long DSC computation thread into several shorter ones. Each of these threads are "pushed up" or scheduled to run as early as possible, subject to the constraint that all dependences must be respected. These threads spread out parallel computations as they hop out to the remote nodes on the network. The DPC implementation of the example is listed in Fig. 5(b). Each computation of j becomes a thread that is "injected" or spawned by another thread running the outer loop of j (lines (1), (1.2), and (6)). The code for each thread, lines (1.1) through (4.1), remains almost the same as the DSC code listed in Fig. 5(a). The only difference is the insertion of two new lines of event handling, to synchronize the accesses to the entry $a[1]$. Each thread waits at line (2.2) until the previous thread is done accessing $a[1]$, and at line (3.1) it notifies all other threads on the node that it has finished accessing $a[1]$. In this way, the threads organize themselves into a pipeline when they access $a[1]$: the thread computing $a[j]$ runs immediately after the thread computing $a[j − 1]$. Because MESSENGERS uses non-preemptive FIFO scheduling, and because threads hopping from the same source node to the same destination node preserve their ordering, the pipeline remains intact throughout the entire computation: migrating threads do not pass each other in the mobile pipeline. Each computation migrates through the pipeline, progressively visiting the successive stages (the elements $a[i]$ that it successively incorporates into its computation). Note that the code works correctly irrespective of how the array $a[.]$ is distributed.

There are three advantages to building a mobile pipeline: (1) In programming a DSC, we follow the principle of *pivot-computes*. This principle says that computation should occur on the node containing the largest amount of data to be used by the computation (the *pivot node*), so that a small amount of data is carried to meet with a large amount of data rather than the other way around. In the present example, this principle says that the computation of $a[j]$ should happen on the nodes that host the $a[i]$'s. As the computation of $a[j]$ proceeds, the pivot node changes. Assigning the computation of $a[j]$ statically to any single node would cost more than our DSC does because it requires more data communication. (2) We use concurrent threads to explore parallelism. For algorithms that exhibit pipelining opportunities, we simply insert multiple DSC threads to

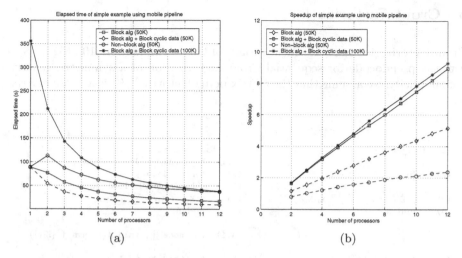

Fig. 7. Performance of the simple problem. (a) Elapsed time; (b) Speedup.

form a mobile pipeline, and synchronize them using events. Because threads are not allowed to access data remotely, all synchronization events are local to a node and hence efficient. (We note that when data parallelism is present, we can use multiple concurrent DSC threads to exploit this data parallelism [8], but this is not the focus of this paper.) (3) The NavP code as listed in Fig. 3(b) and Fig. 5(b) work for arbitrary data distribution. All that changes is the contents of the $node\_map[.]$ and $l[.]$ arrays. This provides considerable flexibility, because the programmer can experiment with different data distribution patterns using exactly the same code. For better performance, we can use a block algorithm (listed in Fig. 6(a)) so the granularity is coarse. Figure 6(b) shows the details of the block function $F()$ (called in lines (3) and (5) of Fig. 6(a)). The functions $start(I)$ and $end(I)$ return, respectively, the smallest and largest global indices of array entries stored in block I. The block pipeline code is similar to the code listed in Fig. 5(b) and therefore omitted. Transforming from the original sequential algorithm to the corresponding block algorithm can be automated using loop tiling techniques [9].

Asymptotically, if this algorithm is run on P processors, each processor will hold N/P array entries. The thread that starts on processor k (for $k = 1, \ldots, P$) will hop to processor 1, then processor 2, and so forth, ending up at processor k, for a total of k hops. On each hop it will carry N/P array entries. Hence the total communication costs of all the threads is $N/P \cdot \sum_{k=1}^{P} k$, which is $O(N \cdot P)$ as stated at the beginning of Sect. 2. Since on any particular processor, a thread hops away as the next thread is executing, the additional storage required on each processor is $O(N/P)$. We can further improve performance by using a block cyclic data distribution. This allows all the processors in the pipeline to get involved in the computation earlier and hence increases parallelism. As shown in Fig. 7, the block algorithm and block cyclic data distribution both help improve performance dramatically.

3   Crout Factorization

Crout factorization is a convenient variant of Gauss elimination [10]. Figure 8(a) lists the pseudocode for sequential Crout factorization of a symmetric matrix. For simplicity, we assume that the matrix being factorized, K, is a dense matrix.

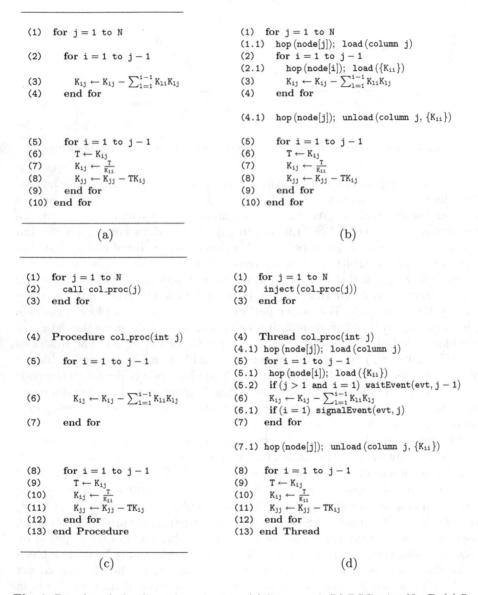

Fig. 8. Pseudocode for Crout factorization. (a) Sequential; (b) DSC using NavP; (c) Sequential re-written (with procedure call); (d) Pipelining using NavP.

This algorithm is left-looking because the updating of the jth column uses all the columns to its left from 1 to $j - 1$.

Figure 8(b) lists the pseudocode of DSC Crout factorization. Three $hop()$ and load/unload compound statements are inserted at lines (1.1), (2.1) and (4.1). The columns of matrix K are distributed to the nodes in a block fashion. Each column is the basic unit of data distribution and is hence indivisible. In the pseudocode, $node[.]$ provides the node ID of a given column. In the real code, the matrix K is implemented as a 1-D array, and the map $l[.,.]$ from a global index pair $[i, j]$ to a local 1-D index is needed. The detail is omitted in the pseudocode.

Figure 8(d) lists the pseudocode for a pipelined DPC implementation. This is compared side by side with the sequential code re-written with procedure calls, where the inner loop becomes a procedure, listed in Fig. 8(c). Each loop j is now assigned to a thread, and the outer loop becomes a "spawner" thread. In addition to the $hop()$ compound statements, two event handling statements are inserted at lines (5.2) and (6.1). Similar to the simple example in Sect. 2, we utilize the FIFO scheduling of MESSENGERS so that the event handling only happens on the node that hosts column 1 of K. This pipelined NavP code works correctly no matter how the columns are distributed. Similar to the simple example, we a use block cyclic column distribution to exploit parallelism.

4   Performance

Performance data for the simple example is presented in Fig. 7, and for Crout factorization is in Fig. 9 and Table 1. The data was obtained using a network of SUNW Ultra-60's with 450 MHz UltraSPARC-II CPU, 256MB of main memory,

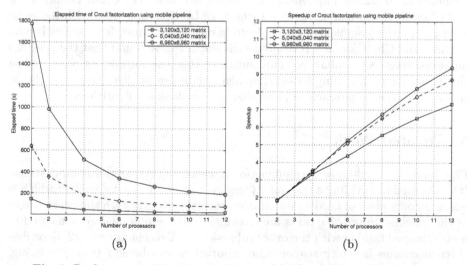

Fig. 9. Performance of Crout factorization. (a) Elapsed time; (b) Speedup.

Table 1. Performance of Crout factorization

Matrix Order	3120		5040		6960	
Num Proc	Time (s)	Speedup	Time (s)	Speedup	Time (s)	Speedup
			Sequential			
1	145.22	1.00	637.69	1.00	1770.74	1.00
			MESSENGERS			
2	78.33	1.85	351.79	1.81	980.00	1.81
4	43.51	3.34	180.15	3.54	510.81	3.47
6	33.16	4.38	125.01	5.10	336.39	5.26
8	26.05	5.57	97.63	6.53	262.48	6.75
10	22.29	6.52	82.42	7.74	215.69	8.21
12	19.80	7.33	73.21	8.71	188.63	9.39

1GB of virtual memory, 100Mbps of Ethernet connection with a collision-free switch, and using the NFS file-sharing system. To keep the presentation simple, we used non-block implementations of Crout algorithm in both the sequential and parallel versions of our algorithms. Thus, even though the sequential implementation is not the fastest possible, the speedup numbers relating our sequential and parallel implementations are based on a fair comparison, and they represent a good indication of the efficiency and scalability of our approach.

We were unable to find a parallel Crout implementation in literature, possibly because of the difficulty of parallelizing left-looking algorithms using conventional approaches. In [8], we compared our speedup numbers with those of the Cholesky factorization implementation in ScaLAPACK [11]. Crout factorization and Cholesky factorization are two variants of LU decomposition with the same asymptotic time complexity. Crout factorization on symmetric matrices is left-looking, and Cholesky factorization is right-looking. We found that the speedup numbers were very similar [8]. This indicates that the techniques presented in this paper for parallelizing left-looking algorithms are as effective as the classic (message-passing) approach to parallelizing right-looking algorithms.

5   Related Work

Pipelining is a well-known technique for parallelizing sequential computations. It is achieved by dividing a task into a sequence of smaller tasks, each of which is executed on a piece of hardware that operates concurrently with the other stages of the pipeline. Successive tasks are streamed into the pipe and get executed in an overlapped fashion with the other subtasks [1]. A recent survey [12] describes three situations in which sequential computations can benefit from pipelining. The examples discussed in this paper fall into the situation when "a series of data items must be processed, each requiring multiple operations." However, the method discussed in [12] for achieving parallelism using pipelining in situations of this type is not directly applicable to our examples. The reason is that this method assumes a regular data distribution with all data items initially residing on the first node initially. (Note that this second assumption is non-scalable.) In our examples, these assumptions do not hold, and once they are removed MP programming becomes significantly harder.

A classical pipeline is the segmentation of a functional unit into different parts, each of which is responsible for partial execution of an operation. It is similar to an assembly line process in a factory. In contrast, a mobile pipeline operates like farm work. The tasks (e.g., weeding, watering, or harvesting) are carried to the large data (the fields) by a mobile pipeline of equipment (e.g., tractors or harvesters) following each other. A mobile pipeline also carries small-sized data (e.g., seeds or fertilizer) that it needs when it carries out its operations.

6   Final Remarks

We have shown that NavP can be used to parallelize a class of sequential programs, namely left-looking programs, that are difficult to parallelize using conventional methods. Our approach can be used for a wide variety of data distributions and adapts automatically to changes in data distribution as long as we update the $node\_map[.]$ and $l[.]$ arrays which are byproducts of data distribution. This is useful for situations where the data distribution pattern is unknown at compile time (e.g. in Grid computing). Our method can be easily extended to algorithms that are neither left-looking nor right-looking (for example Crout factorization on a non-symmetric matrix [1]). This is important, because most algorithms are neither purely left-looking or purely right-looking.

The reason for the effectiveness of our approach can be summarized as follows: supply-driven "pushing" is easier than demand-driven "pulling." Right-looking (producer-driven) algorithms are easy to parallelize using conventional message-passing methods: when data is produced, it is propagated to the consumers, who consume it immediately. In contrast, left-looking (consumer-driven) algorithms based on movement of data require additional processing: once produced, any data that is not consumed immediately must either be replicated to multiple PEs and stored on each PE, or communicated multiple times. In our approach, even though the algorithm is consumer driven, the consumer process does not "pull" the data. Rather, it migrates (i.e., "pushes itself") to the data. This additional flexibility is a fundamental advantage of migrating computations and Navigational Programming over more conventional methods of distributed programming.

The NavP approach is highly mechanical: it requires insertion of $hop()$s and insertion of events. The former is based on data distribution, the latter on dependency analysis. Code transformations are incremental and code structures remain essentially the same throughout the process. These transformations could potentially be semi-automated or perhaps fully automated by a compiler. Investigating this potential is part of our future research.

References

1. Dongarra, J.J., Duff, I.S., Sorensen, D.C., van der Vorst, H.A.: Solving Linear Systems on Vector and Shared Memory Computers. Society for Industrial and Applied Mathematics, Philadelphia, Pa. (1991)
2. Menon, V., Pingali, K.: Look left, look right, look left again: An application of fractal symbolic analysis to linear algebra code restructuring. International Journal of Parallel Programming **32** (2004) 501–523
3. Pan, L., Bic, L.F., Dillencourt, M.B.: Distributed sequential computing using mobile code: Moving computation to data. In Ni, L.M., Valero, M., eds.: Proceedings of the 2001 International Conference on Parallel Processing (ICPP 2001), Los Alamitos, Calif., IEEE Computer Society (2001) 77–84
4. Pan, L., Lai, M.K., Dillencourt, M.B., Bic, L.F.: Mobile pipelines: Parallelizing left-looking algorithms using navigational programming. School of Information & Computer Sciences Technical Report TR# 05-12, University of California, Irvine, Irvine, Calif. (2005)
5. Pan, L., Bic, L.F., Dillencourt, M.B., Lai, M.K.: NavP versus SPMD: Two views of distributed computation. In Gonzalez, T., ed.: Proceedings of the Fifteenth IASTED International Conference on Parallel and Distributed Computing and Systems. Volume 2, Algorithms., Anaheim, Calif., ACTA Press (2003) 666–673
6. Department of Computer Science, University of California, Irvine Irvine, Calif.: MESSENGERS User's Manual (Version 1.2.04). (2005) http://www.ics.uci.edu/~messengr/messengersC/messman1_2_04.ps.
7. Wicke, C., Bic, L.F., Dillencourt, M.B., Fukuda, M.: Automatic state capture of self-migrating computations in MESSENGERS. In Rothermel, K., Hohl, F., eds.: Proceedings, Second International Workshop on Mobile Agents, MA '98. Volume 1477 of Lecture Notes in Computer Science., Berlin, Germany, Springer-Verlag (1998) 68–79
8. Pan, L., Lai, M.K., Noguchi, K., Huseynov, J.J., Bic, L.F., Dillencourt, M.B.: Distributed parallel computing using navigational programming. International Journal of Parallel Programming **32** (2004) 1–37
9. Xue, J.: Loop Tiling for Parallelism. Kluwer Academic Publishers, Boston (2000)
10. Hughes, T.J.R.: The Finite Element Method : Linear Static and Dynamic Finite Element Analysis. Prentice Hall, Englewood Cliffs, N.J. (1987)
11. Blackford, L.S., Choi, J., Cleary, A., D'Azevedo, E., Demmel, J., Dhillon, I., Dongarra, J., Hammarling, S., Henry, G., Petitet, A., Stanley, K., Walker, D., Whaley, R.C.: ScaLAPACK Users' Guide. Society for Industrial and Applied Mathematics, Philadelphia, Pa. (1997)
12. Wilkinson, B., Allen, M.: Parallel Programming: Techniques and Applications Using Networked Workstations and Parallel Computers. 2 edn. Pearson Prentice Hall, Upper Saddle River, N.J. (2005)

Distributed Point Rendering

Ramgopal Rajagopalan, Sushil Bhakar,
Dhrubajyoti Goswami, and Sudhir P. Mudur

Dept. of Computer Science and Software Engineering,
Concordia University, Montreal, Canada
{r_rajago, sushil, goswami, mudur}@cs.concordia.ca

Abstract. Traditionally graphics clusters have been employed in real-time visualization of large geometric models (many millions of 3D points). Data parallel approaches have been the obvious choices when it comes to breaking up the computations over multiple processors. In recent years, programmable graphics hardware has gained widespread acceptance. Today, every processing node in a graphics cluster has two powerful and fully programmable processors - a CPU (Central Processing Unit) and a GPU (Graphics processing unit). It enables distribution of graphics computations targeting an applications's needs in more flexible ways. In this paper we discuss and analyze our implementation of functionality distributed point-based rendering pipeline with impressive performance improvements. To the best of our knowledge, it is the first attempt to devise a functionality distribution scheme for a large data and compute-intensive application. We discuss the merits and limitations of such a distribution scheme by comparing it against traditional data parallel and single node schemes.

1 Introduction

With the advent of 3D scanners and other data capture devices, it has become relatively easy to capture large geometric models. The size of these models, typically many millions of points, poses a serious challenge for real-time rendering. A single node is normally not capable of delivering real-time frame rates when rendering such large models.

Computer graphics applications for visualizing large data-sets have employed many data parallel solutions in the past to achieve divide and conquer. But not much attention has been given to handling data and distribution with regards to functionality of the application as a whole. Traditional data parallel approaches perform data distribution without taking into account any knowledge of the functional complexity of the graphics processing. For example, a highly specialized point based graphics application and a simple polygon renderer of similar geometric complexity would both be distributed exactly the same way over a sort-first configuration (like Chromium [1], a stream processing framework that has popularly been used to implement graphics applications using the data parallel approach). Also most systems which employ data parallel schemes in graphics suffer from a scalability problem on modern day clusters [2]. Such systems could

D.A. Bader et al. (Eds.): HiPC 2005, LNCS 3769, pp. 213–224, 2005.
© Springer-Verlag Berlin Heidelberg 2005

scale better if they break up their functional complexity and suitably distribute functionality as well. Also it is important to exploit the application characteristics to overcome the memory, internal and external bandwidth limitations.

Yet another major development in computer graphics has been the advent of the specialized processors that accelerate 3D graphics computations through hardware implementation of a number of operations in the 3D graphics rendering pipeline. The more recent trend is a programmable graphics processing unit (GPU), which as the name indicates provides programmability and increases the flexibility to control the operations in the 3D graphics pipeline. The programmable GPU has made its mark in the mainstream processing and different non-graphics applications can now take advantage of its raw processing power (about 7 times a CPU in FLOPS) [3]).

We claim that modern day clusters with CPU and GPU processors can be programmed to provide efficient functionality driven distribution strategies. The graphics applications we are targeting are data intensive, demanding real-time rendering performance. Understanding the application domain is crucial to achieve a better distribution. We have chosen to demonstrate functionality distribution with a point based rendering application. The performance improvements are impressive when compared to implementations with only data parallel distribution.

The rest of the paper is organized as follows: Section 2 addresses previous work involving graphics clusters and functionality distribution. It also gives a brief background of point based rendering. Section 3 discusses the important aspects of a point based rendering pipeline. Section 4 discusses how we achieve a distributed point rendering pipeline. Section 5 gives the performance results and compares our configuration against a similar data parallel (sort-first) configuration. Section 6 concludes with a discussion of our plans for future work.

2   Previous Works

The 3D graphics pipeline consists of two principal parts - Object Space operations (transformations and lighting) and Image space operations (e.g. per fragment operations like texture mapping, visibility computations, blending, etc.). Geometric primitives (polygons, lines and points) are sent down this pipeline. The essence of the rendering pipeline is to calculate the effect of each primitive on each display pixel. Due to the arbitrary nature of the modelling and viewing transformations, a primitive can fall anywhere on (or off) the screen. Thus rendering can be also viewed as a problem of sorting primitives to the screen, as noted by Sutherland et al in [4].

2.1   Functionality Distribution in Graphics

In distributed graphics, any strategy involving distribution of graphics primitives over a cluster of nodes has to eventually address sorting them in depth order on the screen. Molnar et al. [5] classified the various possible schemes into three categories: sort-first, sort-middle and sort-last, depending on whether the sorting process takes place before the object space operations, between the object space operations and image space operations (just before rasterization), or after the image space operations respectively. Most of the traditional graphics cluster applications have been built around one of the previous categories.

These sort-based approaches can easily exploit hardware parallelism when the computation can be split into smaller equal weight sub-computations, with good data locality. Such a distribution is also called *regular*. Simple mapping and scheduling algorithms can be used to satisfactorily exploit the hardware parallelism. The same does not happen if the sub-computations do not have similar complexities or do not have good data locality properties. Parallel computations in this category are called *irregular* (e.g. numerical treatment of large sparse matrices). 3D graphics rendering pipeline computations could be viewed as a collection of varied complexity algorithms operating on different data structures. Functionality distribution could be used to create smaller regular sets of computations that best exploit the hardware and data locality.

For example, Govindaraju et al [6] demonstrate solutions to the classic visibility culling problem on a cluster of three GPUs. They generate an occlusion representation on one node, cull away occluded objects on another and render the geometry on a third different node. They further extend their distributed visibility culling technique to perform interactive shadow generation over a cluster [7]. The significant point to note is that in every frame, each of the three nodes above implement different algorithms (e.g. level of detail selection and occluder rendering, frustum culling and hierarchical Z-buffering, and rendering of visible nodes) which operate on specialized data structures (Z-buffer, scene graph and queue of visible nodes). Hence they manage to break the irregularity of the computations into smaller manageable regular sets of computations on three nodes.

Isard et al [8] perform distributed soft shadow rendering by programming every GPU on a cluster to calculate the contributions of a disjoint set of light sources for every object and finally compose the result over a Sepia 2a compositing network [9] for display. Heirich et al [10] demonstrate the need for parallelizing the iterative multi-grid solver routines of a CFD in order to visualize the pressure field of a developing steady-state solution. Zara et al [11] simulate cloth animation over a cluster of 100 nodes by exploiting data as well as task parallelism. Zhe et al [12] use a cluster of 30 GPU nodes to perform parallel flow simulation using the lattice Boltzmann model (LBM). Their application virtualizes the cluster as a 2D grid which facilitates communication sideways and diagonally. Kipfer et al [13] demonstrate a distributed lighting network by distributing radiosity, ray tracing and photon mapping.

All the above mentioned schemes exploit application characteristics like data locality and functional complexity to break the irregular computations into smaller regular sets of computations. They cannot be categorized purely on the basis of sorting classification.

While the approach in each of the above cases is largely that of providing a specific distributed architecture suited to the needs of the computations in a single application, in our research we have investigated distribution of computations and data in the more general setting of a graphics cluster to take into account the differential capabilities of CPUs and GPUs. It is accepted that optimal functionality distribution cannot take place without adequate knowledge of the graphics application under consideration. To that extent, we have been able to identify various architectures that can augment data parallel approaches with functionality distribution to provide distribution scalability [14]. We have also specified a set of system and application parameters to model the performance of a system which can help in making the appropriate choice for distribution in a given application [15].

2.2   Background of Point-Based Graphics

The use of points as a display primitive for continuous surfaces was introduced by Levoy and Whitted [16]. In 1989, Westover [17] introduced splatting for interactive volume rendering. In splatting, each projected voxel (a unit of volume data) is represented as a radially symmetric interpolation kernel (e.g. Gaussian) giving the appearance of a fuzzy ball. Grossman [18] investigated the use of point sampled representations as an alternative to triangles for rendering. One of the first point based rendering systems was QSplat [19], where in a multi-resolution hierarchy, based on bounding spheres, is employed for the representation and progressive visualization of large models. QSplat used an efficient quantized representation for each hierarchical node using a mere 48 bits. In the same year, Pfister and Zwicker introduced surfels [20] (a short form of surface element), a zero dimensional n-tuple that captures shape and color attributes locally approximating the object's surface at any given point. Zwicker et al. [21] introduce surface splatting wherein they render opaque and transparent surfaces from sampled point representations. Their approach is based on a screen space formulation of the Elliptical Weighted Average (EWA) filter [22]; Disc-shaped splats in object-space project to elliptical splats (stored as textures) with Gaussian intensity distribution in image-space. It results in high-quality anti-aliased rendering but the number of point samples rendered is less.

Of late, there has been a growing interest in using programmable graphics hardware to accelerate point rendering. Dachsbacher et al. [23] present a hierarchical LOD structure that is suitable for GPU implementation. They can process 50M low quality points per second.

Although extremely fast, a GPU's on-board memory is currently rather limited in terms of data storage. It is inevitable to employ a PC cluster for larger data models since a PC cluster provides a scalable memory model. Also for larger complex scenes a single GPU can not handle all the processing, so distribution is inevitable. In [24] Hubo and Bekaer have described a sort-first point rendering configuration and noted that in the absence of topological information, point-rendering is ideally suited for sort-first rendering. They demonstrate a peak performance of splatting 1.5 million points per second per node. In comparison, our functionality distributed pipeline splats over 3.5 million points per second per node (see Table 2).

3 Point Based Rendering

In this section we discuss the stages involved in point data processing with the help of a standard point rendering pipeline. Just as a good understanding of the computations involved in an application is essential for achieving optimal functionality distribution, the choices for representation, access mechanism, organization, and storage of graphics application data are important factors that can considerably affect the overall performance of a distribution scheme. A substantial part of our experimental investigation efforts have been devoted to the analysis of these factors in our point data processing application. We discuss these in brief below.

We broadly categorize the stages (Figure 1) involved in point based rendering into two major phases - selection and rendering. The selection phase consists of a set of view dependent algorithms which decide on the candidate points to be rendered to obtain a hole-free image (Figure 2). The rendering phase feeds the candidate points through the GPU for generating an image on the display. Each phase has multiple sub-stages, discussed in detail below.

3.1 Point Representation and Organization

A point may possess several attributes depending upon the application. For a simple watertight rendering we need coordinates, surface normal, neighbour

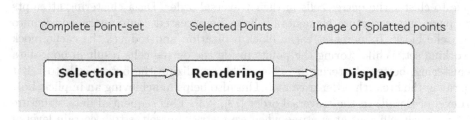

Complete Point-set   Selected Points   Image of Splatted points

Selection $\longrightarrow$ **Rendering** $\longrightarrow$ **Display**

Fig. 1. Two major phases of a point processing pipeline

Fig. 2. Illustration of hole filling in Stanford Buddha model. Holes in the rendering of a point sampled model are filled with increasing spat size (from left to right).

points, color, texture and splat size. Some of these properties could be represented as a dedicated structure, also referred to as a surfel [20], and others could be calculated at runtime: the decision as to which attributes are put in the surfel structure and which ones are to be calculated at runtime is an application dependent issue. An attribute, expensive to compute may be precomputed. However, with the growing relative memory latency, computing some of the desired point attributes at runtime may be beneficial as compared to storing them in memory and incurring an access cost at runtime [25]. We henceforth refer to the collection of surfels representing surface of a point sampled model as point-set.

We employ octrees to organize our point data-set, a well established data structure based on regular subdivision of the cube in 3D space. The simplicity and uniformity of the octree naturally lends itself to efficient queries and traversal. Also the construction time is minimal as compared to other space partitioning schemes. Figure 3 illustrates visually the recursive construction of an octree for a point-set.

Fig. 3. Octree construction of a point sampled model (Stanford Bunny)

Traversing the octree hierarchically adds overheads of the recursive function calls and pointer dereferencing. Further, to achieve better cache hit, we would like to cluster the octree cells in their traversal order. Data clustering attempts to pack data structure elements likely to be accessed contemporaneously into a cache block. It increases cache block utilization and reduces the cache block working set. While storing the points inside the octree cells result in poor data clustering, better clustering is achieved by serializing the octree cells to a flat array in the breadth order traversal. This also helps in achieving an implicit LoD (Level of Detail) in the traversal order [19], [23]. This sequential data structure is extremely efficient at runtime when we restrict ourselves to a certain level of the octree depending upon the results of the selection algorithms. Also as the different octree cells at the same level of detail are located contiguously it helps in data clustering. We call the sequential structure a Sequential Octree [15].

The performance gains achieved with a sequential octree can be seen by comparing rows 2 and 3 of Table 1 (30% over the hierarchical structure). However, we still end up with only one-third of the performance by rendering out of a flat unorganized point-array. (Compare rows 1 and 2 of Table 1.) This is largely due to the unavoidable cache thrashing when we toggle between the octree (for selection) and the point-set (for rendering) at runtime. We shall present an innovative scheme to alleviate this problem when we discuss functionality distribution in the section 4.

Table 1. Rendering performance readings for different point organization schemes

Stanford Bunny model rendered from:	Frames/Second
Flat unorganized point-set array	724.3
Seq. traversal of Seq. Octree (with point data clustering)	246.5
Hier. traversal of Seq. Octree (with point data clustering)	192.4
Hier. traversal of Hier. Octree (no point data clustering)	64.3

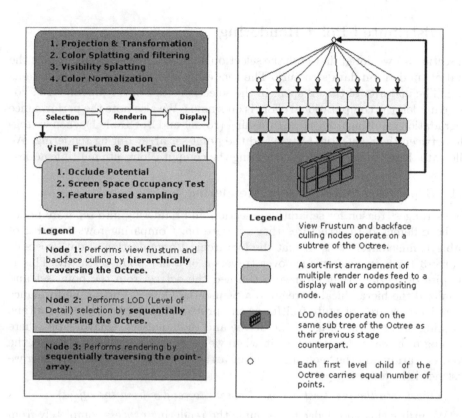

Fig. 4. Left Side: Distributed Point Rendering Pipeline. The legend outlines the assignment of the respective computations and data structures of the point rendering pipeline to the different nodes of the cluster. Right Side: Scaling the functionality distribution of point rendering for larger models by employing data parallelism in the individual stages of the pipeline.

3.2   Point Selection and Rendering

Selection algorithms (Figure 4) decide on the points which are splatted in a given rendering pass. This decision depends on a number of factors like features present in a region, screen space projection of an octree cell and visibility. Details of these algorithms [15] are not presented here as they do not form part of the core theme of this paper.

The final step in any graphics data processing is rendering. In our case it corresponds to pushing the selected points through the GPU to obtain an image. The selection algorithm picks a smaller subset of points for rendering. Figure 4 (Left side) captures the rendering tasks involved. Again for details the reader is referred to [15].

4   Distributed Point Rendering

In section 3.1 we noted that both the selection and rendering stages work with the spatial subdivision data structure. The rendering stage needs the point data as well. The majority of time is spent in the selection stage of the pipeline [19]. For large models the computational effort is enormous. We also note the performance degradation resulting from cache thrashing (rows 1 and 2 of Table 1) when we toggle between the Sequential Octree and the point-set array during rendering. We alleviate these bottlenecks by distributing the computations into multiple nodes.

4.1   Separating Selection from Rendering

The strongest reason for separating selection and rendering into separate nodes is the different data structures they operate on. Comparing rows 1 and 2 of Table 1 immediately tells us that the hierarchical organization of data costs us nearly 3 times in performance over traversal of a flat point-set array. This is in spite of the fact that we have decoupled the octree from the point-set and serialized the hierarchical octree into a Sequential Octree. Further performance optimizations are not quite fruitful as the problem lies in the inherent runtime coupling between the usage of the octree and the point-set. Unavoidably, there is going to be cache thrashing again when we toggle back and forth between the point-set array and the Sequential Octree array. We notice that for rendering we just need the following pair:

{ *Offset into the point-set array, # points to render from the offset*}

We utilize this knowledge to decouple the rendering process completely from the octree. Hence we perform rendering from the point-array on a separate node. In fact the renderer node works just with the point-array. After constructing the point-array during the preprocessing phase, it destroys the octree hierarchy as it no longer needs it. For every frame, the selection nodes send across the aforementioned pair in a network optimized packet. We notice a performance improvement of more than 2 times employing this clever strategy of separating the selection and rendering phases on separate nodes. Further, the selection nodes destroy the point-array after the construction of the octree, thus releasing memory.

4.2   Distributing the Computations

Figure 4 (Left side) illustrates how the computations are split among different nodes of a cluster. The nodes are connected in a pipelined fashion. Each node operates concurrently on an incremental frame. This means when node 3 is rendering frame f, node 2 is computing $f + 1$ and node 3 is working on $f + 2$. Although this causes a 2-frame delay in the rendering node, we observe that the benefit we achieve in overall frame rate significantly offsets this delay. The point-set data is replicated on each node to avoid expensive geometry data distribution at runtime.

4.3 Incorporating Data Parallelism

As the size of the 3D models grow, incorporation of data parallel approaches become necessary. Each of the selection stages operating on the octree is split into multiple nodes as shown in Figure 4 (Right side). The first node of our distribution performs visibility culling. It traverses the octree hierarchically and generates visibility culling information for each Octree cell (whether the Octree cell and its contents are visible from the given viewpoint). To distribute the task of this node different sub-trees of the octree are assigned to multiple nodes. Depending upon the data size we distribute the first level of the octree cells or the second (the latter case is rare). In Figure 4 (Right side) the visibility culling calculations on the first level sub-trees are distributed.

To achieve a fair distribution, on application instancing the first level octree subdivision is created such that each octree cell receives equal number of points. As visibility calculations are view dependent runtime load balancing will give better results.

The results from each visibility node are fed into a point sampling node. The point sampling node works with a sequential version of the same sub-tree as the previous visibility culling node. It calculates the point subsets that need to be rendered for the given sub-tree. The point subsets are represented as an offset into the point-set array and a count of the number of points to be rendered from therein. This information is fed to the next stage for rendering. It should be noted that the same data structures are available on each node as the point-set data is replicated on each node at load time. This replication is done to avoid expensive geometry data transmission at run-time.

The rendering stage can be arranged in a sort-first configuration [1], [2] (refer to section 2.1). The images generated by the render nodes are fed to a display wall or can be composed into a single node.

5   Implementation and Results

In this section we compare the performance of our functionality distribution with a sort-first only implementation. We also outline the minimal model sizes needed to achieve performance benefits from the distribution by comparing the results with a single node implementation. We use a cluster of 3 nodes; each node is a Pentium-4 2.8 GHz, 512 K L2 Cache, 1 GB RAM with ATI Raedon 9800 128 MB graphics card, Catalyst 4.2 driver running Windows 2000. The backbone is supported by 100Mbps Ethernet connectivity. We have employed MPI over the cluster for communication among the nodes.

Table 2. Rendering performance in frames per second for different data models over a single node renderer, functionality distributed renderer and sort-first renderer

Model	Size (#Points)	Octree (#Cells)	Single Node FPS	Functionality Distr. FPS	Sort-First Distr. FPS
G Tech Blade	882,954	65,209	8.0	12.5	3.5
Stanford Dragon	437,645	32,347	5.03	11.60	4
Stanford Buddha	543,652	38,351	9.8	20.0	NA
Stanford Bunny	35,947	12,626	83	70	NA

Table 2 shows the frame rates achieved from rendering models of different complexities. It compares the performance of a point rendering pipeline for a single node, functionality distributed and a sort-first rendering respectively. For the sort-first rendering we use the recently reported results from a similar configuration [24]. We note (from rows 1 and 2 of Table 2) that our optimized functionality aware distribution gives over three times the performance benefit over a traditional sort-first configuration [24].

We obtain twice the performance over a single node with functionality distribution. As the model size reduces the benefit of a performance distribution is offset by the overhead in communication. As we haven't employed out of core strategies so our models are relatively small (less than a million points as reported in Table 2).

Another interesting point to note is the maximum number of points allowed in the leaf node of an Octree. The smaller this number is the greater the height of the octree, which might be needed to get a better sampling of a region on the surface. But it results in larger data packets getting transmitted over the network from the point sampling node to the renderer node per frame. We chose 10-20 points as an optimal tradeoff for our experiments. We find that it gives us good image quality as well adequate rendering speed. The single node performance drops too as the height of the hierarchical octree increases. It must be recalled that we have to perform a hierarchical traversal on the single node because the visibility calculation makes it mandatory.

6   Conclusion

Today's GPU can outperform the CPU in most workstations by a factor of seven or more. The programmability that has been introduced in GPUs, a recent trend in graphics hardware, now makes it possible to offload application specific computational functionality to the GPU, thus enabling functionality distribution among the CPUs and GPUs available in a graphics cluster. The focus of our research has been to study this type of functionality distribution for large graphics applications. While there have been quite a few attempts to program GPUs with special algorithms, ours is the first research investigation that has tried to address the problem of functionality distribution in a more general setting of a graphics cluster. It is this investigation that has led us to formulate our conclusion that functionality distribution achieved by programming multiple GPUs combined in effective ways with traditional sort based data parallel approaches provide high scalability to an existing data parallel scheme.

Functionality distribution becomes advantageous primarily due to the flexibility provided by programming the GPUs of a cluster and by organizing data for better cache hit. In section 4.2 we analyze these issues for our point based rendering application and present a simple sequential organization, which demonstrates very well the advantages of a cache-conscious organization.

The effectiveness of functionality distribution is aptly demonstrated in Sections 4 and 5, wherein we show that with just 3 nodes we clearly outperform a sort-first configuration by a factor greater than 3. In future we intend to extend our implementation to incorporate the data parallelism discussed in section 4.3 and develop out of core strategies.

Acknowledgement

This research has been supported by Natural Sciences and Engineering Research Council (NSERC) of Canada and Concordia University Faculty Research grants.

References

1. Humphreys, G., Houston, M., Ng, R., Frank, R., Ahern, S., Kirchner, P.D., Klosowski, J.T.: Chromium: a stream processing framework for interactive rendering on clusters. In: SIGGRAPH. (2002) 693–702
2. Muller, C.: The Sort-First Rendering Architecture for High-Performance Graphics. In: Symposium on Interactive 3D Graphics. (1995)
3. Buck, I., Foley, T., Horn, D., Sugerman, J., Fatahalian, K., Houston, M., Hanrahan, P.: Brook for GPUs: Stream Computing on Graphics Hardware. In: SIGGRAPH. (2004)
4. Sutherland, I.E., Sproull, R.F., Schumacker, R.A.: A Characterization of Ten Hidden Surface Algorithms. ACM Computing Surveys 6 (1974) 1–55
5. Molnar, S., Cox, M., Ellsworth, D., Fuchs, H.: A Sorting Classification of Parallel Rendering. IEEE Computer Graphics and Algorithms (1994) 23–32
6. Govindaraju, N.K., Sud, A., Yoon, S.E., Manocha, D.: Interactive visibility culling in complex environments using occlusion-switches. In: Symposium on Interactive 3D Graphics. (2003) 103–112
7. Govindaraju, N.K., Lloyd, B., Yoon, S., Sud, A., Manocha, D.: Interactive Shadow Generation in Complex Environments. In: ACM SIGGRAPH. (2003)
8. Isard, M., Shand, M., Heirich, A.: Distributed rendering of interactive soft shadows. In: Parallel Graphics and Visualization, EGPGV. (2002) 71–76
9. Heirich, A., Moll, L.: Scalable Distributed Visualization Using Off-the-Shelf Components. In: IEEE Parallel Visualization and Graphics Symposium. (1999)
10. Moll, L., Heirich, A., Shand, M.: Sepia: Scalable 3D Compositing Using PCI Pamette. In: IEEE Symposium on Field Programmable Custom Computing Machines. (1999)
11. Zara, F., Faure, F., Vincent, J.M.: Physical cloth simulation on a PC cluster. Parallel Graphics and Visualisation (2002)

12. Fan, Z., Qiu, F., Kaufman, A., Yoakum-Stover, S.: GPU Cluster for High Performance Computing. In: Proceedings of ACM/IEEE Supercomputing Conference, Pittsburgh PA, USA (2004)
13. Kipfer, P., Slusallek, P.: Transparent Distributed Processing for Rendering. In: Parallel Visualization and Graphics Symposium (PVG), San Francisco (1999)
14. Rajagopalan, R., Goswami, D., Mudur, S.: Functionality Distribution for Parallel Rendering. In: IEEE IPDPS. (2005)
15. Rajagopalan, R.: Functionality Distribution in Graphics. Master's thesis, Concordia University, Canada (2005)
16. Levoy, M., Whitted, T.: The use of points as display primitives. Technical report, CS Departement, University of North Carolina at Chapel Hill (1985)
17. Westover, L.: Interactive Volume Rendering. In: Chapel Hill Workshop Volume Visualization. (1989) 9–16
18. Grossman, J.P.: Point Sample Rendering. Master's thesis, Dept. of Electrical Engineering and Computer Science, MIT (1998)
19. Rusinkiewicz, S., Levoy, M.: QSplat: A Multiresolution Point Rendering System for Large Meshes. In: SIGGRAPH. (2000)
20. Pfister, H., Zwicker, M., Baar, J.V., Gross, M.: Surfels: Surface elements as rendering primitives. In: SIGGRAPH. (2000) 335–342
21. Zwicker, M., Pfister, H., Baar, J.V., Gross, M.: Surface splatting. In: SIGGRAPH. (2001) 371–378
22. Ren, L., Pfister, H., Zwicker, M.: Object space EWA surface splatting: A hardware accelerated approach to high quality point rendering. In: Eurographics 2002. (2002) 461–470
23. Carsten, D., Christian, V., Marc, S.: Sequential Point Trees. In: SIGGRAPH. (2003)
24. Hubo, E., Bekaer, P.: A Data Distribution Strategy for Parallel Point-Based Rendering. In: WSCG. (2005)
25. Chilimbi, T.M., Hill, M.D., Larus, J.R.: Making Pointer-Based Data Structures Cache Conscious. IEEE Computer (2000)

An Intra-task DVS Algorithm Exploiting Program Path Locality for Real-Time Embedded Systems

G. Sudha Anil Kumar and G. Manimaran

Real-Time Computing and Networking Laboratory,
Dept. of Electrical and Computer Engineering,
Iowa State University, Ames, IA 50011, USA
{anil, gmani}@iastate.edu

Abstract. In this paper, we present a novel intra-task Dynamic Voltage Scheduling (DVS) algorithm based on the knowledge of frequently executed paths in the control flow graph for real-time embedded systems. The basic idea is to construct a common path composing all the frequently executed paths (hot-paths) and perform DVS scheduling based on this common path, rather than the most probable path. We compare the performance (energy consumption) of our algorithm with a recently proposed algorithm. Our simulation results show that the proposed algorithm performs better than the existing algorithm for most of the simulated conditions. We also identify interesting research problems in this context.

1 Introduction

Portable embedded devices, such as personal digital assistants, mobile phones and palmtops have become extremely popular in the recent past. These devices rely on batteries for power supply and their operation is limited by the available battery life. Therefore, efficient utilization of energy is one of the key challenges in the design and operation of embedded devices. Most of the embedded processors are based on CMOS technology, where the energy dissipated per cycle is directly proportional to the square of the supply voltage,V_{dd} [1]. A widely used technique that exploits this characteristic is the DVS, whose goal is to minimize the energy consumption by choosing the supply voltage and operating frequency as per the performance level required by the tasks. Several energy aware DVS algorithms have been proposed for real-time systems [1, 2, 3, 4, 5].

The existing real-time DVS (RT-DVS) algorithms can be broadly classified into intra-task and inter-task DVS algorithms based on the granularity at which the voltage scaling is performed. The intra-task voltage scaling algorithms [2, 3, 4, 5] adjust the supply voltage within a task boundary. The inter-task voltage scaling algorithms [1] perform voltage scaling on a task by task basis.

Intra-task DVS algorithms typically work with the control flow graph (CFG) of the real-time programs. CFG represents the block level control flow

D.A. Bader et al. (Eds.): HiPC 2005, LNCS 3769, pp. 225–234, 2005.

structure of the program. Each node in the CFG denotes a basic block of computation. The edges in the CFG indicate the control dependency between the blocks.

The objective of an intra-task voltage scheduling algorithm for real-time programs is to assign proper clock frequency to each of the basic blocks so as to minimize the total energy consumption while meeting the task deadline. Ideally, each basic block can be operated at any voltage point which lies in the operational range of the processor. However, current commercial processors supply a fixed number of discrete voltage (and corresponding frequency) levels [6]. Therefore, each basic block needs to be operated in one of the discrete supply voltage levels. In this paper, we assume the processor supports a fixed number of supply voltage and corresponding frequency levels.

2   Related Work and Motivation

Lee et. al. [2] introduced the basic idea of intra-task voltage scheduling. Shin et.al. [4] extended this work with a worst case execution path based scheme which does not consider the likelihood of different possible execution paths. However, programs typically display a high degree of path locality, that is, only a small fraction of total possible paths execute most of the time [7].

Seo et al.[5] take the path locality into account by considering the branch probabilities of the CFG. Based on the branch probabilities, the proposed algorithm achieves optimal average energy savings. However, obtaining all the branch probabilities for a large program (with varying degree of path locality) is impractical. On the other hand, the less detailed information like the most frequently executed paths (hot-path information) is much easier to obtain as it incurs less profiling.

Shin et. al. [3] proposed a hot-path information based intra-task DVS scheme (RAEP), which chooses one of the hot-paths (where a hot-path is a path that exhibits high execution locality) and perform voltage scaling at each basic block that gives the best possible energy savings when the chosen path is executed. However, this heuristic scheme does not always achieve the minimum energy consumption, because there could be more than one hot path.

2.1   Motivation

The optimal frequency with respect to a particular execution path in the CFG depends on its length, where path length is defined as the execution time of the path when operated at the maximum frequency. Therefore, operating at a frequency based on the lengths of the hot-paths results in best energy savings.

The RAEP algorithm takes the above approach by considering one of the hot-paths (the most probable hot path). It operates at a frequency closest to the optimal frequency of the chosen path. However, there could be several hot paths of varying lengths. For example, commercial programs like MPEG-4 video decoder & encoder, compress and gcc typically have more than 15 hot-paths [7]. In such programs, the non-chosen hot-paths may together contribute a higher probability of execution and therefore RAEP which chooses just the most probable path cannot be very effective in minimizing the energy consumption.

Consider the following example with three hot-paths (CFG shown in figure 1.(a), the thick edges represent hot-paths): $p_1(B1, B2, B8)$, $p_2(B1, B3, B8)$, $p_4(B1, B5, B8)$. Let the respective execution probabilities be 0.35, 0.30 and 0.30. The probabilities for the other paths are unknown. For this example, the RAEP considers path p_1 only though it contributes less to the total energy savings as compared to executing paths p_2 and p_4 together. Therefore, considering just the most probable hot-path may not be effective in maximizing energy savings.

In this paper, we present an intra-task DVS algorithm which considers all the hot-paths together. The proposed algorithm constructs a common hot-path composing all the hot-paths and performs DVS scheduling based on this common hot-path. We have presented the preliminary idea of this paper in [8].

The proposed intra-task DVS algorithm can handle all possible CFG structures. To demonstrate its wide applicability, we follow the branch graph CFG model introduced in [9] which is typical in expressing structured programming constructs. In this model, a CFG is modeled as a branch graph consisting of a collection of components in series and/or parallel combinations. Each component is a basic fan graph with $n + 2$ $(n > 1)$ nodes defined as follows: A basic fan graph is a directed acyclic graph consisting of n independent nodes with one common parent and one common child. An example of the basic fan graph with 8 nodes is shown in figure 1(a).

The rest of the paper is organized as follows. In section 3, we present the algorithm for the basic fan graph with an illustrative example. In section 4, we extend the basic fan graph algorithm and demonstrate its working on more complex CFGs. In section 5, we present our simulation results. Finally, in section 6, we make several concluding remarks.

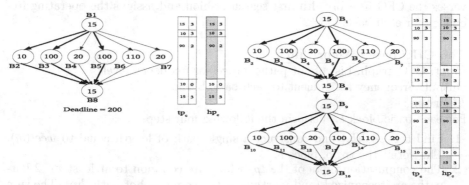

(a) A basic fan graph with its hp_c (b) A Series Fan graph with its hp_c

Fig. 1. Working of the CHP algorithm

3   Common Hot Path (CHP) Based Intra-task Algorithm

The proposed CHP algorithm considers all the hot paths together. The basic idea is to combine all the hot-paths into a single common path which represents a (virtual) hot-path that is common in length to majority of the hot-paths.

The common hot path is formed by first composing all the paths into a single path of computation, called the *common-total-path*, tp_c, which represents the longest path that the program can ever take. Each computation unit in the tp_c is contributed by one or more paths in the CFG. In figure 1.(a), the first 15 units of the tp_c are contributed by all the paths, while the last 10 units of the middle 110 units in the tp_c are contributed by the path $p_5(B1, B6, B8)$ alone.

The computational units contributed by majority of the hot paths constitute the common-hot-path, hp_c. In figure 1.(a), the number of contributing hot-paths (known as the hot-path count) is shown adjacent to the corresponding computation units. For example, the first 15 units are contributed by all the three hot-paths and therefore, the corresponding hot-path count is three; where as the last 10 units are not contributed by any of the hot-paths and therefore, the corresponding hot-path count is zero. The computational units with hot-path count greater than or equal to two (majority in this case) are marked to constitute the hp_c. In figure 1.(a), the highlighted 130 units of computation, forms the hp_c. The hp_c represents the path that is common to majority of the hot paths. Therefore, performing DVS scheduling based on this common hot path length would be beneficial to all the hot paths rather than based on a single hot-path.

We use the following notations in the rest of the paper:

- D: deadline of the task.
- t_c: current time.
- t_l: time remaining until the deadline, $(D - t_c)$.
- $l(p_i)$: length of a particular path p_i.
- f_i: frequency at which the basic block b_i is operated.
- $wcet(b_i)$: remaining WCET of the task starting from block (b_i).

Following is the detailed description of the CHP algorithm. The algorithm traverses the CFG in a breadth first search fashion and assigns the operating frequency for each basic block.

The CHP based Algorithm

Input: CFG graph, List of hot-paths, processor frequency levels
Output: Frequency assignment to each basic block.
Algorithm:
For each basic block b_i, perform the following four steps:

1. Find the $wcet(b_i)$ and construct a single path of length equal to $wcet(b_i)$. This forms the tp_c of block b_i.
2. The computation units of the tp_c which are common to at least $n_h/2$ hot-paths are recognized (and marked) as the common-hot-path, hp_c. The sum of all the marked computational units forms the path length of the hp_c.
3. The operating frequency for b_i is chosen so as to operate the common-hot-path at its minimum possible frequency(normalized), which is given by:

$$f_i = \frac{l(hp_c)}{t_l - (l(tp_c) - l(hp_c))} \tag{1}$$

4. The smallest discrete frequency level which is greater than (or equal to) f_i is chosen as the b_i's operating frequency.

The time complexity of this algorithm is $O(v + e)$, where v and e represent the number of basic blocks and number of edges in the CFG respectively.

3.1   Illustrative Example

Consider the CFG shown in figure 1.(a) with three hot paths. Paths $p_1(B1, B2, B8)$, $p_2(B1, B3, B8)$, and $p_4(B1, B5, B8)$ are the hot paths with $p2$ being the most probable hot path. The execution probability of the paths p_1, p_2 and p_4 are 0.35, 0.30 and 0.30 respectively. The probabilities of the other paths are unknown. In this example, we assume the processor can operate at any of the ten equally spaced discrete frequency levels (normalized with respect to the maximum frequency) in the range $[0.1, 1.0]$. The numbers in each basic block represent the computation time of the block when operated at the maximum frequency.

The RAEP calculates the frequency of each basic block based on the most probable path [3]. The operating frequency of block B1 is calculated as follows:

$$\frac{l(p_1)}{t_l} = \frac{40}{200} = 0.20 \tag{2}$$

Operating $B1$ at this frequency results in operating at the maximum frequency on the execution of longer paths (say p_2 or p_4) as shown in figure 2. Since paths p_2 and p_4 together constitute a higher probability of execution, the RAEP executes at the maximum frequency for most of the program runs. This results in a high average energy consumption.

The proposed CHP scheme calculates the operating frequency of each basic block by considering all the hot paths. The following is the step by step execution of the CHP algorithm for basic block $B1$:

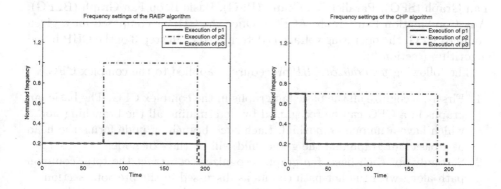

Fig. 2. Frequency settings of the two algorithms for the basic fan graph

1. The worst case path of the CFG rooted at $B1$ is $(B1, B6, B7)$ with a path length of 140 computation units. Therefore a single path with $l(tp_c) = 140$ is constructed as shown in figure 1.(a).
2. In step 2, the computation units that are common to majority (two in this case) of the hot-paths are marked. This is done in a breadth first search fashion considering just the hot-paths. The block $B1$ is common to all the hot-paths, so the first 15 units (equal to the size of $B1$) get marked in the tp_c. In the next level, two of the hot paths (p_2 and p_4) will execute 100 computational units. Therefore, 100 units are marked as the common (in length) computation units. The block $B8$ is again common to all the hot-paths and hence 15 more units get marked in tp_c. Therefore, $l(hp_c) = 130$. The hp_c is shown shaded in figure 1.(a).
3. The operating frequency for block $B1$ is 0.68 calculated using equation (1).
4. The smallest operational frequency level which is greater than 0.68 is 0.7, hence $B1$ is operated at 0.7.

The program continues to execute at the same frequency if it executes one of the two hot-paths p_2 and p_4. On the other hand, it reduces the frequency when it executes path p_1. Since, both paths p_2 and p_4 together execute with a higher probability, CHP consumes lesser energy for most of the times the program is run. This results in a lower average energy consumption compared to the RAEP scheme. For this example, CHP shows an improvement of 40% over RAEP.

4   CHP on Complex CFG Structures

CFGs of typical application programs will have more complex structure than the basic fan graph. A complex CFG with two or more fan graphs may be viewed as either series or parallel arrangement (or a combination of both) of the basic fan graphs. A CFG can also have loops in addition to the above combinations. The common hot-path formation technique is non-trivial for such complex CFGs and therefore, we demonstrate CHP formation techniques for the complex CFGs. In particular, we present the technique for the following complex CFGs: Series Fan Graph (SFG), Parallel Fan Graph (PFG), Basic Loop Fan Graph (BLFG), Any Combination Fan Graph (ACFG). Once the CHP is composed for a given complex CFG, the operating voltage is determined as in step 3 of the CHP based algorithm (section 3).

The following *generalized_CHP* procedure is applied to the complex CFGs:

1. Firstly, recognize all the basic fan graphs in the complex CFG. The basic fan graphs in a CFG can be recognized by determining all the branching nodes which have a unique grandchild. Each such branching node forms the head of a basic fan graph and the grandchild will be the exit node.
2. Secondly, for each basic fan graph recognized, construct the total common path along with the hot-path counts as discussed in the previous section.
3. Thirdly, combine all the tp_cs by taking hot-path counts into consideration to form the final hp_c. The working of this step depends on structure of the

CFG, that is, whether the basic fan graphs are in series or parallel. This step is elaborated in detail for each of the above four complex CFGs.

4.1   CHP Formation for an SFG

An SFG has two or more basic fan graphs one followed by the other. Figure 1.(b) shows an SFG with two basic fan graphs. Each hot-path in an SFG can be visualized as a concatenation of k partial paths, where the ith ($i \leq k$) partial path is a part of the ith basic fan graph. In the 2-SFG ($k = 2$) shown in figure 1.(b), the hot-path $p_1(B_1, B_2, B_8, B_9, B_{10}, B_{16})$ has two partial paths: $p_{11}(B_1, B_2, B_8)$ and $p_{12}(B_9, B_{10}, B_{16})$. Similar to every hot-path, the CHP (yet to be formed) will have k concatenated partial CHPs each formed independently from each of the k basic fan graphs. Therefore the final hp_c of the SFG can be obtained by concatenating the tp_cs obtained as a result of the first two steps of the *generalized_chp* algorithm and marking the computational units which have hot-path counts greater than $n_h/2$.

In the following three subsections, we present the basic ideas for handling PFG, BLFG and ACFG. We skip the working details for each of them due to space constraints.

4.2   CHP Formation for a PFG

A PFG has two or more basic fan graphs as alternatives following a branching basic-block. The procedure to find the final hp_c of a PFG is little more involved. The basic idea is to find the final tp_c (the longest of all tp_cs) and update its hot-path counts by considering every other tp_c. Once the final hot-path counts are available the algorithm marks the computational units contributed by majority of the hot-paths to obtain final hp_c.

4.3   CHP Formation for a BLFG

A BLFG has one basic fan graph within a loop. The general procedure to handle a loop is to find the tp_c of the CFG ignoring the loop. Once the tp_c for the basic fan graph is found, the loop has to be unrolled. Since a straightforward loop unrolling can be very expensive, the loop is unrolled twice in a fashion that captures the effect of iterations.

4.4   CHP Formation for an ACFG

An ACFG consists of several basic fan graphs arranged in a complicated fashion. Typical application programs fall into this category. Interestingly, any complicated CFG can be viewed as a series-parallel combination of basic fan graphs. Therefore, the hp_c of an ACFG can be obtained by recognizing all the basic fan graphs and applying the series-parallel CHP formation techniques appropriately.

5   Simulation Studies

We have compared the performance of the proposed CHP scheme with the existing RAEP scheme and the clairvoyant algorithm. The performance metric is the
normalized average energy consumption (normalized with respect to the DVS
unaware scheduler). The clairvoyant algorithm by definition knows the exact
path the program will execute and hence operates at the corresponding optimal frequency. Therefore, clairvoyant algorithm represents the theoretical lower
bound of the energy consumption. We have simulated the average energy consumption of the above schemes on randomly generated ACFGs. Each ACFG
was generated with n_h hot paths and n_l non hot-paths. All the n_h hot paths
together execute with a probability of 0.95. The most probable path executes
with a probability p_m and has a path length equal to (l_1) while the remaining
hot paths execute with equal probabilities and each has a length of $(1 + \alpha)l_2$.
Each of the non hot-paths have a path length equal to $(1 + \beta)l_3$. Where α and β
are uniformly chosen from $[0, 1]$. We assumed $l_1 = 1000$ for all our performance
studies and varied the following parameters:

- Hot-path length ratio: $lr_1 = \frac{l_1}{l_2}$; • Non hot-path length ratio: $lr_2 = \frac{l_1}{l_3}$
- Slack factor: $s_f = \frac{D - wcet(B_1)}{D}$; • p_m: probability of the most probable path

5.1   Results and Discussions

Effect of the Path Length Ratio: Figure 3(a) shows the relative performance
of the CHP and RAEP schemes varying the hot-path length ratio (lr_1). This
graph shows the effect of path length variations and has three disjoint regions
of interest defined by the value of lr_1. The region with the value of lr_1 very
close to one (unity region) corresponds to the case where all the hot-paths have
approximately the same path length. In this region, performing DVS scheduling
based on the most probable path (or any single path) will be very effective.
Consequently, RAEP performs slightly better than CHP in the unity region.

The region which is left to the unity region (left region) corresponds to the
case where most of the hot-paths are much longer than the most probable path.
In this region, RAEP which considers the (shorter) most probable path alone
performs aggressive voltage down scaling in the beginning and ends up increasing
the voltage when the other (longer) hot-paths execute. On the other hand, CHP
which considers all the hot-paths together performs conservative voltage scaling
considering the fact that majority of the hot-paths are long in length.

Similarly, the region which is right to the unity region (right region) corresponds to the case where most of the hot-paths are much shorter than the most
probable path. In this region, RAEP which considers the (longer) most probable
path alone performs conservative voltage up scaling in the beginning and ends
up decreasing the voltage or even leaving the slack unutilized when the other
(shorter) hot-paths execute. On the other hand, CHP which considers all the
hot-paths together performs aggressive voltage scaling considering the fact that
majority of the hot-paths are short in length.

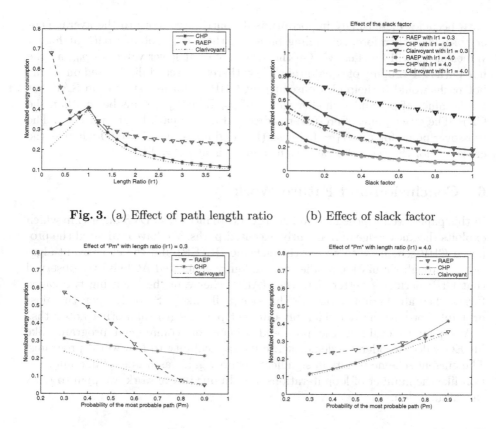

Fig. 3. (a) Effect of path length ratio   (b) Effect of slack factor

Fig. 4. (a) & (b) Effect of the path probability (P_m) on energy consumption

Effect of the Slack Factor: Figures 3(b) shows the relative performance of the two schemes varying the slack factor (s_f) for different values of lr_1 corresponding to the left and right regions discussed above. We have chosen $p_m = 0.3$ for this set of experiments. In general, with the increasing slack factor both the schemes operate at relatively lower frequencies and hence consume less energy. CHP performs consistently better than RAEP throughout the range. It shows an improvement of 21% at $s_f = 0$ and an improvement of 67% at $s_f = 1.0$ for $lr_1 = 0.3$ case (left region). Similarly, CHP shows an improvement of 27% at $s_f = 0$ and an improvement of 50% at $s_f = 1.0$ for $lr_1 = 1.0$ case (right region).

Effect of the Probability of the Most Probable Path: Figures 4(a) & 4(b) show the relative performance of the above two schemes varying the probability of the most probable path (p_m) for different values of lr_1 corresponding to the left and right regions discussed in the previous result. We have chosen $s_f = 0.5$ for this set of experiments. CHP performs better than RAEP at lower values of p_m, while RAEP performs better at higher values of p_m. This is due to the following reason: as the probability (p_m) increases, the most probable path becomes increasingly important as it contributes more to the average energy savings; therefore, scheduling based on the most probable path at higher values of p_m will be effective. On the other hand, at lower values of p_m, all the hot-paths are roughly of equal probability; therefore, scheduling based on all the hot-paths would be helpful. Consequently, CHP performs better than RAEP at lower values of p_m and at higher values of p_m RAEP performs better than the CHP. The exact point of crossover is dictated by the path length ratio lr_1. The crossover point for the $lr_1 = 0.3$ case (left region) is at $p_m = 0.60$ whereas the crossover for the $lr_1 = 4.0$ case (right region) is at $p_m = 0.75$.

6   Conclusion and Future Work

In this paper, we proposed a novel energy aware intra-task DVS algorithm which exploits the knowledge of frequently executed paths. We have evaluated the proposed CHP scheme with an existing scheme (RAEP) and the clairvoyant algorithm through simulation studies on randomly generated ACFGs. We observed that CHP performs better than the RAEP scheme in the following two cases: First, when all the hot-paths are almost equally likely. Second, when the most probable hot-path has considerably different path length than other hot-paths.

We plan to evaluate the proposed scheme on commercial programs like MPEG video decoders to demonstrate its applicability to real world programs. The current scheme results in significant energy gains assuming offline information like the number of loop iterations, etc. In our future work, we plan to relax this assumption.

References

1. P. Pillai and K. G. Shin, "Real-Time Dynamic Voltage Scaling for Low-Power Embedded Operating Systems," *ACM Symp. on O.S Principles*, 2001, pp.89-102.
2. S. Lee and T. Sakurai, "Run-Time Voltage Hopping for Low-Power Real-Time Systems," *in Proc. of ACM Design Automation Conference(DAC).*, 2000, pp.806-809.
3. D. Shin, J. Kim, and S. Lee. "Intra-task voltage scheduling on DVS-enabled hard real-time systems," *IEEE transactions on CAD*, Vol. 24, Issue 9, Sep. 2005.
4. D. Shin, J. Kim, and S. Lee. "Intra-task Voltage Scheduling for Low-energy Hard Real-Time Applications," *IEEE D & T of Comp.*, Vol. 18, No. 2, 2001, pp.20-30.
5. J. Seo, T. Kim, K. S. Chung, "Profile-Based Optimal Intra-task Voltage Scheduling for Hard Real-Time Applications," *in Proc. of ACM Design Automation Conference(DAC)*, June 2004, pp.87-92.
6. Transmeta Corporation. Crusoe Processor. http://www.transmeta.com, June 2000.
7. T. Ball, P. Mataga and M. Sagiv, "Edge Profiling versus Path Profiling: The Showdown" *in Proc. of ACM SIGPLAN-SIGACT symposium on principles of programming languages*, January 1998, pp.134-148.
8. G. S. Anil Kumar and G. Manimaran, "An intra-task DVS algorithm exploiting path probabilities for real-time systems" *SIGBED Review, special issue on 11th IEEE RTAS Work-in-Progress*, Vol. 2, No. 2, April 2005.
9. H. El-Rewini and H. H. Ali, "Static Scheduling of Conditional Branches in Parallel Programs," *Journal of parallel and Distributed Comp.*, Vol. 24, Jan. 1995, pp.41-54.

Advanced Resource Management and Scheduling of Workflow Applications in JavaSymphony*

Alexandru Jugravu[1] and Thomas Fahringer[2]

[1] Institute of Scientific Computing, University of Vienna,
Nordbergstr. 15/C/3, A-1090 Vienna, Austria
[2] Institute for Computer Science University of Innsbruck,
Technikerstr. 21 A, A-6020 Innsbruck, Austria

Abstract. JavaSymphony is a high-level programming model for performance oriented distributed and parallel Java applications, which allows the programmer to control parallelism, load balancing, and locality at a high level of abstraction. In this paper, we describe an extension of JavaSymphony that deals with distributed workflow applications as graphs of software components, which can be executed on a distributed set computers. Workflows are not limited to DAGs, but also cover complex control flow including loops. Furthermore, we introduce a novel approach for workflow scheduling based on the HEFT algorithm and resource brokerage for a heterogeneous set of computers. We demonstrate the effectiveness of our approach with two real-world applications and compare our techniques against the widely known DAGMan Condor scheduler.

1 Introduction

The workflow model has emerged as a very promising paradigm for programming distributed applications. Workflow-based applications have become a fashion topic in the Grid research community. Commonly, a static scheduling strategy is used to build a schedule for a DAG (Directed Acyclic Graph) based workflow, which is known as a NP-complete optimisation problem. However, static scheduling is not appropriate for dynamic distributed environments such as the Grid, in which resources may randomly become unavailable or unsuitable or may change their runtime behaviour during the execution of distributed applications. At the same time, repetition of parts of the application until convergence criteria are met, cannot be modelled by DAG-based models, and existing DAG-based schedulers commonly do not address the non-deterministic behaviour due to data available only at runtime.

In previous work [4], we have introduced a workflow model with conditional branches and loops. The execution plan associated with these workflows can be

---

* The work described in this paper is partially supported by the Austrian Grid Project, funded by the Austrian BMBWK (Federal Ministry for Education, Science and Culture) under contract GZ 4003/2-VI/4c/2004.

D.A. Bader et al. (Eds.): HiPC 2005, LNCS 3769, pp. 235–246, 2005.

adjusted to dynamic changes of the underlying execution environment or non-deterministic application behaviour (e.g. convergence criteria for until-loops).

In order to deal with such dynamism of the execution environment and work-flow applications, we have implemented a novel scheduling strategy, which is applied as part of the JavaSymphony Workflow Management System. In addition, the JavaSymphony Workflow Management System provides a graphical user interface to design and control the execution of workflow applications. The activities and the resources may be associated with performance parameters and a build-in specification language is used to describe the workflow.

This paper presents important extensions to our scheduling approach. We introduce a new algorithm that manages loops and conditional branches of a workflow by combining a classical list scheduling algorithm [12] with our dynamic scheduling technique [5]. Furthermore, we describe a new theoretical framework, which models the resource broker functionality for controlling computer resources as part of a heterogeneous computing environment. We evaluate our approach with one real-world application, and we compare our scheduler performance against the widely known Condor DAGMan scheduler.

The rest of this paper is organised as follows: Section 2 presents preliminaries notions, which include a short description of the workflow model and the dynamic scheduling technique. Sections 3, 4 introduce the new dynamic scheduling algorithm, respectively the theoretical framework for the resource brokerage, whilst Section 5 demonstrates our scheduling technique in an experiment. Section 6 discusses related work. Finally, some concluding remarks are made and future work is outlined in Section 7.

2   Background

JavaSymphony is a high-level programming model for performance-oriented distributed and parallel Java applications, which allows the programmer to control parallelism, load balancing, and locality at a high level of abstraction. The JavaSymphony programming paradigm [3] offers high-level constructs to manage distributed resources and to access their static/dynamic system parameters. It offers constructs to create, map or migrate distributed objects. The communication between the objects of a distributed application is based on several types of remote method invocation. Mechanisms for distributed synchronization and distributed events are provided. The JavaSymphony programming paradigm allows flexible implementation of a large range of distributed applications, such as meta-task applications or workflow applications. However, the developer usually has to manage the resources, build Java objects, and control the mapping of the objects onto resources. In order to improve the performance, the developer has to use a scheduling strategy adapted to his particular application. All these issues require a significant programming effort. In order to alleviate the programming effort for distributed workflow applications, the JavaSymphony Workflow Management System, built on top of JavaSymphony runtime system, has been introduced to support automatic resource allocation and scheduling.

2.1   The Workflow Model

A workflow application can be seen as a collection of computing activities (computational tasks) that are processed in a certain order. Between two computing activities there may be: (1) a control flow dependency, which means that one activity cannot start before its predecessors finished or (2) a data dependency, which means that one activity needs input data that is produced by the other. In previous work [4], we have presented a new workflow model which extends the classical workflow that is limited to DAGs of tasks, with loops and conditional branches. Each workflow has an associated workflow graph, which has vertices for the workflow basic elements: activities, dummy activities, conditional branches, initial and final states. Between the vertices of the workflow graph, there are edges associated with the control-/data-dependencies between the elements of the workflow, with the sequential loops and with the parallel loops of the workflow.

2.2   Scheduling Workflows with Loops and Conditional Branches

Commonly, for DAG-based workflows, the activities of the workflow are scheduled before the execution begins. Static scheduling, however, is mostly unsuitable for graph-based workflows that are being executed on a dynamically changing execution environment. We have introduced a scheduling strategy [5] to transform the workflow associated with the application into one with no conditional branches or loops, and recursively find a schedule, by using one of the many existence algorithm for static scheduling of DAG-based workflows. According to this strategy, several transformations are recursively applied to the workflow graph, which include eliminations of parallel loops, conditional branches, initial and final states, and sequential loop transformations.

2.3   The JavaSymphony Workflow Management System

To build a JavaSymphony workflow application, one has to first design the workflow graph, by using the specialized graphical user interface. The developer puts together the activities, dummy activities, initial and final states of the workflow and connects them using control links, data links, loops and parallel loops, according to the model presented in [4]. The result is an easy-to-understand graphical representation of the workflow (Figure 2), based on the UML Activity Diagram, which can be stored in a file by using the specific XML-based specification language. Behind the graphical representation, each element (vertices and edges of the graph) is associated with relevant workflow information. Some of this data is mandatory (e.g. class names for activities, activity ids, input parameters, files to be transferred, termination conditions or the number of iterations for the loops, the number of iterations for parallel loops, branch conditions, etc.). Other information is optional (e.g. performance characteristics of the computational activities or communication, resource constraints for mapping the activities or for communication network, performance contracts). However, the latest may be needed by the build-in scheduler to improve the performance of the whole application or to match the user preferences.

Within the same scheduling process, the workflow specification is analyzed, a resource broker determines which resources are suitable for each activity of the workflow, a scheduler computes the execution plan of the workflow, and an enactment engine manages the execution of the activities according to the execution plan.

3   A List-Scheduling Algorithm for Workflows

In this section, we present a new scheduling algorithm that manage dynamic workflows with loops and conditional branches. The new algorithm (Figure 1) combines the dynamic scheduling strategy introduced in Section 2.2 with a classical list-scheduling algorithm for DAGs of tasks.

3.1   List-Scheduling for DAG-Based Workflows

The general DAG scheduling problem was extensively studied and many research efforts proposed heuristics to solve this problem, both for homogeneous and for heterogeneous domains [7,10]. A significant number of the proposed heuristics are based on the *list scheduling* technique. The algorithms of this type are known to perform well, at relatively low cost. However, all of them propose a static approach, in which the schedule is computed at compile time, and do not address the problem of scheduling conditional branches and loops.

We have chosen HEFT (Heterogeneous Earliest Finish Time) [12], which is a DAG scheduling algorithm that supports a bounded number of heterogeneous processing elements and is considered an important representative of the list-scheduling algorithms for heterogeneous systems [2]. The HEFT algorithm associates activities with priorities, based on the so-called activity *upward rank*:

$$rank_u(t) = \overline{exec(t)} + \max_{s \in succ(t)} \left(\overline{comm(t,s)} + rank_{u(s)} \right)$$

1. Apply all possible transformations to produce WF_t.
2. Perform steps 3-8, whenever a new scheduling event occurs
3. **begin**
4.   (Re)Compute $U(WF_t)$, $S(WF_t)$ and $DAG(WF_t)$
5.   Eliminate finished tasks from $DAG(WF_t)$
6.   Apply HEFT strategy
        - Determine exit nodes of the reduced $DAG(WF_t)$
        - Recursively compute $rank_u(t)$, starting from the exit nodes.
        - Build a list of activities, sorted by descending order of $rank_u$ values.
        - while the list is not empty
        - **begin**
        -        Remove t, the first element from the list
        -        Compute $ct(t/m)$ for each m and assign t to m_t that minimizes it.
        - **end**
7.   Start the activities on the assigned machines, ordered by the $st(t/m)$ values
8. **end**

Fig. 1. HEFT-based algorithm for scheduling workflows

where $succ(t)$ is the set of immediate successors of task t and $\overline{exec(t)}$ and $\overline{comm(t,s)}$ are *the average execution cost* of task t, respectively *the average communication cost* of edge (t,s), defined as:

$$\overline{exec(t)} = 1/|M| \sum_{m \in M} exec(t/m) \text{ and } \overline{comm(t_1,t_2)} = data(t_1,t_2)/\overline{rate}$$

with M - the set of available machines, $exec(t/m)$ - the execution time of t running onto m, $data(t_1,t_2)$ being the size of data sent from t_1 to t_2, and $\overline{rate}$ the average transfer rate between the machines in the domain. The *upward rank* is computed recursively, starting from the exit node(s). It is clear that each task is ranked higher than its successors. Therefore, the tasks are processed in descending order of their ranks. For each task, the machine which gives the best completion time $ct(t/m)$ is chosen, and the start time $st(t/m)$ is calculated accordingly.

3.2 Managing Workflow Conditional Branches and Loops

We use the technique introduced in Section 2.2, the HEFT strategy and the notations described above to create a new algorithm for scheduling workflows. We iteratively build a transformed workflow as follows: Initially (pre-scheduling), all possible transformations, except branch elimination, are applied. The workflow application is scheduled/executed until a conditional branch is reached. Upon this event, the branch elimination and the subsequent possible transformations are applied to obtained a new workflow graph.

We use the notation $WF \longmapsto WF_t$ to express that WF_t is obtained from WF by applying the above-mentioned transformations. At each scheduling iteration, the activities are separated into two sets: $U(WF_t)$ - the set of **unsettled activities** comprises activities for which the scheduling/execution decision depends on data that is not yet available (e.g. activities subsequent to a conditional branch, for which the associated condition cannot be evaluated); $S(WF_t)$ - the set of **settled activities** comprise the rest of the workflow activities. The activities in $S(WF_t)$ build up a DAG of activities, denoted by $DAG(WF_t)$.

The new scheduling algorithm is summarized in Figure 1. Note that the algorithm recursively computes a partial DAG and a partial schedule. The activities in $U(WF_t)$ are not scheduled. The schedule is dynamically updated, if necessary, at runtime, on scheduling events (e.g. termination of activities - successful or with error, performance contract violation, or user intervention). The termination of activities and evaluation of Boolean expression associated with conditional branch or loops are mainly responsible for the recalculation of $DAG(WF_t)$ in step 4. Step 6 of the algorithm determines a static schedule by applying the classical HEFT algorithm. The algorithm finishes when all the activities in $DAG(WF_t)$ have finished and no other scheduling event occurs.

4 The Resource Broker

A resource broker is essential for scheduling distributed applications in heterogeneous environment. A resource broker determines which resources are available and suitable for a workflow activity, and may support more advanced features like reservations. In the previous sections, we have assumed that the scheduling algorithms use a static set of resources and determine a near-optimal mapping of the activities onto these resources. However, in real life, this is hardly true, since resources may crash, or become available at random times. Moreover, resource performance may largely vary in time, such as suitable resources become unsuitable and vice versa, thus affecting the scheduling performance. In this section, we introduce a new theoretical framework to describe the functionality of the resource broker.

4.1   Modelling the Resources

We consider $M = \{m_1, m_2...m_{n_M}\}$ the set of all resources that may be used. The resources are associated with a set of attributes $Att = \{att_1, att_2, ...att_n\}$. Each attribute att_i associates to each resource an attribute value (numeric or string) in the values set. If the attribute is a dynamic attribute (e.g. system load, idle times, available memory), this value varies in time and the attribute is defined as a function of the machine and the time: $att_i : M \times T \rightarrow Values, att_i(m, t) = v$. If the attribute is static (e.g. machine name, operating system, peak performance parameters), the attribute is defined as a function over M only: $att_i : M \rightarrow Values, att_i(m) = v$

4.2   Modelling the QoS for Workflow Activities

The workflow activities are associated with a set of constraints, as defined in the workflow specifications. We denote the set of constraints by $\mathbf{C} = \{c_1, c_2, ...\}$. Each constraint c_i is uniquely associated with a resource attribute, which we denote by $att(c_i)$. A constraint is a Boolean function $c_i : M \times Time \rightarrow \{true, false\}$, which determines if a property of the attribute $att(c_i)$ holds or not. Practically, the Boolean value of $c_i(t)$ is determined by comparing $att(c_i)$ with a threshold value. For example, we may have a constraints $c(m, t)$, which takes the value of the predicate "$att(c)(m, t) \geq v_0$", if the associated attribute $att(c)$ takes numeric values. For a workflow WF, each task $T \in Act$ is associated with a (finite) set of constraints denoted by $\mathbf{C}(T) = \{c_{(T,1)}, c_{(T,2)}...\}$. Using these notations, we say that a resource $m \in M$ is **suitable** for the activity $T \in Act$ at time $t \in Time$, if $c_{T,i}(m, t) = true$ for any $c_{T,i} \in \mathbf{C}(T)$. For $m \in M, T \in Act$ and $t \in Time$, the predicate $S(m, T, t) = \wedge_{c_{T,i} \in \mathbf{C}(T)}(c_{T,i}(m, t))$ is called the **suitability-predicate** of the resource m for the activity T (at t).

One of the functions of the resource brokerage process is to find all the suitable resources, for all the activities of the workflow, at any moment in time. In other words, a resource broker provides a function:

$$\mathbf{B}(T, t) = \{m | c_{T,i}(m, t) = true, \forall c_{T,i} \in \mathbf{C}(T)\}$$

that determines at each moment t which resources are suitable for a workflow activity T. It is not feasible to determine the suitability of the resources at each moment: On the one hand, it does not make sense to measure the system parameters continuously, since this may lead to performance problems. Instead, the dynamic system parameters could be updated at regular intervals (as done by the JavaSymphony middleware), and consequently the **resource suitability predicate** would be updated at discrete times, too. On the other hand, a relaxed scheduling policy may not require last updated information on resource suitability. Therefore, we analyze three scheduling scenarios regarding the usage of resource suitability information:

(1) The resource suitability is used only at the start of the scheduling, for the initial mapping. The advantage of this policy is obvious - a set of suitable resources is assigned only once to each tasks, and there is no need for complex monitoring of the resources. However, significant changes of the system dynamic parameters could dramatically deteriorate the performance.

(2) The resource suitability is continuously updated, and the scheduler is notified in case of changes. The scheduling complexity grows, however, adaptive decisions, which prevent the performance deterioration, are possible.

(3) We have adopted a hybrid scheduling policy for JavaSymphony workflow applications. The idea is to use two sets of constraints for each activity: The first set is used to determine the initial suitability of the resource. Optionally, a second set of constraints, which we call **performance contract** is used during the execution of the workflow activities onto resources. If the suitability-predicate for the performance contract of an activity does no longer hold, the scheduler may stop the execution of the activity and migrate it onto another suitable resource.

4.3   The Availability of the Resources

In a dynamic computing system such as computational grids, the computing resources may become unavailable or available randomly. It is important for a distributed application to determine when a resource crashes and to be able to recover and continue the execution after such an incident. Another function of the resource broker would be to monitor the resources in order to determine if they are still available or not. The availability of a resource can be expressed as a function of the resource and the time $Avail(m, t)$, which is *true* if m is alive at t, and *false* otherwise. In combination with the suitability-predicate, we obtain a function, which tells us if a resource m may be used by one activity T, at time t: $S(m, T, t) \wedge Avail(m, t)$ **iif** m may be use by T at t. The function of the resource broker is modified to include the availability of the resources as well:

$$\mathbf{B}(T, t) = \{m | S(m, T, t) \wedge Avail(m, t) = true\}$$

In practice, the resource broker does not calculate such a function, but provides an ordered series of time intervals: $I(T, m) = \{I_1, I_2 ..., I_j..\}$ such as: $I_j = [ts_j, tf_j]$ with $ts_j < tf_j$, $Avail(m, t) = true$, and $S(m, T, t) = true$ for any $t \in I_j$. Moreover, the resource broker is not able to forecast future intervals, but updates the $I(T, M)$ sets on scheduling events (e.g. resource becomes (un)available, or resource does no longer fulfil the suitability-condition).

5 Experiments

In order to demonstrate the usefulness of JavaSymphony workflow scheduling technique, we have implemented one JavaSymphony workflow application and used the JavaSymphony Workflow Management System to build the application workflow, and to schedule and execute the application onto a set of distributed resources.

The workflow application is built on top of WIEN2k [8] package, which is a program package for performing structure calculations of solids using density functional theory, based on the full-potential (linearised) augmented plane-wave ((L)APW) and local orbitals (lo) method. The components of the WIEN2k package can be organized as a workflow (Figure 2): The *lapw1* and *lapw2_TOT* tasks can be solved in parallel by a fixed number of so-called *k-points*. This is modelled by two parallel loops in the workflow graph. Various files are sent from one workflow activity to another (supported by the JavaSymphony runtime system), which determine complex data dependencies between the activities (Figure 2(b)). At the end of the main sequence of the activities, a *testconv* activity performs

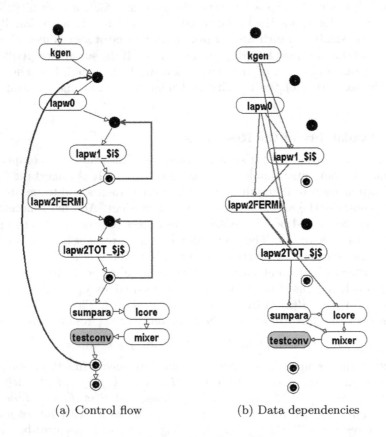

(a) Control flow   (b) Data dependencies

Fig. 2. Wien2k workflow

a convergence test to determine if the calculation needs to be repeated. This is modelled by the main sequential loop.

We have used the HEFT-based dynamic scheduling algorithm presented in Section 3, to schedule and run this application onto a set of workstations. Moreover, we have enhanced the JavaSymphony enactment engine to export the workflow application as an input file for Condor DAGMan [11]. For each workflow activity that is executed by the JavaSymphony enactment engine, a Condor job submit file is created as well. Condor DAGMan does not offer support for sequential/parallel loops and conditional branches. Therefore, these workflow elements are not present in the output DAG, which comprises only the activities that have finished their execution and their control dependencies.

We compare the scheduling performance of the two schedulers: JavaSymphony dynamically builds a schedule for the workflow application, based on the algorithm presented in Section 3, whilst Condor DAGMan uses the static DAG as built by JavaSymphony workflow enactment engine. Consequently, the two schedulers execute the same sets of activities, restricted by the same control dependencies. Furthermore, the activities are executed on the same set of computing resources, in a Condor pool of workstations (see Figure 3). The workstations are heterogeneous, ranked according to *JavaMFlops* attribute of Condor machine ClassAd, which determines the speed of the machine by using a specific benchmark at the time Condor is started on the machine.

We used artificially controlled execution times for the workflow activities: Each activity *act* is associated with a computing cost *cost(act)* expressed in FLOPs. Consequently, the execution time of one activity *act* on a machine *m* is

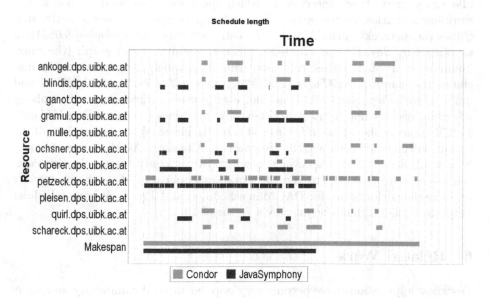

Fig. 3. Gannt Chart for Wien2k workflow execution. JavaSymphony vs. Condor.

determined as $cost(act)/power(m)$, where $power(m)$ is the value of *JavaMFlops* attribute associated with the machine. Therefore, one activity finishes in shorter time if it is mapped onto the machines that are higher ranked. On the other hand, the execution of a specific workflow activity takes the same amount of time, if mapped on the same machine, no matter if the activity is executed by using the Condor or the JavaSymphony scheduler.

The JavaSymphony determines the schedule of the workflow application based on the activity execution times onto the resources. The activities are mapped as Java objects onto the workstations. The Condor DAGMan scheduler uses the DAG file to schedule the DAG of the workflow application. Each activity is associated with a Java Condor universe job submission file, which executes the very same Java class onto the remote resources. The *JavaMFlops* is used as a priority associated with the resources, such that the stronger machines are preferred over the weaker. However, Condor cannot use the estimated execution times to make scheduling decisions. JavaSymphony scheduler, on the other hand, uses the *JavaMFlops* parameters, which are provided by JavaSymphony Runtime System, and the computing costs of the activities, provided in the workflow specification, to estimate execution times for the workflow activities, which are used by the HEFT-based scheduling algorithm (Figure 1).

Figure 3 presents the schedules for the two executions of the Wien2k workflow: by using the JavaSymphony scheduler, respectively the Condor DAGMan scheduler. The experimental run uses 8 *k-points* and the calculation within the main loop is repeated 3 times. This gives us a number of 67 activities in the execution plan, both for JavaSymphony and Condor DAGMan.

Figure 3 shows the timelines for each computing resource, which comprises idle times, respectively intervals in which the resource is used by one of the workflow activities. As we can see, both schedulers prefer to use the better machines (i.e. petzeck- *JavaMFlops*=70.885361, ochsner- *JavaMFlops*=58.082180), as ranked by *JavaMFlops* attribute. Since the width of the graph (the maximum number of activities that may run in parallel) is 8, the slowest machines (i.e. ganot-*JavaMFlops*=27.678907, pleisen-*JavaMFlops*=33.361057 and mulle -*JavaMFlops*=34.130119) are not evenly used. Moreover, JavaSymphony scheduler decides not to use two other slow machines (i.e. quirl-*JavaMFlops*= 37.276459 and ankogel-*JavaMFlops*=34.866943), based on the estimations of the execution times for each activity. In contrast, Condor DAGMan uses the next best available resource whenever a new activity is ready to run. In this case, the use of more resources to run the workflow application does not necessarily improve the performance for DAGMan scheduler, and JavaSymphony scheduler outperforms DAGMan scheduler by a factor of 1.64.

6   Related Work

Workflow applications have become very popular in Grid community and many research and industry groups have proposed new languages to model and develop workflow applications. We do not intend to compete with highly complex workflow definition languages [6,1]. Instead, the JavaSymphony specific XML-based specification language for workflow applications is simple, in order to allow an easy manipulation of the workflow structure by a scheduler. However, complex workflow specification languages are not commonly associated with advanced scheduling techniques for distributed workflow applications. We prefer to use a simpler definition language and a simplified workflow graphical representation, in order to be able to better investigate such advanced scheduling techniques.

On the other hand, most systems for allocating tasks on grids, (e.g. DAGMan [11]), currently allocate each task individually at the time it is ready to run, without aiming to globally optimise the workflow schedule. In addition, they assume that workflow applications have a static DAG-based graph, which may be seen as a too restrictive constraint.

The DAG scheduling problem has been intensively studied in the past, mostly in connection with parallel application compiling techniques. A parallel application is represented by a DAG in which nodes represent application tasks (computation) and edges represent inter-task data dependencies (communication). Numerous scheduling techniques and scheduling heuristics have been developed for both homogeneous and heterogeneous systems [12,7]. However, these heuristics expect a static application graph and statically compute the schedule before the execution is started. Static scheduling of static DAG structures is, however, too restrictive for the new generation of Grid workflow applications. Therefore, we propose a new approach which includes a workflow model with loops and conditional branches, and an extension of the static scheduling with novel dynamic scheduling techniques to accommodate these new constructs.

7   Conclusions and Future Work

JavaSymphony is a system designed to simplify the development of parallel and distributed Java applications on heterogeneous computing resources, ranging from small-scale clusters to large-scale Grid systems. Recently, we have extended the JavaSymphony programming paradigm to support workflow applications. For this purpose, the runtime system has been augmented with an automatic scheduler and an enactment engine for workflow applications, and a simple, yet expressive, workflow specification language has been introduced. This language allows the association of the activities and the resources with performance parameters that can be used by the scheduler/resource broker.

In this paper, we have presented new features of the JavaSymphony Workflow Management System. We have introduced a new scheduling algorithm, which combines dynamic scheduling techniques for workflows with loops and conditional branches, and HEFT, a classical list scheduling algorithm (limited to DAGs in its original form). Moreover, we have described a novel framework for resource brokerage, which analyzes the suitability and availability of resources. These parameters can be derived from the information provided by the JavaSymphony runtime system or from the workflow specification. We have extensively tested our HEFT-based dynamic scheduling algorithm by using two real-world workflow applications. The experiments have demonstrated that the JavaSymphony scheduler significantly outperforms the widely known Condor DAGMan scheduler for DAG-based workflows.

As future work, we plan to evaluate the dynamic scheduling technique with several other DAG-scheduling heuristics, and compare their performance on several workflow applications.

References

1. Tony Andrews, Francisco Curbera, Hitesh Dholakia, Yaron Goland, Johannes Klein, Frank Leymann, Kevin Liu, Dieter Roller, Doug Smith, Siebel Systems, Satish Thatte, Ivana Trickovic, and Sanjiva Weerawarana. Business process execution language for web services (bpel4ws). Specification version 1.1, Microsoft, BEA, and IBM, May 2003.
2. Sanjeev Baskiyar and Prashanth C. SaiRanga. Scheduling directed a-cyclic task graphs on heterogeneous network of workstations to minimize schedule length. In *Proc. of International Conference on Parallel Processing Workshops,Kaohsiung, Taiwan*, Oct. 2003.
3. Thomas Fahringer and Alexandru Jugravu. JavaSymphony: New Directives to Control and Synchronize Locality, Parallelism, and Load Balancing for Cluster and GRID-Computing. In *ACM Java Grande - ISCOPE 2002 Conference*, Seattle, November 2002. ACM.
4. Alexandru Jugravu and Thomas Fahringer. JavaSymphony, A Programming Model for the Grid. *Future Generation Computer Systems (FGCS)*, 21(1):239–246, Jan. 2005.
5. Alexandru Jugravu and Thomas Fahringer. Scheduling Workflow Distributed Applications in JavaSymphony. to appear in European Conference on Parallel Computing (Euro-Par 2005), Aug-Sep 2005.
6. Sriram Krishnan, Patrick Wagstrom, and Gregor von Laszewski. GSFL : A Workflow Framework for Grid Services. Technical Report, Argonne National Laboratory, 9700 S. Cass Avenue, Argonne, IL 60439, U.S.A., July 2002.
7. Yu-Kwong Kwok and Ishfaq Ahmad. Benchmarking and comparison of the task graph scheduling algorithms. *Journal of Parallel and Distributed Computing*, 59(3):381–422, 1999.
8. P.Blaha, K.Schwarz, G.Madsen, D.Kvasnicka, and J.Luitz. *WIEN2k: An Augmented Plane Wave plus Local Orbitals Program for Calculating Crystal Properties.* Vienna University of Technology, 2001.
9. Tan Tien Ping, Gian Chand Sodhy, Chan Huah Yong, and Fazilah Haron andRajkumar Buyya. A Market-Based Scheduler for JXTA-Based Peer-to-Peer Computing System. *Lecture Notes in Computer Science*, 3046:147–157, Apr 2004.
10. Andrei Radulescu and Arjan J.C. van Gemund. Low-cost task scheduling for distributed-memory machines. *IEEE Transactions on Parallel and Distributed Systems*, 13(6), June 2002.
11. The Condor Team. Dagman (Directed Acyclic Graph Manager). http://www.cs.wisc.edu/condor/dagman/.
12. H. Topcuoglu, S. Hariri, and M.-Y. Wu. Task scheduling algorithms for heterogeneous processors. In *Eighth Heterogeneous Computing Workshop*, pages 3–14. IEEE C.S. Press, 1999.

Using Clustering to Address Heterogeneity
and Dynamism in Parallel Scientific Applications[*]

Xiaolin Li[1] and Manish Parashar[2]

[1] Department of Computer Science, Oklahoma State University, OK 74078, USA
[2] Department of Electrical & Computer Engineering, Rutgers University, NJ 08854, USA
xiaolin@cs.okstate.edu, parashar@caip.rutgers.edu

Abstract. The dynamism and space-time heterogeneity exhibited by structured adaptive mesh refinement (SAMR) applications makes their scalable parallel implementation a significant challenge. This paper investigates an adaptive hierarchical multi-partitioner (AHMP) framework that dynamically applies multiple partitioners to different regions of the domain, in a hierarchical manner, to match the local requirements of these regions. Key components of the AHMP framework include a segmentation-based clustering algorithm (SBC) for identifying regions in the domain with relatively homogeneous partitioning requirements, mechanisms for characterizing the partitioning requirements, and a runtime system for selecting, configuring and applying the most appropriate partitioner to each region. The AHMP framework has been implemented and experimentally evaluated on up to 1280 processors of the IBM SP4 cluster at San Diego Supercomputer Center.

Keywords: Parallel Computing, Adaptive Mesh Refinement, Dynamic Load Balancing, Hierarchical Multi-Partitioner.

1 Introduction

Simulations of complex physical phenomena, modeled by systems of partial differential equations (PDE), are playing an increasingly important role in science and engineering. Dynamic structured adaptive mesh refinement (SAMR) techniques [1] are emerging as attractive formulations of these simulations. Compared to numerical techniques based on static uniform discretization, SAMR can yield highly advantageous ratios for cost/accuracy by adaptively concentrating computational effort to appropriate regions at runtime.

Parallel implementations of SAMR-based applications have the potential to accurately model complex physical phenomena and provide dramatic insights. However, while there have been some large-scale implementations [4] [6] [7] [8] [11], these implementations are typically based on application-specific customizations and general scalable implementations of SAMR applications continue to present significant challenges. This is primarily due to the dynamism and space-time heterogeneity exhibited

[*] The research presented in this paper is supported in part by the National Science Foundation via grants numbers ACI 9984357, EIA 0103674, EIA 0120934, ANI 0335244, CNS 0305495, CNS 0426354 and IIS 0430826.

D.A. Bader et al. (Eds.): HiPC 2005, LNCS 3769, pp. 247–257, 2005.
© Springer-Verlag Berlin Heidelberg 2005

by these applications. SAMR dynamism/heterogeneity has been traditionally addressed using dynamic partitioning and load-balancing algorithms, such as the mechanisms presented in [6] [11], that partition and load-balance the domain when it changes. The meta-partitioner approach proposed in [14] selects and configures partitioners at runtime to match the application's current requirements. However, due to the spatial heterogeneity of the SAMR domain, the computation/communication requirements can vary significantly across the domain, and as a result, using single partitioner for the entire domain can lead to decompositions that are locally inefficient. This is especially true for large-scale simulations that run on over 1000 processors.

The objective of the research presented in this paper is to address this issue. Specifically, we investigate an adaptive multi-partitioner framework that dynamically applies multiple partitioners to different regions of the domain, in a hierarchical manner, to match the local requirements of the regions. This research builds on our earlier research on meta-partitioning [14] and adaptive hierarchical partitioning [10] to define an adaptive hierarchical multi-partitioner framework (AHMP). The experimental evaluation of AHMP demonstrates the performance gains using AHMP on up to 1280 processors of the IBM SP4 cluster at San Diego Supercomputer Center.

The rest of the paper is organized as follows. Section 2 presents the problem description. Section 3 presents the AHMP framework and the SBC clustering algorithm. The experimental evaluation is presented in Section 4 . Section 5 reviews related work. Section 6 presents a conclusion.

2   Problem Description

SAMR formulations for adaptive solutions to PDE systems track regions in the computational domain with high solution errors that require additional resolution. SAMR methods start with a base coarse grid with minimum acceptable resolution. As the solution progresses, regions in the domain requiring additional resolution are tagged and finer grids are overlaid on these tagged regions of the coarse grid. Refinement proceeds recursively so that regions on the finer grid requiring more resolution are similarly tagged and refined. It results in a dynamic adaptive grid hierarchy [11].

Parallel implementations of SAMR typically partition the adaptive grid hierarchy across available processors, and each processor operates on its local portions of this domain in parallel. The overall performance of parallel SAMR applications is thus limited by the ability to partition the underlying grid hierarchies at runtime to expose all inherent parallelism, minimize communication and synchronization overheads, and balance load. Communication overheads of parallel SAMR applications primarily consist of four components: (1) *Inter-level communications*, defined between component grids at different levels of the grid hierarchy; (2) *Intra-level communications*, required to update the grid-elements along the boundaries of local portions of a distributed grid; (3) *Synchronization cost*, which occurs when the load is not balanced; (4) *Data migration cost*, which occurs between successive regridding and re-mapping steps.

The space-time heterogeneity of SAMR applications is illustrated in Figure 1 using the 3-D compressible turbulence simulation kernel solving the Richtmyer-Meshkov (RM3D) instability [3]. The figure shows a selection of snapshots of the RM3D

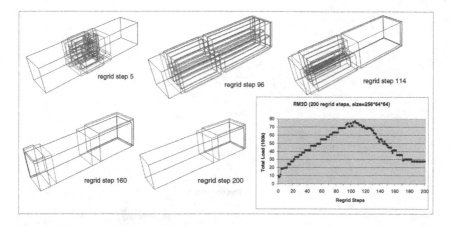

Fig. 1. Spatial and Temporal Heterogeneity and Workload Dynamics for RM3D Simulation

adaptive grid hierarchy as well as a plot of its load dynamics at different regrid steps. Since the grid hierarchy remains unchanged between two regrid steps, the workload dynamics and other features are measured in terms of regrid steps. The workload in this figure represents the computational/storage requirement, which is computed based on the number of grid points in the grid hierarchy. Application variables are typically defined at these grid points and are updated at each iteration of the simulation, and consequently, the computational/storage requirements are proportional to the number of grid points. The snapshots in this figure clearly demonstrate the dynamics and spatial and temporal heterogeneity of SAMR applications - different subregions in the computational domain have different computational and communication requirements and regions of refinement are created, deleted, relocated, and grow/shrink at runtime.

3   Adaptive Hierarchical Multi-partitioner (AHMP) Framework

The operation of the AHMP framework is illustrated in Figure 2. The input of AHMP is the structure of the current grid hierarchy, which is represented as a list of regions and defines the runtime state of the SAMR application. AHMP operation consists of the following steps. First, a clustering algorithm is used to identify clique hierarchies. Second, each clique is characterized and its partitioning requirements identified. Available resources are also partitioned into corresponding resource groups based on the relative requirements of the cliques. Third, these requirements are used to select and configure an appropriate partitioner for each clique. The partitioner is selected from a partitioner repository using selection policies. Finally, each clique is partitioned and mapped to processors in its corresponding resource group. The strategy is triggered locally when the application state changes. State changes are determined using a load-imbalance metric defined below. Partitioning proceeds hierarchically and incrementally. The identification and isolation of cliques uses a segmentation-based clustering (SBC) scheme. Partitioning schemes in the partitioner repository include Greedy Partitioning Algorithm (GPA), Level-based Partitioning Algorithm (LPA), bin-packing

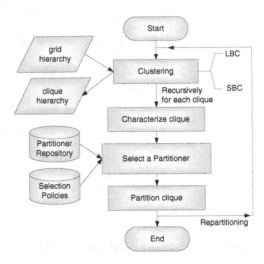

Fig. 2. Flowchart for the AHMP Framework

partitioning algorithm (BPA), geometric multilevel + sequence partitioning (G-MISP+SP), and p-way binary dissection algorithm (pBD-ISP) [10] [14]. AHMP extends our previous work on Hierarchical Partitioning Algorithm (HPA) [10], which applies single partitioner hierarchically, reducing global communication overheads and enabling incremental repartitioning and rescheduling.

The load imbalance factor (LIF) metric is used as the criterion for triggering repartitioning and rescheduling within a local resource group, and is defined as follows:

$$LIF_A = \frac{\max_{i=1}^{A_n} T_i - \min_{i=1}^{A_n} T_i}{\sum_{i=1}^{A_n} T_i / A_n}$$

where A_n is the total number of processors in resource group A, and T_i is the estimated relative execution time between two consecutive regrid steps for processor i, which is proportional to its load. The local load imbalance threshold is γ_A. When $LIF_A > \gamma_A$, the repartitioning is triggered inside the local group. Note that the imbalance factor can be recursively calculated for larger groups as well.

3.1   Clustering Algorithm for Clique Identification

The objective of clustering is to identify well-structured subregions in the SAMR grid hierarchy, called cliques. A clique is a quasi-homogeneous computational sub-domain with relatively homogeneous partitioning requirements.

This section presents the segmentation-based clustering (SBC) algorithm, which is based on space-filling curves (SFC) [12]. The algorithm is motivated by the locality-preserving property of SFCs and the localized nature of physical features in SAMR applications. Typical SAMR applications exhibit localized features and result in localized refinements. Moveover, refinement levels and the resulting adaptive grid hierarchy reflect the application runtime state. SBC hence attempts to cluster subregions with similar refinement levels. SBC defines the load density factor (LDF) as follows:

$$LDF(rlev) = \frac{\text{associated load on the subdomain}}{\text{volume of the subdomain at } rlev}$$

where $rlev$ denotes the refinement level and the volume is for the subregion of interest.

The SBC algorithm first smooths out subregions that are smaller than a predefined threshold, which is referred to as the template size (TS). TS is determined by the stencil size of the finite difference method and the granularity constraint that defines a certain computation communication ratio. SBC then follows the SFC index to extract subregions (defined by rectangular bounding boxes) from the subregion list until the size of the accumulated subregion set is over the template size. It calculates the load density for this set of subregions and computes a histogram of its load density. SBC continues to scan through the entire subregion list, and repeats the above process, calculating the load density and computing histograms. Based on the histogram of the load density obtained, it then finds a clustering threshold θ. A simple intermeans thresholding algorithm [5] is used. Using the threshold obtained, subregions are further partitioned into several clique regions. A hierarchical structure of clique regions is created by recursively calling the SBC algorithm for finer refinements.

Note that this algorithm has similarities to the point clustering algorithms proposed by Berger and Regoutsos in [2]. However, the SBC scheme differs from this scheme in two aspects. Unlike the Berger-Regoutsos scheme, which creates fine grained cluster, the SBC scheme targets coarser granularity cliques. In addition, SBC also takes advantage of the locality-preserving properties of SFCs to potentially reduce data movement costs between consecutive repartitioning phases.

Figure 3 shows the load density distribution and histogram for an SFC-indexed subdomain list. For this example, the SBC algorithm creates three cliques defined by the regions separated by the vertical lines in the figure on the left. The template size in this example is two boxes on the base level. The right figure shows a histogram of the load density. For this example, the threshold is identified in between 1 and 9 using the intermeans thresholding algorithm. While there are many more sophisticated approaches for identifying good thresholds for segmentation and edge detection in image

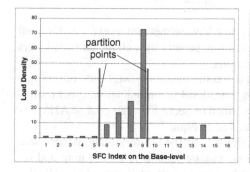
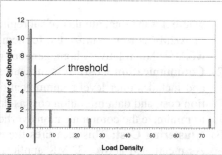

Fig. 3. Load Density Distribution and Histogram

processing [5], this approach is sufficient for our purpose. Note that we assume a predefined minimum size for a clique region. In this example, the subregion indexed 14 does not form a clique as its size is less then the template size. It is clustered with another subregion in its proximity.

3.2 Clique Characterization and Partitioner Selection

The characterization of a clique is based on its computation and communication requirements, and its refinement homogeneity is defined in Section 4.1. Using the characterization of applications and partitioners presented in [14], partitioner-selection policies are defined to select the partitioners. The overall goal of these policies is to obtain better load balance for less refined cliques, and to reduce communication and synchronization costs for highly refined cliques. For example, the policy dictates that the GPA and G-MISP+SP partitioning algorithms be used for cliques with refinement homogeneity below some threshold and partitioning algorithms LPA and pBD-ISP be used for cliques with refinement homogeneity greater than the threshold.

4 Experimental Evaluation

4.1 Evaluating the Effectiveness of the SBC Clustering Algorithm

To aid the evaluation of the effectiveness of the SBC clustering scheme, a clustering quality metric is defined below. The static quality of a clique is measured in terms of its refinement homogeneity and the efficiency of the clustering algorithm. The dynamic quality of the clique hierarchy is measured in terms of its communication costs (intra-level, inter-level, and data migration).

(1) **Refinement Homogeneity:** This measures the quality of the structure of a clique. Let $|R_i^{total}(l)|$ denote the total size of a subregion or a clique at refinement level l, which is composed of $R_i^{ref}(l)$, the size of refined regions, and $R_i^{unref}(l)$, the size of un-refined regions at refinement level l. Refinement homogeneity is recursively defined between two refinement levels as follows:

$$H_i(l) = \frac{|R_i^{ref}(l)|}{|R_i^{total}(l)|}, \quad and \quad H_{all}(l) = \frac{1}{n}\sum_{i=1}^{n} H_i(l), \text{if } |R_i^{ref}(l)| \neq 0$$

where n is the total number of subregions that have refinement level $l+1$. The goal of AHMP is to maximize the refinement homogeneity of a clique as partitioners work well on relatively homogeneous regions.

(2) **Communication Cost:** This measures the communication overheads of a clique and includes inter-level communication, intra-level communication, synchronization cost, and data migration cost as described in Section 2. The goal of AHMP is to minimize the communication overheads of a clique.

(3) **Clustering Cost:** This measures the efficiency of the clustering algorithm itself. As mentioned above, SAMR applications require regular re-partitioning and re-balancing, and as a result clustering cost become important. The goal of AHMP is to minimize the clustering cost.

Partitioning algorithms typically work well on highly homogeneous grid structures. Hence, it is important to have a quantitative measure to specify homogeneity. Intuitively, the refinement homogeneity metric attempts to isolate refined cliques that are potentially heterogeneous. In contrast, unrefined cliques are homogeneous at their finest refinement level.

The effectiveness of SBC-based clustering is evaluated using the metrics defined above. The evaluation compares the refinement homogeneity of 6 SAMR application kernels with and without clustering. These application kernels span multiple domains, including computational fluid dynamics (compressible turbulence: RM2D and RM3D, supersonic flows: ENO2D), oil reservoir simulations (oil-water flow: BL2D and BL3D), and the transport equation (TP2D). The detailed descriptions and characterizations of these applications are presented in [14].

The average refinement homogeneity of 6 SAMR applications without clustering is presented in Table 1. The table shows that the refinement homogeneity $H(l)$ increases as the refinement level l increases. Typical ranges of $H(l)$ are: $H(0) \in [0.02, 0.22]$, $H(1) \in [0.26, 0.68]$, $H(2) \in [0.59, 0.83]$ and $H(3) \in [0.66, 0.9]$. Several outliers require some explanation. In case of the BL2D application, average $H(2) = 0.4$. However, the individual values of $H(2)$ are in the range $[0.6, 0.9]$ with many scattered zeros. Since the refinement homogeneity on level 3 and above is typically over 0.6 and refined subregions on deeper refinement levels tend to be more scattered, the clustering schemes will focus efforts on clustering level 0, 1 and 2. Furthermore, based on these statistics, we set the threshold θ for switching between different lower-level partitioners as follows: $\theta_0 = 0.4$, $\theta_1 = 0.6$, and $\theta_2 = 0.8$, where the subscripts denote the refinement level.

Table 1. Average Refinement Homogeneity $H(l)$ for 6 SAMR Applications

Application	Level0	Level1	Level2	Level3
TP2D	0.067	0.498	0.598	0.6680
RM2D	0.220	0.680	0.830	0.901
RM3D	0.427	0.618		
ENO2D	0.137	0.597	0.649	0.761
BL3D	0.044	0.267		
BL2D	0.020	0.438	0.406	0.316

Table 2. Homogeneity Improvements using SBC

Application	Level0	Level1	Gain on Level0	Gain on Level1
TP2D	0.565	0.989	8.433	1.986
RM2D	0.671	0.996	3.050	1.465
RM3D	0.802	0.980	1.878	1.586
ENO2D	0.851	0.995	6.212	1.667
BL3D	0.450	0.583	10.227	2.184
BL2D	0.563	0.794	28.150	1.813

The effects of clustering using SBC for the 6 SAMR applications are presented in Table 2. In the table, the gain is defined as the ratio of the improved homogeneity over the original homogeneity at each level. The gains for RM3D and RM2D applications are smaller because these applications already exhibit high refinement homogeneity. These results demonstrate the effectiveness of the clustering scheme.

4.2 Performance Evaluation

This section presents an evaluation of the AHMP scheme using the clustering quality metrics defined above.

Clustering Costs: The cost of the SBC clustering algorithm is experimentally evaluated using the 6 different SAMR application kernels on a Beowulf cluster (Frea) at Rutgers University. The cluster consists of 64 processors and each processor has a 1.7 GHz Pentium IV CPU. The costs are plotted in Figure 4. As seen in this figure, the overall clustering time on average is less than 0.01 second. Note that the computational time between successive repartitioning phases is typically in the order of 10's of seconds, and as a result, the clustering costs are not significant.

Overall Performance: The overall performance benefit of the AHMP scheme is evaluated on DataStar, the IBM SP4 cluster at San Diego Supercomputer Center. DataStar has 176 (8-way) P655+ nodes (SP4). Each node has 8 (1.5 GHz) processors, 16 GB memory, and CPU peak performance is 6.0 GFlops. The evaluation uses the RM3D application kernel with a base grid of size 256x64x64, up to 3 refinement levels, and 1000 base level time steps. The number of processors used was between 64 and 1280.

The overall execution time is plotted in Figure 5. The figure plots execution times for GPA, LPA and AHMP. The plot shows that SBC+AHMP delivers the best performance.

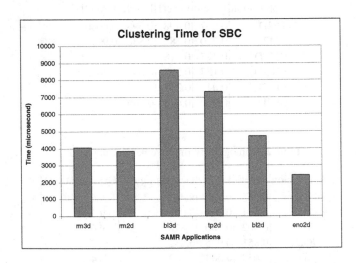

Fig. 4. Clustering Costs for the 6 SAMR Application Kernels

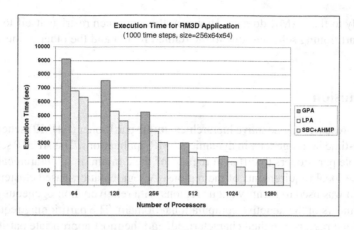

Fig. 5. Overall Performance for RM3D

Compared to GPA, the performance improvement is between 30% to 42%. These improvements can be attributed to the following factors: (1) AHMP takes advantage of the strength of different partitioning schemes matching them to the requirements of each clique; (2) the SBC scheme creates well-structured cliques, which reduces the communication between cliques; (3) AHMP enables incremental repartitioning/redistribution and concurrent communication between resource groups, which extends the advantages of HPA [10].

5   Related Work

Traditional parallel SAMR implementations presented in [6] [11] use dynamic partitioning and load-balancing algorithms. These approaches view the system as a flat pool of processors. They are based on global knowledge of the state of the adaptive grid hierarchy, and partition the grid hierarchy across the set of processors. Global synchronization and communication is required to maintain this global knowledge and can lead to significant overheads on large systems. Furthermore, these approaches do not exploit the hierarchical nature of the grid structure and the distribution of communications and synchronization in this structure. Dynamic load balancing schemes for distributed SAMR applications are proposed in [9], which consist of two phases: global load balancing and local load balancing. However, simplistic partitioning schemes are used without explicitly addressing the spatial and temporal heterogeneity exhibited by SAMR applications. The characterization of SAMR applications presented in [14] was based on the entire physical domain. The research in this paper goes one step further by considering the characteristics of individual subregions. The concept of natural regions was proposed in [13]. Two kinds of natural regions were defined: unrefined/homogeneous and refined/complex. The framework proposed in the paper then used a bi-level domain-based (BLED) partitioning scheme to partition the refined subregions. This approach is one of the first attempts to apply multiple partitioners concurrently to the SAMR domain. However, this approach restricts itself to applying only two partitioning schemes, one to the refined region and the other to the unrefined region.

6 Conclusion

This paper presented the adaptive hierarchical multi-partitioner (AHMP) scheme to address space-time heterogeneity in dynamic SAMR applications. The AHMP scheme applies multiple partitioners to different regions of the domain, in a hierarchical manner, to match the local requirements of the regions. A segmentation-based clustering algorithm (SBC) was used to identify clique regions with relatively homogeneous partitioning requirements in the adaptive computational domain. The partitioning requirements of these clique regions are then characterized, and the most appropriate partitioner for each clique is selected. The AHMP framework and its components have been implemented and experimentally evaluated using 6 SAMR application kernels. The evaluations demonstrated both, the effectiveness of the clustering as well as the performance improvements using AHMP.

References

1. M. Berger and J. Oliger. Adaptive mesh refinement for hyperbolic partial differential equations. *Journal of Computational Physics*, 53:484–512, 1984.
2. M. Berger and I. Regoutsos. An algorithm for point clustering and grid generation. *IEEE Transactions on Systems, Man and Cybernetics*, 21(5):1278–1286, 1991.
3. J. Cummings, M. Aivazis, R. Samtaney, R. Radovitzky, S. Mauch, and D. Meiron. A virtual test facility for the simulation of dynamic response in materials. *Journal of Supercomputing*, 23:39–50, 2002.
4. K. Devine, E. Boman, R. Heaphy, B. Hendrickson, and C. Vaughan. Zoltan data management services for parallel dynamic applications. *Computing in Science and Engineering*, 4(2):90–97, 2002.
5. R. C. Gonzalez and R. E. Woods. *Digital Image Processing*. Prentice Hall, Upper Saddle River, NJ, 2nd edition, 2002.
6. R. D. Hornung and S. R. Kohn. Managing application complexity in the samrai object-oriented framework. *Concurrency and Computation - Practice & Experience*, 14(5):347–368, 2002.
7. L. V. Kale. Charm. URL: http://charm.cs.uiuc.edu/research/charm/.
8. G. Karypis and V. Kumar. Parmetis, 2003. URL: http://www-users.cs.umn.edu/~karypis/metis/parmetis/index.html.
9. Z. Lan, V. Taylor, and G. Bryan. Dynamic load balancing of samr applications on distributed systems. *Journal of Scientic Programming*, 10:4:319–328, 2002.
10. X. Li and M. Parashar.
    Dynamic load partitioning strategies for managing data of space and time heterogeneity in parallel samr applications. In *The 9th International Euro-Par Conference (Euro-Par 2003), Lecture Notes in Computer Science*, volume 2790, pages 181–188. Springer-Verlag, 2003.

11. M. Parashar and J. Browne. On partitioning dynamic adaptive grid hierarchies. In *29th Annual Hawaii Int. Conference on System Sciences*, pages 604–613, 1996.
12. H. Sagan. *Space Filling Curves*. Springer-Verlag, 1994.
13. J. Steensland. *Efficient Partitioning of Structured Dynamic Grid Hierarchies*. PhD thesis, Uppsala University, 2002.
14. J. Steensland, S. Chandra, and M. Parashar. An application-centric characterization of domain-based sfc partitioners for parallel samr. *Ieee Transactions on Parallel and Distributed Systems*, 13(12):1275–1289, 2002.

Data and Computation Abstractions
for Dynamic and Irregular Computations

Sriram Krishnamoorthy[1], Jarek Nieplocha[2], and P. Sadayappan[1]

[1] Department of Computer Science and Engineering,
The Ohio State University, Columbus, OH 43210, USA
{krishnsr, saday}@cse.ohio-state.edu
[2] Computational Sciences and Mathematics,
Pacific Northwest National Laboratory, Richland, WA 99352, USA
jarek.nieplocha@pnl.gov

Abstract. Effective data distribution and parallelization of computations involving irregular data structures is a challenging task. We address the twin-problems in the context of computations involving block-sparse matrices. The programming model provides a global view of a distributed block-sparse matrix. Abstractions are provided for the user to express the parallel tasks in the computation. The tasks are mapped onto processors to ensure load balance and locality. The abstractions are based on the Aggregate Remote Memory Copy Interface, and are interoperable with the Global Arrays programming suite and MPI. Results are presented that demonstrate the utility of the approach.

1 Introduction

The development of scalable application codes is a challenging task. The parallelism in the underlying problem needs to be identified and exposed; the data and computation then must be partitioned and mapped onto processors. Computation partitioning exposes the parallelism in the computation. Data distribution and mapping, together with appropriate mapping of the computation to the processors, can potentially eliminate communication costs, resulting in good scalability. When communication costs cannot be completely eliminated, alternative approaches are taken to minimize the adverse effects of communication on scalability. Data distribution is used to avoid communication *hot-spots*, minimizing node contention. Non-blocking communication primitives are used to overlap communication with computation. In message passing architectures, mechanisms to minimize communication and synchronization have been studied [1].

Effective data distribution and computation mapping of computation involving irregular data structures is a challenging task. The communication patterns of such computations are known only at runtime. This hinders attempts at compile-time analysis and automatic parallelization.

In this paper, we address the twin problems of data distribution and computation mapping in the context of block-sparse matrices. The user is presented with a global view of a distributed block-sparse matrix. The programmer identifies the parallelism in the computation, in the form of independent tasks. He/she also specifies the locality information for each task, in terms of the needed data from the global space. The data

D.A. Bader et al. (Eds.): HiPC 2005, LNCS 3769, pp. 258–269, 2005.

is distributed amongst the processors such that node contention is minimized. The tasks are mapped onto processors so as to maximize locality and minimize communication, while ensuring load-balance. The mechanisms for computation mapping, though presented in the context of computations involving block-sparse matrices, are applicable to other computations that incur non-trivial communication costs.

These abstractions are based on the Aggregate Remote Memory Copy Interface (ARMCI) [2], a distributed-memory one-sided communication mechanism, available as part of the Global Arrays programming suite [3]. The Global Arrays suite provides a variety of programming models, each at a different level of abstraction. The abstraction provided for block-sparse matrices is equivalent to the Global Arrays (GA) abstraction that provides a global view of a distributed dense multi-dimensional array. The primitives provided are inter-operable with the Global Arrays suite, and hence with MPI.

The principal contributions of this paper are as follows:

1. Definition of a high-level model for block-sparse matrices, that facilitates a global view of a distributed block-sparse array.
2. A computation abstraction that allows the user to express the parallelism in the computation. The information presented by the user is used to perform the computation in a locality-aware load-balanced fashion.
3. Performance studies that the demonstrate improved scalability and performance achieved by the proposed mechanisms.

The paper is organized as follows. In Section 2, we discuss the applications that motivated our work. The Global Arrays suite is described in Section 3. The global abstraction for block-sparse arrays is introduced in Section 4. Mechanisms provided for locality-aware load-balancing are presented in Section 5. The proposed model is evaluated by comparing with alternative mechanisms and the results are discussed in Section 6. Related work is detailed in Section 7. Section 8 concludes the paper.

2 Target Applications

2.1 Tensor Contraction Expressions

The development of these primitives is primarily motivated by our work on the The Tensor Contraction Engine (TCE) [4] synthesis system. TCE is a domain-specific compiler for expressing ab initio quantum chemistry models. The TCE takes as input a high-level specification of a computation, expressed as a set of tensor contraction expressions, and transforms it into efficient parallel code. Each tensor contraction expression is comprised of a collection of multi-dimensional summations of products of several block-sparse input arrays. An operation on the indices of the segments that form a block of an array determines if it is non-zero. The wide-ranging sizes of the blocks leads to significant variation in the computation and communication times involved in processing a block. The large sizes of the arrays can significantly increase communication costs, if locality is not taken into account.

2.2   Lennard Jones Energy Minimization Using Force Decomposition

Load balancing is important for force decomposition molecular dynamics algorithms. The array of forces of dimension $N \times N$ is divided into multiple blocks of size $m \times m$, where m is the block size and N is the total number of atoms. Each process owns N/P atoms, where P is the total number of processors, and each processor computes a fixed subset of inter-atomic forces [5]. The forces between atoms farther from each other than the *cut-off distance* need not be evaluated, resulting in unequal processing times for each subset of the force-matrix. This, together with the block decomposition of the force matrix, leads to load imbalance.

2.3   Parallel Dense Matrix Multiplication

Dense matrix multiplication can also benefit from our abstraction for computation mapping. In many scientific applications, the matrix distribution is based on the underlying physical problem and might involve variable block sizes on individual processors, leading to load imbalance. The computation involving these matrices can be partitioned into equal-sized *blocks*, independent of the underlying distribution. The assignment of the logical blocks to individual processors is determined at run-time to achieve load balancing. Taking locality into account can improve performance.

In general, the abstraction presented for computation mapping can benefit applications that:

- Can be partitioned into independent tasks,
- Involve many more tasks than the number of processors,
- Have wide variation in task execution times, and
- Operate on coarse-grain data, and incur communication costs if the task and the data it operates on are not co-located.

Note that computations with data dependences can also benefit from this mechanism, provided there is enough parallelism at any point in the computation. For example, while performing a sequence of block-sparse matrix multiplies, each matrix multiply can be treated as a set of independent tasks and processed using this mechanism.

3   Global Arrays Programming Suite

The Global Arrays programming suite [3] provides a set of inter-operable programming models, each at a different level of abstraction. At the lowest level is MPI, a distributed-memory programming model with message passing for two-sided communication. Though MPI is not part of the suite, it is fully inter-operable with the abstractions provided in the suite, and is an integral part of the hierarchy of abstractions presented to the user.

The Aggregate Remote Memory Copy Interface (ARMCI) library [2] provides a distributed-memory view with one-sided access to remote data. It has a rich set of primitives for non-blocking operations, and contiguous and non-contiguous data transfers optimized to hide latency. ARMCI forms the underlying communication layer for a number of compile/runtime systems, including Co-Array Fortran [6], GPSHMEM [7], and Global Arrays.

The next higher level is the Global Arrays (GA) library. GA exposes a global view of a dense multi-dimensional array distributed amongst the local memories of processors. It provides a shared-memory programming model in which data locality is explicitly managed by the programmer. Explicit function calls are used to transfer data between global address space and local storage. It is similar to distributed shared-memory models in providing an explicit acquire-release protocol, but differs with respect to the level of explicit control in moving blocks of data in multidimensional arrays between remote global storage and local storage. The functionality provided by GA has proved useful in the development of large scale parallel quantum chemistry suites such as NWChem [8] (which contains over a million lines of code), adaptive mesh refinement codes such as NWPhys/NWGrid (www.emsl.pnl.gov/nwphys) and applications in other areas [3].

The Disk Resident Arrays (DRA) model [9] extends the GA programming model to secondary storage. It provides a disk-based representation for multi-dimensional arrays and operations to transfer blocks of data between global arrays and disk resident arrays.

ARMCI, GA, and DRA provide a unified programming model for handling different levels of the memory hierarchy in which the user controls the location of data in the memory hierarchy. This has been shown to achieve high performance, while being a simpler programming model than message passing.

4   Abstraction for Block-Sparse Matrices

The abstraction provided for multi-dimensional block-sparse arrays provides collective functions for creating and destroying arrays and non-collective functions to get/put arbitrary multi-dimensional non-zero regions of the global memory.

The creation and destruction of the arrays is divided into two steps. For each array, an index is first created. The index stores information on the location of the different portions of the data in the distributed memory system. The array is then created, using this index. The decoupling of the creation of the index from the actual creation of the array simplifies creation of multiple aligned arrays. In computations in which memory is dynamically allocated and freed, the index can be computed once, while the actual memory for the array is dynamically allocated and freed.

The abstraction is constructed using the ARMCI library. The one-sided mechanisms provided by the ARMCI library, together with the index, enables the non-collective access functionality.

The arrays can be created by specifying the number of dimensions, the number of blocks, and the actual block sizes. In addition, a *bitmap* can be provided to specify whether a block is zero. Alternatively, a function that takes as argument the block indices and returns whether it is zero, can be provided.

The non-zero blocks of the array are divided into bricks, which are then distributed amongst the processors in a round-robin fashion. This ensures a uniform distribution of the data among all the processors. The user can specify the typical access pattern, to provide hints on the choice of appropriate bricking. A small brick size allows for a more uniform distribution of the data amongst the processors. On the other hand, a large brick size allows for coarse-grained, and possibly more efficient, computation and potential reduction in the communication cost, due to amortization of the communication latency.

The index stores the offsets of all non-zero bricks together with the processor to which it is assigned. The creation of the array allocates a contiguous chunk of memory in each processor. The brick offsets in the index determine the offset of each brick in this contiguous chunk. The index is replicated on all the processors.

The index is constructed using a two-level scheme. Each non-zero block contains a pointer to a dense multi-dimensional array. Each element in this array corresponds to a brick in that block and contains information on the processor to which that brick is assigned, and the offset.

5   Abstraction for Locality-Aware Load-Balancing

5.1   Computation Specification

The abstraction provided to the user enables the specification of a set of independent tasks to be executed in parallel. For each such set, all processes collectively create a *task_pool* object using the *create_task_pool* method.

Each task in the task pool is identified by the routine to be invoked to process that task, identified by a function handle, and the set of *locality elements* it operates upon. In addition, any private data specific to that task can also be specified. Each locality element corresponds to a global data region, identified by its global address, size, and its access mode. Three access modes are supported. Read, write, and access modes allow for put, get, and accumulate of global data.

For dense and block-sparse arrays, the global address is replaced by the array handle and the specification of the data region being accessed.

Tasks are added to a task pool using the *add_task* method. The creation and addition of tasks to the task pool is done by all the processes, in a replicated fashion. Once all the tasks have been added to the task pool, *seal_task_pool* method is used to *seal* the work pool. This method is invoked once for a task pool and is used to perform start-time optimizations.

Subsequently, all the processes collectively invoke the *process_task_pool* method to process the tasks in the task pool. A task pool, once created, can be processed multiple times. The cost of start-time optimizations, performed once, are thus be amortized.

5.2   Implementation

In this section, we discuss the implementation of the locality-aware load-balancing abstraction. Computation mapping is modeled as a hypergraph partitioning problem and a hypergraph partitioning solver, PaToH [10], is used to determine the mapping of the tasks to processors.

Problem Definition. A computation is to be performed on globally addressable data. The data is partitioned into non-overlapping regions and is distributed across the memories of the processors, such that each region is assigned to one and only one processor.

```
TILE_MATMUL(loc_list, private_data) ::
  integer Ni, Nj, Nk, i, j, k
  double *A, *B, *C

  !Actual communication handled outside this routine
  A = get_ptr(loc_list[1]) !Fetch pointer to data
  B = get_ptr(loc_list[2])
  C = get_ptr(loc_list[3])

  Ni = private_data[0] !Tile sizes
  Nj = private_data[1]
  Nk = private_data[3]

  !Matrix multiply for this task
  for i = 0 to Ni-1
    for j = 0 to Nj-1
      for k = 0 to Nk-1
        C[i,j] += A[i,k] * B[k,j]
  !Freeing of buffers and writing/accumulating output data
  !handled outside this routine
```

Fig. 1. Routine to process a single task in block-sparse matrix multiply

The computation is expressible as a set of independent tasks. Each task takes as input a set of data regions and reads, writes and/or updates (accumulates), one or more data regions. The computation cost of each task is also provided.

Note that each task can be executed on any processor. The input data regions associated with the task are brought into local memory and the task is executed. The output data are then written/accumulated into the global regions. If a task is executed on a processor that contains the data regions required by it, no communication is required. In addition, if a set of tasks that require the same data regions are co-located in a processor, communication cost can be significantly reduced by reusing the read-only data across tasks.

We assume that we have enough memory to store all the data required by all the tasks. Thus, given a set of tasks assigned to a processor, the amount of communication performed by that processor is equal to the total size of all the distinct data regions accessed by all the tasks assigned to it.

The objective is to partition the set of tasks among the available processors, such that the amount of communication required is minimized, while maintaining the balance of computational load amongst the processors.

Hypergraph Partitioning. A hypergraph $H = (V,N)$ is defined as a set of vertices V and a set of nets (hyper-edges) N among those vertices. For every net n_j, s_j is equal to the number of vertices it has, i.e., $s_j = |n_j|$. Weights (w_i) and costs (c_j) can be assigned to the vertices ($v_i \in V$) and edges ($n_j \in N$) of the hypergraph, respectively. $P = \{V_1, V_2, ..., V_P\}$ is a P-way partition of H if 1) each part is a nonempty subset of V, 2) the parts are pairwise disjoint, and 3) union of the P parts is equal to V. A partition is said to be balanced if $W_p \leq W_{avg}(1+\varepsilon)$ for $1 \leq p \leq P$, where $Wp = \sum_{v_i \in V_p} w_i$ is the sum of the vertex weights of part V_p, $W_{avg} = (\sum_{v_i \in V} w_i)/P$ denotes the weight of each part under the perfect load balance condition, and ε represents the predetermined maximum imbalance ratio allowed. In a partition P of H, a net that has at least one vertex in a part is said to connect that part. Connectivity λ_j of a net n_j denotes the number of parts connected by n_j. A net n_j is said to be cut if it connects more than one part (i.e. $\lambda_j > 1$). The cut nets are also referred to as external nets and is denoted as N_E. There are various cut-size definitions for representing $\chi(P)$ of a partition P. The relevant, connectivity-1, definition is:

$$\chi(P) = \sum_{n_j \in N_E} c_j(\lambda_j - 1) \tag{1}$$

In equation 1, each cut net n_j contributes $c_j(\lambda_j - 1)$ to the cut-size. The hypergraph partitioning problem can be defined as the task of dividing a hypergraph into two or more parts such that the cut-size is minimized, while a given balance criterion among the part weights is maintained. Algorithms based on the multi-level paradigm, such as hMETIS [11] and PaToH [12], have been shown to compute good partitions quickly for this NP-hard problem.

Modeling Locality-Aware Load-Balancing. We model the problem of locality-aware load-balancing as a hypergraph partitioning problem. Each data region and task in the computation has a corresponding vertex in the hypergraph. A net is introduced in the hypergraph for every data region in the computation. For each data region, the corresponding net connects the vertices corresponding to it and the tasks that access it. The weight associated with each net is the communication cost associated with the data region. We model it to be the size of the data region. The cost of a vertex is zero if it corresponds to a data region, and is the number of operations to be executed if it corresponds to a task.

We can evaluate the hypergraph thus constructed in two ways. It can be used to determine the assignment of both the tasks and data regions to processors. If the data regions are pre-distributed and cannot be remapped, the distribution of the data regions amongst the processors can be pre-specified by constraining each data region to be on a specific processor. The hypergraph is then partitioned to determine the mapping of the tasks to the processors. Given a partition, the cost incurred by a net is the size of the corresponding data region, times the number of remote processors that have been assigned at least one task that accesses this region. The total cost of all the nets is given by the connectivity metric, shown in equation 1.

5.3   Illustration

The work-sharing construct is illustrated using an implementation of block-sparse matrix multiply, shown in Fig. 2. The multiplication is of the form

$$C[i,j] + = A[i,k] * B[k,j]$$

All dimensions are assumed to be divided into *nblocks* segments. The binary operator *op* is applied to the segment indices to determine whether a block is non-zero. Parameters *g_a*, *g_b*, and *g_c* correspond to A, B, and C arrays, respectively. Methods

```
MATMUL(g_c, g_a, g_b, nblocks) ::
  task_pool_t tp
  locality_info_list_t loc_list

  tp = create_task_pool() !Create task pool
  for i=0 to nblocks - 1
    for j=0 to nblocks - 1
      if op(i,j) <> 0  !Non-zero C block
        for k=0 to nblocks - 1
          if op(i,k) <> 0 and op(k,j) <> 0 !Non-zero A/B blocks
            private_data[3] = {bi, bj, bk} !Tile sizes
            loc_list.add(g_c, get_range(g_c,i,j), ACCESS_UPDATE)
            loc_list.add(g_a, get_range(g_a,i,k), ACCESS_READ)
            loc_list.add(g_b, get_range(g_a,k,j), ACCESS_READ)

            !add task to task pool
            add_task(tp, tile_matmul, loc_list, private_data)

  seal_task_pool(tp) !Any start-time optimizations
  for i = 0 to maxiter !An iterative computation
    process_task_pool(tp) !Process all tasks, every iteration
  destroy_task_pool(tp) !Destroy task pool
```

Fig. 2. Block-sparse matrix multiply using the load-balancing abstraction. Each task is processed by the routine in Fig. 1. The task pool is processed *maxiter* times, but is created and sealed once.

get_ranges and *get_size* are used to compute the ranges and sizes of the non-zero block, respectively.

Fig. 1 shows the routine used to process an individual task, matrix-multiply involving a tile from each of the arrays. Note that no explicit communication is involved. The routine assumes that all input data are read into local memory and all output data are written/accumulated into global memory. Fig. 2 shows the implementation of parallel matrix multiply using this routine.

6   Experimental Results

We evaluated the primitives on the Colony2a system in the Pacific Northwest National Laboratory, a twenty-four node cluster with each node being a dual 1GHz Itanium-2 with 6GB memory. We used the Infiniband network available on the cluster for our experiments.

Three alternative load-balancing schemes were implemented for comparison. In the first scheme, henceforth referred to as the *Random* scheme, each processor traverses the entire list of tasks in the same order. For each task in the traversal, each processor generates a pseudo-random number between 0 and $P - 1$, where P is the number of processors. If the random number generated is the processor's rank, the processor executes that task. Since all the processors start with the same random seed, they all generate the same sequence of pseudo-random numbers. This ensures that each task is executed by exactly one process. The randomization results in a uniform distribution of the number and sizes tasks to processors. Note that this scheme balances the number of tasks and not task execution times. In addition, locality is not taken into account.

In the second scheme, one of the locality elements in each task is marked. Each task is executed by the process that "owns" the marked locality element in that task. This scheme is referred to as the *Owner* scheme. This scheme ensures locality for the array used to determine the ownership. Though the round-robin distribution ensures a reasonably balanced distribution of the data and hence the ownership, computational load is not guaranteed to be balanced.

The third scheme is based on dynamic load balancing. In this scheme, referred to as *NextTask*, all the processes enumerates the tasks to be executed in the same order. A global shared counter is used to determine the next task to be executed. Each process, when idle, performs an atomic *fetch-and-add* of the global shared counter. The value obtained by the process specifies the next task to be executed by it. All processes continue this procedure until the counter exceeds the number of tasks to be processed. The strictly increasing counter ensures that no task is executed more than once. It also keeps all the processes busy, till there are no more tasks to be executed. This ensures load balancing. But locality is not taken into account.

Note that this scheme is similar to self-scheduling in OpenMP . This is also the typical model of parallelization used in many applications, including some quantum chemistry codes [13].

Execution times of the following tensor contraction expression, typical of those encountered in quantum chemistry, were measured:

$$a,b,c,d : O$$
$$i : V$$
$$C[a,b,c,d] = A[a,b,i] * B[i,c,d]$$

where O and V correspond to the number of occupied and virtual orbitals, respectively. They are divided into a number of symmetry segments, in turn dividing the matrix into a set of blocks. For example, if O is divided into four symmetry segments, array C would consist of 64 blocks. A block of a matrix is non-zero if a function of its block segment indices is equal to the symmetry value associated with the matrix. Typically, the function is an exclusive OR operator and the symmetry of a matrix is zero. The tensor contraction is, in effect, a block-sparse matrix multiply. The indices were divided into four symmetry segments. The O index was set at 160 with four symmetry segments of length 80, 40, 20, and 20, respectively. The value of V was varied to be a multiple k of O, with k varying from 1 to 16. The number of processors was varied from 2 to 32.

The execution times for four and thirty two processors, which are representative of the trend, are shown in Fig. 3. Other results are not shown due to space restrictions. The three alternative schemes, labeled *Random*, *Owner*, and *NextTask*, and our approach, labeled *Our*, are shown. For our approach the cost is shown including and excluding the overhead of hypergraph partitioning.

For smaller numbers of processors, the communication cost and the hypergraph partitioning overhead are not significant. Hence, the difference in the performance of the various schemes is minimal. Increase in the number of processors increases the communication cost. Our scheme, being locality-aware, performs increasingly better than

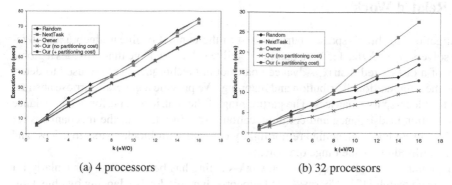

(a) 4 processors   (b) 32 processors

Fig. 3. Execution time, in seconds, of block-sparse matrix multiply for various schemes. Time is shown in y-axis. k=(V/O) is shown along x-axis. Each graph corresponds to a different number of processors.

the other schemes. The NextTask scheme, which completely ignores locality, performs progressively worse. The Owner scheme ensures locality for at least one of the arrays, thus performing better. The Random scheme, performs better than the NextTask and Owner schemes, due to the benefits of randomization.

The cost of hypergraph partitioning increases the cost of our load-balancing mechanism. Though the overhead increases with increase in the number of processors, our mechanism still performs better, even when partitioning overhead is taken into account. Note that in typical applications, the partitioning overhead is amortized over multiple processings of the same task pool.

The speedups obtained by the different schemes, for number of processors varying from 2 to 32, and for k value being 8 and 16, are shown in Fig. 4. The sequential execution times were determined for these problem sizes, and are shown in the figure. Our approach achieves a speed-up of up to 20 on 32 processors, excluding partitioning cost.

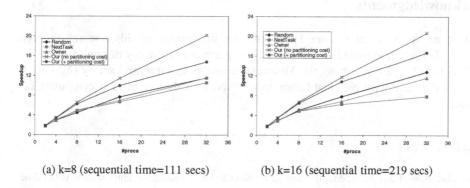

(a) k=8 (sequential time=111 secs)   (b) k=16 (sequential time=219 secs)

Fig. 4. Scalability of the schemes for block-sparse matrix multiply for (a) k=8 and (b) k=16. The number of processors is shown in x-axis. Speedup is shown in y-axis. The corresponding sequential execution times are also shown.

7 Related Work

Abstractions for block-sparse matrices exist in the context of linear algebra and iterative solvers [14] . Aztec [15] is a parallel iterative solver package that provides a global view of a distributed matrix. Advanced partitioning techniques [16] are used to determine the computation distribution and mapping. We provide a general-purpose abstraction for block-sparse matrices. The partitioning of the matrices is performed to balance computation load-balance and communication costs. In addition, the mechanisms for locality-aware load-balancing are not tightly coupled with block-sparse matrices, and can be utilized in a wide range of contexts.

Dynamic load-balancing based on work-stealing has been studied, particularly for state-space search [17] . Charm++ [18] supports dynamic load-balancing by object migration. Cilk [19] supports load-balancing of computations based on work-stealing. OpenMP exploits parallelism at the loop level by distributing different iterations to different processors. Locality is not taken into consideration in any of these schemes. The self-scheduling strategy in OpenMP is similar to the *NextTask* scheme that was evaluated earlier.

Çatalyürek and Aykanat [12] have used hypergraph-partitioning to parallelize sparse matrix-vector multiplications. Chang et al. [20] performed parallel data aggregation based on hypergraphs.

8 Conclusions

We designed and implemented high-level abstractions for manipulating block-sparse matrices. Computation primitives to improve load balancing, by exploiting locality, were presented. The programmer exposes the parallelism in the computation, and the system determines the computation mapping. Our approach consistently performs better than the alternative schemes considered for load-balancing.

Acknowledgments

We thank the National Science Foundation for the support of this research through grants 0121676 and 0403342, and the U.S. Department of Energy through award DE-AC05-00OR22725. We thank the Molecular Sciences Computing Facility (MSCF) at the Pacific Northwest National Laboratory (PNNL) for the use of their computing facilities.

References

1. Lim, A.W., Lam, M.S.: Maximizing parallelism and minimizing synchronization with affine partitions. Parallel Computing **24** (1998) 445–475
2. Nieplocha, J., Carpenter, B.: ARMCI: A Portable Remote Memory Copy Library for Distributed Array Libraries and Compiler Run-time Systems. In: Proc. 3rd Workshop on Run-time Systems for Parallel Programming (RTSPP). (1999)

3. Nieplocha, J., Palmer, B., Tipparaju, V., Krishnan, M., Trease, H., Apra, E.:   Advances, Applications and Performance of the Global Arrays Shared Memory Programming Toolkit. Intern. J. High Perf. Comp. Applications **to appear** (2005)
4. Baumgartner, G., Bernholdt, D., Cociorva, D., Harrison, R., Hirata, S., Lam, C., Nooijen, M., Pitzer, R., Ramanujam, J., Sadayappan, P.: A High-Level Approach to Synthesis of High-Performance Codes for Quantum Chemistry. In: Proc. of Supercomputing 2002. (2002)
5. Plimpton, S.J., Hendrickson, B.A.: Parallel molecular dynamics with the embedded atom method. In: Proc. of Materials Theory and Modelling, MRS Proceedings (1993) 37
6. Coarfa, C., Dotsenko, Y., Mellor-Crummey, J.: A Multi-Platform Co-Array Fortran Compiler. In: Proc. of PACT. (2004)
7. Parzyszek, K., Nieplocha, J., Kendall, R.A.: A Generalized Portable SHMEM Library for High Performance Computing. In: Proc. of the IASTED Parallel and Distributed Computing and Systems. (2000) 401–406
8. High Performance Computational Chemistry Group: NWChem, A Computational Chemistry Package for Parallel Computers, Version 4.6. Pacific Northwest National Laboratory. (2004)
9. Nieplocha, J., Foster, I.: Disk Resident Arrays: An Array-Oriented I/O Library for Out-Of-Core Computations. In: Proc. 6th Symposium on the Frontiers of Massively Parallel Computation. (1996) 196–204
10. Çatalyürek, U.V., Aykanat, C.: PaToH: A Multilevel Hypergraph Partitioning Tool, Version 3.0. Bilkent University, Department of Computer Engineering. (1999)
11. Karypis, G., Aggrawal, R., Kumar, V., Shekhar, S.: Multilevel hypergraph partitioning: Applications in VLSI domain. In: Proc. of 34th Design Automation Conference. (1997)
12. Çatalyürek, U.V., Aykanat, C.: Hypergraph-partitioning based decomposition for parallel spars e-matrix vector multiplication. IEEE TPDS **10** (1999) 673–693
13. Hitara, S.: Tensor contraction engine: Abstraction and automated parallel implementation of configuration-interaction, coupled-cluster, and many-body perturbation theories. J. Phys. Chem. A **107** (2003) 9887–9897
14. Duff, I.S., Marrone, M., Radicati, G., Vittoli, C.: Level 3 basic linear algebra subprograms for sparse matrices: a user-level interface. ACM Trans. Math. Softw. **23** (1997) 379–401
15. Tuminaro, R.S., Heroux, M., Hutchinson, S.A., Shadid, J.N.: Official Aztec user's guide: Version 2.1. Technical report, Sandia National Laboratories (1999)
16. Hendrickson, B., Leland, R.: The Chaco user's guide: Version 2.0. Technical Report SAND94–2692, Sandia National Laboratories (1994)
17. Sinha, A., Kalé, L.: A load balancing strategy for prioritized execution of tasks. In: Seventh International Parallel Processing Symposium, Newport Beach, CA. (1993) 230–237
18. Kalé, L., Krishnan, S.: CHARM++: A Portable Concurrent Object Oriented System Based on C++. In Paepcke, A., ed.: Proceedings of OOPSLA'93, ACM Press (1993) 91–108
19. Randall, K.H.: Cilk: Efficient Multithreaded Computing. PhD thesis, MIT Department of Electrical Engineering and Computer Science (1998)
20. Chang, C., Kurc, T., Sussman, A., Çatalyürek, U.V., Saltz, J.: A hypergraph-based workload partitioning strategy for parallel data aggregation. In: Proceedings of the Eleventh SIAM Conference on Parallel Processing for Scientific Computing, SIAM (2001)

XCAT-C++: Design and Performance of a Distributed CCA Framework*

Madhusudhan Govindaraju, Michael R. Head, and Kenneth Chiu

Grid Computing Research Laboratory (GCRL),
State University of New York (SUNY) at Binghamton, NY 13902, USA
mgovinda@cs.binghamton.edu
head@acm.org, kchiu@cs.binghamton.edu

Abstract. In this paper we describe the design and implementation of a C++ based Common Component Architecture (CCA) framework, XCAT-C++. It can efficiently marshal and unmarshal large data sets, and provides the necessary modules and hooks in the framework to meet the requirements of distributed scientific applications. XCAT-C++ uses a high-performance multi-protocol library so that the appropriate communication protocol is employed for each pair of interacting components. Scientific applications can dynamically switch to a suitable communication protocol to maximize effective throughput. XCAT-C++ component layering imposes minimal overhead and application components can achieve highly efficient throughput for large data sets commonly used in scientific computing. It has a suite of tools to aid application developers including a flexible code generation toolkit and a python scripting interface. XCAT-C++ provides the means for application developers to leverage the efficacy of the CCA component model to manage the complexity of their distributed scientific simulations.

Keywords: CCA, XCAT-C++, component, performance, multi-protocol.

1 Introduction

The software engineering benefits of component based software have been widely described in the literature: components foster code re-usability and provide high level abstractions to shield users from low level details. They provide a manageable unit for software testing, distribution and management, and reduce the complexity of building large scale scientific applications, which often require the integration of multiple numerical libraries into a single application. The plug-and-play characteristic of component architectures provides the ability to reuse components in multiple applications, and serve performance needs by allowing components to be swapped at run-time with others that meet the required Quality of Service (QoS) metrics.

A consortium of university and national laboratory researchers launched the "CCA Forum" [1] in 1998, to develop a Common Component Architecture (CCA) specification for large scale scientific computation. The CCA specification defines the roles and functionality of entities necessary for high performance component-based application

* Supported in part by NSF grants CNS-0454298 and IIS-0414981.

D.A. Bader et al. (Eds.): HiPC 2005, LNCS 3769, pp. 270–279, 2005.

development. The specification is designed from the perspective of the required behavior of software components. However, the design and implementation of the framework, choice of communication protocol, and component discovery mechanisms have not been formally specified. This has facilitated different research groups to design, develop and evaluate the use of the same CCA specification to support a wide variety of applications [2].

XCAT-C++ is tailored for distributed scientific applications and is designed to meet the following goals: (1) the framework should have a modular design so that specialized modules can be easily loaded to extend the capabilities of the system; (2) applications should have the capability of seamlessly and dynamically switching to a suitable communication protocol to maximize effective throughput; (3) the overhead due to component layering should be minimal and not impact the overall performance of the distributed system; (4) each XCAT-C++ component should be capable of interacting with endpoints that are compliant with Grid Web services standards; (5) and a flexible, extensible and powerful code generation toolkit should be provided that can generate the transport protocol specific code and shield away the complexity of the run-time specific details in *stubs* and *skeletons*.

The remainder of this paper is organized as follows. In Section 2 we provide a brief introduction of the CCA specification and highlight its key concepts. In Section 3 we discuss in detail the design and implementation of the key features in XCAT-C++. Section 4 describes some utility modules in XCAT-C++. We present performance of XCAT-C++ in Section 5. Section 6 discusses related work, and we conclude with a summary and pointers to future work in Section 7.

2   The Common Component Architecture

The Common Component Architecture (CCA) [2] specification is an initiative to develop a common architecture for building large-scale scientific applications. CCA places minimal requirements on components to facilitate the integration of existing scientific libraries into a CCA framework and also to minimize the impact of the component layer on performance. The specification does not mandate the use of any specific form of distributed or parallel technology as the underlying communication architecture, thereby ensuring that it does not preclude applicability to serial, parallel, distributed or grid systems. CCA promotes interoperability by requiring all components to define their interfaces via a Scientific Interface Definition Language (SIDL) [3]. The Babel toolkit [4] can be used to generate glue code from SIDL to many programming languages including C, C++, Java, Fortran and Python. SIDL has been specifically designed for high performance scientific applications. It explicitly supports complex numbers, dynamic multi-dimensional arrays, parallel attributes, and communication directives.

Communication between CCA components takes place via their ports, which follow a *uses/provides* design pattern. A *provides* port is the public interface implemented by a component. It can be referenced and used by other components. It can also be viewed as the set of services that are exported by the component. A *uses* port is a connection endpoint that represents the set of functions that it needs to call. Port descriptions for CCA components are provided using the SIDL specification. CCA applications are composed by connecting the *uses* port of one component to the *provides* port of one another. The mechanism by which calls are transferred from the *uses* port to the *provides* port of the connected component is handled differently by each underlying framework.

3   Design and Implementation Features of XCAT-C++

3.1   Mapping CCA Concepts in XCAT-C++

Components in a distributed application often span multiple address spaces and are seldom co-located. Applications are developed by wiring components together into a component assembly. To facilitate this approach, distributed CCA frameworks need to provide support for remote invocation, wherein calls between components seamlessly cross machine boundaries. Components also need the capability to instantiate other components on remote machines. We list a few important CCA concepts and describe how they are designed and implemented in XCAT-C++.

– **Services Object**: Each CCA component contains a *Services* object that is responsible for managing the component's ports, including the ones that are dynamically added during the execution of the distributed application. Application developers can retrieve handles to a component's ports or just inspect its current state via the standard API of the *Services* object. In XCAT-C++, the *Services* object is designed to encapsulate the framework specific bindings for the *provides* and *uses* ports. Whenever a *uses* port is requested, a pointer to a local object is returned, while for a *provides* port a global (serializable) reference is returned that can be sent to components in remote address spaces. The serializable form of the reference contains information necessary to communicate with the provides port from any component. This information includes details such as host name, port number, communication protocol and a globally unique ID for the provides port.
– **ComponentID**: The CCA specification states that each component should design the `ComponentID` as an opaque handle, but does not require any standard format. The motivation for this approach is to allow each framework to design the handle according to its application requirements. In XCAT-C++, the handle has been designed as an object that is serialized to a string format whenever it is transported to another component. The idea is for the remote handle to be compatible with emerging standards in Grid Web services, which have adopted the Web Services Description Language (WSDL) document to represent distributed services. A WSDL document is an XML document that is commonly stored in a string format. This design also allows an XCAT-C++ *ComponentID* to be used for component assembly via work-flow engines [5].
– **Builder Service**: The CCA *Builder Service* presents a standard API for all components to instantiate, connect and disconnect other components. Once a component has been instantiated, the service returns a *ComponentID* to the new component. This *ComponentID* can then be used to directly communicate with the component. In XCAT-C++, the builder service instantiates new components from a set of name-value pairs that encapsulate the remote environment details such as command line

arguments, executable location, target machine name, and creation protocol. Currently, XCAT-C++ supports the use of SSH and we are testing the incorporation of the Grid Resource Allocation and Management (GRAM) service for authenticated launch of components on Grid resources.

- **Component Communication**: In distributed scientific computing, components are instantiated on remote machines and wired together dynamically with running components. As a result the choice of protocol depends on dynamically changing factors including the data type and size that needs to be transferred, security policies, and the list of common protocols supported by a pair of interacting components. Also, a component is typically connected to several components at any given time, each connection probably optimized for a different protocol. We discuss the communication system of XCAT-C++ in detail in Section 3.2.

3.2   XCAT-C++ and Grid Web Services

Web services have emerged as the architecture of choice for grid systems. Standards such as Open Grid Services Architecture (OGSA) [6] and Web Services Resource Framework [7] define a set of Web services based specifications for accessing Grid resources. These standards share many design features with the CCA specification [8]. We briefly discuss how XCAT-C++ components can be used with Grid Web services.

- Two choices for mapping XCAT-C++ components to Web services are (1) every XCAT-C++ component can be a Web service, with the endpoint in the WSDL document for the service pointing to the *ComponentID* of the component; or (2) every *provides* port of a component can itself be a Web service, as it has a well defined interface and endpoint. Unlike in Web services, two ports of the same type belonging to the same CCA component can exhibit semantically different behavior. To keep this flexibility we have chosen to map each *provides* port to a different Web service (and hence a different WSDL document) that can be uniquely identified and separately accessed by Web service clients.
- The Open Grid Services Infrastructure (OGSI) specification (precursor to the WSRF specification) required each Grid service to have a standard Grid Service Port. This requirement can be trivially met in CCA by defining a standard *provides* port with all the operations of the Grid Service Port. We are currently working on mapping the collection of five specifications (WS-Resource Properties, WS-Resource Lifetime, WS-RenewableReferences, WS-ServiceGroup, WS-BaseFaults) of the WSRF framework to standard CCA *provides* ports. The idea is to make these services available by default to all XCAT-C++ components. In the current implementation, each component has access to a *Builder Service* by default.
- Resource Lifetime Management: the life cycle of each component is managed via the *Builder Service* and a standard *go* port. The *Builder Service* design is based on the factory model. The *go* port of each component can be used to stop or kill the current execution of a component. As the *go* port of a component is a *provides* port, it is described in a WSDL document that can be used by clients to manage the component's lifetime.
- Service Handles: Each CCA component has a unique (opaque) handle represented as a *ComponentID* object within the component address space and some native

representation for the on-the-wire format. This concept maps directly to that of a Grid Service Handle (GSH) used by grid services. However, to enable interoperable communication between different CCA frameworks, we have proposed [9] that a standard CCA registry service be defined to convert the GSH of each framework to a WSDL format, which can serve as a service reference pointing to the endpoint of the *provides* port. This idea directly corresponds to the two level naming scheme adopted by grid Web services.

Multi-protocol Approach for Component Interactions.   The imperative for multi-protocol design is clear when we consider the diverse communication characteristics of various distributed applications. There is no single best protocol that can meet the requirements for all data types and communication patterns. We have successfully incorporated the *Proteus* [10] multi-protocol library as the communication substrate for XCAT-C++. Proteus currently has support for two protocols: (1) XBS [11], an efficient streaming binary protocol; and (2) XSOAP, a C++-based implementation of the SOAP specification. We list some features of the multi-protocol approach employed in XCAT-C++:

- For communication between two XCAT-C++ components, both enabled by Proteus, communication can switch to an optimized communication protocol on a per-call basis. Proteus provides an API, which the XCAT-C++ framework can call, to select the communication library to be used for subsequent calls. Also, communication modules can be dynamically loaded for each component. An example scenario is shown in Figure 1, in which three entities (A, B and C) are connected to form a distributed application. For communication between components A and B, the efficient XBS protocol is used. When interaction with the Web Service (entity C) is required, the components can automatically switch to XSOAP, as SOAP is the only protocol supported by the Web service.
- The use of Proteus can serve as the basis for interoperability with components running in a different CCA framework. For example, to communicate with a compo-

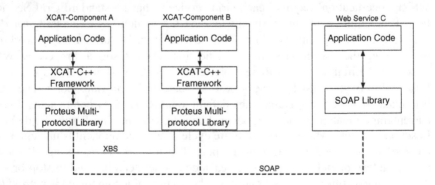

Fig. 1. The Figure shows a simple architecture of XCAT-C++ components using the Proteus Multi-protocol communication library. For communication between two XCAT-C++ components, the XBS (a streaming binary serializer) protocol is used, while for interaction with Web services, the XSOAP protocol is used.

nent running on the Legion framework [9], a common-denominator protocol can be used to negotiate the use of the most optimized protocol available with both the frameworks. The negotiation and switch to the appropriate protocol can be handled by the framework and remain transparent to the application.

– The use of a multi-protocol approach provides a fail-safe mechanism for data transfer in XCAT-C++. If a particular protocol results in errors or has an unexpected loss in performance, the framework can dynamically switch to another communication protocol. Moreover, it allows error reporting to take place via a protocol that is different from the one that generated it.

A central issue in any system that relies on multiple protocols is finding break-even points – to know when one method is preferable to another. In numerical computing, such approaches are called *polyalgorithms*, and the break-even points can often be specified in terms of a few parameters giving problem characteristics independent of the computing environment. In communication systems, however, the issue is significantly more complex because of dependence on hardware, networks, and software implementations. For scientific applications, with widely varying communication characteristics, an extensive testing framework is required for each application.

3.3 Wormhole Routing

In wormhole routing, a message is divided into a sequence of (fixed size) data units, called flits. As the header flit moves, the remaining flits follow in a pipeline fashion. As opposed to the store-and-forward policy, worm-hole routing allows parts of a message to be forwarded to the next node even before the entire message has been received. All parts of a single message follow the same route. The overlapping of transmission with reception of data sets, when done for fixed sized chunks tuned for each system, can also maximize the benefits of cache hits. The wormhole routing feature is included as part of the Proteus communication library [10]. The use of wormhole routing in XCAT-C++ components will allow them to efficiently function as gateways for some distributed applications that just require efficient streaming of large data sets via network storage depots [12]. The XCAT-C++ gateway components do not have to store the entire data in memory at any given time and can start forwarding data chunks even before the entire data set has been received.

4 Utility Modules for Application Development

In this section we briefly describe a few tools that facilitate in providing ease-of-use for application scientists. While these don't directly address performance requirements they contribute to the rich experience of using component based technology, wherein each component is a binary unit of composition, and just the interfaces are needed for plug-and-play application development.

4.1 A Flexible Code Generation Toolkit

The use of the Proteus multi-protocol library with XCAT-C++ allows applications to be built by composing components living in disparate heterogeneous environments using

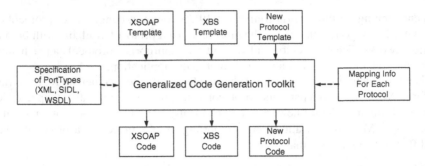

Fig. 2. The Figure shows the architecture of a flexible code generation toolkit that can generate the required glue code for many different communication libraries. The toolkit needs to be provided with the interface description and a template for the required communication library. Additionally, mappings can be provided to steer the code generation process.

various communication protocols. Each communication substrate has low level details that are shielded from the user by isolating them in a library that is generated by a specialized code-generator. However, the use of several code generators to compose a distributed application is tedious and inconvenient. It imposes a burden on the user. It is desirable to have a single code generation toolkit that can be used for all communication protocols available to the framework.

Figure 2 shows a simplified design of our code generation toolkit that can be used to generate *stubs* and *skeletons* for a wide-variety of communication protocols. The same code generator can also be used to generate framework specific code. The common patterns in the generated code for distributed object systems are captured in a grammar, that can be used to specify *templates* for each communication protocol. Apart from these design patterns, the template can specify control structures and mappings for variables in code-templates. These templates need to be written only once by the designer of the communication library. A user needs to specify a CCA port type interface in XML (we will add support for SIDL in the near future) and pick a template from the available protocol-templates with the toolkit. The code generation toolkit understands the grammar used to define the templates, and can generate the required code accordingly.

4.2 Composition Model

In scientific computing, one often needs to run a distributed computation multiple times with minor variations. This makes a scripting language interface for building such applications invaluable. To accomplish this, we have developed a simple Python interface to XCAT-C++ by using the Simple Wrapper Interface Generator (SWIG) [13] to translate calls between Python and the XCAT-C++ library. Scientists can use the CCA API directly from the python script and do not have to be concerned about the details of the XCAT-C++ implementation. The design of XCAT-C++ does not preclude the use of other composition models such as Matlab or Graphical Uses Interfaces (GUIs). We plan to add a convenient GUI interface so that users can visually drag and drop components from a repository to compose an application. This will also allow end-users to save configurations of successful runs as python scripts.

5   Performance

Our test environment consisted of two dual processor machines, each configured with 2.0 GHz Pentium 4 Xeon with 1GB DDR RAM and a 15K RPM 18GB Ultra-160 SCSI drive running Debian Linux 3.1 ("sarge") with the 2.4.26 kernel. The machines were connected by Gigabit Ethernet. Two XCAT-C++ components with compatible ports were launched on different nodes of the cluster. The port communication involved sending and receiving one and two dimensional arrays of various sizes. The code was compiled with gcc version 3.3.5. Our results reflect the average of multiple measurements for each reported data point.

The primary aim of our tests was to measure the overhead imposed by the XCAT-C++ component layering on data types commonly used by scientific applications. XCAT-C++ was run by selecting the high performance streaming XBS parser [11] from the Proteus library. Figure 3 shows the performance comparison of raw TCP, Proteus/XBS and XCAT-C++ on top of Proteus/XBS. The performance of raw TCP is better than that of Proteus/XBS. This is expected as Proteus/XBS has additional overhead above pure transmission. The performance of XCAT-C++ closely matches that of XBS for all data sizes. The design of XCAT-C++ ensures that all application calls are transferred to the Proteus communication module without buffer copying and the component layering cost is restricted to just a few virtual method calls. The second plot of Figure 3 compares the performance of XCAT-C++ and Proteus/XBS for two dimensional arrays, and once again for large data sizes both XCAT-C++ and Proteus/XBS achieve an average of 900Mbps on a Gigabit switched network.

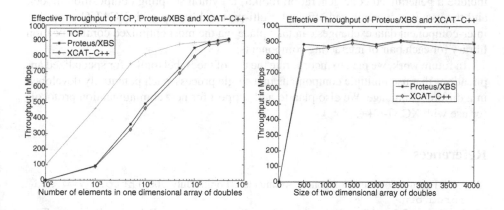

Fig. 3. The figures compare the performance of raw TCP, Proteus/XBS and XCAT-C++ for one- and two- dimensional arrays. XCAT-C++ is layered on top of Proteus and XBS is one of the communication protocols of Proteus. The plot for raw TCP serves as a standard with which we can judge the overhead of the other protocols. For very large arrays of floating point data (doubles), the performance of Proteus/XBS approaches that of TCP. The plot for XCAT-C++ shows that the overhead due to component layering is minimal. For large data sizes, XCAT-C++ achieves 900 Mbps on a Gigabit switched network.

6 Related Work

The most widely used component models developed by the industry include CORBA Component Model (CCM), Distributed Component Object Model (DCOM) and Enterprise Java Beans (EJB). These component models have not been explicitly designed to meet the challenges of scientific computing. In particular, scientific applications require the component models to encapsulate parallel and distributed programs sending large, complex, and rapidly changing data objects.

Many CCA systems have been developed for different application domains. SCIRun2 [14] is specialized for parallel-to-parallel remote method invocation in a distributed memory environment. SCIRun2 has mainly been used for visualization applications. XCAT-Java [8], [15] is a Java framework that uses the Web services model as its basic architecture and supports the SOAP communication protocol. LegionCCA [16] uses the Legion object model and run-time system to launch applications on Legion-based grids. The CCAFFEINE [17] framework is specialized for parallel computing and supports both single program/multiple data (SPMD) and multiple program/multiple data (MPMD) models.

7 Summary and Future Work

We presented performance results to show that overhead of component layering on applications is minimal and for large arrays of floating point data XCAT-C++ delivers very high throughput. The features provided by the framework can shield users from the details of managing the scale and complexity of their scientific applications. These include a generalized code generation toolkit, a Python scripting composition model, and the use of a dynamic and efficient multi-protocol library with XCAT-C++ so that inter-component data exchanges can take place via the most optimized communication library for each pair of interacting components.

In future work, we plan to incorporate the use of the Babel toolkit for specialized applications that use multiple components in a single process, each potentially developed in a different language. We also plan to add support for new communication protocols for use with XCAT-C++.

References

1. CCA Forum: Common Component Architecture Forum (July, 2005) http://www.cca-forum.org.
2. Bernholdt, D.E., Allan, B.A., Armstrong, R., Bertrand, F., Chiu, K., Dahlgren, T.L., Damevski, K., Ewasif, W.R., Epperly, T.G.W, Govindaraju, M., Katz, D.S., Kohl, J.A., Krishnan, M., Kumfert, G., Larson, J.W., Lefantzi, S., Lewis, M.J., Malony, A.D., McInnes, L.C., Nieplocha, J., Norris, B., Parker, S.G., Ray, J., Shende, S., Windus, T.L., Zhou, S.: A Component Architecture for High Performance Scientific Computing. International Journal of High Performance Computing Applications, ACTS Collection Special Issue (2005)
3. Kohn, S., Kumfert, G., Painter, J., Ribbens, C.: Divorcing Language Dependencies from a Scientific Software Library. In: Proceedings of 10th SIAM Conference on Parallel Processing, Portsmouth, VA. (March 12-14, 2001)

4. Elliot, N., Kohn, S., Smolinski, B.: Language Interoperability for High-Performance Parallel Scientific Components. In: International Symposium on Computing in Object-Oriented Parallel Environments (ISCOPE 1999), San Francisco, CA. (September 29 - October 2nd)
5. Gannon, D., Ananthakrishnan, R., Krishnan, S., Govindaraju, M., Ramakrishnan, L., Slominski, A.: 9, Grid Web Services and Application Factories. In: Grid Computing: Making the Global Infrastructure a Reality. Wiley (2003)
6. Foster, I., Kesselman, C., Nick, J., Tuecke, S.: Grid Services for Distributed System Integration. Computer 35(6) (2002)
7. Globus Alliance: The WS-Resource Framework (2004) http://www.globus.org/wsrf/.
8. Govindaraju, M., Krishnan, S., Chiu, K., Slominski, A., Gannon, D., Bramley, R.: Merging the CCA Component Model with the OGSI Framework. In: 3rd IEEE/ACM International Symposium on Cluster Computing and the Grid. (May 12-15, 2003, Tokyo, Japan.)
9. Lewis, M.J., Govindaraju, M., Chiu, K.: Exploring the Design Space for CCA Framework Interoperability Approaches. In: In Workshop on Component Models and Frameworks in High Performance Computing. (June 2005)
10. Chiu, K., Govindaraju, M., Gannon, D.: The Proteus Multiprotocol Library. In: Proceedings of Supercomputing 2002. (November 2002.)
11. Chiu, K.: XBS: A streaming binary serializer for high performance computing. In: Proceedings of the High Performance Computing Symposium 2004. (2004)
12. Swamy, M.: Improving Throughput for Grid Applications with Network Logistics. In: Proceedings of Supercomputing Conference. (2004)
13. SWIG: Simplified Wrapper and Interface Generator (1997) http://www.swig.org.
14. Zhang, K., Damevski, K., Venkatachalapathy, V., Parker, S.: SCIRun2: A CCA framework for high performance computing. In: Proceedings of the 9th International Workshop on High-Level Parallel Programming Models and Supportive Environments (HIPS 2004), Santa Fe, NM, IEEE Press (2004)
15. Krishnan, S., Gannon, D.: XCAT3: A Framework for CCA Components as OGSA Services. In: Proceedings of HIPS 2004: 9th International Workshop on High-Level Parallel Programming Models and Supportive Environments. (April 2004)
16. Govindaraju, M., Bari, H., Lewis, M.J.: Design of Distributed Component Frameworks for Computational Grids. Proceedings of International Conference on Communications in Computation (June 2004) 160–166
17. Allan, B.A., Armstrong, R.C., Wolfe, A.P., Ray, J., Bernholdt, D.E., Kohl, J.A.: The CCA Core Specification In A Distributed Memory SPMD Framework. CCPE **14** (2002) 323–345

The Impact of Noise on the Scaling
of Collectives: A Theoretical Approach

Saurabh Agarwal, Rahul Garg, and Nisheeth K. Vishnoi

IBM India Research Lab,
Block-1, IIT Delhi, Hauz Khas, New Delhi 110016, India
{saurabh.agarwal, grahul, nvishnoi}@in.ibm.com

Abstract. The performance of parallel applications running on large
clusters is known to degrade due to the interference of kernel and daemon activities on individual nodes, often referred to as *noise*. In this
paper, we focus on an important class of parallel applications, which repeatedly perform computation, followed by a collective operation such
as a barrier. We model this theoretically and demonstrate, in a rigorous way, the effect of noise on the scalability of such applications. We
study three natural and important classes of noise distributions: The exponential distribution, the heavy-tailed distribution, and the Bernoulli
distribution. We show that the systems scale well in the presence of exponential noise, but the performance goes down drastically in the presence
of heavy-tailed or Bernoulli noise.

1 Introduction

Motivation. It is well known that many parallel applications do not scale well
on large high-performance computing systems [1,2,3]. The per-node performance
degradation is more pronounced in systems with more than 1K nodes, running a
multi-tasking operating system such as Unix. In order to build high-performance
computing systems that are capable of very high and sustained performance, it
is important to understand the reasons for such performance degradation.

It is increasingly becoming evident that one of the main causes of performance
degradation is the *noise* in the system; in the form of daemons and interrupts,
see [1,2].

A detailed study of the *noise* and its impact on performance was done by
Petrini *et al.* [3] on the 8192 processor ASCI Q machine. It was observed that
the overheads due to noise were mostly in the range 0.5% to 2.5% (see Figure 9
in [3]). However, this noise had a large impact on the system performance. By
reducing the *intensity* of noise in the system it was possible to get a factor
of 13 improvement in the performance of a micro-benchmark that repeatedly
calls *barrier* with no intervening computation. Similarly, Kramer and Ryan [4]
concluded that the performance variability in EP (of NAS parallel benchmarks)
was due to the noise in systems.

D.A. Bader et al. (Eds.): HiPC 2005, LNCS 3769, pp. 280–289, 2005.
© Springer-Verlag Berlin Heidelberg 2005

Our Contribution. The main contribution of this paper is to initiate the study of the impact of noise on the scaling of parallel applications in a *formal manner*. We focus on a particularly important class of parallel applications which often arise in scientific computations. Here, typically, each node in the cluster is repetitively involved in a computation stage, followed by a collective operation; such as a barrier computation. We model this theoretically and demonstrate the effect of noise on the performance of such parallel applications. We study three natural and important classes of noise distributions: The exponential distribution, the heavy-tailed distribution, and the Bernoulli distribution. We show that the systems scale well in the presence of an exponential noise, but their performance goes down drastically in the presence of a heavy-tailed or a Bernoulli noise. Though our model is very simple, it is powerful enough to predict the effect of noise on scaling. We believe that this study will also be extremely useful in identifying and improving bottlenecks in the scalability of systems in a more systematic way, for instance, by designing scheduling policies, which take into account the nature of the noise, to improve the overall system performance. To the best of our knowledge, this is the *first attempt* to explain the impact of noise with a mathematical model.

Related Work. One way to reduce the impact of noise on scalability is to reduce the intensity of noise itself. This can be done by removing several system daemons, dedicating a spare processor to absorb the noise, and decreasing the frequency of daemons [3].

Another approach is to *synchronize* the noise across the nodes of the system. This may be done by either periodically adjusting the scheduling priorities of the processes, or by changing the scheduler in the kernel, see [5,6,7,8,9].

Though these methods have resulted in a reduced impact of noise on the performance of the respective systems, a general solution is more desirable both with regards to scalability and applicability. Our work provides a structured approach to understand the impact of noise on the overall system performance. Using the insights from our results, it might be possible to further enhance all of the above approaches, thereby advancing the frontier of scalability and yielding better resource utilization in the present high-performance computing systems.

Organization. Section 2 presents the theoretical model of a typical scientific parallel application with noise. Therein, we also justify the assumptions about the model. In Section 3, we analyze the proposed model and present the results obtained when the noise is distributed according to the exponential, heavy-tailed or Bernoulli distribution. In Section 4 we briefly discuss the implications of our results. Due to space limitations, the proofs of the theorems in Section 3 and a detailed discussion will appear in the full version of this paper [10].

2   The Model

In this section we describe a general model that captures the case of a compute intensive program with periodic synchronization. We assume that the program has perfectly balanced load and it carries out minimal I/O and message exchanges. However, it carries out periodic synchronization using a collective operation. A footprint of such a program is typically present in many parallel applications, in particular in those which involve scientific computations.

2.1   Modeling the Computation

Consider a parallel program with N threads running on a system which has N processors. We assume, for simplicity of analysis, that $N = 2^k - 1$ for some positive integer k.

The Communication and the Three Stages. We assume that the barrier is implemented using message passing along a complete binary tree. A thread is associated to each node of the binary tree. There is a special node called the *root* which initiates the *post-barrier* stage and the *pre-barrier* stage ends at it. In the post-barrier stage, the root thread starts by sending a message to both its children to start with the *compute* stage. Whenever this message reaches a thread, it forwards the message to both its children in the tree (unless the node is a leaf) and starts the computation assigned to it. After finishing its computation, a leaf node sends a message to its parent indicating the end of its computation stage. This starts the pre-barrier stage. The parent, after finishing its computation and receiving the message from both its children, sends the message to its parent indicating the end of computation at every node in its subtree. This stage ends when the root finishes its computation and receives a message from both its children indicating the same. An iteration of the loop would, thus, consist of a compute stage, followed by a pre-barrier and a post barrier stage. For simplicity, we assume that each message transmission between a parent and a child node takes time τ. Again, for simplicity of analysis, we consider a *phase* which consists of a sequence of a post-barrier, a compute and a pre-barrier stage. The program consists of M such phases. Now, we model various aspects of one such phase.

A Phase. Let t_{ij}^s represent the time instant when the i-th thread begins the computation stage in the j-th phase. Let t_{ij}^f represent the time instant when the i-th process ends the computation stage in the j-th phase. Let W_{ij} represent the amount of work (say the number of operations) carried out by thread i in the compute stage of the j-th phase. If the system is noiseless, the time required by processor i to finish work W_{ij} in its j-th phase will be a constant, say w_{ij}, which typically depends on the characteristics of the processor, such as clock frequency, architectural parameters, and the state of the node (such as cache contents) just before the j-th phase is entered. Therefore, $t_{ij}^f - t_{ij}^s = w_{ij}$. Due to the presence of system level daemons that get scheduled arbitrarily, the wall-clock time taken by processor i to finish the work W_{ij} is typically not a constant. There will be a variable component that represents the time consumed to service the daemons and other asynchronous events. We capture this by a random variable δ_{ij}. More precisely, $t_{ij}^f - t_{ij}^s = w_{ij} + \delta_{ij}$, where δ_{ij} is a random variable that captures the overheads incurred by processor i in servicing the daemons and other asynchronous events during the j-th phase. Note that δ_{ij} also includes the context switching overheads, as well as, the time required to handle additional cache or TLB misses that arise due to cache pollution by background processes. The mean value of δ_{ij} depends on the time taken to do work W_{ij} and the system load on processor i during the j-th phase. Let $f_{ij} \in [0,1]$ be the fraction representing the system overhead for the processor. We may write the wall-clock time taken by processor i for the compute stage of the j-th phase as $t_{ij}^f - t_{ij}^s = w_{ij}\left(1 + \frac{f_{ij}}{1-f_{ij}}\eta_{ij}\right)$, where η_{ij} is the normalization of δ_{ij} such that $\mathrm{E}[\eta_{ij}] = 1$.

2.2   The Assumptions and Justifications

In this section we state and justify the assumptions we make about the model. The underlying principle in making the assumptions is to present an *ideal* model which captures the impact of the noise on typical parallel programs for scientific applications, and which is at the same time, susceptible to a rigorous theoretical analysis. We show, in a formal manner, that even in this ideal setting, the nature of noise may impact the system performance considerably.

Balanced Load: $W_{ij} = W$ **for all** i, j. Application programs try to divide the load equally among its threads. Best performance is obtained when the load across every thread in a compute phase is equal (i.e. $W_{ij} = W_j$ for all i).

Identical Processors: $w_{ij} = w$ **for all** i, j. If the processors are heterogeneous, the performance of the parallel application will be dictated by the performance of the slowest processor in the system. Best performance is obtained (with perfectly balanced load) when the processors are identical[1]. In addition to this, we make two more assumptions: (1) The application starts with all its threads in identical states. (2) The time taken by a computation does not depend on the input data. Together, these assumptions imply that the time taken by a compute phase is same across all the processors. The second assumption will not be true in general, because, due to cache effects, the time taken to carry out a set of operations also depends on the order in which the operations are carried out. However, it can be verified that this is the most optimistic assumption that will give the best program performance.

Stationary and Balanced Overheads: $f_{ij} = f$ **for all** i, j. In typical HPC systems, the processors are allocated to an application for the lifetime of the application. Running any other application on the node is avoided. Thus, all the interference is due to the background processes or daemons. The amount of daemon activity is not expected to change over time. Thus we may assume

---

[1] We do not consider programs running on heterogeneous clusters that distribute the load across multiple nodes depending on the relative speed of the nodes.

$f_{ij} = f_{ij'}$, for all i, j and j'. The daemons and overheads may be classified into *intrinsic* and *extrinsic* processes. The intrinsic processes run on every node and carry out book-keeping activities for the node. The system overhead due to intrinsic processes is expected to remain the same across all the nodes. However, the overheads of extrinsic processes are expected to vary across nodes. A detailed analysis may be carried out along the same lines while taking into account the activities of the extrinsic processes as well. Therefore, we assume $f_{ij} = f_{i'j}$, for all i, i' and j.

Identical Noise: $\eta_{ij} \sim \eta$ for all i, j.   Due to homogeneity of nodes and the fact that we choose to ignore the effect of extrinsic processes, we may assume that the nature of noise associated to the intrinsic processes is the same across the nodes and phases.

Spatial Independence: $\{\eta_{ij} : i \in [1 \ldots N]\}$ are independent for each j. This assumption is the key to all of our results. In a typical cluster environment, there is no co-ordinated scheduling policy to synchronize processes across different nodes. Some HPC systems may deploy different scheduling policies to alleviate the daemon problems [7,8,9], as discussed earlier in Section 1. However, our analysis is restricted to systems that do not employ a co-ordinated scheduling policy. Note that we do not assume the random variables $\{\eta_{ij} : j \in [1 \ldots M]\}$ to be independent. In fact, many of the daemons are periodic, and we do expect complex correlation pattern between these random variables. In general, the nature of noise η_{ij} may depend on the quantum of work w_{ij} carried out in the phase. To analyze this, the compute phases may be grouped into quanta of work w_j and the same analysis may be independently carried for each quantum (with its associated noise). Due to linearity of expectation, the expected run time for each quantum of work can be added up to give the expected running time of the application.

3   Analysis

The Ideal Noiseless Case. Figure 1 illustrates the sequence of events in the ideal noiseless case. It is clear that in this case the time taken by each thread in a phase is $w + 2\tau(k-1)$, where $N = 2^k - 1$. (The figure is for the case when $k = 3$.) In terms of N, this is $w + 2\tau(\log(N+1) - 1)$. This will be the benchmark performance we will use for comparison with the noisy case.

The Ideal Noisy Case. Now, we no longer assume that f_{ij} are 0. We refer to this as the *ideal noisy* case. In this case, $t_{ij}^f - t_{ij}^s$ is randomly distributed. An example of this scenario is presented in Figure 1. The post-barrier phase of the communication is still the same as in the case of the ideal noiseless case.

Let a_i denote the time it takes the message to reach thread i from the root in the post-barrier stage. Further, let b_i denote the time it takes the message from thread i to reach the root in the pre-barrier stage. The time taken to

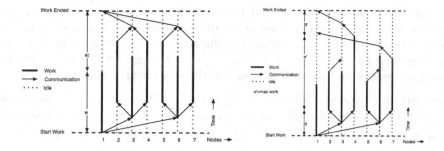

Fig. 1. An ideal noiseless (left) and noisy (right) barrier computation cycle

complete the j-th phase then is at-least $\max_{i=1}^{N}(a_i + t_{ij}^f - t_{ij}^s + b_i)$. Notice that since the pre and post-barrier stages are done via communicating through a binary tree, for the leaves of this binary tree, which are $2^{k-1} = \frac{N+1}{2}$ in number, $a_i = b_i = \tau(k-1) = \tau(\log(N+1) - 2)$. Let us just restrict our attention to these leave threads. Since the noise is independent across the threads, the maximum of $a_i + b_i + w_{ij}$, for i restricted to these threads is a lower bound to the time taken to complete the j-th phase. Let Y_η^r denote the maximum of r random variables which are independent and identically distributed according to η. Hence, the expectation of $t_{ij}^f - t_{ij}^s$ is at-least $2\tau(\log(N+1) - 2) + w\left(1 + \frac{f}{1-f}\mathrm{E}\left[Y_\eta^{\frac{N+1}{2}}\right]\right)$. Therefore, we have the following theorem.

Theorem 1. *The expected time taken per phase is bounded by*

$$\mathrm{LowerBound} : w\left(1 + \frac{f}{1-f}\mathrm{E}\left[Y_\eta^{(N+1)/2}\right]\right) + 2\tau(\log(N+1) - 2)$$

$$\mathrm{UpperBound} : \quad w\left(1 + \frac{f}{1-f}\mathrm{E}\left[Y_\eta^N\right]\right) + 2\tau(\log(N+1) - 1)$$

We call the term $2\tau(\log(N+1) - 2)$ in the above expression as the *latency component* which is an indication of time spent in barrier due to the communication latency. The term $w\frac{f}{1-f}\mathrm{E}\left[Y_\eta^{(N+1)/2}\right]$ is called the *noise component* as it represents the slow-down due to the presence of asynchronous daemons. The expected time taken by a phase can be decomposed into the *work component* w, the *latency component* and the *noise component*. The daemons start playing a significant role as soon as the noise component becomes comparable to the latency component. Now, we examine different types of noise distributions and prove a lower bound for the expected time taken to complete a phase.

The Exponential Case. This distribution arises as the continuous limit of the discrete geometric random distribution and occurs very often in practice as a description of the time elapsing between unpredictable events, such as, telephone calls, radioactive emission, arrivals of buses. Being one of the most natural and important distribution to model such events, in this section we consider the case when the noise η_{ij}-s are also distributed according to the exponential distribution. An exponential distribution $X_{\exp}$ with mean 1 has the following distribution: $\forall x \geq 0, \quad \Pr[X_{\exp} \leq x] = 1 - \exp(-x)$. In this case, the following lower bound shows the impact of the noise being exponential.

Theorem 2 (Exponential Noise). *If $\{\eta_{ij} : i \in [1 \ldots N]\}$ are independently and identically distributed according to $X_{\exp}$, then the expected time taken per phase is at-least $w \left(1 + \frac{f}{1-f}\left(\ln(N+1) - \Theta(1)\right)\right) + 2\tau(\log(N+1) - 2)$.*

The lower bound has the form $c \log N + d$, where $c = \frac{wf}{(1-f)\log e} + 2\tau$. This is a linear function of $\log N$, similar to the ideal noiseless case. When $\frac{wf}{(1-f)\log e}$ is comparable to (or less than) 2τ, the performance is close to the ideal noiseless case. Hence, only when $\frac{wf}{(1-f)\log e}$ is large compared to 2τ, this model of noise impacts the performance by a constant factor of $\frac{wf}{2\tau(1-f)\log e}$ compared to the ideal noiseless case.

The Heavy-Tailed Case: The Pareto Distributions. In this section we consider the case when the noise has a heavy tail. This is unlike the exponential case and the noise looks more like the uniform distribution. A natural and very popular way to model data which has heavy tail is the so-called Pareto distribution. The Pareto random variable X_{par}^a with parameter a has the following distribution: $\forall x \geq 1, \quad \Pr[X_{\text{par}}^a \leq x] = 1 - \frac{1}{x^a}$. The Pareto distribution has mean $\frac{a}{a-1}$. To make this random variable with unit mean, we let η be $\frac{a-1}{a} X_{\text{par}}^a$.

Theorem 3 (Pareto(a) Noise). *If $\{\eta_{ij} : i \in [1 \ldots N]\}$ are identically and independently distributed according to $\frac{a-1}{a} X_{\text{par}}^a$, then the expected time taken per phase is at-least $w \left(1 + \frac{f}{1-f}\left(\frac{N+1}{2}\right)^{1/a}\left(\frac{a-1}{a}\right)^{1-1/a}\right) + 2\tau(\log(N+1) - 2)$.*

The theorem shows that, in this case, the scalability of the parallel systems suffers far more than in the exponential case or the ideal noiseless case. The scaling becomes worse as the value of a goes lower. Hence, fixing w, τ, f and a, and letting N increase, the term that will dominate here is $N^{1/a}$. We refer to this as *polynomial* scaling, and such a scenario is extremely undesirable, especially for small values of a.

The Bernoulli Case. This is parameterized by a probability p and a time T. In this setting, each thread takes time $w + T$ with probability p, and time w with probability $1 - p$. The Bernoulli distribution models the expected scaling behavior of collectives in the presence of in-frequent and bursty noise. This model can also be thought of as a first order and discrete approximation of a heavy-tailed noise, where the size of the tail can be controlled by varying pT.

Theorem 4 (Bernoulli Noise). *If $\{\eta_{ij} : i \in [1 \ldots N]\}$ are identically and independently distributed according to the Bernoulli distribution, then the expected time taken by a phase j is at-least $w + T(1 - (1-p)^{(N+1)/2}) + 2\tau(\log(N+1) - 2)$.*

When $\frac{pN}{2}$ is small compared to 1, the first term in the above lower bound is essentially $w(1 + \frac{pT}{2w}N) = w\left(1 + \frac{f}{1-f}N\right)$. Hence, in this range, the system is expected to show *linear* scaling. For very large values of N, the maximum slow-down in the performance is approximately a factor of $1 + T/w$.

4   Discussion

In this section, we discuss the implications of our results by plugging in the values for w, τ and f, which are typical to the HPC systems, see [3].

We use the weak scaling model to measure the scalability of the system in presence of different noise distributions. In the weak scaling model, the work per processor is kept fixed and the performance is studied as the number of processors is increased. Define $\mathcal{N}_{1/2}$ to be the minimum number of processors with which the program takes twice as much time as with one processor. For the ideal noiseless case, this happens when $w \approx 2\tau \log(N + 1)$, or $N + 1 \approx 2^{w/(2\tau)}$. This parameter gives an indication of how well a program scales in the presence of noise. Subsequently, we also discuss the values of $\mathcal{N}_{1/2}$ for different noise distributions.

The Exponential Case. Figure 2 shows the expected time needed for one phase of computation in our model when η is distributed according to the exponential distribution (see Theorem 2). When $w = 10\mu s$, the noise intensity f has little impact on the performance, whereas when $w = 1ms$, f has a significant impact.

In this case, $\mathcal{N}_{1/2}$ may be approximated as $\mathcal{N}_{1/2} \approx \exp\left(\frac{1}{f/(1-f)+2\tau/(w\ln 2)}\right)$. With 1% exponentially distributed noise, $\mathcal{N}_{1/2} \approx 2.3 \cdot 10^{27}$ and with 10% noise, $\mathcal{N}_{1/2} \approx 5196$. This shows that in the presence of an exponentially distributed noise, the programs are expected to scale well. However, unlike the the ideal noiseless case, the time taken by collectives may be dominated by the noise component (when $wf > \tau$) as opposed to the latency component.

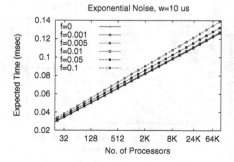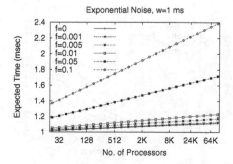

Fig. 2. Expected time taken by a phase in the presence of exponential noise

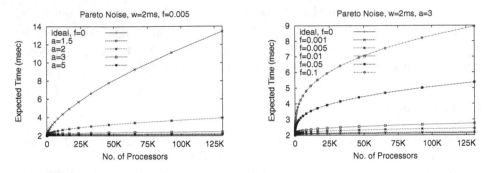

Fig. 3. Expected time taken by a phase in the presence of Pareto noise

The Pareto Case. Figure 3 shows the expected time needed for one phase of computation as a function of N when η is distributed according to the Pareto distribution (see Theorem 3). In the first plot, different lines represent different values of a, while the value of f is kept fixed at 0.005. In the second plot, different lines represent different values of f, while a is kept fixed at 3.

In this case, $\mathcal{N}_{1/2}$ may be approximated as $\mathcal{N}_{1/2} \approx \min\left(2 \cdot \left[\frac{1-f}{fc_a}\right]^a, 2^{w/(2\tau)+2}\right)$, where $c_a = \left(\frac{a-1}{a}\right)^{1-1/a}$. If f is kept fixed at 0.005, then $\mathcal{N}_{1/2} \approx 35.4 \cdot 10^6$ for $a = 3$, $\mathcal{N}_{1/2} \approx 158,404$ for $a = 2$, and $\mathcal{N}_{1/2} \approx 9724$ for $a = 1.5$. Similarly, for $a = 2$, $\mathcal{N}_{1/2} \approx 39,203$ when $f = 0.01$, and $\mathcal{N}_{1/2} \approx 9604$ when $f = 0.02$. This shows that scaling behavior is sensitive to the Pareto parameter a, as well as the noise intensity f.

The Bernoulli Case. Figure 4 shows the expected time taken by a phase as a function of number of nodes for different values of p and T (with $w = 2$ms). Note that the x-axis is in logarithmic scale. For small values of N, the total time varies linearly with N (with a slope of $pT/(2w)$). For large values of N, it saturates to $w + T$.

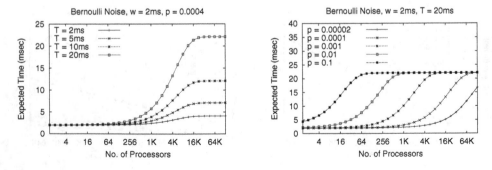

Fig. 4. Expected time taken by a phase in the presence of Bernoulli noise

The $\mathcal{N}_{1/2}$ in the presence of Bernoulli noise may be approximated as $\mathcal{N}_{1/2} \approx 2/f$. For $f = 0.01$, $\mathcal{N}_{1/2} \approx 200$. This indicates that systems with Bernoulli noise are expected to have very poor scaling properties.

References

1. R. Gioiosa, F. Petrini, K. Davis, and F. Lebaillif-Delamare, "Analysis of System Overhead on Parallel Computers," in *The 4th IEEE International Symposium on Signal Processing and Information Technology (ISSPIT 2004)*, (Rome, Italy), Dec. 2004.

2. T. R. Jones, L. B. Brenner, and J. M. Fier, "Impacts of Operating Systems on the Scalibility of Parallel Applications," Tech. Rep. UCRL-MI-202629, Lawrence Livermore National Laboratory, Mar. 2003.

3. F. Petrini, D. J. Kerbyson, and S. Pakin, "The Case of the Missing Supercomputer Performance: Achieving Optimal Performance on the 8,192 Processors of ASCI Q," in *ACM/IEEE Conference on Supercomputing (SC'03)*, (Phoenix, Arizona, USA), Nov. 2003.

4. W. T. C. Kramer and C. Ryan, "Performance Variability of Highly Parallel Architectures," in *International Conference on Computational Science (ICCS 2003)*, (Melbourne, Australia), Jun. 2003.

5. E. Frachtenberg, F. Petrini, J. Fernandez, S. Pakin, and S. Coll, "STORM: Lightning-Fast Resource Management," in *ACM/IEEE Conference on Supercomputing (SC'02)*, (Baltimore, Maryland, USA), Nov. 2002.

6. A. Hori, H. Tezuka, and Y. Ishikawa, "Highly Efficient Gang Scheduling Implementation," in *ACM/IEEE Conference on Supercomputing (SC'98)*, (Orlando, FL, USA), Nov. 1998.

7. T. Jones, S. Dawson, R. Neely, W. Tuel, L. Brenner, J. Fier, R. Blackmore, P. Caffrey, B. Maskell, P. Tomlinson, and M. Roberts, "Improving the Scalability of Parallel Jobs by adding Parallel Awareness to the Operating System," in *ACM/IEEE Conference on Supercomputing (SC'03)*, (Phoenix, Arizona, USA), Nov. 2003.

8. E. Frachtenberg, D. Feitelson, F. Petrini, and J. Fernández, "Flexible Coscheduling: Mitigating Load Imbalance and Improving Utilization of Heterogeneous Resources," in *International Parallel and Distributed Processing Symposium 2003 (IPDPS03)*, (Nice, France), Apr. 2003.

9. S. Agarwal, G. S. Choi, C. R. Das, A. B. Yoo, and S. Nagar, "Co-ordinated Coscheduling in Time-Sharing Clusters through a Generic Framework," in *IEEE International Conference on Cluster Computing (CLUSTER'03)*, (Hong Kong), Dec. 2003.

10. S. Agarwal, R. Garg and N. Vishnoi, "The Impact of Noise on the Scaling of Collectives: A Theoretical Approach", Tech. Rep. RI-05003, *IBM Research Report*, Feb. 2005.

Extensible Parallel Architectural Skeletons

Mohammad Mursalin Akon[1], Ajit Singh[1],
Dhrubajyoti Goswami[2], and Hon Fung Li[2]

[1] Department of Electrical and Computer Engineering,
University of Waterloo, Ontario, Canada
{mmakon, a.singh}@ece.uwaterloo.ca
[2] Department of Computer Science,
Concordia University, Montreal, Canada
{goswami, hfli}@cs.concordia.ca

Abstract. Complexity of parallel application development has been one
of the major obstacles towards the mainstream adoption of parallel programming. In order to hide some of these complexities, researchers have
been actively investigating the pattern-based approaches to parallel programming. As reusable components, patterns are intended to ease the
design and development phases of parallel applications. Parallel Architectural Skeleton (PAS) is one such pattern-based parallel programming
model which describes the architectural aspects of parallel patterns. Like
many other pattern-based parallel programming models and tools, the
benefits of PAS were offset by the difficulties in extending PAS. *EPAS* is
an extension of PAS that addresses this issue. Using EPAS, a skeleton designer can design new skeletons and add them to the skeleton repository
(i.e., extensibility). EPAS also makes the PAS model more flexible by
defining composition of skeletons. In this paper, we describe the model
of EPAS and also discuss some of the recent usability and performance
studies. The studies demonstrate that EPAS is a practical and usable
parallel programming model and tool.

1 Introduction

Parallel application design and development is often a complex process. There
are several approaches to parallel programming for hiding some of the complexities. This research focuses on one such approach, which is based on the idea
of (frequently occurring) design patterns in parallel computing. In the domain
of parallel computing, (parallel) design patterns specify recurring parallel computational problems with similar structural and behavioral components, and
their solution strategies. Several parallel programming systems have been built
with the intent to facilitate rapid development of parallel applications through
the use of design patterns as reusable components. Some of these systems are
Enterprise [1], *DPnDP* [2], *COPS* [3], *PAS* [4], and *ASSIST* [5].

Unlike the *algorithmic skeletons* [6] research, which deals with the behavioral
aspects of patterns, Parallel Architectural Skeletons (PAS) [7,4] focus on the
architectural or structural aspects of message-passing parallel patterns. Each

D.A. Bader et al. (Eds.): HiPC 2005, LNCS 3769, pp. 290–301, 2005.
© Springer-Verlag Berlin Heidelberg 2005

architectural skeleton in PAS encapsulates the various structural attributes of
a pattern in a generic (i.e., pattern- and application-independent) fashion. An
architectural skeleton can be considered as a pattern-specific virtual machine
with its own communication, synchronization and structural primitives. A developer, depending upon the specific needs of an application, chooses the appropriate skeletons, supplies the required parameters for the generic attributes,
and finally fills in the application-specific code. An architectural skeleton supply
most of the code that is necessary for the low-level and parallelism-related issues.
Consequently, there exists a clear separation between application dependent and
application independent issues (i.e., *separation of concerns*).

Though re-usability is an obvious benefit, the lack of extensibility and the lack
of support for pattern composition are some of the major concerns associated
with many of the pattern-based parallel programming systems, including PAS.
Most existing systems support a limited and fixed set of patterns that are hard-
coded into those systems. Generally, there is no provision for adding a new
pattern without understanding the entire system (including its implementation)
and writing the pattern from scratch (i.e., lack of extensibility). Consequently,
if a required parallel computing pattern demanded by an application is not
supported, generally the designer has no alternate but to abandon the idea of
using the particular approach altogether (lack of flexibility).

EPAS is an extension of the PAS system and it addresses the drawbacks mentioned previously. An earlier discussion of EPAS emphasizing more on the user
interface part (Skeleton Description Language) appeared in [8]. In this paper,
we describe the complete model of EPAS along with some of the recent usability
studies. Using EPAS, a skeleton designer can extend PAS by adding new skeletons to an existing skeleton repository. EPAS also makes the PAS system more
flexible by defining the composition of skeletons, whereby two or more existing
skeletons can be composed into a new skeleton.

In the next section, we introduce the necessary preliminaries. We elaborate
the model of EPAS in Section 3. The subsequent section describes the sum-
mery of usability and performance test results. Section 5 discusses the related
works and compares EPAS with them briefly. Finally, Section 6 concludes our
discussion.

2   Preliminaries

Parallel Architectural Skeletons (abbreviated as PAS) [4, 7] generically encapsulate the structural/architectural attributes of message-passing parallel computing patterns. Each PAS skeleton is parameterized where each parameter is
associated with some attribute. The value of a parameter is determined during
the application development phase. A PAS skeleton with unbound parameters
is called an *abstract skeleton* or an *abstract module*. An abstract skeleton becomes a *concrete skeleton* or a *concrete module*, when the parameters of the
skeleton are bounded to actual values. A concrete skeleton is yet to be filled
in with application-specific code. Filling a concrete skeleton with application-

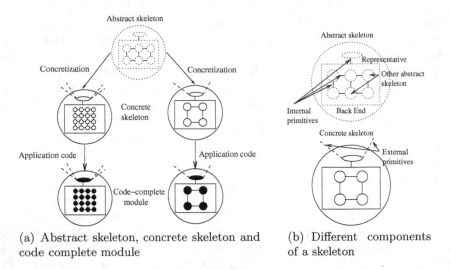

(a) Abstract skeleton, concrete skeleton and code complete module

(b) Different components of a skeleton

Fig. 1. PAS skeletons and their components

specific code results in a *code-complete parallel module* or simply a *module*. Various phases of an application development using PAS are roughly illustrated in Fig. 1(a). The figure shows that different parameter bindings to the same abstract skeleton can result in different concrete skeletons.

Each abstract skeleton (or abstract module) consists of the following set of attributes: (i) *Representative* of a skeleton represents the module in its action and interactions with other modules. The initial representative is empty and is subsequently filled with application-specific code during application development. (ii) The *back-end* of an abstract module A_m can be formally represented as $\{A_{m1}, A_{m2}, \ldots, A_{mn}\}$, where each A_{mi} is itself an abstract module. The type of each A_{mi} is determined after the abstract module A_m is concretized. Note that collection of concrete modules inside another concrete module results in a (tree-structured) hierarchy. Consequently, each A_{mi} is called a *child* of A_m, and A_m is called the *parent*. The children of a module are *peers* of one another. In this paper, the children of a module are also referred as *computational nodes* of the associated skeleton or patterns. (iii) *Topology* is the logical connectivity between the children inside the back-end as well as the connectivity between the children and the representative. (iv) *Internal primitives* are the pattern-specific communication, synchronization or structural primitives. Interactions among the various modules are performed using these primitives. The internal primitives, the inherent properties of the skeleton, capture the parallel computing model of the associated pattern as well as the topology. Fig. 1(b) diagrammatically illustrates attributes of an abstract and a concrete 2-D Mesh skeleton.

There are pattern-specific parameters associated with some of the previous attributes. For instance, if the topology is a Mesh, then the number of dimensions of the mesh is one parameter, and the nature of the connectivities among the nodes at the edges (i.e., toroidal or non-toroidal) is another parameter. Binding these parameters to actual values, based on the needs of an application, results in a concrete module. A concrete module C_m becomes a code-complete module when: (i) the representative of C_m is filled in with application-specific code, and (ii) each child of C_m is code-complete.

All of the attributes of an abstract skeleton are inherited by the corresponding concrete skeletons as well as the code-complete modules. In addition, we define the term *external primitives* of a concrete or a code complete module as the set of primitives using which the module (i.e., its representative) can interact with its parent (i.e., representative of the parent) and peers (i.e., representatives of the peers). Unlike internal primitives, which are inherent properties of a skeleton, external primitives are adaptable, i.e., a module adapts to the context of its parent by using the internal primitives of its parent as its external primitives. While filling in the representative of a concrete module with application-specific code, the application developer uses the internal and external primitives to interact with other modules in the hierarchy.

A parallel application developed using PAS is a hierarchical collection of (code-complete) modules. Conceptually, each concrete module can be considered as a pattern-specific virtual machine with its own communication, synchronization and structural primitives. A user fills in these virtual machines with application-specific code, starting bottoms-up in the hierarchy, to create the complete parallel application. The root of the hierarchy, i.e. a code-complete module with no parent, represents a complete parallel application. Each non-root node of the hierarchy represents a partial parallel application. Each leaf of the hierarchy is called a *singleton module* (and correspondingly, a *singleton skeleton* for the abstract counterpart).

3   The EPAS Model

EPAS is targeted for two categories of users: skeleton designers and application developers. Often a user can fall into both the categories. EPAS provides a set of virtual processors, interconnected to form a specific topology (described in the next subsection), and a rich set of basic communication and synchronization primitives for interactions among these processors. In the process of designing a new skeleton, a skeleton designer first decides about the constituents of the new skeleton (i.e., back-end components, topology, primitives, etc). Subsequently she needs to map the children (i.e., abstract modules in the back-end) of the newly designed (abstract) skeleton to the virtual processors provided by EPAS, and define the topology and the communication-synchronization primitives of the new skeleton on top of the basic primitives and the specific topology provided by EPAS. A Skeleton Description Language (SDL) facilitates this design phase. The discussion in the following sub-sections elaborates these issues.

3.1   The Virtual Processor Grid

The previously discussed virtual processors of EPAS are arranged into a set of multidimensional grids. Each node of such a grid is a virtual processor. Consequently we will use the term *virtual processor grid* (*VPG*) to describe such a grid, and use the terms *node of a grid* and *virtual processor* interchangeably. The topologies of the newly designed skeletons are embedded into these grid topologies. Usually one skeleton is embedded into one VPG. Each multidimensional VPG is equipped with its own communication and synchronization primitives. These primitives include primitives for synchronous and asynchronous peer-to-peer communication, collective communication and synchronization primitives. We chose to make the VPG primitives a super set of the basic communication-synchronization primitives supported in some of the prominent parallel programming environments. Our choice is influenced by the research article [9], MPI standard, PVM documentations, and our experiences with PAS and other pattern-based systems.

3.2   Mapping a Skeleton Topology into a VPG

In the process of designing an abstract skeleton, the skeleton designer needs to embed the topology of the newly designed skeleton to the existing grid topology provided by EPAS, and consequently map each of its children (i.e., abstract modules of the back-end) into a VPG node. Fig. 2 shows one such mapping where the designer wants to design an abstract *Wavefront* skeleton, a structural implementation of the *Wavefront* pattern. Fig. 2(a) is the visualization of the Wavefront skeleton where the topology of its constituents is shown. The visualization helps the designer to make several design decisions. At first, she decides about the *parameters* of the skeleton. For example, in this case, the structure of a Wavefront skeleton becomes generic if the the number of rows (or the number of columns) of the skeleton is considered to be a parameter rather than a constant. In this example, we name this parameter as *size*.

In the case of a Wavefront, the choice of a two dimensional VPG (Fig. 2(b)) is obvious because it facilitates a one-to-one mapping of the children (of the skeleton) into the nodes of the VPG. Fig. 2(c) shows such a mapping. Note that

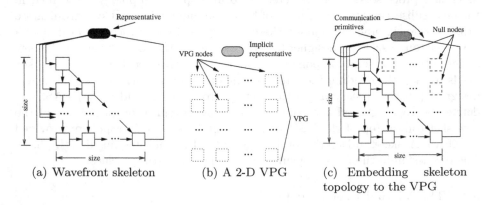

(a) Wavefront skeleton   (b) A 2-D VPG   (c) Embedding   skeleton topology to the VPG

Fig. 2. Mapping *Wavefront* skeleton components into a 2-D VPG

each VPG has an implicit representative node, to which the representative of the skeleton is mapped onto. From the figure, it can be found that even after limiting the height and width of the VPG (to the parameter *size*), there are virtual processors to which no child of the skeleton are mapped onto. Those virtual processors are called *null virtual processors* or *null nodes*.

The embedding of a skeleton into a VPG is complete when the associated communication, synchronization and structural primitives of the skeleton are defined. These primitives are defined on top of the existing EPAS-provided primitives for the VPG (in this case, a 2-D VPG). In Fig. 2(c), some of the channels along which communication primitives are defined are marked. Examples of some communication primitives for a Wavefront skeleton are: (1) receive a message from the representative, (2) send a message to the left neighbor, etc. Examples of some structural primitives are the operations: (1) to check whether a child is located at the nth column, (2) to check whether a child is located on the diagonal, etc. The discussion in the following elaborates it further.

In the EPAS model, the topology of an abstract skeleton (as in Fig. 2(a)) is called the *abstract topological space* ($\mathcal{AT}$) of the skeleton. The abstract topological space of a skeleton is constituted of zero or more nodes (i.e., children of the back-end) along with their connectivities. Connectivity among the nodes of $\mathcal{AT}$ is represented by a connectivity function $\mathcal{T}$. The mapping function, $\mathcal{M}$, maps nodes of $\mathcal{AT}$ to the virtual processors of the *VPG space* (designated as $\mathcal{VPG}$). The embedding of an $\mathcal{AT}$ into a $\mathcal{VPG}$ results in an *abstract mapped space*, designated as $\mathcal{P}$, which is a subgraph of the *VPG space* and is constituted of only the non-null nodes of $\mathcal{VPG}$. Note that for a composite skeleton, discussed in the next subsection, there can be more than one abstract mapped spaces.

Provided that $\mathcal{M}$ and $\mathcal{T}$ are already defined, it is easy to express the connectivities among the non-null virtual processors of the $\mathcal{VPG}$ (i.e., nodes of $\mathcal{P}$) as the composite function: $\mathcal{M}.\mathcal{T}.\mathcal{M}^{-1}$. As mentioned before, skeleton-specific primitives are defined on top of the VPG-primitives. Once $\mathcal{M}$, $\mathcal{T}$, and consequently $\mathcal{M}.\mathcal{T}.\mathcal{M}^{-1}$ are known, it is a rather mechanical procedure to define any skeleton-specific communication-synchronization and structural primitive in terms of a sequence of EPAS-provided VPG-primitives. Further discussion in this direction is omitted due to space constraints.

3.3   Composition

EPAS model supports the idea of composition of abstract skeletons. Composition is the way to combine simpler abstract skeletons into a complex one. In the following discussion, we describe the motivation behind incorporating the idea of composition into EPAS.

Motivation Behind Composition. A large-scale parallel application is often a composition of multiple patterns. Sometimes it is more desirable to have a single composite skeleton rather than a collection of smaller skeletons, if the composite skeleton will be used frequently later on. Another reason for having a composite skeleton is performance. Let us consider the example in Fig. 3(a),

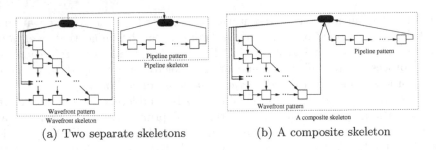

(a) Two separate skeletons   (b) A composite skeleton

Fig. 3. Composing skeletons towards performance

where a Wavefront and a Pipeline skeleton are shown. The output of the right-most child of the Wavefront is sent back to the representative; the representative routes it to the representative of the Pipeline skeleton, which in-turn again routes to the first stage of the Pipeline. Composition of these two skeletons is shown in Fig. 3(b). It is evident from the figure that composition, in this case, reduces the number of routing requirement by 1 as compared to the case in Fig. 3(a).

Readers should note that composition is different from the construction of skeleton hierarchy during concretization. Composition is performed on abstract skeletons to create a composite abstract skeleton. Composition is performed by the skeleton designers, whereas the skeleton hierarchy is constructed by the application developers. Composition may require an in-depth knowledge of the *abstract mapped spaces*, whereas the creation of the skeleton hierarchy does not require that. Composition of two or more skeletons during concretization increases the height of the skeleton hierarchy during application development. On the contrary, use of composite abstract skeletons during application development reduces the height of the skeleton hierarchy.

Model of Composition. In simple words, composition of skeletons S_i and S_j into a skeleton S_k results in the union of the parameters, primitives, and the abstract mapped spaces of S_i and S_j. Moreover, newer primitives may be defined for a composite skeleton. Say, constituents of skeleton S_i are mapped onto abstract mapped spaces $\mathcal{P}_{1i}, \mathcal{P}_{2i}, \ldots, \mathcal{P}_{mi}$; and those of S_j are mapped onto $\mathcal{P}_{1j}, \mathcal{P}_{2j}, \ldots, \mathcal{P}_{nj}$; then constituents of S_k will be mapped onto $\mathcal{P}_{1i}, \mathcal{P}_{2i}, \ldots, \mathcal{P}_{mi}, \mathcal{P}_{1j}, \mathcal{P}_{2j}, \ldots, \mathcal{P}_{nj}$. We define the *extended mapped space* $(\mathcal{E})$ of a skeleton as a space which is exactly big enough to hold all the abstract mapped spaces of that skeleton. Formally, let us assume that a skeleton S consists of the abstract mapped spaces $\mathcal{P}_1, \mathcal{P}_2, \ldots, \mathcal{P}_N$. Assuming that the abstract mapped space $\mathcal{P}_i$ is a k_i dimensional space (i.e., result of mapping the abstract topological space of the skeleton into a k_i dimensional $\mathcal{VPG}$), the extended mapped space, $\mathcal{E}$, would be of dimension $K = \max\{k_i \mid 1 \leq i \leq N\} + 1$.

To create the extended mapped space $\mathcal{E}$, an abstract mapped space $\mathcal{P}_i$ is extended from k_i dimension to $K - 1$ dimension. While extending the dimension, the higher $K - 1 - k_i$ dimensions are made limited to length 1 to ensure consistency. The length of the K-th dimension of the extended mapped space,

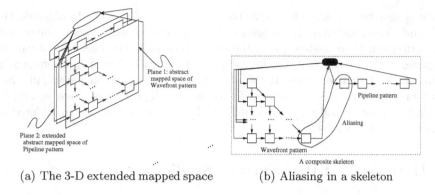

(a) The 3-D extended mapped space   (b) Aliasing in a skeleton

Fig. 4. Extended mapped space and aliasing

$\mathcal{E}$, is N and the extended abstract mapped space of $\mathcal{P}_i$ is placed on the i-th entry of the K-th dimension of $\mathcal{E}$. Fig. 3(b) is redrawn in Fig. 4(a) to reflect the idea of the extended mapped space. The extended mapped space, as is shown in the figure, includes the abstract mapped spaces of the Wavefront skeleton and the Pipeline skeleton. Among the two abstract mapped spaces, the mapped space for the Wavefront is of the highest dimension, which is two. As a result, the extended mapped space is 3-D. As is shown in the figure, the first plane of the extended mapped space includes the mapped space of the Wavefront skeleton, and the second plane includes the mapped space of the Pipeline skeleton.

In order to achieve more flexibility, EPAS provides an *aliasing* facility. Aliasing is the way to combine two nodes from two different abstract mapped spaces of a particular skeleton. The idea is shown in Fig. 4(b). Aliases in EPAS are expressed using an *aliasing function* (designated as $\mathcal{A}$). EPAS provides two types of aliasing: (1) *fusion* paradigm, and (2) *linkage* paradigm. In the linkage paradigm, two aliased nodes are connected via a channel and both the nodes remain as separate entities. On the other hand, in the fusion paradigm, two aliased nodes are unified into one node. As a result, that unified node becomes members of both of the abstract mapped spaces, where the original nodes belonged to. To describe the idea formally, let us assume that $\mathcal{E}_S$ is the extended mapped space of skeleton S, and $\mathcal{P}_i$, $\mathcal{P}_j$ and $\mathcal{P}_k$ are three abstract mapped spaces of S. Lets $P_l \subseteq \mathcal{P}_l$ where $l \in \{i, j, k\}$. The aliasing function is defined as, $\mathcal{A} : P_m \rightarrow P_n$ where $m, n \in \{i, j, k\} \wedge m \neq n$. Say, $\mathcal{A}(p_i) = p_j$ and $\mathcal{A}(p_j) = p_k$, where $p_l \in P_l \wedge l \in \{i, j, k\}$. In the fusion paradigm of the model, those two aliases imply that $\mathcal{A}(p_i) = p_k$. However, this implication is not true for the linkage paradigm.

3.4   Labelling

In PAS, during the concretization phase, a labelling function $\mathcal{L}$ labels each child of a concrete skeleton, CS, with an abstract skeletons type. In other words, labelling can be considered as specifying the types of the children inside the back-end. These children are subsequently concretized, and consequently concretization can be considered as a top-down process in the skeleton hierarchy. Suppose, in the presence of aliasing, $\mathcal{A}(p_i) = p_j$ and $\mathcal{L}(p_i) = AS_s$ where AS_s is an an abstract skeleton type. If the aliasing follows the fusion paradigm, then $\mathcal{L}(p_j)$ must also be AS_s. However, if linkage paradigm is used then p_j can be labelled without considering the labelling of p_i, as p_i and p_j are considered to be two separate entities.

4   Usability and Performance Studies

A Skeleton Description Language (SDL), a usable form of the EPAS model, facilitates both the skeleton designers and the application developers. EPAS generates C++ object frameworks for a given skeleton hierarchy and the run-time system uses MPI-2 as a system software. A discussion of the SDL and EPAS tools are out of the scope of this paper and can be found in [8].

To conduct our usability tests, we chose a group of twelve students, enrolled in an introductory graduate-level course on parallel and distributed computing. Students were asked to compare their experiences with MPI and EPAS. The study pointed out five important points: (1) the learning curve for the EPAS model is more than the MPI model. On the average, the time to learn the EPAS model and the SDL is approximately 30% more than that of the MPI model and its API; (2) developing parallel applications is significantly easier and less time consuming, if the required abstract skeletons already exist in the repository. In the case of EPAS, it took approximately 50% less time and coding effort as compared to MPI; (3) the EPAS system becomes more beneficial with increased complexity of the given problem, i.e., if the problem structure is simple, it is better to use MPI, provided that the required abstract skeletons are not available in the repository; (4) the object-oriented interface and skeleton-specific primitives for communication-synchronization are easier to use as compared to the generic primitives provided by MPI.

To test the performance of EPAS system, we developed two image processing applications. The first application convolutes a series of images and the second one finds the contours of objects in images of maps of buildings and roads. The applications were developed using both MPI and EPAS. The run-times of the applications were measured as an average of at least 10 runs. The results showed that the performance degradation using EPAS as compared to MPI is rather negligible (less than 1%). It is also found that the initial environment initialization phase for a EPAS application is much more complex and hence more time consuming than that of a similar MPI application. However, it should be noted that this initialization takes place only once during the life time of the application. Though the initialization time grows proportional to the complexity of the skeleton hierarchy, it becomes rather insignificant if the application has a relatively long execution time.

5   Related Research Works

Patterns in parallel computing have often been employed not only at the design level but also at the implementation level, i.e., parallel design patterns are often pre-implemented as reusable components. This is analogous to the idea of a *framework* in conventional software engineering. Beginning with the late 80s, several parallel programming systems have been built with the intent to facilitate rapid development of parallel applications through the use of pre-implemented, reusable components. Some of the earlier systems include *Frameworks* [10], *Enterprise* [1], *Code2* [11], and *HeNCE* [11]. Some of the recent systems are *Tracs* [12], *DPnDP* [2], *COPS* [3] and *ASSIST* [5]. Around the same time period of late 80s, another approach to explore the use of parallel design patterns emerged, which focused on the idea of substituting explicit parallel programming by the selection of a variety of pre-packaged parallel algorithmic forms popularly known as *algorithmic skeletons*. Algorithmic skeletons are described as higher order polymorphic functions and are, in practice, realized using various functional and logic programming languages [6,13]. A comprehensive survey of some these approaches can be found in [14].

The significance of extending a given library of parallel patterns has been recognized for about a decade now. Early systems such as *Frameworks* [10], *Enterprise* [1], and *HeNCE* [11] did not have any support for extensibility and composition. The next line of parallel design patterns based systems like *Tracs* [12], *DPnDP* [2], *COPS* [3] and *ASSIST* [5] have some support for extensibility and composition. However, the support is often quite limited in several ways. For example, patterns in TRACS are essentially parallel program segments without any application code. Also, these patterns are not parameterized and there is not much support for creating applications that employs multiple patterns. Similarly, support for extensibility in *DPnDP* [2] is limited to the structural aspects and, thus, it is not possible to describe the behavioral aspects of a pattern and, therefore, a programmer has to fill in all the parallel code to get the right behavior. The COPS system does not support pattern composition. Although, ASSIST system supports composition, the system is limited to only three types of topologies among the virtual processors, i.e., *multi-dimensional array*, *none* (they work independently of each other) and *one* (sequential component with features like non-determinism, etc.). The *multi-dimensional array* topology can easily express data parallelism whereas the *none* topology can easily express independent task parallelism. But, in real life, a parallel application is a complex composition of both of them. In ASSIST, it is difficult to describe such complex compositional structures. Neither COPS nor ASSIST supports reusable components with parametric structures (for example, a k dimensional mesh rather than a two dimensional mesh). Note that *parametric templates* supported by COPS are more from the algorithmic or behavioral perspective. ASSIST demands explicit declaration of connectivities among concurrent modules. As a result, the user of the patterns either needs to design her solution space conforming to the restricted structure of the available pattern (in the repository) or needs to design new patterns according to the needs of her application. The EPAS model allows skeletons with parametric structures and topologies, and the user just needs to tailor the required skeleton according to the application requirements by specifying proper values for the associated parameters.

6   Conclusion and Future Work

EPAS is a step towards making PAS more flexible and usable by supporting both extensibility and skeleton composition. In this paper, we describe the EPAS model for designing abstract PAS skeletons. We also extend the model for supporting composition of abstract skeletons to design new abstract skeletons. Recent usability studies have demonstrated that EPAS might ease the development process for big and complex applications. We have also found that there is no significant performance degradation (less than 1%) while using EPAS.

Perhaps, the usability of the EPAS system could be further enhanced by designing a suitable graphical user interface. Moreover, the associated subsystems for *performance modeling and profiling* need to be included into the system to provide a complete parallel programing environment. Currently we are investigating these aspects. We are also working on the issues of *static and dynamic optimizations* and *fault tolerance* aspects of applications developed using EPAS, and these issues will be reported in our future works.

Acknowledgment

This research has been supported by a grant from the Natural Sciences and Engineering Research Council (NSERC) of Canada.

References

1. Schaeffer, J., Szafron, D., Lobe, G., Parsons, I.: The enterprise model for developing distributed applications. IEEE Parallel and Distributed Technology: Systems and Applications **1** (1993) 85–96
2. Siu, S., Singh, A.: Design patterns for parallel computing using a network of processors. In: 6th International Symposium on High Performance Distributed Computing (HPDC '97), Portland, OR (1997) 293–304
3. MacDonald, S., Szafron, D., Schaffer, J., Bromling, S.: From patterns to frameworks to parallel programs. Parallel Computing **28** (2002) 1663–1683
4. Goswami, D., Singh, A., Preiss, B.R.: From design patterns to parallel architectural skeletons. Journal of Parallel and Distributed Computing **62** (2002) 669–695
5. Vanneschi, M.: The programming model of assist, an environment for parallel and distributed portable applications. Parallel Computing **28** (2002) 1709–1732
6. Cole, M.: Algorithmic Skeletons: Structured Management of Parallel Computation. MIT Press, Cambridge, Massachusetts (1989)
7. Goswami, D.: Parallel Architectural Skeletons: Re-Usable Building Blocks for Parallel Applications. PhD thesis, University of Waterloo, Canada (2001)

8. Akon, M.M., Goswami, D., Li, H.F.: A model for designing and implementing parallel applications using extensible architectural skeletons. In: The Eighth International Conference on Parallel Computing Technologies (To appear in LNCS), Russia (2005)

9. Chan, F., Cao, J., Sun, Y.: High-level abstractions for message passing parallel programming. Parallel Computing **29** (2003) 1589–1621

10. Singh, A., Schaeffer, J., Green, M.: A template-based tool for building applications in a multicomputer network environment. In: Parallel Computing 89. (1989) 461–466

11. Browne, J.C., Hyder, S.I., Dongarra, J., Moore, K., Newton, P.: Visual programming and debugging for parallel computing. IEEE Parallel and Distributed Technology **3** (1995)

12. Bartoli, A., Corsini, P., Dini, G., Prete, C.A.: Graphical design of distributed applications through reusable components. IEEE Parallel and Distributed Technology **3** (1995) 37–50

13. Darlington, J., Field, A.J., Harrison, P.G.: Parallel programming using skeleton functions. In: Lecture Notes in Computer Science. Volume 694., Munich, Germany (1993) 146–160

14. Singh, A., Schaeffer, J., Green, M.: A template-based tool for building applications in a multicomputer network environment. In: Parallel Computing 89, North-Holland, Amsterdam (1989) 461–466

An Efficient Distributed Algorithm for Finding Virtual Backbones in Wireless Ad-Hoc Networks

B. Paul, S.V. Rao, and S. Nandi

Department of Computer Science & Engineering,
Indian Institute of Technology Guwahati,
Guwahati - 781039, Assam, India
{bpaul, svrao, sukumar}@iitg.ernet.in

Abstract. A minimum connected dominating set is an efficient approach to form a virtual backbone for ad-hoc networks. We propose a tree based distributed time/message efficient approximation algorithm to compute a small connected dominating set without using geographic positions. The algorithm has $O(n)$ time, $O(n \log n)$ message complexity, and has an approximation factor of eight. The algorithm is implemented using dominating set simulation program, which shows that our method gives smaller connected dominating set than the existing methods.

Keywords: independent set, connected dominating set, MANET.

1 Introduction

An ad-hoc network can be modeled as a unit disk graph (UDG) $G = (V, E)$ [1], where V represents the wireless mobile hosts and E represents the connectivity among them. Two hosts are connected if distance between them is at most one unit. A dominating set D is a subset of V in which, each node of V is either in D or adjacent to at least one node in D. If the graph induced by D is connected then it is called a connected dominating set (CDS). The nodes in CDS are known as dominator/gateway hosts and others are called dominatee/non-gateway nodes. A independent set (IS) I is a subset of V such that no two vertices of I are adjacent. A maximal independent set (MIS) is an independent set in which no node can be added with out destroying the IS property. If a dominating set is independent then it is called independent dominating set (IDS).

CDS based routing is an efficient approach for routing, since it simplifies the routing process through a smaller sub-network. In proactive approach, only gateway hosts need to keep routing information and in reactive method, route search space is reduced to CDS. So the efficiency of routing depends largely on the size of CDS and its maintenance. But, unfortunately finding a minimum connected dominating set (MCDS) for most graphs is a NP-complete problem [1]. So, several researchers proposed various approximate algorithms for computing CDS. We briefly discuss related work.

One class of algorithms are based on independent dominating set. Gerla and Tsai [2] proposed two localized algorithms, one using node-ID and other using

D.A. Bader et al. (Eds.): HiPC 2005, LNCS 3769, pp. 302–311, 2005.
© Springer-Verlag Berlin Heidelberg 2005

node degree respectively as a rank. Using both node-ID and degree as a rank, Chen and Stojmenovic [3] proposed a distributive algorithm for IDS. The rank idea is generalized by Basagni [3] such that any meaningful parameter can be used as rank to exploit the network properties.

Main draw back of IDS based approach is maintenance of IS property under the mobility condition. So, another class of algorithms are proposed on the bases of CDS. A centralized approximation algorithms for CDS is proposed by Guha and Khuller [4] which have asymptotically optimal approximation ratios of $O(\log \Delta)$ where Δ is the maximum vertex degree of the input graph. Ad-hoc networks need distributed algorithms. A distributed version of the algorithm proposed in [5]. A localized distributed algorithm is proposed by Wu and Li [6] for finding CDS. But the approximation ratio of this algorithm can be as large as $n/2$, where n is the number of nodes in the network [7]. Another distributed algorithm for CDS is proposed in [8]. Wan et al. proposed many algorithms [9, 10, 11, 7] for finding connected dominating sets in unit disk graphs. Two distributed algorithm for CDS is proposed in [9]. The first algorithm uses node-ID as a rank and the second algorithm is a tree based approach using level and node-ID. The size of CDS is further improved for level based algorithm and discussed in [11, 7]. Another node-ID based algorithm is proposed [10], which is a message optimal algorithm, but the approximation ratio of the CDS can be bounded above by 192. Wu et al. added rule-k [12] to their previous algorithm [6] to get a smaller CDS. The performance comparison of these algorithms are tabulated in the Table 1.

Table 1. Performance comparison of various algorithms: n is number of nodes, m is number of edges, and Δ is the maximum node degree

	[5, 13, 14]	[6]	[8]	[11, 7]	[9]	[10]
Approx factor	$\Theta(\log n)$	$O(n)$	$n/2, n$	≤ 8	$8 - 12$	≤ 192
Msg. Complexity	$O(n^2)$	$\Theta(m)$	$O(n^2)$	$O(n \log n)$	$O(n \log n)$	$O(n)$
Time Complexity	$O(n^2)$	$O(\Delta^3)$	$\Omega(n)$	$O(n)$	$O(n)$	$O(n)$
Non-trivial	Yes	No	No	Yes	Yes	Yes
Ngh. Knowledge	2 hop	2 hop	1 hop	1 hop	1 hop	3 hop

In this paper we propose a new distributed algorithm for small connected dominating set. Our algorithm uses *level*, *degree*, and *ID* as a rank. The time and message complexity of the algorithm are respectively in $O(n)$ and $O(n \log n)$. The approximation size factor of the CDS is 8. To the best of our knowledge, tree based CDS maintenance of mobile nodes has not been discussed in the literature, which is an important issue in ad-hoc networks. In this paper we propose a method of maintaining the CDS under mobility.

The rest of the paper is organized as follows: next section describes our distributed algorithm for constructing a small connected dominating set. Section 3 we present a method for node maintenance. Our simulation results are shown in the fourth section. We conclude in the last and final section.

2   The Proposed Algorithm for Finding Virtual Backbone

Each node in the network is identified uniquely by its rank. In our approach, we use $(level, degree, ID)$ as the rank of node. We explain each term in the rank. The *level* of a node is the number of hop distance from the root of an arbitrary spanning tree rooted at a particular node called leader/root. The *degree* of a node is the number of neighbor nodes. The *id* of a node is the node's unique ID.

Our algorithm works in two phases. First phase constructs a maximal independent set using a spanning tree. In the second phase, we connect nodes in IS using the other nodes to form a CDS. The detailed description of these phase are given in the following subsections.

2.1   Maximal Independent Set Formation

We first elect a leader node x and construct a spanning tree (ST) rooted at node x using the distributed algorithm described in [15]. One can use any criteria in [15], like ID or $(degree, ID)$, to elect the leader. We use $(degree, ID)$ criteria, since by selecting a node with maximum degree as a dominator covers maximum number of nodes. Next step is to find level of each node in ST. This process starts at the root. The root node sets its level to zero and sends this information to its children. Each child sets its level one more than the level of its parent and propagates this information to its children. Each node informs its parent after all nodes complete their level marking in the subtree rooted at it. Once the root receives such message, all nodes in the ST are marked. During the construction of ST, each node collects information about ID, level, and number of neighbor of all its neighbors. This can be done by single broadcasting by each node. After construction of ST, independent set formation starts.

We discuss the algorithm using colors WHITE, GRAY, and BLACK. Initially, each node is marked WHITE. After completion of the IS formation algorithm, each node colored either GRAY or BLACK, such that the set of the BLACK nodes forms an IS. The GRAY nodes are later used to connect BLACK nodes to form a CDS. In addition to *level* and ID, each node maintains a sorted list of BLACK neighbors in ds, which is initialized to null. As we mentioned earlier, we use $(level, degree, ID)$ to rank of nodes. We can order nodes as per their ranks. For example, let (l_i, d_i, ID_i) and (l_j, d_j, ID_j) be ranks of nodes n_i and n_j respectively. The node n_i is higher rank than n_j if $(l_i < l_j)||((l_i = l_j)\&\&(d_i > d_j))||((l_i = l_j)\&\&(d_i = d_j)\&\&(ID_i > ID_j))$.

The root node marks it self BLACK and sends its color information to its neighbors. All other nodes mark as per the following algorithm.

- When a WHITE node receives the message from a BLACK node, it marks itself GRAY and stores the sender's ID in ds. Then sends its color information to its neighbors.
- When a WHITE node receives the message from GRAY node for the first time, it marks itself BLACK, if its rank is higher than its unmarked neighbors. Then sends its color information to neighbors.

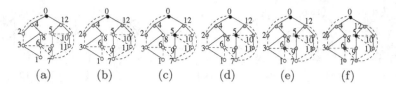

Fig. 1. Independent set construction

- Subsequently, any node receives the message from BLACK node, adds the sender's ID in ds.
- When a leaf node is marked, it send a message to its parent stating marking is complete. Each non-leaf node of ST wait till it receives this message from each of its children and then forward it up the tree.

By the time when root gets the marking complete message, all nodes in the tree ST have been marked with either BLACK or GRAY and thus root moves to the construction of CDS. To show our algorithm gives smaller CDS, we consider the example discussed in [7]. The Fig. 1[1] illustrates the IS construction process.

2.2 CDS Formation

Main idea of this phase is to form a dominating tree (DT), by connecting BLACK nodes with the help of some GRAY nodes, of which all the non-leaf nodes form a CDS. This dominating tree is another spanning tree of the set of nodes. In dominating tree construction, each node maintains root, parent, list of children, and list of gateway neighbors respectively in $root$, $parent$, $child$, and $gate$. These are initialized to NULL for each node. We also use another variable b_degree like [7], to indicate total number of black neighbors. But unlike in [7], we initialize b_degree of root to one.

Among the root of ST and its neighbors, we choose the node with highest number of BLACK neighbors as the root of DT. This can be achieved by the root of the ST by broadcasting CHECK message. Gray nodes receiving the CHECK message, sends their number of black neighbor's in the form of a STATUS message. Upon receiving the STATUS messages from all its neighbors, root node set the variable $root$ with the ID of node with highest number of BLACK neighbors. Then sends a START message to the node, whose ID is in $root$ for construction of DT. The Fig. 2(a)- 2(c) illustrates the process of finding root of DT. Upon receiving the START message, this new root node starts the construction of DT, by setting its $level$ to zero and broadcast the CONSTRUCT message. Each node act according to the following principle.

- BLACK nodes having $parent = NULL$ receiving CONSTRUCT message, sets its $parent$ to sender's ID, and $level$ to one more than the level of sender, send an ACCEPT message to sender, and broadcast an INITIATE message.

---

[1] In figures, tree and backbone edges are respectively denoted by solid and bold solid/dashed edges. Message passing direction is indicated by arrows.

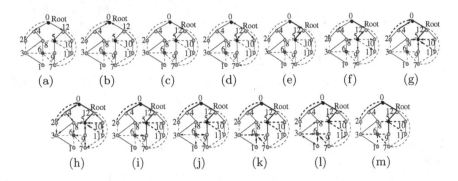

Fig. 2. CDS construction process

- GRAY nodes having *parent = NULL* receiving the first INITIATE message, sets its *parent* to sender's ID and *level* to one more than the level of sender, send an ACCEPT message to sender, and broadcast a CONSTRUCT message.
- Upon receiving the ACCEPT message, each node adds ID of the sender to its child list and then broadcast the GATEWAY message.
- Upon receiving the GATEWAY message, each node adds the sender ID to *gate* list.

The Fig. 2 shows a possible execution scenario of CDS construction process. The size of the CDS can be further improved by using the following rule. We assume every parent and child exchange periodic HELLO messages for finding each other activeness. This rule can also be applied during node movement.

Improvement Rule: Each non-gateway node tries to reduce its level by connecting to a lesser level gateway neighbor node. This can be done by searching for a node in its *gate* list, and if found, it establishes a parent-child relation. In the process of changing the parent, if any gateway node looses all its children, it becomes a non-gateway node by broadcasting a NONGATEWAY message. All nodes receiving the NONGATEWAY message, removes the sender ID from its *gate* list.

The distributed spanning tree algorithm [15] is in $O(n)$ time and $O(n \log n)$ message complexity. In our proposed algorithm, each node sending constant number of messages in IS formation, hence the time and message complexity is in $O(n)$. Similarly, one can see the time and message complexity of the second phase also in $O(n)$. Hence, follows the theorems.

Theorem 1. *The time and message complexity of CDS construction is in $O(n)$ and $O(n \log n)$ respectively.*

Theorem 2. [7] *The size of the CDS can be bound above by $8 * opt - 2$, where opt is the size of MCDS.*

3 Backbone Maintenance

In ad-hoc network, the communication links and backbone may get disconnected due to nodes switch on/off or movement. It is necessary to maintain the backbone such that the whole network is connected. In this section, we propose a strategy for backbone maintenance under switch on/off cases and mobility environment.

3.1 Switch On/Off

In ad-hoc network, any node can go down or come up due to limited resources. The backbone gets disconnected if any gateway node goes down. All the children of it connect themselves to some other gateway node whose level is less than its level, so that the backbone is again connected. We assume that the root node can move but can not go down. If the root goes down, we reconstruct a CDS from the beginning, which is initiated by the level one nodes. We describe the handling process case wise.

Switch On. Initially, when a node comes up, it notifies its existence by broadcasting a NOPARENT message. We handle different cases separately as follows.

New Node Near to Gateway Nodes: If any gateway node receives the NOPARENT message, it sends a AVAIL message to the sender in response. If this new node receives AVAIL message first time, it sets its parent to sender ID, level to one more than the sender level, adds the sender ID in its *gate* list, and sends a OK message to the sender. Otherwise, it simply adds sender ID in its *gate* list. When gateway node receives this OK message, it adds the sender ID in its child list and sends a DONE message for confirmation.

New Node Near to Non-gateway Nodes: If within specific interval, the new node does not get AVAIL message, it broadcasts the NOPARENT2 message to connect through some non-gateway nodes. This time non-gateway nodes which are in its range reply with the AVAIL2 message. Upon receiving the first AVAIL2 message, the new node sets its parent to the sender ID, level to one more than the sender level, and sends a OK message in reply. When non-gateway node receives this OK message, adds the sender ID in its child list and broadcasts a GATEWAY message. Each node receiving this GATEWAY message, adds the sender ID in their respective *gate* lists.

Switch Off. If any non-gateway node goes down, its parent gateway node does not receive periodic message, so its parent removes its entry from *child* list. By doing this, if it does not have any more child, it broadcasts NONGATEWAY message. Each node receiving the NONGATEWAY message removes the sender from their respective *gate* lists.

When any gateway node goes down, its children searches for another parent. If the child is a non-gateway node, it broadcast the NOPARENT message and the remaining process of finding a parent is same as discussed in the section 3.1.

If the child is a gateway node, it searches for a gateway node whose level is less than its level. This process of searching initially starts in 1-hop neighbors in the form of a NOPARENT3 message. If successful, it establishes the connection and changes its level in consistence with the new parent gateway. If unsuccessful, searches in 2-hop gateway neighbors and so on till it finds a gateway node with lesser level. Let assume such a node y is found at a distance of i-hops through a node path $y_1, y_2, \ldots, y_{i-1}$. This path becomes a part of the backbone. These nodes changes their level inconsistent with the level of y and propagates the level change to their respective subtrees rooted at them in DT. We show an example of this process in node movement section 3.2.

3.2   Node Movement

We deal the cases non-gateway and gateway nodes movement separately in the following sub sections.

Non-gateway Node Movement. When a non-gateway node moves out of its parent vicinity, two cases may arise. First, it moves near to other gateway nodes and second, near to non-gateway nodes. In these case, the earlier parent of this node does not receive any periodic message from this child. Hence, the parent gateway node removes the child from its child list. When any child non-gateway node does not receive periodic message from its parent, the child broadcasts the NOPARENT message. The next process of finding its parent is similar to the method described in the section 3.1.

Gateway Node Movement. If gateway node moves from the vicinity of any child, the child need to find a new parent for maintenance of the backbone. The process of finding is very similar to the method discussed in the section 3.1. Consider an example given in the Fig. 3(a) where node 12 is the root. Assume node 9, at level two, moves from its original position to the new position as shown in Fig. 3(b). Since, node 9 does not get periodic messages from its parent node 5, searches for a gateway node at level zero or one, by broadcasting a NOPARENT3 message with $hop = 1$. But it is unsuccessful, so it broadcasts again NOPARENT3 message with $hop = 2$ after some fixed time interval. This time, it finds a gateway node 12 with level zero through the node 7. The node 12 responds with AVAIL3 message to node 9 through node 7. The nodes 9-7, and 7-12, establishes the child-parent relation by sending OK2 and DONE2 messages.

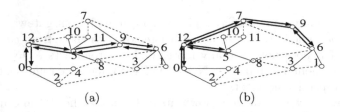

(a)   (b)

Fig. 3. (a) Initial backbone.   (b) The updated backbone after movement of node 9.

Once the path 9, 7, 12 is established, node 12 sends NEWLEVEL messages to node 7. All nodes in the subtree rooted at 7 changes their level in consistence with the level of node 12, forwarding the NEWLEVEL message. The Fig. 3(b) shows the updated backbone of the network.

4   Simulation Results

Theoretical comparison of various algorithms is given in the Table 1. These bounds give an upper bound on the size of CDS, but it does not give actual difference in size. Simulation gives more realistic approximation to real scenario than theoretical bounds. So, we have simulated our algorithm to show the size difference with other methods. We have compared only with tree based approaches, since they give smaller backbone than other approaches [12].

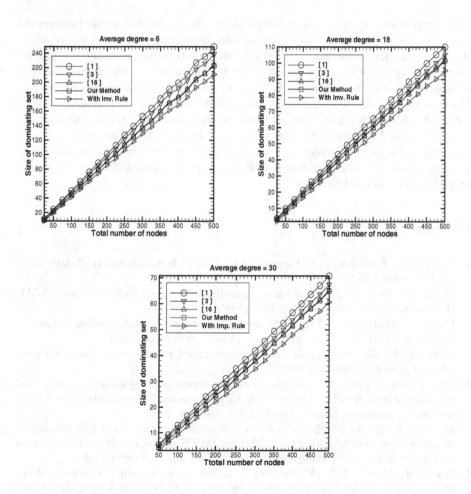

Fig. 4. Simulation results for different values of d

We have used the **ds** custom simulator [17]. We assume, the network is connected with bidirection links. We have consider a 50×50 square units area, with 10 units transmission range. For a fixed transmission range, average degree, d, of nodes increase as network density increases. Network density increase as the number of nodes, N, increases in a fixed confined area. So, we change the average degree and number of nodes in the network to cover all possible scenarios.

We have chosen $d \in \{6, 18, 30\}$, and $N \in [25, 500]$. For given d and N, the average of 250 readings are plotted in the Fig. 4. From this simulation, we can conclude that our algorithm without improved rule gives smaller backbone than existing tree based algorithms. The size of the backbone reduce further by using improved rule.

5   Conclusion

In this paper we presented a distributed algorithm for finding a small connected dominating set. The algorithm consists of two phases: first, maximal independent set construction, and second, CDS formation. The time and message complexity is respectively in $O(n)$ and $O(n \log n)$. It has been shown by simulation that our algorithm constructs smaller CDS compared to prior heuristics method of MCDS in ad-hoc network. A simple node maintenance scheme also incorporated in the proposed algorithm. The increase in the size of the CDS, due to node movement, can be reduced by improvement rule. Note that the same rule can also applied for gateway nodes without level consistency. It would be interesting to study other network parameters like delay, jitter, packet delivery ratio, etc using the constructed backbone.

References

1. Clark, B.N., Colbourn, C.J., Johnson, D.S.: Unit disk graphs. Discrete Mathematics. 86 (1990) 165–177
2. Gerla, M., Tsai, J.T.C.: Multicuster, mobile, multimedia radio network. ACM-Baltzer Journal of wireless networks 1 (1995) 255–265
3. Chen, G., Stojmenovic, I.: Clustering and routing in mobile wireless networks. Technical report, University of Ottawa, Computer Science (1999)
4. Guha, S., Khuller, S.: Approximation algorithms for connected dominating sets. Algorithmica 20 (1998) 374–387
5. Bharghavan, V., Das, B.: Routing in adhoc networks using minimum connected dominating sets. In: IEEE International Conference on Communications (ICC'97). Volume 1., Montreal, Canada (1997) 376 – 380
6. Wu, J., Li, H.: On calculating connected dominating sets for efficient routing in adhoc wireless networks. In: Proc. of the 3rd Int'l Workshop on discrete algorithms and methods for mobile computing and communications. (1999) 7–14
7. Wan, P.J., Alzoubi, K.M., Frieder, O.: Distributed construction of connected dominating set in wireless adhoc networks. In: Twenty-First Annual Joint Conference of the IEEE Computer and Communications Societies. (2002) 1597 – 1604

8. Stojmenovic, I., Seddigh, M., Zunic, J.: Dominating sets and neighbor elimination based broadcasting algorithms in wireless networks. IEEE Transactions on Parallel and Distributed Systems (2002) 14 – 25
9. Alzoubi, K.M., Wan, P.J., Frieder, O.: Distributed heuristics for connected dominating set in wireless adhoc networks. IEEE ComSoc / KICS Journal of communications and networks 4 (2002) 22–29
10. Alzoubi, K.M., Wan, P.J., Frieder, O.: Message-optimal connected dominating sets in mobile adhoc networks. In: The Third ACM Int'l Symposium on mobile adhoc networking and computing. (2002) 157–164
11. Alzoubi, K.M., Wan, P.J., Frieder, O.: New distributed algorithm for connected dominating set in wireless adhoc networks. In: Proceedings of the 35th Annual Hawaii International Conference on System Sciences. (2002) 3849 – 3855
12. Dai, F., Wu, J.: An extended localised algorithm for connected dominating set formation in ad-hoc wireless networks. IEEE Transactions on Parallel and distributed systems 15 (2004)
13. Das, B., Sivakumar, R., Bharghavan, V.: Routing in adhoc network using a spine. In: Int'l Conference on computers and communications networks '97, Las Vegas, NV (1997) 34 – 39
14. Sivakumar, R., Das, B., Bharghaban, V.: An improved spine-based infrustructure for routing in adhoc networks. In: IEEE Symposium on Computers and Communications '98, Athens, Greece (1998)
15. Cidon, I., Mokryn, O.: Propagation and leader election in multihop broadcast environment. In: the 12th Int'l Symposium on DIStributed Computing (DISC98), Greece (1998) 104–119
16. Marathe, M.V., Breu, H., III, H.B.H., Ravi, S.S., Rosenkrantz, D.J.: Simple heuristics for unit disk graphs. Networks 25 (1995) 59–68
17. Dai, F.: Dominating set simulation program. http://www.cse.fau.edu/ fdai/adhoc (2001)

A Novel Battery Aware MAC Protocol for Minimizing *Energy × Latency* in Wireless Sensor Networks

M. Dhanaraj, S. Jayashree, and C. Siva Ram Murthy*

Department of Computer Science and Engineering,
Indian Institute of Technology, Madras 600036, India
{dhanaraj, sjaya}@cs.iitm.ernet.in, murthy@iitm.ac.in

Abstract. Wireless Sensor Networks (WSNs) possess highly constrained energy resources. The existing Medium Access Control (MAC)
protocols for WSNs try to either minimize the energy consumption or the
latency, which are conflicting objectives, or find a trade-off between them.
They fail to achieve the minimum *energy × latency*, which ensures that
transmission should occur such that both the energy consumption and latency are minimized. We propose a novel Battery-aware Energy-efficient
MAC protocol to minimize the Latency (BEL-MAC) that exploits the
chemical properties of the batteries of the sensor nodes, in order to increase their lifetime. Our protocol reduces the latency of the packets in
an efficient manner without compromising on the lifetime of the network.
We compare our protocol with the SMAC, DSMAC, TMAC, and IEEE
802.11 MAC, in terms of throughput and latency and show that our protocol outperforms these existing protocols, in terms of *energy × latency*.

1 Introduction

A Wireless Sensor Network (WSN) consists of a large number of distributed
sensor nodes that organize themselves into a multi-hop wireless network. They
possess one or more sensors, embedded processor(s), and low-power radios, and
are normally battery operated. The main task of the sensor nodes in a sensor
field is to detect events, perform quick local data processing, and then transmit
the processed information to the data sink.

Sensor nodes are battery-driven and hence operate on an extremely low energy
supply. In most applications, once deployed, battery replacement is not preferred
because of a larger number of nodes and hostile and remote environment they
are deployed in. Hence, battery power is a precious resource that has to be
used efficiently. The nodes of a WSN have the following three modules: sensing,
processing, and communication. All these modules support various operating
modes for power management purposes. The amount of energy consumed by
the sensor nodes mainly depends on the operating modes of the modules. For

---

* Author for correspondence. This work was supported by the Department of Science
and Technology, New Delhi, India.

D.A. Bader et al. (Eds.): HiPC 2005, LNCS 3769, pp. 312–321, 2005.

example, the communication module consists of *Transmit, Receive, Idle*, and *Sleep* modes. Power consumption for communication is several times higher than that for other activities, such as sensing and processing. Hence, the radio should be completely shut off whenever possible, in order to save energy. Experimental results indicate that the energy consumed by the wireless devices, while in idle state, is almost equal to that in the transmission or receiving state [1]. Recent works in [2] - [5], aim only at switching off the communication module, which they achieve by applying a wakeup/sleep schedule for the radios. In addition, the batteries of the sensor nodes tend to possess an interesting chemical property and the maximum lifetime is achieved by switching off the battery periodically [6], [7]. This paper attempts to exploit this property of the battery, in order to increase its lifetime.

2   Related Work

In order to save energy and prolong the network lifetime, switching between the operating modes of the sensor node radio can be applied. Several works [2] - [5] have been proposed in this context. SMAC protocol [2] periodically switches the radio between the sleep and listen modes. This listen mode, also called the active period, is known as the basic active slot. Since the sleep time duration is chosen statically without considering the network traffic and is much higher than the active time, SMAC introduces a large latency. On the other hand, the PAMAS [3] and STEM [1] protocols use a dual channel radio setup. The data channel always remains in the sleep mode and it is activated for the data transmission by sending a wakeup signal in the control channel, which becomes on periodically. However, all the above-mentioned protocols have a fixed sleep duration and do not consider the network traffic, while designing the wakeup/sleep schedule.

DSMAC [4] improves the latency over SMAC by appropriately adding or removing the active slots. Though DSMAC protocol reduces the sleep time duration of the nodes compared to SMAC, it, at the maximum, uses only 40% of the frame duration for the data transmission, and thus, achieves lesser throughput in the presence of a high network traffic. In TMAC protocol [5], after a node gains access to the channel, based on the network traffic, it transmits a burst of packets by increasing the active slot time duraion. However, this continuous transmission of packets leads to a decrease in the nodes' battery charge and thus reduces the network lifetime [6]. Though DSMAC and TMAC protocols attempt to dynamically add/remove the active slots based on the network load, they do not achieve the minimum possible latency for a given fixed lifetime. To the best of our knowledge, there exists no MAC protocol in WSNs that considers chemical behavior of the batteries of the nodes to arrive at an efficient wakeup/sleep schedule. Hence, we propose a novel MAC protocol, which dynamically alters the number of active slots and achieves the minimum possible latency without compromising on the network lifetime by taking the chemical behavior of the battery into consideration.

3   Our Work

Since the WSNs deal with real world processes, it is often necessary for the communication to meet real-time constraints, in the form of end-to-end latency targets. Making nodes sleep too often may considerably save energy, but may also increase latency. The existing works aim at reducing either of these conflicting requirements. Hence, it is required to design a generalized wakeup/sleep schedule, which reduces both the energy consumption and latency. We propose a novel Battery-aware Energy-efficient MAC protocol to minimize the Latency (BEL-MAC). The main goal of our protocol is to attain the minimum achievable end-to-end latency and maximum lifetime by using an appropriate dynamic wakeup/sleep schedule by exploiting the chemical behavior of the batteries of the sensor nodes. We, in this section, discuss the basic battery characteristics, working principle of our protocol, and the dual-battery setup.

3.1   Basics of Batteries

The following definitions and notations for batteries are taken from [6]. Battery capacity can be expressed in either of the three ways: (1) *Theoretical capacity (T):* This value is based on the amount of energy stored in the battery, and is an upper bound on the total energy that can be extracted in practice, (2) *Nominal capacity (N):* This value is the energy that can be extracted when it is discharged under standard load conditions, which are specified by the manufacturer, and (3) *Actual capacity:* This value is the amount of energy that the battery delivers under a given load, and is usually used as a metric to judge the battery efficiency for constant loads.

Two important effects that decide the battery performance are: (1) *Rate-capacity effect:* As the intensity of the discharge current increases, an insoluble component develops between the inner and outer surfaces of the cathode of the batteries. The inner surface becomes inaccessible and as a result of this phenomenon, the cell (battery) becomes unusable even while a sizable amount of active materials still exists. (2) *Recovery effect:* This effect is concerned with the recovery of charges under idle conditions. Due to this effect, on increasing the idle time of the batteries, the theoretical capacity can be completely utilized.

A battery can be represented by a tuple $< N_i, T_i >$, where N_i and T_i represent its nominal and theoretical capacities at any time instant i. Actual capacity may be higher than N, but cannot exceed T. The recovery process depends on the idle time duration of the battery, its N_i, and T_i. Whenever the battery is discharged, it goes from state $< N_i, T_i >$ to $< N_{i-k}, T_{i-k} >$ where k is the number of time units for which a constant discharge current is applied to the battery. If a battery is continuously discharged for N time units, it goes from state N to 0, also called the dead state, after which no more recovery is possible. Using a pulsed discharge, the battery alternates between the discharge and recovery process and thus it recovers one charge unit in one unit of idle time with a probability (R_{N_i, T_i}) [6] and goes from state $< N_i, T_i >$ to $< N_{i+1}, T_i >$, where

$R_{N_i,T_i} = e^{-g \times (N-N_i) - \phi(T_i)}$ if $(1 \leq N_i \leq N, 1 \leq T_i \leq T)$ and otherwise $R_{N_i,T_i} = 0$. Here, $g = 0.05$ is the battery constant [6], and $\Phi(T_i)$ is a piecewise constant function of T_i, which are specific to the battery's chemical properties. Hence, N_i of the battery should be maintained close to N, in order to increase its recovery capabilities.

3.2 BEL-MAC Protocol

Basic Assumptions: The basic assumptions, which facilitate the working of our protocol, are as follows. (1) Sensing and computing (if any) are carried out periodically. (2) We assume the presence of a dual-battery setup, a setup with a primary and a smaller secondary battery, as explained at the end of this Section. (3) Sleep state of the nodes corresponds to a complete shut-off of the primary battery during which the battery can recover its nominal capacity.

Working of BEL-MAC Protocol: Here, we introduce a new wakeup/sleep schedule, which reduces the latency without compromising on the lifetime. During the sleep time of the radio, the primary battery is switched off for a particular period of time by using the dual-battery setup. In the discussions that follow, battery of a sensor node represents its primary battery, unless otherwise mentioned. The recovery of a battery decreases as N_i decreases [6]. The wakeup/sleep schedule of a node is designed so as to maintain N_i close to its initial value (N) by providing enough idle times for charge recovery.

In our scheme, each frame has a number of slots, also called the dynamic active slots added, in addition to the basic active slot. Figure 1 shows the wakeup/sleep schedule of SMAC, DSMAC, TMAC, and BEL-MAC protocols. Each node's primary battery is turned on at the beginning of every slot by the SBS system [8] and the nodes attempt to grab the channel. As followed in the SMAC protocol,

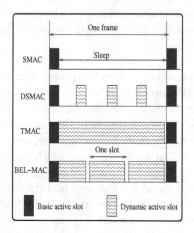

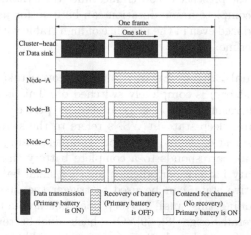

Fig. 1. Wakeup/sleep schedule of different protocols

Fig. 2. BEL-MAC – Data transmission for 1-hop network

BEL-MAC also uses 802.11 DCF for resolving contentions. After the contention is resolved, except the sender and receiver, all the other nodes enter the idle state by turning off their primary batteries to enable recovery. The sender node, on the other hand, transmits its packets continuously, until all of its packets are transmitted or the slot ends. Since the SMAC and DSMAC protocols add only a limited number of active slots in a complete frame, these protocols are meant for low and medium traffic. The proposed BEL-MAC aims at adding optimum number of additional active slots and each of the active slots is used by different nodes, which have higher battery capacity. Since our protocol does not allow a continuous transmission by a node for consecutive slots, it enables recovery of the battery charges, and thus, achieves the maximum lifetime. However, the latency of the packets is also minimized in our protocol, because the nodes that gain access to the channel are allowed to transmit packets continuously for the complete duration of the slot.

Figure 2 shows the data transmission for 1-hop network, where all the nodes transmit their packets to the cluster-head/data sink. The nodes are placed in any one of the following states: 1) *Transmission* (Transmit/receive), 2) *Recovery*, and 3) *Idle* (contend for channel). When a node remains in the *Transmission* state, its battery is assumed to discharge by two units for every packet transmission, that is, both N_i and T_i values decrease by 2 [6]. When the battery is switched off during *Recovery* state, the recovery of charges increases its N_i by one unit with a probability R_{N_i, T_i}. Thus, a longer sleep time for the battery may allow it to attain the maximum value of N. This helps to utilize the maximum value of T. In our protocol, Only those nodes which have packets to transmit and $N_i = N$ are allowed to contend for the channel. Hence, the complete recovery is provided for the nodes in the network. Since we add additional active slots in addition to the basic active slots, the packets can be transmitted at any active slot instead of waiting for the basic active slots. Thus, the latency is minimized in the presence of low traffic. Since the dynamic active slots can be extended and a complete frame duration can be used for packet transmissions, the latency is reduced even in the presence of high traffic without compromising on the network lifetime.

Dual-Battery Setup: Battery recovery takes place only when the power source (battery) is completely switched off. This is quite possible in sensor networks, where information is generated by periodically sensing the environment. Till date, sleep state in WSNs refers to the switching off of the radio alone, with the rest of the modules (for example, sensing) *on*, whereas, in our scheme, sleep state corresponds to a complete battery shut off. In order to wakeup the primary battery periodically, another small battery, which remains always in the active state, can be used to trigger the sleeping primary battery to turn on. Adding this small battery does not account for much space and cost in the overall design [8], [9]. Our basic dual-battery setup contains a primary battery, secondary battery (non-rechargeable button cell), timer, and the SBS for controlling both the batteries. The primary battery is periodically made active using SBS.

4   Theoretical Analysis

We now analyze the lifetime and latency of the SMAC, DSMAC, BEL-MAC, TMAC and IEEE 802.11 MAC protocols.

The lifetime of a node is estimated by calculating the maximum number of packets that can be transmitted (P_m) by that node throughout its lifetime. Since SMAC protocol has enough sleep time and BEL-MAC protocol allows for complete recovery of the battery charges, they achieve the maximum lifetime $(T/2)$ independent of the network load. This is the maximum possible lifetime with a battery of theoretical capacity (T) as each transmission consumes 2 charge units. Since the 802.11 MAC protocol does not have sleep slot, the battery charges never get recovered and hence, P_m is $N/2$. In the presence of heavy traffic, DS-MAC and TMAC protocols achieve only the minimum lifetime $(N/2)$. During a heavy traffic, the recovery of the battery charge may not be possible, whereas in the case of light traffic, the recovery is possible. Hence, for DSMAC/TMAC protocols, $N/2 \leq P_m \leq T/2$.

The latency denotes the sum of the propagation, queuing, and transmission delays encountered by a packet. The notations used for the latency estimation are given as follows: DC - Number of active slots per frame (Basic DC + Dynamic DC), L_{total} - Total latency for transmitting all the packets, L_{avg} - Average latency per packet, n_t - Number of nodes trying to access the channel at time t, T_a - Inter arrival time of the packets, T_f - Time duration of one frame, T_c - Time taken to resolve a collision, T_d - Time taken to transmit a data packet, L_i - Latency of packet i, L_{qi} - Queuing delay for packet i, L_{ci} - Channel access delay for packet i, and L_{ti} - Time taken for transmitting packet i after gaining access to the channel.

For SMAC and DSMAC protocols, the latency is estimated as follows. The number of active slots (DC) is 1 and 4 for SMAC and DSMAC, respectively. Latency of any packet i is given by, $L_i = L_{qi} + L_{ci} + L_{ti}$. We now calculate L_i for the first packet in the queue of a node. If a packet i, on its arrival, has x_i previous packets in the queue, queuing delay for packet i is the time taken for the previous x_i packets to get transmitted. Since the first packet has no previous packets in the queue, $L_{q1} = 0$. Let L_{ti} be T_d for all i. In order to calculate the channel access delay, we use the delay analysis of IEEE 802.11, proposed in [10]. In [10], the backoff time or L_{ci} gives the channel access delay (total waiting time to get access to the channel for a node). Hence,

$$L_{c1} = \frac{\alpha(W_{min}\beta - 1)}{2q} + \frac{(1-q)}{q}t_c \quad where, \quad \beta = \frac{q - 2^m(1-q)^{m+1}}{1 - 2(1-q)} \quad (1)$$

Here, if q_i is the probability that a packet is successfully transmitted at the end of ith stage, we assume, according to [10], $q_1 = q_2 = \dots$ W_{min} and W_{max} are the minimum and maximum contention windows, respectively. If m is the maximum back-off stage, $W_{max} = 2^m W_{min}$. If i, c, and s denote the events, idle state, collision of packets, and successful transmission of a neighbor, respectively, when a node is in the back-off state, $\alpha = p_i t_i + p_c t_c + p_s t_s$, p_i and t_i are the probability of channel to be idle and the corresponding time duration, when a node is in the back-off state. p_c, t_c, p_s, and t_s follow corresponding notations for the other two events. In the case of SMAC, DSMAC, BEL-MAC, and TMAC protocols, $\alpha = T_f/DC$. This is because, unlike 802.11 protocol, in which the nodes attempt to transmit continuously, in all the other protocols, the nodes contend for the channel only during the beginning of the time slot. Hence, in SMAC protocol, $L_{q1} = 0$; $L_{t1} = T_d$

$$L_{c1} = \frac{\frac{T_f}{DC}(W_{min}\beta - 1)}{2q} + \frac{(1-q)}{q}t_c \; ; \quad L_i = (\sum_{j=1}^{x_i} L_j) + (L_{ci} + L_{ti}) \qquad (2)$$

$$L_{total} = \sum_{i=1}^{T/2} L_i = \sum_{i=1}^{T/2}[(\sum_{j=1}^{x_i} L_j) + L_{ci} + L_{ti}]; \quad L_{avg} = \frac{L_{total}}{T/2} \qquad (3)$$

Hence, for any i^{th} packet in the queue, which has x_i preceding packets to be transmitted when it arrives, latency is given in the above equation. Since, this L_{avg} is directly proportional to $\frac{T_f}{DC}$, it is higher for SMAC ($DC = 1$) than the DSMAC ($DC = 4$) protocol.

For TMAC and BEL-MAC protocols, latency is estimated as follows. TMAC protocol uses the same DC value of 1 as that of SMAC protocol, whereas DC value is 3 for BEL-MAC protocol. Then,

$$L_{total} = \begin{cases} \sum_{i=1}^{T/2}(L_{ci} + L_{ti}) + \sum_{i=1}^{T/2} Reco(i) & \text{:for low traffic} \\ \sum_{i=1}^{T/2}[\frac{(\sum_{j=1}^{x_i} L_j)+L_{ci}}{\frac{T_f}{DC \times T_d}} + L_{ti}] + \sum_{i=1}^{T/2} Reco(i) & \text{:for heavy traffic} \end{cases} \qquad (4)$$

where, $Reco(i)$ is the time spent by the battery in recovering its charges completely, before transmitting the ith packet. For TMAC, $Reco(i)$ value is 0 always.

5   Simulation Results

We have carried out extensive simulations for measuring the performance of our protocol and compared our results with that of SMAC, DSMAC, TMAC, and IEEE 802.11 MAC protocols using GloMoSim simulator. The various parameters used in our simulation are listed in Table 1. In our simulations, we consider a cluster, where the nodes communicate with the cluster-head in a single-hop path.

Table 1. Simulation Parameters

Description	Value	Description	Value
Simulation area	200m × 200m	Frame duration (T_f)	3000 ms
Number of nodes	10	Active slot duration (T_b)	300 ms
Channel bandwidth	20 Kbps	Theoretical capacity of the battery (T)	200
Packet size	512 bytes	Nominal capacity of the battery (N)	25
Transmission range	300m	Battery parameter (g)	0.05

However, our protocol can be extended for multi-hop networks. Throughout the simulations, we assume that each packet transmission discharges its battery charge by two units. All the simulation results in this section have been obtained at 95% confidence level.

The average end-to-end latency of the packets by varying the network load is measured for all the protocols and the results are shown in Figure 3. The latency of all the protocols reduces as the load reduces, because the number of packets in the queue is lesser and thus the average queuing delay reduces. In the case of TMAC, 802.11 MAC, and BEL-MAC protocols, the latency is lesser than that of SMAC and DSMAC protocols, because SMAC and DSMAC do not use the complete frame duration for the transmissions. In the case of SMAC, the latency is 4 times higher than that of DSMAC and hence, this is not shown in the figure. As the inter arrival time decreases, the latency in the case of 802.11 MAC and TMAC becomes lesser than that of BEL-MAC, because the number of packets transmitted in the case of 802.11 MAC and TMAC is lesser and we calculate the latency only for the transmitted packets. The latency of the packets which are not transmitted because of the death of a node is not used in the latency calculation. In the case of BEL-MAC, the number of packets transmitted remains higher which increases the queuing delay. In the presence of a medium/low network traffic, the latency of BEL-MAC is lesser than that of SMAC, DSMAC, and TMAC, because it transmits a set of packets continuously and adds additional active slots. However, the latency of 802.11 MAC is lesser than that of all other protocols, because it does not wait for active slot to start the transmission. The simulation and theoretical results (Figures 3 and 4) are compared, in terms of the average latency of the packets. The theoretical and simulation results are found to be almost close to each other. In the simulations, the chances of getting access to the channel are random. Since a node that has a few packets may get more number of chances to transmit, it leads to a reduction in the amount of channel utilization. Hence, the TMAC and BEL-MAC protocols have higher latency in the simulation results and slightly lesser latency in the theoretical results.

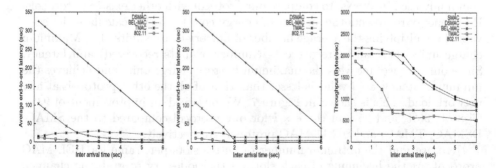

Fig. 3. Varying load Vs Latency   **Fig. 4.** Theoretical result - Varying load Vs Latency   **Fig. 5.** Varying load Vs Throughput

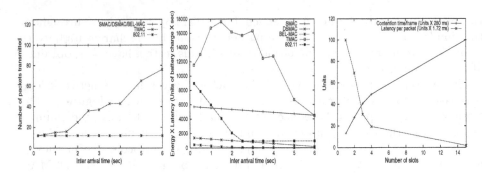

Fig. 6. Varying load Vs Lifetime   **Fig. 7.** *Energy × Latency* for varying load   **Fig. 8.** Optimum number of slots per frame

Since SMAC and DSMAC have a fixed number of active slots, the throughput remains constant irrespective of the network load as shown in Figure 5. The throughput of 802.11 MAC and BEL-MAC protocols remains higher, since they transmit packets continuously. However, the throughput of 802.11 MAC is higher than that of BEL-MAC, because BEL-MAC waits for the active slot period to start the transmission and thus there exists a delay between the packet transmissions. In the case of 802.11 MAC, the packets are transmitted continuously without any delay. The throughput of TMAC protocol is lower than that of 802.11 MAC and BEL-MAC protocols, because TMAC transmits only during the basic active slot. In addition, the complete frame is used for the transmission only if the node has enough packets for transmission. As shown in Figure 6, SMAC, DSMAC and BEL-MAC achieve the maximum lifetime independent of the network load, since SMAC and DSMAC protocols have enough sleep slots and BEL-MAC provides enough sleep time for the battery recovery. 802.11 MAC does not have the sleep slot and the lifetime always remains lesser. In the case of TMAC, the lifetime increases as the network load reduces, because the nodes get more sleep time and thus enables recovery of the nodes' battery.

We use $e \times l$ (*energy × latency*) metric, which should be minimized and is commonly used in WSNs, to compare our protocol with other existing protocols. Here, energy represents the amount of energy consumed. We calculate the $e \times l$ value by multiplying the average number of battery charge units (1 + Maximum charge units that can be recovered - Number of units recovered) and latency. Since our protocol recovers the maximum battery charge units and achieves the minimum latency, $e \times l$ value is lesser than that of all the other protocols at any network load. This is shown in Figure 7. We observed an improvement of 93%, 65-85%, 95%, and 95% in the $e \times l$ for our protocol compared to the SMAC, DSMAC, TMAC, and 802.11 MAC protocols, respectively.

Here, we find the optimum number of active slots per frame for BEL-MAC protocol. At the beginning of each slot, all the nodes try to grab the channel, which leads to a collision. This causes additional energy consumption. Thus, as the number of slots added increases, the total energy consumption increases.

However, higher number of slots reduces the duration of the sleep time, and thus reduces the waiting time for a slot (latency). As the traffic increases, both the latency per packet and the contention time durations increase, leading to a corresponding shift of both the curves. Figure 8 shows the contention time per frame and the average latency per packet for varying number of active slots through the simulation. Thus, we find the optimum number of slots/frame to be 3 for different traffic scenarios.

6 Conclusion

In this paper, we proposed a novel Battery-aware Energy-efficient MAC protocol to minimize the Latency (BEL-MAC) for WSNs. Our protocol exploits the battery characteristics to achieve the maximum lifetime and minimize the latency to the maximum extent possible. Our protocol increases the network lifetime by allowing the nodes to recover their charges completely, before attempting to transmit the packets. The analytical and simulation results proved that our protocol achieves the minimum possible latency without compromising on the network lifetime. In addition, we showed that our protocol performs better than all the other existing protocols at varying network load conditions by using *energy* × *latency* metric.

References

1. C. Schurgers, V. Tsiatsis, and M. B. Srivastava, "STEM: Topology Management for Energy Efficient Sensor Networks", in *Proc. IEEE Aerospace Conference 2002*, vol. 3, pp. 1099-1108, September 2002.
2. W. Ye, J. Heidemann, and D. Estrin, "Medium Access Control with Coordinated Adaptive Sleeping for Wireless Sensor Networks", *IEEE/ACM Transactions on Networking*, vol. 12, no. 3, pp. 493-506, June 2004.
3. C. S. Raghavendra and S. Singh, "PAMAS-Power Aware-Access Protocol with Signaling for Ad Hoc Networks", *ACM Computer Communication Review*, vol. 28, no. 3, pp. 1209-1213, September 2000.
4. P. Lin, C. Qiao, and X. Wang, "Medium Access Control with a Dynamic Duty Cycle for Sensor Networks", in *Proc. IEEE WCNC 2002*, pp. 1522-1527, March 2004..
5. T. V. Dam and K. Langendoen, "An Adaptive Energy-Efficient MAC Protocol for Wireless Sensor Networks", in *Proc. ACM SenSys 2003*, pp. 171-180, Nov 2003.
6. S. Jayashree, B. S. Manoj, and C. Siva Ram Murthy, "On Using Battery State for Medium Access Control in Ad hoc Wireless Networks", in *Proc. ACM MobiCom 2004*, pp. 360-373, October 2004.
7. C. F. Chiasserini and R. R. Rao, "Pulsed Battery Discharge in Communication Devices", in *Proc. ACM MobiCom 1999*, pp. 88-95, August 1999.
8. Smart Battery System Implementers Forum (http://www.sbs-forum.org).
9. K. Lahiri, A. Raghunathan, S. Dey, and D. Panigrahi, "Battery-Driven System Design: A New Frontier in Low Power Design", in *Proc. International Conference on VLSI Design 2002*, pp. 261-267, January 2002.
10. M. M. Carvalho and J. J. Garcia-Luna-Aceves, "Delay Analysis of IEEE 802.11 in Single-Hop Networks", in *Proc. IEEE ICNP 2003*, pp. 146-155, November 2003.

On the Power Optimization and Throughput Performance of Multihop Wireless Network Architectures*

G. Bhaya[1], B.S. Manoj[2], and C. Siva Ram Murthy[3]

[1] Department of Computer Science and Engineering,
University of Washington, WA 98195, USA
[2] Department of Electrical and Computer Engineering,
University of California at San Diego, CA 92093, USA
[3] Department of Computer Science and Engineering, IIT Madras, India
gbhaya@cs.washington.edu, bsmanoj@ucsd.edu
murthy@iitm.ac.in

Abstract. With the emergence of powerful processors and complex applications, wireless communication devices are increasingly power hungry. While there exist several solutions to provide transmission power management in cellular wireless networks and ad hoc wireless networks, it remains an open problem in recently proposed hybrid wireless networks. The Multihop Cellular Network (MCN) and Multi Power Architecture for Cellular network (MuPAC) are instances of hybrid wireless networks, which are proposed to increase the system throughput and spectrum reuse by infusing multihop radio relaying mechanism into the infrastructure-based wireless networks. This paper proposes a novel variable power optimization scheme for the hybrid wireless network architectures such as MCN and MuPAC in order to optimize the power consumption at a mobile node without losing the throughput advantage gained by the multihop scheme. Extensive simulation results show 10% to 15% improvement in power consumption and system throughput which is significant in case of power constrained mobile nodes.

1 Introduction

The mobile terminals in the wireless domain lack significant advantages that their counterparts in the wired domain have. These include the computing resources, energy storage, potentially large bandwidth and of course the threats posed by the wireless domain such as high bit error rates and security issues. With the fast developing computing technology, the problem of limited computing resources gets alleviated to some extent, but applies a greater pressure on other scarce resources provided by the network. The increasing number of applications on the mobile devices places an additional demand on the limited

---

* This work was supported by Infosys Technologies Ltd., Bangalore, India and Department of Science and Technology, New Delhi, India.

D.A. Bader et al. (Eds.): HiPC 2005, LNCS 3769, pp. 322–332, 2005.

bandwidth and available battery power. A significant amount of power is consumed at the mobile nodes during transmission and reception, display, and I/O access. Earlier attempts for design of power saving protocols for mobile ad hoc networks that allow mobile hosts to switch off to a low power sleep mode are suggested in [1]. An attempt is made in [2] to solve the problem of adjusting transmit power in an ad hoc network to create the desired network topology. A power control multiple access protocol (PCMA) is proposed in [3] which uses variable transmission power. PCMA uses a separate channel, to indicate busy state of channel, over which the receiver transmits a busy signal during data reception. Though this scheme gives a significant power improvement, it does not account for mobility of the sender and the receiver and it also consumes additional power for the signaling channel. The presence of mobility and interference might cause fluctuations in the strength of the busy signal that is crucial for determining transmission power. Vaidya et. al. [4] show that using variable power for transmission, in fact leads to a reduction in throughput. However, their study is based on increasing power for RTS/CTS while retaining and normal power level for the data and ACK transmission. However, by doing so they prevent some nodes which could have otherwise been permitted to send data from transmission. The MCN architecture though primarily proposed with the aim of improving the throughput, showed a significant improvement in power utilization per successfully delivered packet as compared to the Single-hop Cellular Networks (SCNs) [5]. The MCNs proposed in [6] and [7] are examples of this performance improvement, which show a significant power optimization (*i.e.*, average power consumption per successfully delivered packet) in addition to very high throughput as compared to the SCN [5]. But, these network architectures require mobile nodes to expend power for relaying traffic generated by other nodes. Hence we propose a new power optimization scheme for these architectures. The MuPAC architecture proposed in [8] extends the idea of MCN to use multiple transmission ranges. This provides flexibility to the mobile nodes to use a lower transmission power when the receiver is nearer. But the choice of the transmission range was restricted to a limited set of data channels decided *a priori*. We extend this idea further to incorporate variable power transmission ranges. This leads to a better power utilization and an improvement in the throughput over the existing MCN and MuPAC architectures.

The organization of the rest of this paper is as follows. We first briefly describe the MCN and the MuPAC architectures in Section 2. Section 3 describes our solution for transmission power management. In Section 4 we discuss the performance results. Finally, Section 5 concludes this paper.

2   Related Work

The MCN [6] was originally designed to provide spacial reuse of the channel by decreasing the transmission power of the mobile nodes without having to pay the penalty for a large number of Base Stations (BSs). The MCN architecture uses a data channel transmission range that is a fraction *1/k* of the cell radius, *R*.

This means that, in the same cell, up to a maximum of k^2 nodes can transmit simultaneously. The analysis in [6] proved that the hop count increases linearly with k. So, in effect, the throughput increase is expected to follow linearly with k. The extension of MCN architecture and a unicast routing protocol proposed in [7] used a single data channel of transmission range $R/2$ and a control channel with the transmission range of R. The control channel is used for delivering the topology information to the BS. The registration and *Route Request* messages use the control channel for communication with the BS. The MuPAC [8] is a multi-power scheme for packet data cellular networks. This scheme was primarily proposed to overcome the limitations posed by the MCN architecture and to enhance the throughput further. The n-channel MuPAC uses (n) data channels with different transmission ranges and a control channel. The available bandwidth is divided among n data channels and a control channel, thus denoted as $(n, r_1, b_1, ..., r_n, b_n)$, where r_i is the transmission range of the ith data channel and b_i is the bandwidth of the ith data channel. The transmission range of the control channel is R, the cell radius. In case of a route established using multiple hops, higher channels serve as backups to avoid link breaks, therefore, if the node on the next hop which was reachable using the ith data channel is no longer reachable using the same then MuPAC upgrades the transmission to the $(i+1)^{th}$ data channel. MuPAC chooses the transmission power as follows: For a given route the data is transmitted to the next hop using the data channel given by $r_i > \alpha * d_{approx}$; where $\alpha > 1$ is the safety factor and d_{approx} is the distance estimated using the received power of the *Hello* beacons. The MuPAC-2 refers to specific case of the MuPAC architecture in which the number of data channels is two, and similarly MuPAC-3 has three data channels. From the simulation studies in [8], for the MuPAC-2, the optimal performance is obtained when r_1 is $R/3$ and r_2 is $R/2$ with $b_1 = b_2 = 2.5$Mbps when the bandwidth for control channel is 1Mbps. MuPAC-3 uses $r_1 = 170$m, $r_2 = 220$m and $r_3 = 250$m for a cell radius (R) of 500m. $b_1 = 0.75$Mbps, $b_2 = 1.25$Mbps, and $b_3 = 3$Mbps. Henceforth MuPAC-2 and MuPAC-3 mean the division with the above characteristics and we shall use the same for our study.

The popular MAC protocols such as the IEEE 802.11 may not work efficiently with variable power transmission system. This is because IEEE 802.11 uses the Request-to-Send/Clear-to-Send (RTS-CTS) mechanism for gaining access to the channel. In this mechanism the sender senses the channel to be idle, sends an RTS message to the receiver. The receiver replies back with a CTS message. Neighbor nodes avoid transmission during this period so as to avoid collisions. However, when multiple transmission powers are used this effect will be nullified as illustrated with an example in Figure 1 (a). Consider a wireless network shown in Figure 1 (a). Node A wishes to send data to Node B. Node A finds the channel idle and hence sends an RTS message to Node B. Since Node B is close enough to Node A, A uses a lower transmission power sufficient enough for Node B to hear. Node B replies back with the CTS with the same power. Both the RTS and CTS are not heard by Nodes C and D who are relatively further apart. Thus the transmission between Nodes A and B is initiated. Now, Node C wishes to send

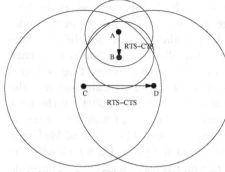

(a). Problem in using variable power in IEEE 802.11

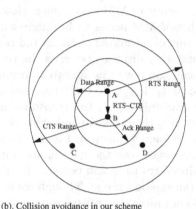

(b). Collision avoidance in our scheme

```
SendHelloMessage()
begin
        if (the time since last broadcast is greater
              than  MIN_HELLO_INTERVAL)
              Broadcast Hello Message
end

HelloMessageReceived()
begin
        Update the neighbour table
        Inform the changes to the BS
end

SendMessage(Message m)
begin
        if (the next hop of m is reachable)
        begin
              Estimate the distance to the next hop
              Apply the mobility margin correction to
              the estimated distance
              Calculate the required transmission
              power based on the above distance
              Estimate the transmission power by
              MuPAC scheme
              if (MuPAC transmission power is less)
                    Send the packet using MuPAC scheme
              else
              begin
                        Send the RTS/CTS with power
                            calculated using MuPAC
                        Send the data using reduced power
              end
        end
        else
              Send a route error to the BS
end
```

Fig. 1. (a) Problem in using variable power directly on MAC protocols like IEEE 802.11 and (b) the collision avoidance in the proposed power optimization scheme

Fig. 2. The proposed algorithm

data to Node D, also finds the channel idle, but Node D being relatively further apart Node C uses a higher transmission power. This transmission is audible at Node B and hence a collision occurs at Node B. Thus, the origination of a new packet transmission causes interference with an existing packet transmission.

3   Our Work

We propose a transmission power management scheme for hybrid wireless architectures. We assume that the available bandwidth is divided into n different channels with different transmission ranges similar to the MuPAC architecture. We do not consider MCN separately because it can be considered as a special case of the MuPAC with one data channel. Thus we shall use the same representation as MuPAC to represent this division: $(n, r_1, b_1,..., r_n, b_n)$, where r_i is the transmission range of the ith data channel and b_i is the bandwidth of the

i^{th} data channel. The r_n is used to exchange the *Hello* messages as in the Mu-PAC architecture. Using the *Hello* message and received power at every node, an approximate distance to the sender can be calculated as in the MuPAC. The MuPAC uses α as a safety factor to calculate this distance. In addition to this, we define Mobility Margin (MM) to be the mobility safety factor. This is necessary because the factor α only compensates for the error in the calculation of the distance. But, mobility may cause two nodes to move further apart till the next *Hello* message is transmitted. This is accounted by MM. MM may be transmitted by the BS to every node at the time of registration. The value of MM may be defined based on the desired need for power consumption (in which case the value needs to be less) and the mobility of the network (in which case the value needs to be more). Also, the value of MM depends on the frequency of neighbor updates. In case of high update frequency a low value of MM may suffice. However, in case of low update frequency the value of MM needs to be increased. The transmission power is estimated from the sum of estimated distance and the MM value. This is done as follows: Let the estimated distance between the two nodes be x m. Let s be the MM. Thus a transmission power needed to transmit up to a distance of radius $x+s$ is calculated to be t_x. Let r_i be the transmission range of the channel selected by MuPAC for transmission. If the transmission power t_x is greater than the transmission power of r_i then the transmission takes place as in MuPAC, without any change. Otherwise, as shown in Figure 2 (in function SendMessage()), the RTS-CTS are exchanged over the $r_i{}^{th}$ channel but the data transmission takes place with the reduced transmission power. The acknowledgment is also sent over the reduced transmission range. Though we are affecting the transmission range of data, we do not alter the transmission power of the RTS and CTS messages. Hence, we may expect a higher throughput due to increased reusability of bandwidth as opposed to the study by Vaidya et. al. in [4] which increases the transmission range of RTS and CTS to the maximum possible value. Hence, our scheme prevents only the nodes that pose a threat to the ongoing transmission from transmitting. In fact, in our scheme, the same number of nodes that are blocked by the non-optimized version of the scheme are only prevented. For example, in Figure 1 (b), Nodes C and D hear the RTS sent by Node A or the CTS sent by Node B though they may not hear the actual transmission. Thus Nodes C and D are prevented from transmitting data on the same data channel as Nodes A and B for the transmission interval specified by RTS-CTS. Hence the chances of collision during the data transmission interval is reduced. In our proposed optimization to the MCN and MuPAC architectures, we depend on the estimated distance between the nodes for the calculation of the reduced transmission range. We assume that it is possible to estimate the approximate distance between two nodes. This can be done in many ways. If the GPS information is available the receiver can convey this to the sender in the CTS. Alternatively, the sender can translate the signal strength of the *Hello* messages received form the neighbors, thus estimating the distance. An example for such a physical layer system that is capable of estimating the distance between the transmitter and receiver is the Ultra Wide Band physical layer. In other physical layers, the transmission power information may be included in the packets by all the nodes in their transmissions. Further to do away with the variable noise levels a node may average the received power for the last few receptions from the same sender. In this work we assume that provisions exist for approximate calculation of distance between nodes.

3.1 Theoretical Analysis

We analyze the improvement in the power consumption of the above proposed architectures for MCN and MuPAC. Let α, β, and γ be the traffic generated on the data, neighbor, and the control channel respectively. Let δ denote the MM. By traffic, we refer to bytes of data generated and not the number of packets. We do not take into account the reception energy because it is a constant and unaffected in all the schemes studied in this paper. In case of MCN, the mobile node uses a transmission range of $R/2$ for transmitting data to any other mobile node within a distance of $R/2$. Regions 1, 2, and 3 in Figure 3 (a) correspond to the area covered by the transmission range of $R/2$. However, in our proposed optimization, the mobile node uses a transmission range r for transmitting to the nodes in Region 2. Region 2 corresponds to a circular ring of thickness dr and radius r. In the calculations below we integrate over r to find the average transmission power per data transmission. The average power consumed per node per transmission in the MCN architecture for transmitting data alone over one hop is proportional to

$$\int_0^{R/2} \frac{2\pi r dr (R/2)^2}{\pi R^2/4} = \frac{R^2}{4} \tag{1}$$

Similarly, the average power consumed in case of the power optimized MCN architecture for transmitting data over one hop is given by

$$\int_0^{R/2} \frac{2\pi r dr (r+\delta)^2}{\pi R^2/4} = \frac{R^2}{8} (ignoring\ \delta) \tag{2}$$

Here δ is the MM. Hence the ratio of the total power consumed by power optimized MCN to MCN without power optimization is given by

$$\frac{\alpha R^2/8 + \beta R^2/4 + \gamma R^2}{\alpha R^2/4 + \beta R^2/4 + \gamma R^2} \tag{3}$$

Here, the terms from left to right (in both the numerator and the denominator) refer to the power consumed in transmissions over data channel, neighbor exchanges, and the transmissions over the control channel, respectively. These can be compared directly owing to the fact that the routes chosen for data transmission in both the schemes will be the same and no additional traffic is generated by the power optimized version of the protocol.

In case of MuPAC-2, the mobile node uses a transmission range of $R/3$ for transmitting data to any other mobile node within a distance of $R/3$ but a transmission range of $R/2$ for transmitting to mobile nodes beyond a distance of $R/3$

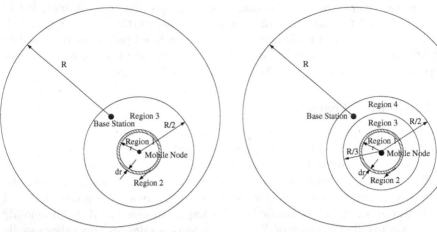

r = Transmission range used in the optimized version
    of MCN for transmissions to nodes in Region 2.
R/2 = Transmission range used in MCN for transmissions
      to nodes in Regions 1, 2, and 3.

(a). Power optimized MCN

r = Transmission range used in the optimized version
    of MuPAC–2 for transmissions to nodes in Region 2.
R/2 = Transmission range used in MuPAC–2 for transmissions
      to nodes in Region 4.
R/3 = Transmission range used in MuPAC–2 for transmissions
      to nodes in Regions 1, 2, and 3.

(b). Power optimized MuPAC

Fig. 3. Analysis for power optimization: (a) MCN and (b) MuPAC

but within a distance of $R/2$. Regions 1, 2, and 3 in Figure 3 (b) correspond to the former and Region 4 corresponds to the latter. In our proposed optimization, a transmission range of r is used for the nodes in Region 2. The value of r varies up to $R/2$. The average power consumed per node per transmission in the MuPAC architecture for transmitting data alone over one hop is proportional to

$$\int_0^{R/3} \frac{2\pi r dr (R/3)^2}{\pi R^2/9} + \int_{R/3}^{R/2} \frac{2\pi r dr (R/2)^2}{\pi R^2/4 - \pi R^2/9} = \frac{13 \times R^2}{36} \tag{4}$$

Similarly, the average power consumed in case of the power optimized MuPAC architecture is given by

$$\int_0^{R/2} \frac{2\pi r dr (r + \delta)^2}{\pi R^2/4} = \frac{R^2}{8} (ignoring\ \delta) \tag{5}$$

Hence the ratio of the total power consumed by power optimized MuPAC to MuPAC without power optimization is given by

$$\frac{\alpha R^2/8 + \beta R^2/4 + \gamma R^2}{\alpha 13 R^2/36 + \beta R^2/4 + \gamma R^2} \tag{6}$$

Here, the terms from left to right (in both the numerator and the denominator) refer to the power consumed by data transmission, neighbor exchanges, and the transmissions over the control channel, respectively. In both the Equations 3 and 6 the numerator is proportional to the power consumed by the power optimized version of MCN and MuPAC, respectively while the denominator refers to the power consumed by the MCN and MuPAC without power optimization. From the above we see that the numerator in both the MCN and the MuPAC cases remains the same but the denominator is more in case of MCN. Hence we expect more saving in case of MCN than in case of MuPAC. Comparing the above analysis for the data channel alone (*i.e.*, the terms with α alone) we see up to 50% improvement in case of power optimized MCN and 34% in case of power optimized MuPAC-2. When we consider the power consumed by all the interfaces this saving percentage reduces. However, the values of the quantities in Equations 3 and 6 determined from the simulations in Section 4 show about 10% to 15% improvement in power consumption.

4   Performance Analysis

In order to evaluate the performance of the power optimization scheme we simulated the MCN and MuPAC architectures using GloMoSim. Each simulation result is averaged over 8 runs. The bandwidth of the system was fixed at 6 Mbps (1 Mbps for the control channel and the rest for data channels). The cell radius, R is set to 500m. For the MCN architecture we used the data channel transmission ranges as $R/2$. We evaluated the performance of MCN, MuPAC-2, and MuPAC-3 with and without power optimization on the basis of throughput obtained and the transmission power expended at various values of UDP load. The UDP load generated at each node varies from 2 packets/sec/node to 10 packets/sec/node. The maximum mobility of the nodes is restricted to 10 m/s and the value of MM is fixed at 4m. The results were studied under two traffic locality (L) scenarios: L = 1 and L = 0. (Here L = 1, corresponds to the case when all the mobile nodes in a cell have data meant for nodes in the same cell.) At a mobility margin value of 4m, the power consumption in all the power optimized MuPAC is around 10% to 15% lower (see Figure 4). MCN performed slightly better than MuPAC. The MCN and MuPAC-3 schemes give almost the same throughput with or without power optimization. MuPAC-2 with optimization however gives a slightly poorer performance at lower values of load but outperforms its counterpart without optimization at higher loads. But when the MM was increased to 6m or 8m the performance in terms of throughput matches MuPAC-2 without power optimization (see Figure 5). We compared the MuPAC-2 without power optimization and the power optimized version at MM of 4m, 6m, and 8m when traffic locality is zero (see Figures 6 and 7). The power consumption shows minor differences across various values of MM. We simulated the system at various maximum mobility values, from 2 m/s to 18 m/s, at varying values of MM from 2m to 10m. This was at a fixed UDP load of 5 packets/sec/node and at L=1. For the power optimized MuPAC-2 the throughput was not significantly different from its non-optimized counterpart but the power consumed showed an improvement of 10% to 15% as shown in Figure 8. The increase in MM did not affect the power consumption significantly but the throughput improved with the MM, more so at high mobility values. The same conditions when applied to

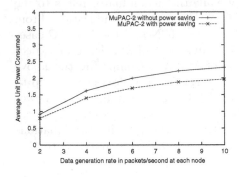

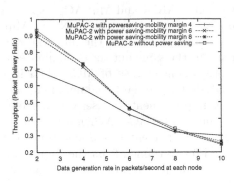

Fig. 4. Power consumption vs Network load for MuPAC-2 (Mobility=10m/s, MM=4m, L=1)

Fig. 5. Packet Delivery Ratio vs Network load for MuPAC-2 (Mobility=10m/s, L=1, Nodes=160)

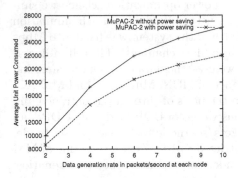

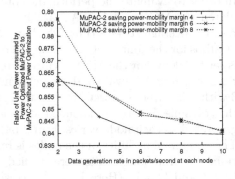

Fig. 6. Power Consumption vs Network Load for MuPAC-2 (Mobility=10m/s, MM=4m, L=0)

Fig. 7. Power Consumption vs Network Load for MuPAC-2 (Mobility=10m/s, L=0, Nodes=160)

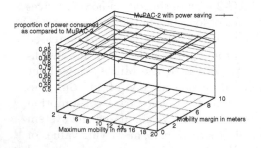

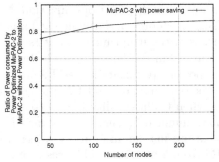

Fig. 8. Power Consumption vs Mobility vs MM for MuPAC-2 (Locality=1, Load=5 pkts/sec/node, Nodes=160)

Fig. 9. Power consumption vs Number of nodes (L=1, Load=5 pkts/sec/node, Mobility=10m/s, MM=4m)

MCN architecture though the power gain was significantly higher, about 20% to 25%, the throughput remained slightly lower than the non-optimized version for low mobility values (about 2% to 5% lower). We simulated the system at varying sizes of the network ranging from 40 nodes to 240 nodes, at a fixed locality of 1, load of 5 packets/sec/node and a maximum mobility of 10m/s. In case of low node density the power optimized MuPAC-2 performed significantly better both in terms of throughput and power consumption and the improvement was about 25%. Increase in the network size was not found to be improving the throughput, whereas the power consumption increased but seemed to stabilize at about 12% gain as compared to MuPAC-2 without power optimization (see Figure 9).

5 Conclusions

In this paper we proposed a transmission power management scheme for the hybrid wireless network architectures such as MCN [7] and MuPAC [8] to improve power consumption performance and throughput. Though the main aim of the scheme was to conserve power, up to 10% improvement of the throughput is achieved. The main cause of improvement of throughput is that some collisions caused as a result of mobility are avoided because the data transmissions affect a smaller portion of the network. On the other hand the factor that has a negative effect on the throughput is mobility margin. From our studies we noticed 10% to 15% improvement in power consumption which is a significant saving for power constrained mobile nodes. Theoretical analysis shows a significant saving in power consumed. Considering the data channels alone in MCN and MuPAC we get almost 50% and 34% reduction of power consumption.

References

1. Y. Tseng, C. Hsu, and T. Hsieh, "Power Saving Protocols for IEEE 802.11-Based Multihop Ad Hoc Networks", *in Proc. of IEEE INFOCOM 2002*, June 2002, Vol. 1, pp. 200-209.
2. R. Ramanathan and R. Rosales-Hain, "Topology Control of Multihop Wireless Networks using Transmit Power Adjust", *in Proc. of IEEE INFOCOM 2000*, March 2000, pp. 404-413.
3. J. P. Monks, V. Bharghaven, and W. M. W. Hwu, "A Power Controlled Multiple Access Protocol for Wireless Packet Networks", *in Proc. of IEEE INFOCOM 2001*, April 2001, pp. 219-228.
4. E. S. Jung and N. H. Vaidya, "A Power Control MAC Protocol for Ad hoc Networks", *in Proc. of ACM MOBICOM 2002*, June 2002, pp. 36-47.
5. K. J. Kumar, B. S. Manoj, and C. Siva Ram Murthy, "On the Use of Multiple Hops in the Next Generation Wireless Architectures", *in Proc. of IEEE ICON 2002*, August 2002, pp. 283-288.
6. Y. D. Lin and Y. C. Hsu, "Multihop Cellular: A New Architecture for Wireless Communications", *in Proc. of IEEE INFOCOM 2000*, March 2000, pp. 1273-1282. 2002, August 2002, pp. 283-288.

7. R. Ananthapadmanabha, B. S. Manoj, and C. Siva Ram Murthy, "Multihop Cellular Networks: The Architecture and Routing Protocols", *in Proc. of IEEE PIMRC 2001*, October 2001, Vol. 2, pp. 78-82.
8. K. J. Kumar, B. S. Manoj, and C. Siva Ram Murthy, "MuPAC: Multi Power Architecture for Wireless Packet Data Network", *in Proc. of IEEE PIMRC 2002*, September 2002, Vol. 4, pp. 1670-1674.

A Novel Solution for Time Synchronization in Wireless Ad Hoc and Sensor Networks*

Archana Sekhar[1], B.S. Manoj[2], and C. Siva Ram Murthy[3]

[1] McKinsey & Company Inc., Express Towers,
Nariman Point, Mumbai 400021, India
[2] Department of Electrical and Computer Engineering,
University of California at San Diego, CA 92093, USA
[3] Department of Computer Science and Engineering,
IIT Madras, Chennai 600036, India
archana_sekhar@mckinsey.com, bsmanoj@ucsd.edu
murthy@iitm.ac.in

Abstract. Time synchronization is an important aspect of distributed computer systems and networks. Nodes must be synchronized to a common clock to determine slot durations for a TDMA based transmission scheme. Most efficient slot-assignment algorithms apportion the TDMA slots with the underlying assumption of a reasonably accurate global synchronization of the network. In this paper, we propose a novel synchronization protocol for ad hoc, sensor, and other dense multi-hop infrastructure-less wireless networks. The protocol performs a random leader election to achieve global network synchronization. We have analyzed the variation of synchronization time and error with different node densities and mobility speeds, by simulating the protocol. Expressions have been derived reflecting the worst case synchronization error, and the maximum synchronization time, for a network with uniform distribution of nodes. Simulation results show that out-of-band and piggybacked signaling have good accuracy of synchronization, and that a considerable bandwidth saving occurs with piggybacking on data or acknowledgment packets.

1 Introduction

Sensor networks form a class of ad hoc wireless networks, where the nodes are low-cost, lightweight, and highly power-constrained. They are deployed in very large numbers to collect data about the environment or any physical event like an intrusion, aggregate the information, and convey parameters of interest to monitor nodes either on demand or periodically. Typical scenarios of interest include seismic monitoring, power plant or nuclear reactor control, and military usage to sense the enemy territory. The nodes are organized into hierarchical clusters to reduce long range message transfers. Messages are short and bursty

---

* This work was supported by Infosys Technologies Ltd., Bangalore, India and Department of Science and Technology, New Delhi, India.

D.A. Bader et al. (Eds.): HiPC 2005, LNCS 3769, pp. 333–342, 2005.

in nature, and spaced apart, to optimize power consumption. A TDMA schedule can be chalked out so that power loss due to collisions is minimized and sensors can be in standby mode when they are not scheduled to transmit. Synchronization is essential to support TDMA-based multihop communication. Sensor networks also require synchronization to determine temporal ordering of events and discard duplicate reports.

2   Related Work

Many existing synchronization algorithms rely on the time information in the Global Positioning System (GPS) to provide coarse time synchronization. In the worst case, with only one observed satellite, GPS offers an accuracy of 1 μs [1]. But, GPS is not a suitable choice for ad hoc and sensor networks, because GPS receivers cannot be used inside large buildings and basements, or underwater, or in other satellite-unreachable environments where sensor or ad hoc networks may have to be deployed. In [2], Yoram Ofek proposed a global synchronization protocol, with the assumption that the total number of nodes in the network is known *a priori*, and a leader can be elected by majority, but this may not always be known accurately in ad hoc and sensor networks due to mobility and power drain. In [3], precision of upto 1 ms is achieved based on local computation and communication. But the synchronization achieved is localized and short-lived. In [4], authors discuss a method for randomized initialization and leader election in ad hoc wireless networks. Two synchronization algorithms are discussed in [5], which offer synchronization accuracy of the order of $10\mu s$. Reference-Broadcast Synchronization (RBS) proposed in [6] is a receiver-receiver synchronization scheme. In RBS scheme, nodes send reference beacon packets to their neighbors. The time of arrival of a reference beacon is exchanged by its receivers for using as a reference point for comparing their clocks. A survey of existing synchronization mechanisms is provided in [7].

3   Our Work

In this paper, we propose a time synchronization protocol that ensures global synchronization of a connected network, or synchronization within connected partitions of a network. Any two nodes are considered to be synchronized if their clocks have a time difference small enough to be accounted for by the guard bands of a TDMA scheme. The synchronization is intended to be long-lasting, i.e., last for as long as the network operates. We have preferred a random leader election protocol over optimal leader choice using the knowledge of the topology in order to avoid the expensive overhead of topology discovery. The network synchronization is maintained through periodic beaconing. We have studied synchronization under the following scenarios: (i) out-of-band synchronization where the synchronization packets are sent over a dedicated control channel, (ii) piggybacking on data packets where the control information is piggybacked onto outgoing data packets, and (iii) piggybacking on acknowledgments. The piggybacking on acknowledgements scheme is explicitly designed for use in sensor networks where data usually flows from all sensors to the monitor, which is a fixed node with greater computing and power resources than the sensors. If the monitor is forced to be the leader, the synchronization information moves in the reverse direction, i.e., along the link-level acknowledgments sent by the nodes for each hop of the data packets.

In the basic synchronization protocol, each node in the network maintains its own local clock, and a virtual clock, to keep track of its leader's clock. A unique leader is elected for each partition in the network, and virtual clocks are updated to match the leader's real clock. On power-up, every node makes an attempt to either locate a leader in its partition, or claim to be a leader itself. The node decides, with a small probability, to stake a claim for leadership, by transmitting a *LeaderAnnouncement* packet, which contains a random number generated by the claimant, its transmission power, and a timestamp referring to its real clock. The transmission power field is to account for the possibility that each node may use a different transmission power. The algorithm used by the nodes is explained through the flowchart in Figure 1. As soon as any node receives a *LeaderAnnouncement*, it takes the following actions: (i) it does not stake a claim to become a leader, (ii) it checks the last *LeaderAnnouncement* it has received, (iii) the node discards the new *LeaderAnnouncement* if its random number is higher than the earlier packet, otherwise, it resynchronizes to the new leader, (iv) in the highly unlikely event that two different nodes have generated the same random number, the 'clash' is resolved on the basis of preferring the lower node address, and (v) if the *LeaderAnnouncement* packet still has a non-zero time-to-live (TTL), the node relays the packet with its own time stamp and transmission power. The node receives the packet at a particular reception power. Assuming a suitable pathloss model, the distance d between source and receiver can be estimated. Given the transmission and reception power, and the antenna characteristics, the distance between receiver and sender can be estimated. This is used to calculate propagation delay for the packet transfer. When added to the time stamp on the packet, this gives the new clock value to which the receiving node should change its virtual clock. The error involved in this estimation is mainly due to the error in the estimation of propagation delay and receive time. Propagation delay is of the order of $1\mu s$ for 300m distance for electromagnetic waves, hence it does not affect the accuracy of the clock significantly. In reality, the pathloss may drop inversely as d^2 to d^6. Hence, the error in distance estimation is usually positive, i.e., the receiving nodes advance their clock earlier than the actual leader clock. The receive delay has been experimentally analyzed in [6], and found to follow a Gaussian distribution, of the order of a bit-reception time. Randomness in queuing and medium access delay has been circumvented by time-stamping the outgoing packets at the physical layer. We also make an assumption that there is no buffering at the physical layer, so the time stamp accurately reflects the actual commencement of packet transmission. The TTL is measured in terms of number of hops traversed since

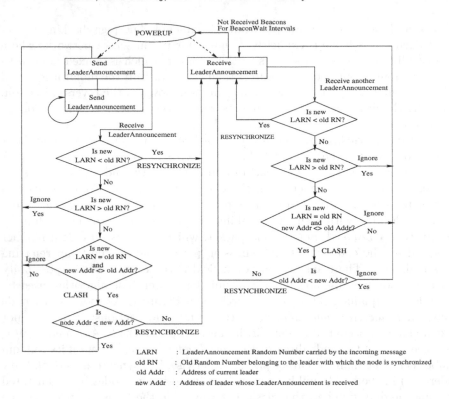

LARN   : LeaderAnnouncement Random Number carried by the incoming message
old RN   : Old Random Number belonging to the leader with which the node is synchronized
old Addr   : Address of current leader
new Addr : Address of leader whose LeaderAnnouncement is received

Fig. 1. Synchronization algorithm

originating from the leader. This is used as an indicator so that a *LeaderAnnouncement* packet does not get broadcast indefinitely to all nodes, and a false presence of a leader is not reported in a partition even after change in topology. The value of TTL must be set depending on the expected network diameter so that the *LeaderAnnouncement* reaches all nodes. Every node which has sent out a *LeaderAnnouncement* continues to send periodic *LeaderAnnouncements*, until it receives a *LeaderAnnouncement* with a lower random number than its own. In order to adapt to topology changes, a node starts its synchronization process if it misses *LeaderAnnouncement* beacons for more than a number of intervals specified by *BeaconWait*.

The synchronization protocol discussed requires an independent channel for the transmission of control packets. We have explored the possibility of piggybacking the synchronization information along with data packets. Sensor networks could piggyback synchronization information onto the periodic data updates sent to the monitor. The effectiveness of piggybacking depends on the duration of buffering allowed for control packets. A small latency does not utilize the advantage of piggybacking, while a very large latency will delay the synchronization of the network. In the piggybacked version of the synchronization protocol, a node which has to send *LeaderAnnouncements* includes the information with all the data packets that are sent from it. For any intermediate node which has to relay the *LeaderAnnouncements*, it buffers the packet and waits for *MaximumPiggybackWait* units of time for any data packet to be sent before it creates an independent control packet and relays the synchronization information. The number of packets buffered at a node can be decided depending on the expected traffic generated from a node. Piggybacking offsets the overhead of extra message size by power saving through fewer transmission startup attempts. Synchronization information can also be piggybacked effectively on the link-level acknowledgment packets in sensor networks, when the fixed monitor is the leader.

3.1   Theoretical Analysis

A bound on the maximum number of hops required for synchronization of the network is derived. We assume that the diameter of the network is k. In the worst case, this will be one less than the number of nodes in the network. We model the network as a uniform distribution of n nodes in a circular region of radius R. We assume that there are relaying nodes on the straight line joining any two nodes. We first calculate the probability of a 'clash' occurring during the leader election i.e., two nodes claiming to be the leader generating the same random number. A leader packet needs atmost k hops to reach all the nodes. We define a slot as the minimum time for a 1-hop transmission to be performed. Let $L_{LeaderAnnouncement}$ be the length of the *LeaderAnnouncement* packet in bits and the bandwidth of the synchronization channel be B bits per second. The propagation delay depends on the distance between sender and receiver. Under the condition of uniform distribution, it is a random variable τ. If the expected value of propagation delay be $E(\tau)$, then the slot time is given by

$$\frac{L_{LeaderAnnouncement}}{B} + E(\tau) \tag{1}$$

This is the minimum time at which a packet can be received by a neighboring node, assuming zero queuing and medium access delay. Any node which is i hops away from the leader gets i slots to generate its own leader packet. If a node stakes a claim to becoming a leader with a probability p, then the probability that it does not generate a leader packet in any of these i slots is $(1-p)^i$. Hence, probability of staking a claim before seeing the other leader packet is $1 - (1-p)^i$. Assuming that random numbers are generated in the range 1 to N, the two claimants can generate the same random number with a probability $1/N$. Hence, $prob(clash\ with\ a\ node\ i\ hops\ away) = \frac{1-((1-p)^i)}{N}$. Therefore, the probability of a clash is given by

$$1 - \prod_{i=1}^{n-1}[1 - \frac{1 - ((1-p)^{i_p})}{N}]. \tag{2}$$

Consider the general case of the leader being at a distance a from the center of the region. We divide the entire region into zones of i-hop reachability. Since the whole radius R can be reached in k hops, each hop distance is roughly R/k. Let

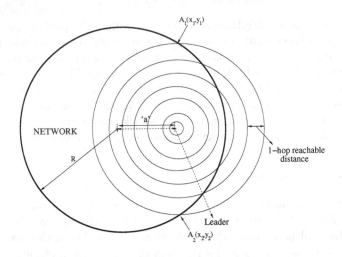

Fig. 2. Geometric representation of network

A_i represent the common area between the network, and the i-hop reachable region. As Figure 2 indicates, for smaller i, A_i is a circular annulus, and for larger i, it is the annulus of the intersection of two circles. For the smaller values of i, where the i-hop zone is a circle,

$$\forall i \le k(1 - \frac{a}{R}), \; A_i = \pi(\frac{iR}{k})^2 \tag{3}$$

Represent the network as a circle of radius R, centered at $(0,0)$. The leader is located at $(a,0)$, and its i-hop zone is a circle of radius iR/k. The two circles can be represented by the equations $x^2 + y^2 = R^2$ and $(x-a)^2 + y^2 = (\frac{iR}{k})^2$. The points of intersection $A_1(x_1, y_1)$ and $A_2(x_1, y_2)$, where say y_1 is the positive root, are calculated as

$$x_1 = \frac{R^2(1 - \frac{i^2}{k^2}) + a^2}{2a} \tag{4}$$

$$y = \sqrt{R^2 - x^2} = \frac{1}{2a}\sqrt{[(R+a)^2 - (\frac{iR}{k})^2]} \times \sqrt{[(\frac{iR}{k})^2 - (R-a)^2]} \tag{5}$$

The total area of intersection is given by $\int_{(a-\frac{iR}{k})}^{R} y\,dx$, which can be rewritten as

$$A = 2[\int_{(a-\frac{iR}{k})}^{x_1} \sqrt{(\frac{iR}{k})^2 - (x-a)^2}\,dx + \int_{x_1}^{R} \sqrt{R^2 - x^2}\,dx] = 2(I_1 + I_2) \tag{6}$$

We use $\int \sqrt{R^2 - x^2}\,dx = R^2 \sin^{-1}(\frac{x}{R}) + \frac{x\sqrt{R^2-x^2}}{2}$ to obtain

$$I_1 = (\frac{iR}{k})^2[sin^{-1}\frac{(x_1 - a)}{iR/k} + sin^{-1}\frac{a}{iR/k}]$$

$$+ \frac{(x_1 - a)}{2}\sqrt{(\frac{iR}{k})^2 - (x_1 - a)^2} + \frac{a}{2}\sqrt{(\frac{iR}{k})^2 - a^2} \tag{7}$$

and

$$I_2 = R^2(\frac{\pi}{2} - sin^{-1}\frac{x_1}{R}) - \frac{x_1 y_1}{2} \tag{8}$$

This expression for area A is valid for $\frac{iR}{k} \leq (R + a)$, i.e., $\forall i \leq k(1 + \frac{a}{R})$, $A_i = 2(I_1 + I_2)$ where $A_i - A_{i-1}$ gives the area reachable by the ith hop. The number of nodes in this region is calculated, by the assumption of uniform distribution, as

$$n_i = (A_i - A_{i-1})\frac{n}{\pi R^2} \tag{9}$$

where there are n nodes in the network. The average hop-count from the leader to a node is

$$\sum_{i=1}^{k} \frac{i \times n_i}{n} \tag{10}$$

Also, the expression for probability of a clash can be simplified to

$$1 - \prod_{i=1}^{k}[1 - \frac{1 - ((1-p)^i)}{N}]^{n_i} \tag{11}$$

The maximum time required for synchronization of the network is thus a function of the diameter of the network. The maximum possible diameter of a network of n nodes is n, when all are arranged linearly. Hence, the network is synchronized in finite time. The protocol relies on calculations using the two-ray pathloss model which assumes a variation inversely proportional to d^4. We try to place a bound on the maximum error that can be present in delay calculation. We represent the transmission power as P_t and the received power as P_r. As the ratio, P_r/P_t decreases i.e., the distance increases, the error between the different distance estimates increases. This is the reason for higher error in estimates with greater hop distance. Also, the effects of multi-path interference and fading lead to unpredictability in the propagation delay. Let the maximum possible transmission power be Tx_{max} and minimum received power required be Rx_{min}. Then the largest $\frac{P_t}{P_r}$ ratio is $\frac{Tx_{max}}{Rx_{min}}$. The worst case error in distance estimation is given by Equation 12, where α is a proportionality constant.

$$\delta r \leq \alpha \times ((\frac{P_t}{P_r})^{1/2} - (\frac{P_t}{P_r})^{1/4}) \leq \alpha \times ((\frac{Tx_{max}}{Rx_{min}})^{1/2} - (\frac{Tx_{max}}{Rx_{min}})^{1/4}) \tag{12}$$

We now analyze the worst case scenarios for a network of n nodes. If all the nodes are arranged linearly, there are at most $n - 1$ hops to be traversed by the *LeaderAnnouncement* packet to reach all nodes. Therefore, the maximum error in synchronization is given by $(n - 1) \times \delta t$. Also, referring to our definition of a slot in Equation 1, the minimum time for complete synchronization with the worst-case diameter is given by $(\frac{L_{LeaderAnnouncement}}{B} + E(\tau)) \times (n - 1)$ assuming zero delay at each node.

4   Results

The proposed synchronization protocol was implemented on GloMoSim simulation platform. We have chosen a simulation area of $2000m \times 2000m$ with number of nodes ranging from 40 to 120 and the node mobility model used was random way-point. The BeaconWait duration was taken as $5 \times slot$ where the slot duration is 50ms. Also, the MaximumPiggybackWait duration was taken in the range 1ms to 100ms. The channel bandwidth is 2Mbps. The transmission powers used were 9 dBm, 15 dBm, and 18 dBm and we used a two-ray radio propagation model. IEEE 802.11 DCF is used as the MAC protocol. The protocol's parameter *BeaconWait* was set as 5 slots, where each slot width is 50 ms. The probability p of claiming to be a leader was set at 10%, to ensure that even if the network is highly partitioned, there is a high probability of some node claiming to be a leader. The value of TTL was set at 10, comparable to the network diameter. Statistics were collected for the different synchronization information delivery mechanisms - out-of-band signaling and piggybacking on data and acknowledgments. Each set of simulation parameters was run on several seeds, and the results were averaged.

Figure 3 plots the variation of standard deviation of the clock times of nodes in a partition, with respect to node density and mobility. It is observed that there is a marginal increase in standard deviation with higher node density, as the *LeaderAnnouncement* packets need to travel through greater number of hops, and the errors in propagation delay estimates increase with higher hop count. The maximum difference between any two *NodeClocks* in the partition is observed to be less than 1 μs, at the transmission power of 15 dB. As Figure 4 shows, the dispersion of *NodeClocks* does not show a specific dependence on node density or mobility, due to the inherent randomness in the propagation delay and receive error estimates, the major contributors to the error in clock synchronization. Similarly, the average difference between *NodeClocks* and the *LeaderClock* is found to be less than 1 μs. The accuracy of synchronization relies on the accuracy of estimation of distance, and hence propagation delay between any two nodes. Additionally, the receive delay error contributes to a random error of about a bit-width.

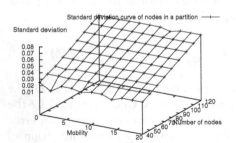 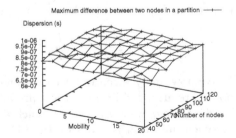

Fig. 3. Standard deviation of *NodeClocks* with separate control channel

Fig. 4. Difference between *NodeClocks* with separate control channel

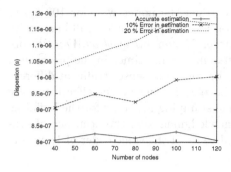

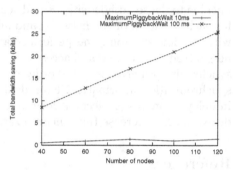

Fig. 5. Difference between *NodeClocks* with error in estimation

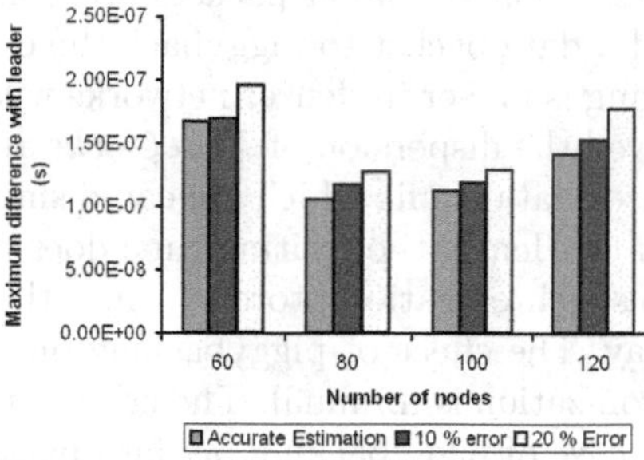

Fig. 6. Difference between *NodeClocks* and *LeaderClock* with error in estimation

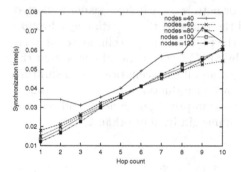

Fig. 7. Synchronization time Vs hop count

Fig. 8. Average bandwidth saving Vs *MaximumPiggybackWaits*

Figure 5 shows the effect of a deliberate inaccurate estimation on the maximum difference between any two *NodeClocks* in a partition. Though the dispersion does increase with larger errors, the maximum dispersion is still $1.2\mu s$ with 20 % error in estimation, which is of an acceptable order. Similar increasing differences are indicated between the *NodeClocks* and *LeaderClock* in Figure 6. Figure 7 analyzes the dependency of synchronization time on hop count from the leader. At all node densities, the synchronization time shows an upward trend with greater hop count, due to the increased queuing delays experienced by the *LeaderAnnouncement* packets at the intermediate nodes. The synchronization control information was piggybacked onto the data packets being sent from the nodes. The saving per piggybacked control packet is the difference between the overhead introduced by extra fields in the data packet, and sending a dedicated control packet. The minimum size of a packet, with only the headers and zero payload, is 160 bytes. The actual control information carried is only 25 bytes. As indicated earlier, the effectiveness of piggybacking depends on the *MaximumPiggybackWait*. Figure 8 clearly indicates that a larger number of

LeaderAnnouncement packets are buffered when the nodes wait for 100 ms to find a data packet to piggyback the control information. The overall bandwidth saving is higher in denser networks with larger *MaximumPiggybackWait*. We observed the dispersion of *NodeClocks* at different waiting times in the presence of sensor data traffic which appeared similar for voice and sensor traffic at different delays. Hence, the waiting time does not have an impact on the effectiveness of the synchronization protocol, since the time-stamping accounts for the buffering delay. The effect of piggybacking on acknowledgments on the accuracy of synchronization is minimal. The error is found to be of the same order of $1\mu s$, as in the case of piggybacking on data packets.

5   Conclusions

We have proposed a novel distributed global time synchronization protocol applicable to dense ad hoc wireless and sensor networks. A unique leader is elected for each partition of the network, and all the nodes in the partition synchronize with it. Time-stamping was performed at the physical layer to eliminate randomness introduced in send and access times by MAC layer. It was observed that the synchronization error is of the order of $1\ \mu s$, in all the scenarios of transmitting synchronization information. Even deliberate introduction of error of upto 20% in delay estimates, to account for randomness in propagation delay and receiver delay, did not increase the synchronization margin by more than $2\mu s$.

References

1. A. Ebner, H. Rohling, R. Halfmann, and M. Lott, "Synchronization in Ad Hoc Networks Based on UTRA TDD", in *Proc. of IEEE PIMRC 2002*, September 2002, Vol. 4, pp. 1650-1654.
2. Y. Ofek, "Generating a Fault-Tolerant Global Clock Using High-Speed Control Signals for the MetaNet Architecture", *IEEE/ACM Transactions on Networking*, Vol. 3, No. 2, (1995) pp. 169-180.
3. K. Romer, "Time Synchronization in Ad Hoc Networks", in *Proc. of ACM MobiHoc 2001*, October 2001, pp. 173-182.
4. K. Nakano aand S. Olariu, "Randomized Initialization Protocols for Ad Hoc Networks", *IEEE Transactions on Parallel and Distributed Systems*, Vol. 11, No. 7, (2000) pp. 749-759.
5. P. Blum and L. Thiele, "Clock Synchronization Using Packet Streams", in *Proc. of 16th International Symposium on Distributed Computing 2002*, October 2002.
6. J. Elson, L. Girod, and D. Estrin, "Fine-Grained Network Time Synchronization using Reference Broadcasts", in *Proc. of Operating System Design and Implementation 2002*, December 2002, pp. 147-163.
7. F. Sivrikaya and B. Yener, "Time Synchronization in Sensor Networks: A Survey", *IEEE Network Magazine*, Vol. 18 , No. 4, (2004) pp. 45-50.

An Algorithm for Boundary Discovery in Wireless Sensor Networks

Jitender S. Deogun, Saket Das, Haitham S. Hamza, and Steve Goddard

Department of Computer Science & Engineering,
University of Nebraska-Lincoln, Lincoln, NE 68588-0115, USA
{deogun, sdas, hhamza, goddard}@cse.unl.edu

Abstract. Wireless Sensor Networks (WSNs) consist of a large number of nodes networked via wireless links. In many WSN settings, sensor nodes are deployed in an ad hoc manner. One important issue in this context is to detect the boundary of the deployed network to ensure that the sensor nodes cover the target area. In this paper, we propose a new algorithm that can be used to discover the boundary of a randomly deployed WSN. The algorithm does not require the sensor nodes to be equipped with positioning devices and is scalable for large number of nodes. Simulation experiments are developed to evaluate the performance of the proposed algorithms for different network topologies. The simulation results show that the algorithm detects the boundary nodes of the network with high accuracy.

1 Introduction

Advances in wireless technologies, low-power computing, and embedded systems have enabled new and exciting applications for Wireless Sensor Networks (WSNs) [3]. A WSN consists of a large number of sensor nodes networked via wireless links. Numerous applications, such as efficient information sharing, military surveillance, disaster aid and risk management, benefit from the ability to efficiently deploy a (temporary) wireless network.

Typical applications of WSN systems may require a random deployment of sensor nodes over a large target area. Moreover, military surveillance systems, may also require detection of any activities around the boundaries of the target area under surveillance. Thus, the system should be capable of detecting and identifying any object that enters or leaves the monitored area. Hence, the development of mechanisms by which the *boundary nodes* of the network can be identified is important and a challenging problem. The nodes that represent the perimeter of the target area under surveillance are called boundary nodes.

The WSN systems are naturally dynamic systems. A WSN infrastructure keep evolving, because, nodes can fail, due to power-depletion, or displacement due to some natural phenomenon or nodes can be redeployed. This dynamic nature of WSNs makes the identification of the network boundary important as well challenging. Detecting the boundary nodes of a WSN is usually not the primary goal of deploying a WSNs. It is, therefore, required that the detection process

D.A. Bader et al. (Eds.): HiPC 2005, LNCS 3769, pp. 343–352, 2005.

does not tax node resources so that the network can perform its designated functions. While equipping the nodes in the network with positioning systems such as Global Position System(GPS) can simplify the problem, it does not provide a cost-effective solution. Moreover, GPS can drain a considerable amount of node energy over time leading to networks with short life-time.

In this paper, we develop an algorithm for detecting the boundary nodes of a WSN. The algorithm *does not* require the sensor nodes to be equipped with GPS. In addition, the algorithm relies on minimum communication between a node and its neighboring nodes, making the algorithm suitable for large-scale WSNs. The algorithm accuracy is evaluated for various network topologies and different parameters. Our simulation results show that our algorithm performs with high accuracy, even for random deployments with more than 500 nodes.

The remainder of the paper is organized as follows. The boundary detection problem is formally defined in Section 2. The proposed algorithm is described and analyzed via simulations in Sections 3 and 4, respectively. Concluding remarks are presented in Section 5.

2 Problem Overview

In this section, we describe the main assumptions, present definitions and notations, and develop a formal definition of the boundary detection problem.

2.1 Assumptions

In this work, we consider a WSN with homogeneous nodes, where each node has a unique global coordinate. The network is sufficiently dense so that each node in the network has at least 3 neighbors. In addition, we do not require the nodes to be equipped with GPS; however, nodes can determine their distance to their neighbors. For simplicity, we consider a network with no communication holes. The following is a summary of the main assumptions in this work:

1. Nodes can determine its distance to some of its neighbors.
2. Each node has a distinct global coordinate.
3. The network is sufficiently dense so that each node in the network has *at least* three neighboring nodes from which it can calculate its distance.
4. There exist no communication holes inside the network.
5. Communications between the nodes are bidirectional and symmetric.

2.2 Notations and Definitions

A sensor network with n nodes can be modeled as a graph $G = (V, E)$, where $n = |V|$ represents the set of nodes in the network, and E is the set of *undirected* edges $e(i, j)$, where $i, j \in V$ and node j is within the communication range of node i.

Let $p \in V$ be a node in the network, define $\mathcal{N}^k(p)$ to be the set of all the nodes that are within k hops from node p. Formally, $\mathcal{N}^k(p) = \{i : i \in V, d(p, i) \leq k\}$, where $d(p, i)$ represents the distance between nodes p and i. Thus, $\mathcal{N}^k(p)$

represents the kth order closed neighborhood of p in G. The set $\mathcal{N}^1(p)$, or simply $\mathcal{N}(p)$, contains all nodes that are within the communication range of node p. Formally, $\mathcal{N}(p) = \{i : i \in V, (p, i) \in E\}$.

Definition 1 (Interior Node). A node $p \in V$ is said to be an *interior* node if there $\exists$ at least three nodes $a, b, c \in V$ such that $a, b, c \in \mathcal{N}(p)$ and the nodes a, b, c forms a triangle that encloses the node p.

Definition 2 (Boundary Node). A node $p \in V$ is said to be a boundary node if p is not an interior node.

Define B to be the set of *boundary* nodes in the network; and I to be the set of *interior*, i.e. non-boundary, nodes in the network.

Definition 3 (Network Boundary). The imaginary line that connects the boundary nodes of the sensor network is defined as a network boundary. The network boundary defines the perimeter of the entire network.

2.3 Problem Definition

Given a randomly distributed network with a set of nodes with unknown position coordinates, and a mechanism by which a node could determine its distance to its neighbors, our objective is to devise a method for identifying the boundary nodes of the network.

The problem, thus can be stated formally as follows. Given a graph $G = (V, E)$, where nodes of G are embedded in a two dimensional space. For a node $p \in V$, determine if there exists a set S such that $S \subset \mathcal{N}(p)$, and $|S| \geq 3$, where there exist at least three nodes $a, b, c \in S$ that enclose the node p in a triangle.

3 Proposed Algorithm

In this Section, we describe the proposed algorithms for detecting the boundary of the network. The approach consists of two algorithms: the *InteriorPoint* al-

---

INTERIORPOINT(IP) ALGORITHM:
Input: Four points P, A, B, C and $d_{AP}, d_{BP}, d_{CP}, d_{AB}, d_{BC}, d_{CA}$
Output: Whether or not P is inside $\triangle ABC$

1. Find the areas of $\triangle$s ABC, PAC, PBC, and PCA using the equation(Heros formula):
   $Area = \sqrt{[s(s-a)(s-b)s-c)]}$
   where $s = (a+b+c)/2$ and a, b, c are length of three sides of the $\triangle$.
2. IF $Area \, \triangle \, ABC = Area \, (\triangle \, PAC + \triangle \, PBC + \triangle \, PCA)$
   RETURN P is inside the $\triangle \, ABC$
   ELSE RETURN P is out side the $\triangle \, ABC$

Fig. 1. The InteriorPoint (IP) Algorithm

gorithm (IP), and the *ChooseGoodNeighbors* algorithm (CGN). Figures 1 and 3 give the details of the IP and CGN algorithms, respectively.

In the IP algorithm we verify whether or not a given node is enclosed inside its three chosen neighbors.

It should be noted, however, that the accuracy of this algorithm depends on how we choose the three neighbors A, B and C of node P. Because possibly each node will be have more than three neighbors. It may be possible that the node P will be interior or exterior node depending on the three neighboring nodes we choose. A situation is depicted in Figure 2. If we choose the neighbors A, B

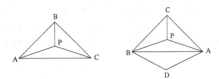

Fig. 2. Steps of CGN Algorithm

CHOOSEGOODNEIGHBORS(CGN) ALGORITHM:
Input: Node P.
Output: Four good neighboring nodes A, B, C and D of P

1. Find a node from $\mathcal{N}(P)$ which at the minimum distance from P. Name it as node A.
2. $SearchRadius = d_{PA}$;
3. WHILE($CommunicationRange \neq SearchRadius$) FOR all the nodes within $SearchRadius$ of P
   Find a node B, such that it is in both $\mathcal{N}(P)$ and $\mathcal{N}(A)$, and distance d_{AB} is maximum.
   ROF
   IF node B is found
   EXIT WHILE   —
   ELSE
   $SearchRadious + +$;
   END-WHILE
4. Find a node C such that it is in $\mathcal{N}(A)$, $\mathcal{N}(B)$ and $\mathcal{N}(P)$, and $|d_{AC} - d_{BC}|$ is minimum and d_{PC} is minimum.
5. Find a point D such that, D is in $\mathcal{N}(A)$, $\mathcal{N}(B)$, $\mathcal{N}(C)$, $\mathcal{N}(P)$ and $|d_{AD} - d_{BC}|$ is minimum and d_{DC} is maximum and $d_{DC} > d_{DP}$ and $d_{DC} > d_{PC}$
6. IF (INTERIORPOINT (P, A, B, C))
   RETURN 'P is an interior point.'
   ELSEIF (INTERIORPOINT (P, A, B, D))
   RETURN 'P is an interior point.'
   ELSE RETURN 'P is a boundary point'

Fig. 3. The ChooseGoodNeighbors(CGN) Algorithm

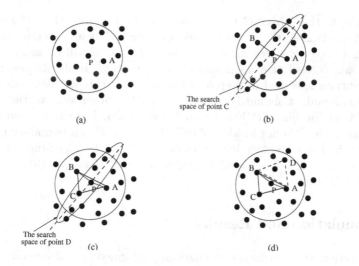

(a)

(b)

(c)

(d)

Fig. 4. Steps of CGN Algorithm

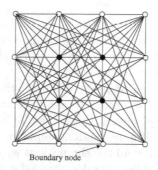

Boundary node

Fig. 5. Ideal case

and C as the three neighbors of the P, then P is an interior node as it could be enclosed inside the $\triangle ABC$ (Figure 2(a)). But incase we choose the $\triangle ABD$ then the node P is an exterior node(Figure 2(b)).

Hence we present our second algorithm (Figure 3) which intelligently selects four neighbors, A, B, C and D, of a node P (Figure 2(b)). The four nodes have the property that these are pairwise neighbors among themselves. The node P is detected as an interior node if it is either inside the $\triangle ABC$ or $\triangle ABD$.

Figure 4 explains how the CGN algorithm works. This algorithm tries to discover neighbors close to the given node, which could possibly surround it. So first, a node A closest to the given node P is found (Figure 4(a)). Then a node B is searched, such that it would possibly lie on the opposite side, to side of A to P(Figure 4(b)). If such a node is found, a third node C, is searched which is equidistant from A and B, but at a minimum distance from P. So it could lie anywhere inside the search space shown in Figure 4(b). So if C is found on the other side, to the side of C to AB shown in Figure 4(c), then P could not be enclosed inside $\triangle ABC$. Hence P could be detected as an exterior node, even if it is not. Hence we search for another node D, which will be again equidistant from A and B, and at a maximum distance from C. It should also satisfy the two conditions specified in Step 5 of Figure 3. By this we make sure that, if a node D is found, it should lie somewhere on the other side, to the side where node C is of the line APB(shown in Figure 4(d)). Note that if no node D is found and node P is not inside $\triangle ABC$, then node P is a boundary node.

Figure 5 shows the interior nodes detected by the algorithm in black and others nodes detected as boundary nodes in white. Hence, in this case both the FD and MD is 0 %.

4   Simulation and Results

In this section we describe the simulations performed to evaluate our algorithms and analysis of the results obtained.

4.1   Simulation Set-Up

We used MATLAB to simulate all our algorithms. We assumed the communication range of all the nodes in the network to be of fixed radius and all the communication channels to be symmetric.

For our simulation, we imagined our area of deployment to be divided into unit grids. In that area, two types of sensor network deployment were considered: *grid deployment* and *random deployment*. In grid deployment, the nodes are placed exactly at the grid points. But in random deployment, sensors are randomly placed inside one unit grid. For all the simulations with grid deployment, except for simulations on various number of nodes, we used network of 78 nodes deployed in $10{\times}10$ grid, on a square terrain of dimension 10 m × 10 m. Nodes are placed on 78 randomly selected grid points out of 100 such points. Similarly, for a random deployments, nodes are randomly placed inside 78 unit grids out of 100 such possible unit grids.

We assume that each node in the network knows all its one-hop neighbors and its distance to them. In addition, it must know the list of one-hop neighbors of each of its one-hop neighbors and their distances to them. Each node can do that locally by querying all its one-hop neighbors and acquiring the one-hop neighbor list from each of them.

4.2   Simulation Parameters

Three different evaluation parameters were considered for the simulation- *number of nodes, node communication range* and *sensor network deployment*.

- **number of nodes.** In this scenario we evaluated the algorithm's performance with increasing network size (number of nodes). We consider: $6{\times}6$

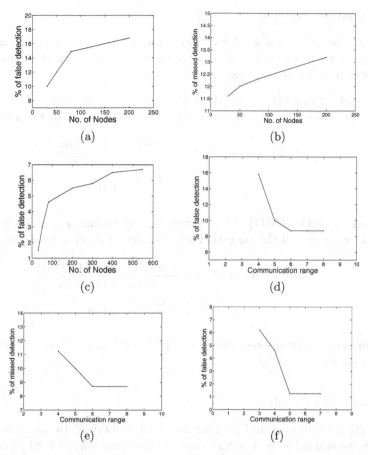

Fig. 6. Simulation results for different network size and different communication range

grid with 30 nodes, 10×6 grid with 50 nodes, 10×10 grid with 78 nodes, 15×16 grid with 200 nodes, 18×20 grid with 300 nodes, 22×22 grid with 400 nodes and 30×22 grid with 564 nodes. For all these network sizes, we simulated both grid and random deployment.

- **communication range.** In this scenario we evaluated the algorithm's performance with increasing communication range of each node. Different communication ranges for each node that were considered: 3, 4, 5, 6, 7 and 8 units. This defines a communication range of radius 3, 4, 5, 6, 7 or 8 units for each node and all the other nodes within that radious are the one-hop neighbors for that node. For these simulations we kept the network size fixed with 10×10 grid with 78 nodes.
- **network deployment.** In this scenario we compared the algorithm's performance with two above defined network deployments: grid deployment and random deployment. We considered networks with 10×10 grid and random topology with 78 nodes.

4.3   Evaluation Metrics

To measure the effectiveness of our algorithm we considered two evaluation metrics- *% of false detection* and *% of missed detection*. Let $D =$ the set of nodes detected as boundary nodes by the algorithm.

- **False detection(FD).** False detection represents the nodes that are not boundary nodes, but they are detected as boundary nodes by the algorithm.

$$\% \text{ of false detection} = \frac{\# \text{ of false detection}}{\text{Total } \# \text{ of nodes}}$$
$$= \mid \frac{I \cap D}{N} \mid \times 100$$

- **Missed detection(MD).** Missed detection represents the nodes that are boundary nodes, but the algorithm failed to detect them as boundary nodes.

$$\% \text{ of missed detection} = \frac{\# \text{ of missed detection}}{\text{Total } \# \text{ of nodes}}$$
$$= \mid \frac{B \setminus D}{n} \mid \times 100$$

All the simulation parameters defined in Section 4.2 were evaluated using the above two metrics.

4.4   Simulation Results

In this section we discuss and analyze the results obtained by the simulation. *Ideal case.* We started with an ideal sensor network deployment with 16 nodes in 4×4 grid. So all the nodes are placed at one of the grid points and each having a communication range of 4 units. Hence all the nodes in the network are within the communication range of each other.

Network size. Next we considered the scenario of varying network size. Figures 6-a, 6-b, and 6-c summarizes the results for both the grid and random deployment. The first two subplots describe FD and MD for the random deployment respectively and the third one shows FD for grid deployment. It is evident from the three plots that the FD and MD increases with increasing network size. The maximum value of FD in case of grid deployment is 6.7 % for 546 nodes, where as it is 16.8 % for random deployment with same number of nodes. It is interesting to note that for all the different network sizes in case grid deployment the MD remained 0 % (no missed detection). Thus, the algorithm is highly scalable.

Communication range. In our next experiment, we increased the communication range of each node in the network keeping the size of the network fixed (10×10 grid with 78 nodes). Figures 6-d, 6-e, and 6-f shows that in all the cases the FD or MD deceases with increasing communication range. But after certain value of

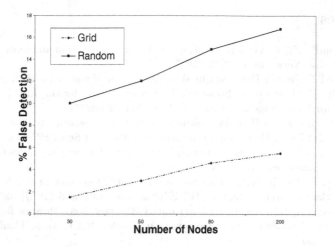

Fig. 7. Grid vs. Random deployment

communication range the values of FD and MD remains constant. Thus, when each node in the graph has reached the threshold on the number of neighboring nodes, the algorithm fails to improve its performance with further increase in the number of neighboring nodes. Hence the FD or MD value remains constant.

Network deployment. Figure 7 shows the comparison of FD for both grid and random deployment with increasing network size. The first bar (black) of each pair of bars shows the FD for grid deployment and the other (gray) is for random deployment. It is clear that for all the cases the FD is more incase of random deployment, but the increase in the value of FD grid deployment with network size, is proportionate to that of random deployment.

5   Conclusion and Future Work

In this paper, we proposed an algorithm that can be used to discover the boundary nodes of a randomly deployed WSN. The algorithm rely on limited communication between a node and its neighbors. This feature makes the algorithm suitable for networks with large number of nodes. Simulation experiments of different network topologies confirm the accuracy of the algorithm in detecting the boundary of the network, even for random networks with more than 500 nodes.

Future work includes the extension of the algorithm to detect the *coverage boundary* of the network. We define the coverage boundary of a network as the boundary encompassing the areas that could be monitored or covered by the sensor nodes inside the network. This defines the entire area that is within the communication range of the sensor nodes of a WSN. The extension of the algorithm to detect the network boundary in the presence of communication holes (i.e. unconnected sensor network) is our current research topic.

References

1. Chintalapudi, K., Govindan, R. :Localized Edge Detection in Sensor Network. IEEE Ad Hoc Networks J. (2003) 59-70.
2. Distributed Perimeter Detection in Wireless Sensor Network (July 2, 2004).
3. Mainwaring,A., Polastre, J., Szewczyk, R., Culler, D., Anderson, J.: Wireless sensor networks for habitat monintoring. ACM WSNA 02 (2002).
4. Nowak, R., Mitra, U.: Boundary estimation in sensor networks: theory and methods, Proc. 1st Inter. Workshop on Information Proc. in Sensor Netwks, 2003.
5. Savvides,A. , Fang, J., Lymberopoulos, D.: Using mobile sensing nodes for dynamic boundary estimation. Proc. WAMES 04, (2004).
6. Clouqueur, T., Saluja, K.K., Ramanathan, P.: Fault-tolerance in collaborative sensor networks for target detection. IEEE Tran. Computers, 3 (2004) 320-333.
7. Krishnamachari, B., Iyengar, S.: Distributed Bayesian algorithm for fault-tolerant event region detection in wireless sensor netowrks. IEEE Tran. Computers, 53 (2004) 241-250.
8. Ding, M., Chen, D., Xing, K., Cheng, X.: Localized fault-tolerant event boundary detection in sensor networks. IEEE INFCOM (2005).
9. Zhao, Y., Govindan, R., Estrin, D.: Residual scans for monitoring weireless sensor networks. IEEE WCNC (2003) 78-89.
10. Meguerdichian, S., Koushanfar, F., Potkonjak, M., Srivastava, M.B. : Converage problems in wiereless ad-hoc sensor networks. IEEE INFOCOM 01, (2001) 1380-87.
11. Huang, C-F., Tseng, Y-C.: The coverage problem in a wireless sensor network. Proc. WSNA (2003).

A Low-Complexity Issue Queue Design with Speculative Pre-execution*

Won W. Ro[1] and Jean-Luc Gaudiot[2]

[1] Department of Electrical and Computer Engineering,
California State University, Northridge
wro@csun.edu
[2] Department of Electrical Engineering and Computer Science,
University of California, Irvine
gaudiot@uci.edu

Abstract. Current superscalar architectures inherently depend on an instruction issue queue to achieve multiple instruction issue and out-of-order execution. However, the issue queue is implemented as a centralized structure and mainly causes globally broadcasting operations to wake up and select the instructions. Therefore, a large issue queue ultimately results in a low clock rate along with a high circuit complexity. This paper proposes Speculative Pre-Execution Assisted by compileR (SPEAR), a low-complexity issue queue design. SPEAR is designed to manage the small issue queue more efficiently without increasing the queue size. To this end, we have first recognized that the long memory latency is one of the factors which demand a large queue, and we aim at achieving early execution of the miss-causing load instructions using another hierarchy of an issue queue. We speculatively pre-execute those miss-causing instructions as an additional prefetching thread.

1 Introduction

For the past ten years, the superscalar architecture model has been adopted for general purpose microprocessors. The superscalar architecture model ultimately aims at expediting program execution by finding parallelism among instructions; this fine-grain parallelism is called Instruction Level Parallelism (ILP). In general, exploiting ILP in superscalar architectures is based on the two runtime techniques: *multiple instruction issue* and *out-of-order execution*, and both techniques are achieved through dynamic scheduling. Dynamic scheduling requires the storing of multiple decoded instructions in a pool of instructions (the pool is often called *issue queue* or *instruction window*) and issues "ready" instructions from the queue (where ready instruction means the instruction for which the source registers have been computed so that there is no data hazard in issuing the instruction).

---

* This paper is based upon work supported in part by NSF grants CCR-0234444 and INT-0223647. Any opinions, findings, and conclusions or recommendations are those of the authors and do not necessarily reflect the views of NSF.

D.A. Bader et al. (Eds.): HiPC 2005, LNCS 3769, pp. 353–362, 2005.
© Springer-Verlag Berlin Heidelberg 2005

In essence, the execution of an instruction i is decided by the previous instruction k which produces the value for the source registers of instruction i. This implies that instruction k should broadcast the register name and value across the issue queue to *wake up* and *select* any ready instruction. In conventional superscalar designs, this operation cannot be pipelined and represents a one-cycle operation [1]. Therefore, as far as the size of the issue queue is concerned, more instructions within an issue queue mean more communication overhead (*wire delay*) for the wake up and select operations within a single clock cycle. In fact, the wires tend to scale poorly compared to semiconductor devices, and the amount of states that can be reached in a single clock period eventually ceases to grow [2]. As a consequence, there is a scaling problem regarding the issue queue size in any superscalar design. Larger issue queue and faster clocks, while both necessary, become paradoxically antagonistic in the design of future superscalar processors [2].

This paper introduces our low-complexity SPEAR (Speculative Pre-Execution Assisted by compileR) architectural technique as an alternative to conventional superscalar designs. Instead of using a large issue queue, we implement an additional separate issue queue which is dedicated to the execution of *performance-critical* instructions. The performance-critical instructions include cache miss instructions and instructions on which the cache miss instructions have data dependencies. Indeed, the increasing performance gap between processor and main memory imposes a high burden on the dynamic scheduling capability of a small issue queue since it causes a significant amount of pipeline stalling at cache misses. Therefore, the proposed architecture aims at reducing pipeline stalling in the main program flow by an early scheduling of those instructions.

Our performance analysis is based on a cycle-time simulator derived from the SimpleScalar 3.0 tool set [3]. The simulation results demonstrate that the proposed SPEAR architecture yields high degrees of ILP, comparable to what a superscalar architecture with a large issue queue (256 entries) would achieve. In the next section, we describe background research and related work. Section 3 presents the detailed description of the proposed SPEAR architecture. Section 4 includes experimental results and performance analysis.

2   Motivation and Background

2.1   Design Motivation

The increasing demands for a large issue queue are caused by the long latency operations which occupy the queue slots for considerable amounts of time. It consequently means that all the instructions which also need the data from those long latency operations should remain inside the queue (Fig. 1-(a)). In fact, this observation means a reduction of the available queue slots and accordingly requires a large issue queue. One of the most critical long latency operations in current microprocessors is memory access where a cache miss can reach hundreds of cycles. Therefore, these long memory access latencies strongly imply the need for a large issue queue.

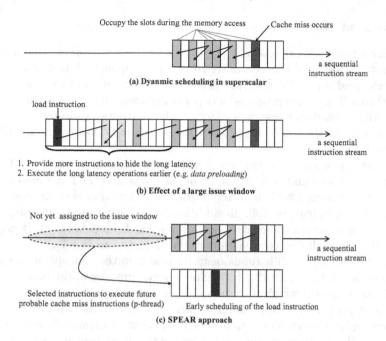

Fig. 1. Basic concept of the SPEAR approach

More specifically, a large issue queue hides the long access latency in two ways (Fig. 1-(b)): first, it provides more issue slots during the memory accesses. More slots mean more opportunities to uncover independent *ready* instructions to hide the long access latency. Another advantage of the large queue is the potential for *dynamic preloading* of the necessary data. Preloading operations would take place when future load instructions are executed early. In a dynamic scheduling scheme, any load instruction inside the issue queue can be executed as long as the data dependencies have been resolved. Therefore, the large queue increases the chances to execute load instructions even earlier. However, despite the above advantages, a large issue queue has an inherent critical scaling problem. As previous research indicates [1], the critical path delay shows a quadratic dependency on the queue size.

As an alternative approach, we here propose our SPEAR model which is designed to achieve the early scheduling of future cache miss instructions without increasing the size of the issue queue. Our main idea is to extract frequently miss-causing instructions and make them into prefetching threads. These instructions can be scheduled through another issue queue to achieve data prefetching (Fig. 1-(c)). The prefetching thread runs as a stand-alone thread and is expected to update the cache blocks before the main thread accesses them. Therefore, the prefetching thread suitably reduces the number of cache misses on the main thread. Consequently, fewer cache misses result in a reduction of the number of load instructions occupying the issue queue due to memory accesses. A more detailed description of our SPEAR technique will be given in Section 3.

2.2   Related Work

There have been several research projects which have attempted to solve the complexity problem of the centralized instruction queue. Palacharla *et al.* [1] have performed an initial analysis of the complexity of wide issue superscalar architectures. They have proposed a dependence-based instruction queue design as a solution. Based on these dependencies, the instructions are sent to separate FIFO queues. At the issue stage, only the head instructions of each FIFO are considered for issuing.

Another approach called *clustering scheme* has also been developed. Clustering is strongly based on the partitioning of the queue and the functional units. In general, the clustering technique separates the instruction streams considering the register dependencies [1,4]. In addition, speculative multithreading which finds parallelism across the control flow limits, has been developed. It normally uses single-chip multiprocessor architectures as platforms [5,6,7].

Our SPEAR approach is fundamentally based on the concept of speculative pre-execution [8,9,10,11,12,13,14,15]. Speculative pre-execution is a promising prefetching technique which uses an auxiliary prefetching thread in addition to the main program flow. The prefetching thread (*p-thread*) includes the frequently miss-causing load instructions (as also referred to as *delinquent loads*) and backward slices (the group of instructions on which the delinquent loads have register dependencies). The p-thread runs earlier than the main program, on the spare hardware context, and can achieve timely data prefetching. The speculative pre-execution is first motivated by the fact that most of the cache misses are caused by a small number of load instructions. Those load instructions (*d-loads*) are normally identified through profiling.

3   Complexity-Effective Issue Queue Design with SPEAR

3.1   Description of the Basic Pipeline

The structure of the SPEAR pipeline is depicted in Fig. 2. It is mainly based on an SMT model; our SPEAR design requires the support for the simultaneous execution of the main thread and of the p-thread. A unique feature of our proposed model is that the p-thread is a strict subset of the main program instruction stream, and it is not required to store the p-thread instructions in separate memory locations. Instead, those instructions that belong to the p-threads are simply marked with appropriate *p-thread indicators* during the pre-decoding operation. When a p-thread is triggered, the instructions marked as belonging to the p-thread are extracted from the Instruction Fetch Queue (IFQ).

Three additional units called the p-thread detector (PD), the p-thread table (PT), and the p-thread extractor (PE) have been designed and implemented to facilitate the execution of the prefetching thread. First, the PD is designed to examine whether the instructions currently being fetched belong to the p-thread or not. For that purpose, it looks up the PT with the PC of the instructions being fetched. Indeed, the PT contains the PCs of all the instructions in the

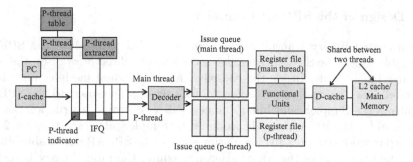

Fig. 2. Hardware description of the SPEAR

p-thread. If the instruction in the pre-decoding stage is identified as a p-thread instruction, the PD sets the p-thread indicator of the instruction (recall that, as described in Fig. 2, each slot of the IFQ includes a one bit p-thread indicator). The third structure, PE, will be explained along with the p-thread execution in the following subsection.

3.2   Extraction and Execution of the P-Thread Instructions

If an instruction being fetched is detected as a d-load, the PD changes the machine state to *pre-execution enabled* mode. After that, the *extraction* and *delivery* operations for the p-thread instructions are controlled by the p-thread extractor (PE). First, to guarantee a deterministic state has been reached before the live-in values are copied, the PE should wait until all instructions which have been decoded retire the commit stage. After that, the live-in values are copied from the main thread context to the p-thread context; the register names for the live-in values are also provided by the PT. Finally, the PE starts extracting p-thread instructions and delivering them to the decoder.

During the pre-execution enabled mode, the PE looks up each entry starting with the head of the IFQ in order to extract the p-thread instructions. As the PE finds instructions for which the p-thread indicator is "on," it extracts them and sends them to the decoder. In fact, it only copies the instruction to the input field of the decoder and leaves the instruction in the IFQ for the main thread to execute. This is because, although the instruction has been selected and delivered to the decoding logic as a p-thread instruction, it also needs to be executed as part of the main thread as well.

After the execution of the p-thread has begun, the processor should operate in a multithreaded mode; the p-thread is executed as a thread running along with the main thread. Every operation of an instruction is tagged with a dedicated *thread id*. In our SPEAR model, *0* is assigned to the main program thread as a thread id, while *1* is assigned to the p-thread. Indeed, the thread id *1* is assigned to the p-thread instructions when the PE sends the instruction to the decoder. Since the goal of the p-thread execution is to pre-execute the d-load, after the d-load is retired from the reorder buffer at the commit stage, the pre-execution mode is finished and the processor returns to the normal mode.

3.3 Design of the SPEAR Compiler

An automated binary translator has been developed to produce the SPEAR executable code. The SPEAR compiler is designed to directly work on the SimpleScalar binary code. Inside the compiler, four individual modules have been developed based on the *SimpleScalar-3.0* tool-set [3]. The input to the SPEAR compiler is the SimpleScalar binary named PISA (Portable Instruction Set Architecture) [3]; the PISA binary is produced by SimpleScalar targeting gcc-2.6.3. The output produced after all compilation steps is the SPEAR executable binary.

At the beginning of the compilation procedure, the input binary is sent to the control flow graph (CFG) drawing tool and the profiling tool. The CFG drawing tool has been developed to create the control-flow graph and identify the loop-region. The second one (*profiling tool*) has been designed to find frequently miss-causing load instructions (*delinquent loads*) and also to collect runtime information; different input data sets have been used between profiling and the performance evaluation. After these two modules, the program slicing tool collects the *program structure information* and *dynamic information* from the previous two modules and constructs the p-threads.

Our slicing method is applied to the static program structure with the control-flow graph which is drawn by the CFG drawing tool. In the slicing tool, each *static* instruction has its own data structure. In addition to that, the data structures for basic blocks are defined and pointed to by the corresponding instructions. The control-flow is defined by identifying the target address of each conditional/unconditional jump instruction. The procedures are also defined by identifying jump instructions to the function calls.

On the other hand, the profiling tool derives the data-flow graph among instructions. The dynamic instances of instructions are analyzed and the dependencies are examined by the source/destination-register names. Furthermore, the access addresses of each store and load instructions are analyzed to find the address dependencies. Finally, the p-threads are defined and constructed by the backward chasing on the data-flow graph. The last module is the attaching tool which attaches the p-thread information to the SPEAR binary. This information needs to be loaded into the p-thread table (PT) at program execution time.

4   Experimental Results and Analysis

4.1 Benchmarks Description and Simulation Parameters

The set of benchmarks includes 12 applications: six benchmarks selected from the SPEC2000 suite (*bzip2, gzip, vortex, vpr, art,* and *equake*), four applications chosen from the Atlantic Aerospace Stressmark suite (*matrix, neighborhood, pointer,* and *update*), and two benchmarks from the Atlantic Aerospace Data-Intensive Systems Benchmarks suite (*data management* and *ray tracing*). The SPEC benchmarks have been compiled at peak optimization level and tested with the reference input set.

The execution-driven architectural simulator for SPEAR has been designed by modifying *sim-outorder* simulator which is provided by the SimpleScalar 3.0 tool set [3]. The baseline superscalar architecture for performance comparison has a 256-entry issue queue and the issue and commit widths are 8. The SPEAR also has been tested with various sizes of issue queues: 32, 64, and 128. For all simulation results, the performance is measured in terms of IPC.

4.2   Performance Results and Analysis

Since our SPEAR aims at achieving performance comparable to that of super-scalars with a large issue queue, we first compare it to a superscalar model with a 256-entry issue queue. Throughout all results, the diagrams show the normalized performance in terms of IPC; although we do not quantify the expected clock rates of our design, we claim that the small queues in our design would eventually contribute to higher clock rates.

The simulation results for the SPEAR models are shown in Fig. 3. The first SPEAR model has been tested with a 128-entry instruction fetch queue, shared functional units between two threads, and two 128-entry issue queues for both threads. The second bars in each benchmark show the normalized performance of the initial SPEAR (*SPEAR-128*), whereas the first bars show the normalized performance of the baseline superscalar architecture. In all twelve benchmarks, the performance of *SPEAR-128* remains better than a 10% degradation of the baseline performance; recall that the baseline performance is measured with a twice large issue queue. More particularly, *SPEAR-128* produces an even better performance in the six benchmarks.

Since the p-thread is spawned from the IFQ, the size of IFQ is believed to affect the prefetching performance. To show how the IFQ size affects the overall performance, we have changed the IFQ size from 128 to 256 entries. The performance results for the longer fetch queue configuration (denoted as *SPEAR.lfq-128*) are shown in the third bars in each benchmark. As the diagram shows, the long range fetch queue considerably improves the SPEAR performance in some applications. More particularly, matrix shows a 44.2% performance improvement compared to the previous configuration. This is due to the fact that the matrix

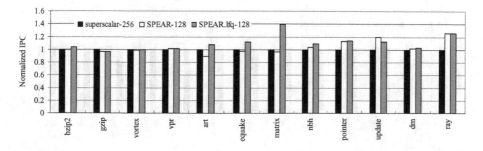

Fig. 3. Performance results of the SPEAR model compared to the baseline superscalar

benchmark contains a high branch hit ratio (0.9942) and could yield significant benefits from the long range fetch queue. On the contrary, the update benchmark shows a very low branch hit ratio (0.8865) and consequently does not gain any improvement from the longer fetch queue. On average, the long fetch queue achieves a 7.25% performance gain from the short queue configuration.

As explained earlier, our p-thread stream is extracted from the main thread as a subset instruction stream. Therefore, the execution behavior of both threads would contain similar patterns, and simultaneous accesses to the same functional units are inevitable. Our next enhancement targets the elimination of this resource conflict by assigning dedicated resources for each thread. Fig. 4 shows the performance enhancement by assigning dedicated resources to the p-thread. The second bars in each benchmarks show the performance of the shared resource configuration, and the third bars indicate the performance of the dedicated resource model (which is denoted as *SPEAR.lfq.dr-128*). The performance improves in most benchmarks with dedicated functional units. On average, a 3.6% performance improvement has been achieved compared to the shared resource model. Over twelve benchmark programs, the average performance improvement of *SPEAR.lfq.dr-128* is 14.4% compared to the 256-entry queue superscalar.

In this part, the issue queues of the SPEAR approach have been further reduced to 32 and 64 (with *long fetch queue* and *dedicated resources*). The performance results of the 32 and 64 queue sizes are shown in Fig. 5 (denoted

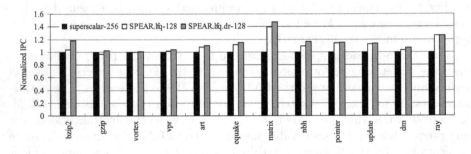

Fig. 4. Performance results of the SPEAR model compared to the baseline superscalar

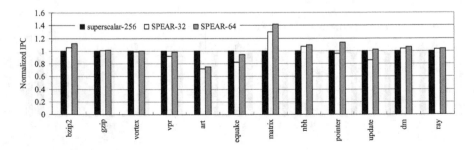

Fig. 5. Performance results of 32 and 64-entry issue queue SPEAR

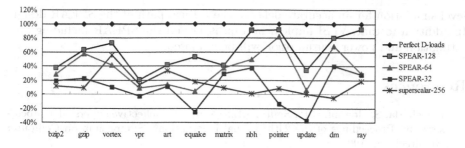

Fig. 6. Latency hiding effect of each architectural model

SPEAR-32 and *SPEAR-64*, respectively). The performance of *SPEAR-64* is higher than the baseline superscalar architecture in 8 benchmarks; note that the baseline superscalar has a four times larger issue queue. On average, *SPEAR-64* achieves a 4.7% improvement compared to *superscalar-256*. In fact, the average performance of *SPEAR-32* over 12 benchmarks reaches 98.1% of the baseline performance even with one eighth of queue size.

Fig. 6 shows the effectiveness of latency hiding for each architecture model. Compared to the latency hiding of the perfect d-load case, the latency hiding of each architecture model has been calculated. In the three benchmarks (*nbh*, *pointer*, and *ray*), the *SPEAR-128* achieves more than 90% of latency hiding compared to the perfect d-load prefetching. On an average over twelve benchmarks, *SPEAR-128* hides delinquent load latency up to 59.9% of time.

Regarding the smaller queue configurations, the *SPEAR-64* configuration hides 35.5% of the delinquent load latency compared to the perfect d-load results. *SPEAR-32* only hides 9.5% of the delinquent load latency in terms of time, and *superscalar-256* hides 14.3% of the delinquent load latency. However, in half of the benchmarks, *SPEAR-32* shows better performance than the superscalar model with a 256-entry issue queue. In fact, *superscalar-256* shows a more stable performance than *SPEAR-32*. This is because the small issue queue size of *SPEAR-32* restricts the parallelism too much in some applications.

5   Conclusions and Future Work

In this paper, we have introduced a low-complexity issue queue design with our SPEAR approach which implements another separate issue queue for data prefetching instead of implementing a large issue queue. Compared to a superscalar architecture with a 256-entry issue queue, the SPEAR technique achieves an even higher performance in some benchmarks with issue queue only half the size. On average, SPEAR with the long fetch queue and the dedicated resource achieves a 14.4% improvement compared to the baseline superscalar.

As future research, more study on possible clock rates will be investigated. Quantifying possible clock speed will provide a better evaluation of the actual performance gain of the proposed architecture. We will also perform a circuit level simulation for an accurate delay of the critical path of the SPEAR design. In addition to that, VLSI implementation issues on the SPEAR including area estimation and power consumption will be explored.

References

1. Palacharla, S., Jouppi, N.P., Smith, J.E.: Complexity-effective superscalar processors. In: Proceedings of the 24th Annual International Symposium on Computer Architecture. (1997)
2. Agarwal, V., Murukkathampoondi, H.S., Keckler, S.W., Burger, D.C.: Clock rate versus IPC: The end of the road for conventional microarchitectures. In: Proceedings of the 27th Annual International Symposium on Computer Architecture. (2000)
3. Burger, D., Austin, T.: The SimpleScalar tool set. Technical Report CS-TR-97-1342, University of Wisconsin-Madison (1996)
4. Farkas, K.I., Chow, P., Jouppi, N.P., Vranesic, Z.: The multicluster architecture: Reducing cycle time through partitioning. In: Proceedings of the 30th Annual International Symposium on Microarchitecture. (1997)
5. Krishnan, V., Torrellas, J.: A chip-multiprocessor architecture with speculative multithreading. IEEE Transactions on Computers **48** (1999)
6. Marcuello, P., González, A.: Clustered speculative multithreaded processors. In: Proceedings of the 13th International Conference on Supercomputing. (1999)
7. Sohi, G.S., Breach, S.E., Vijaykumar, T.N.: Multiscalar processors. In: Proceedings of the 22nd Annual International Symposium on Computer Architecture. (1995)
8. Annavaram, M., Patel, J.M., Davidson, E.S.: Data prefetching by dependence graph precomputation. In: Proceedings of the 28th Annual International Symposium on Computer Architecture. (2001)
9. Collins, J.D., Wang, H., Tullsen, D.M., Hughes, C., Lee, Y.F., Lavery, D., Shen, J.P.: Speculative precomputation: Long-range prefetching of delinquent loads. In: Proceedings of the 28th Annual International Symposium on Computer Architecture. (2001)
10. Kim, D., Yeung, D.: Design and evaluation of compiler algorithms for pre-execution. In: Proceedings of the 10th International Conference on Architectural Support for Programming Languages and Operating Systems. (2002)
11. Luk, C.K.: Tolerating memory latency through software-controlled pre-execution in simultaneous multithreading processor. In: Proceedings of the 28th Annual International Symposium on Computer Architecture. (2001)
12. Liao, S.S.W., Wang, P.H., Wang, H., Hoflehner, G., Lavery, D., Shen, J.P.: Post-pass binary adaptation for software-based speculative precomputation. In: Proceedings of the Programming Language Design and Implementation. (2002)
13. Ro, W.W., Gaudiot, J.L.: SPEAR: A hybrid model for speculative pre-execution. In: Proceedings of the 18th International Parallel and Distributed Processing Symposium. (2004)
14. Roth, A., Sohi, G.S.: Speculative data-driven multithreading. In: Proceedings of the 7th International Symposium on High Performance Computer Architecture. (2001)
15. Zilles, C.B., Sohi, G.S.: Execution-based prediction using speculative slices. In: Proceedings of the 28th Annual International Symposium on Computer Architecture. (2001)

Performance and Power Evaluation
of an Intelligently Adaptive Data Cache[*]

Domingo Benítez[1], Juan Carlos Moure[2], Dolores Isabel Rexachs[2],
and Emilio Luque[2]

[1] DIS Department & IUSIANI, University of Las Palmas G.C.,
35017 Las Palmas, Spain
dbenitez@dis.ulpgc.es
[2] Computer Architecture and Operating Systems (CAOS) Department,
Universidad Autónoma de Barcelona, 08193 Barcelona, Spain
{JuanCarlos.Moure, Dolores.Rexachs, Emilio.Luque}@uab.es

Abstract. We describe the analysis of an on-line pattern-recognition algorithm to
dynamically control the configuration of the L1 data cache of a high-performance
processor. The microarchitecture achieves higher performance and energy saving
due to the accommodation of operating frequency, capacity, set-associativity, line
size, hit latency, energy per access, and chip area to program workload and ILP.
We show that for the operating frequency 4.5 GHz, the execution time is always
reduced with an average measure of 12.1% when compared to a non-adaptive
high-performance processor. Additionally, the energy saving is 2.7% on average,
and t1he product time-energy is reduced on average by 14.9%. We also consider a
profile-based reconfiguration of data cache, which allows picking different cache
configurations but only one can be chosen for each program. Experimental results
indicate that this approach yields a high percentage of the performance improvement and energy saving achieved by the on-line algorithm.

1 Introduction

The effects of caches on processor performance, area cost, energy consumption and
power dissipation are correlated. Cache memory reduces average memory access
time, and this is one of the reasons why some new families of processors provide
features that include an increase of on-chip memory. Although the power density of
SRAM memory is an order of magnitude smaller than logic [11], a large on-chip
cache may be responsible for a significant part of the energy and power dissipated by
the entire processor [12]. Power analysis is currently important since as power density
of processors rises, die temperature increases and long-term reliability can be compromised. Additionally, energy consumption analysis is specially a determining factor
in choosing a processor for battery-powered systems.

The average memory access time, power dissipated and energy consumed by the
on-chip cache depend on the number of hits and misses. However, these numbers

---

[*] This work was supported by the MCyT-Spain under contract TIN 2004-03388, the Gobierno de Canarias,
the Generalitat de Catalunya (SGR-00218), and the HiPEAC European Network of Excellence.

D.A. Bader et al. (Eds.): HiPC 2005, LNCS 3769, pp. 363–375, 2005.

change in different applications and even in different execution phases of a single process [17]. A processor that fixes its cache organization at design time cannot adapt to these changing characteristics. Instead, adaptive processors, depending on the runtime and/or off-line behavior of a process, may rearrange their microarchitecture to increase performance and energy saving ([4], [8]).

This work describes and evaluates an adaptive L1 data cache designed for high-performance and energy-efficient processors. It can be used in high-performance processors and embedded processors. The total amount of memory used, line size, and set-associativity of the cache are reconfigurable after the chip has been fabricated. The cache's hit time (in cycles) and the processor's power consumption per memory access vary depending on the selected cache configuration and on the processor clock frequency, which our design allows to vary dynamically.

This paper studies two types of adaptation methods. *Dynamic Adaptation* is a mechanism that, for each execution interval, selects the cache configuration that tries to provide the best performance. We propose the *Real Dynamic Adaptation* which uses a control methodology that is inspired by pattern-recognition algorithms, to tune the cache organization with the running application. Measures taken at runtime provide feedback to decide if the current configuration should be changed and which configuration to use instead. The second type of proposed adaptation is called *Static Adaptation*, which determines the best cache configuration for the whole execution of a program. It is a software-based and lower cost method, which is inspired by the real dynamic adaptation. We demonstrate that the proposed realistic methods can reduce both the execution time of the processor and the energy consumption of the on-chip cache when compared to non-adaptive high-performance processors. We also evaluate the advantage of our design over other adaptive proposals for the L1 data cache.

The rest of the paper is organized as follows. Section 2 contains discussion of related work. Section 3 presents the processor with both adaptive methodologies for the L1 data cache. In Section 4, we describe our simulation methodology. In Section 5, we present a performance and power analysis of several baseline configurations. Section 6 evaluates our control methodology for real systems with dynamic and static adaptive caches. Section 7 shows our concluding remarks.

2 Related Work

Configuration management algorithms are used to find the configuration that best suits the characteristics of a given program. Work on the subject has explored three basic properties of these algorithms ([3], [7]): (1) efficiency on detecting a phase boundary during execution of a process, (2) the tuning overhead, and (3) the reconfiguration overhead. Additionally, we have observed that the set of tunable configurations is required to be analyzed as it is one of the keys to success.

Semeraro et al. have proposed the Multiple Clock Domain (MCD) architecture with multiple domains (one of them is the on-chip cache memory) for which frequency and voltage can be reduced independently [15]. They maintain the same cache access latency during the voltage/frequency scaling. Balasubramonian et al. proposed an adaptive cache that was tuned at every instruction interval [3], with fixed clock frequency and varying cache latency (in cycles). Our adaptive L1 data cache proposal differentiates from [3] in the following ways: we don't use the same tuneable cache configurations for all programs; we propose a new methodology to know the most efficient configurations for each program; the operating frequency of the processor/cache system can be varied; our methodology can be used for a wide range of frequencies and chip areas; the runtime overhead is significantly lower and independent of the number of configurations, as we don't prove all the possible cache configurations each time a program phase change is detected; and finally, our proposal improves performance and reduces energy consumption at the same time. Most of the adaptive techniques that have been proposed for energy saving in cache memory reduce energy consumption or the time-energy product, but also reduce performance ([2], [3], [10]).

Off-line profiling and instrumentation of the application can be used to alternatively implement the adaptation control. It provides a more global view of the program than with a hardware solution and in some cases can achieve better results [13]. Our static adaptation is a control methodology that relies on an off-line profiling which determines for each application the cache configuration with the highest performance. This cache configuration is activated just before the application runs.

3 Adaptive Cache

We propose an Adaptive L1 Data Cache with Intelligent Closed-Loop Control that is integrated into an out-of-order superscalar processor with a two-level memory hierarchy, where the L1 data cache may adopt different configurations during runtime. The cache can be reconfigured either dynamically (on-line), after each instruction interval, or statically (off-line), before a different program gets started.

3.1 Hardware Support

The hardware support for the adaptive cache is shown in Fig. 1. Firstly, a few internal microarchitectural registers called *Sensors* are required to measure frequency of instruction basic blocks (*BB Sensor*), and furthermore to measure execution time (*cycle Sensor*). Secondly, a hardware coprocessor (called *Copro*) integrated in the same chip with the instruction processor is required for the on-line control tasks. At the end of each instruction interval, Copro reads the BB sensor and decides whether or not to change the cache configuration. Additionally, Copro contains two small tables: a *Pattern Table* with data of each program phase and the best cache configuration associated to that pattern; and a *Configuration Table* that contains the necessary control information to reconfigure the data cache, including the operating frequency at which the processor communicates with the cache memory. Thirdly, the adaptive cache memory is conceived as a set of modular blocks of SRAM-based circuits that are selectively connected through the use of a fixed routing architecture with programmable switch-boxes at possible junction points, in much the same way as a Field-Programmable Gate Array [6]. The cache configuration is determined by the content of an integrated configuration memory which provides the cache change mechanism. The cache reconfiguration can be performed either on-line by the coprocessor or off-line by software. From an architectural point of view, the reconfigurable cache is characterized in this study by: (a) the set of possible cache configurations, in which capacity, set-associativity, line size, load-use latency and operating frequency can be individually selected; (b) the reconfiguration time; (c) the energy consumed in each hardware reconfiguration; (d) the maximum operating frequency of data cache (f_{limit}, the same as processor clock speed); and (e) the maximum chip area devoted to data cache (A_{limit}). Many cache configurations may not need to use all available chip area devoted to data cache. The decommissioned portion of cache is considered to be powered down. When a different cache configuration is picked, all cache lines are invalidated.

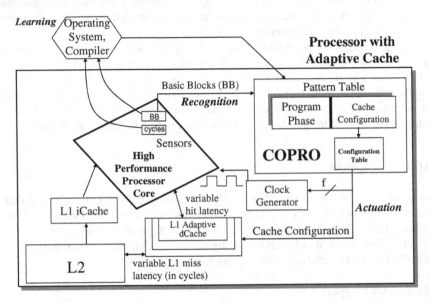

Fig. 1. Microarchitecture of the Processor with Adaptive L1 Data Cache

3.2 Intelligent Control for On-Line Adaptation

We propose an intelligent control system for dynamically managing the reconfigurable cache. Like other approaches used in human-like systems [9], three stages are required: Learning, Recognition, and Actuation. The Learning stage is proposed to identify patterns/phases of program behavior and associate each of them with a configuration of our adaptive cache that provides the highest performance. Firstly, fingerprints of instruction basic blocks are gathered for small intervals of the program's execution. A fingerprint collects the frequency and size of each executed basic block in the form of a *Basic Block Vector* (BBV) [16]. The BB sensor in our design holds the BBV that is being built for the current execution interval. Each BBV represents an execution interval in a multidimensional space. Vectors that are close together in that space represent intervals with similar behavior, i.e. a program pattern/phase. In a second learning step, the K-means clustering algorithm (used in [17] for another purpose) applies by software on BBVs collected from the execution of a large number of instruction intervals. The BBV vectors are grouped into a set of clusters called "*Sim-Point (SP) Classes*", where each SP class represents a different program pattern/phase. The task of identifying these SP classes is performed off-line either by the operating system, compiler, or an independent software layer [7]. In the last learning stage, associating each SP class with an optimal cache configuration involves executing the program and monitoring the times and SP classes of several execution intervals for each cache configuration. This task and the periodic activation of the cache reconfiguration are also implemented by software. Since we are assuming that the best cache configuration is the same for all the intervals belonging to the same class, it is not necessary to execute the full program for every available cache configuration. A single program execution will suffice. Since an SP class should be composed of many intervals, the learning time is small in comparison with the total execution time. Additional executions of the same application amortize the learning time overhead. Each entry in the Pattern Table contains an association between the set of close BBV vectors of an SP class and a cache configuration.

The Recognition stage is active on-line for the whole program execution. It detects if the current cache configuration does not provide the highest performance for the running program pattern/phase, and determines what different cache configuration should be used instead. The coprocessor performs the recognition task by firstly reading the BBV vector from the BB sensor after each instruction interval. Next, the vector position in the representation space used in the learning stage makes it possible to recognize the SP class of the interval. The recognition stage can be executed in parallel with the instruction flow and does not modify the critical execution path. So, this control stage does not provide performance degradation. The number of clock cycles to determine both the SP class of the previous instruction interval and whether the cache configuration requires to be changed depends on the execution time of each interval. This means that for example a processor with maximum IPC equal to 2 must execute the recognition phase of a 10^5 instruction interval in at most $5 \cdot 10^4$ cycles.

The Actuation stage is activated by the coprocessor when the SP classes of three consecutive instruction intervals are assigned to the same cache configuration and this configuration is different from the current one. When an actuation is fired, the instruction flow is stalled and the appropriate entry in the configuration table is read to obtain the data required for the reconfiguration process, including the operating frequency. After reconfiguring the hardware, the instruction flow is restarted. When a different program gets started, the pattern and configuration tables are previously updated with the information derived from the respective learning stage, and the internal registers of Copro with the SP classes of recent instructions intervals are cleared.

The Recognition and Actuation stages provide a closed-loop control mechanism since the actuation stage changes the cache configuration when this is recognized not to provide the highest performance for the current program phase. The actuation stage requires flushing of the cache contents, and introduces an overhead. It is dependent on the reconfigurable technology and we assume that each cache reconfiguration takes 1 μs and consumes 5 μJ. However, since actuations are fairly infrequent, this overhead only very slightly reduces performance. So, our control policy adds minimum overhead to the processor execution when the program enters different phases.

3.3 Static Adaptation

In the type of adaptation called *Static Adaptation*, the program's phase behaviour is not considered. For each program, the configuration of the cache memory and clock frequency can be different but is not changed during its run-time. This is done by software in two phases: Learning and Actuation. The *Learning* stage is executed either during a profiling time or when the application is running normally. The software calculates the performance from the measures made by the cycle sensor after a long interval of retired instructions. This task is repeated for each available cache configuration. After that, each program is associated with the cache configuration that provides the highest performance. In the *Actuation* stage, this configuration is picked and the hardware is reconfigured before running the program.

4 Experimental Methodology

This section describes the experimental methodology used to evaluate our proposals. We have used the Simplescalar-Alpha-3.0 tool set [5] to generate the dynamic instruction trace of the first 2 billion instructions for some programs of the SPEC benchmark suites. Table 1 shows the selected benchmarks and their inputs (Alpha ISA, cc DEC 5.9, –O4). These programs were chosen to demonstrate how our proposed hardware/software methodology can outperform both highly efficient non-adaptive approaches and other adaptive systems on SPEC benchmarks, and additionally, because they represent different program domains (integer, floating-point, and multimedia). Accurate cycle-by-cycle simulation was performed using KScalar [14], a CPU simulator based on Simplescalar, to subsequently calculate for each available L1 data cache configuration the execution time, energy consumption, power dissipation and product time-energy during each instruction interval. Table 2 lists the parameters used for the simulated processor. The standard SimPoint-1.1 Toolkit [16] was used to extract

Table 1. Selected SPEC benchmarks

Benchmark	input	SPEC	Benchmark	input	SPEC	Benchmark	input	SPEC
bzip	source	int00	parser	Ref.	int00	mgrid	Ref.	fp00
eon	cook	int00	facerec	Ref.	fp00	ijpeg	Ref.	int95
gcc	200	int00	lucas	Ref.	fp00	mcf	Ref.	int00
gzip	graphic	int00						

Table 2. Microarchitecture parameters for the cycle-accurate simulations

General Out-of-Order Microarchitecture	Adaptive L1 Data Cache
Up to **8 instructions** renamed, dispatched, issued and retired per cycle	**Set-Associativity**: 1-, 2-, 4-way, full associative
Fetch Queue: 16 instructions	**Size**: 1KB, 2KB, .., 128KB
Branch Predictor: perfect. **Issue Queue**: 48 instructions	**Line Size**: 8, … , 64 bytes
Reorder Buffer: 256 instructions	2 read/write **ports**
Operation **latencies** like Pentium 4	Perfect **memory disambiguation**
Load/Store Queues: 64/32 instructions	Store to Load **forwarding**
I-Cache: perfect, 2-cycle load-use latency	**Load-use Latency**: 1, 2, 3, 4 clock
L2-Cache: perfect, 4.6 ns access time (load-use latency depends on operating frequency, f)	cycles (depends on cache organization and operating frequency, f)
L1-L2 interface: 16GB/s	

BBV vectors and classify execution intervals into simpoint (SP) classes. Benchmark traces were analyzed using intervals of 10^5 instructions per 1 billion instructions executed, after a warming-up of 1 billion instructions. We also used CACTI-3.2 [18] to estimate the access time (T_{access}), energy consumed in each memory access, and chip area of each L1 data cache organization, for a CMOS technology with $\lambda = 100$ nm. CACTI was conceived for on-chip caches whose organization is determined in design-time. However, its results are pessimistic when compared with the cache circuits that regularly are presented at the *IEEE ISSCC conferences* [1]. So, we have considered that the values provided by CACTI characterize to cache configurations of our adaptive cache memory. Additionally, we reduced the access times and energy consumptions of the baseline configurations (described below) that were predicted by CACTI by 5% and 10% respectively. To relate the cache load-use latency in cycles (n) with the operating frequency of the cache and processor (f), we use the equation: $f = n/T_{access}$.

Our experiments consider cache configurations with different operating frequencies (f), and then different load-use latencies (n), and requiring a variable amount of chip area (A). To make fair comparisons, we set up a maximum frequency (f_{limit}) and maximum chip area available for the data cache (A_{limit}). We have performed simulations exploring the design space of 48 different initial conditions (f_{limit}, A_{limit}) for the microarchitecture design: $f_{limit}=\{1, 1.5, .., 4.5$ GHz$\}$, $A_{limit}=\{0.5, 1, .., 3$ mm$^2\}$. These initial conditions restrict the number of available cache organizations to those that achieve the following restrictions: $f \leq f_{limit}$ and $A \leq A_{limit}$ (from 11 tunable cache configurations for $f_{limit}= 1$ GHz and $A_{limit}= 0.5$ mm^2, to 322 for $f_{limit}= 4.5$ GHz and $A_{limit}= 3$ mm^2). In this paper, all the results reported consider $A_{limit}=1.5$ mm^2.

This paper reports results for four architectural metrics: execution time, energy consumption of the L1 data cache, power dissipation (obtained by dividing the energy consumption by the execution time), and time-energy product. The negligible energy consumed by sensors and the tables integrated into the coprocessor were not considered since they are physically very small.

5 Reference Configurations

This section describes four reference configurations used in evaluating performance and energy consumption. The first configuration is called *Perfect Dynamic Adaptation* and is an ideal mechanism that for each instruction interval uses future knowledge to select the cache configuration that provides the highest performance.

One of the conventional non-adaptive microarchitectures that we used as baseline design to estimate the potential of perfect dynamic cache adaptation is similar to that used by Balasubramonian et al. to evaluate their proposal of adaptive memory hierarchy [3]. The processor is as described in Table 2. The L1 data cache has 256 KB, 64-byte lines, is direct-mapped, has two read/write ports and a load-use latency of 2 clock cycles. CACTI predicts ($\lambda=100$ nm) that its maximum operating frequency is 1.44 GHz, its energy consumption is 0.72 nJ per access, and its chip area is 8 mm^2. We assume 15-cycle load-use latency for the perfect L2 cache, and an energy cost of 1.83 nJ per L1 miss. In order to make a fair comparison in a first experiment, the clock speed of the perfect dynamic adaptation was set to f_{limit}= 1.44 GHz. However, following the trend of current high performance processors, the chip area of the L1 data cache was limited to A_{limit}= 1.5 mm^2 (81% less chip area than baseline's L1 data cache). So, the largest capacity explored in Perfect Adaptation was 32 KB.

We also evaluated a third reference configuration called *Average-Base*. It is a non-adaptive microarchitecture with the L1 data cache configuration that provides the best performance of the second 1 billion instructions interval of all the benchmarks. Results presented in this section for the average-base configuration also assume f_{limit} = 1.44 GHz and A_{limit} = 1.5 mm^2. In this case, the selected reference configuration has a capacity of 32 KB, is direct-mapped, its line size is 8 bytes, load latency is 1 clock cycle, and the load-use latency of the perfect L2 cache was 7 clock cycles.

We evaluated a fourth type of reference configuration called *Adaptive-Base*. It consists of an adaptive L1 data cache with fixed clock speed (f=1.44GHz), and varying cache latency and energy consumption that resemble the proposal of Balasubramonian et al. [3]. The differences are that we simulate inclusive caches instead of exclusive caches, our design has an 8-way superscalar core, and we consider perfect branch prediction, perfect L2 cache and 15 cycles for L1 miss latency.

As can be seen in Fig. 2, Perfect Dynamic Adaptation achieves on average a performance improvement of 15.9%, energy reduction of 68.4%, power reduction of 59.7%, and time-energy reduction of 74.6% with respect to baseline system. The average-base configuration achieves on average a performance improvement of 11.5%, energy reduction of 65.8%, power reduction of 61.4%, and time-energy reduction of 76.1% with respect to baseline system. And the adaptive-base configuration only achieves a 4.4% performance improvement with respect to baseline system.

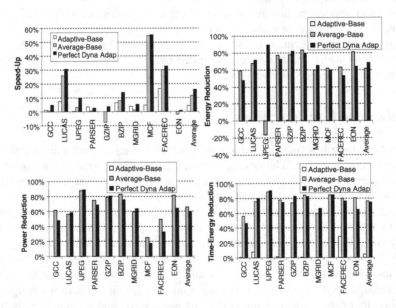

Fig. 2. Improvement of execution time, power dissipation, energy consumption and time-energy of three reference configurations with respect to baseline for f_{limit} =1.44 GHz. (Note: Most of the results for Adaptive-Base system were close to 0%).

The design-time conditions influence on the cache configurations with the highest performance and/or lowest energy consumptions. This justifies not adopting a unique baseline configuration. So, the criteria used in defining the Average-Base system have allowed determining baseline configurations for different operating frequencies and chip areas. Additionally, they are consistently better than our initial baseline configuration (see Fig. 2). For all these reasons, they provide appropriate levels of comparison, and we will use them as our new baseline configurations for the rest of the paper.

6   Results

6.1   Real Dynamic Adaptation

This section evaluates the potential of our Real Adaptive Cache Memory with Intelligent Control. We simulate the learning stage described in Section 3 by combining for each 10^5 instruction interval the caches selected in Perfect Dynamic Adaptation with SimPoint (SP) classes of instruction intervals obtained by using SimPoint. Once the learning stage has initialized the contents of the pattern tables, the recognition and actuation stages of the Real Dynamic Adaptation can be simulated. After the execution of each instruction interval, the BBV vector provided by the processor core and the contents of the pattern table are used to find the corresponding SP class. We used different data sets for training (first billion instruction interval) and recognition (second billion instruction interval). A reconfiguration is fired only when three consecutive instruction intervals are assigned to the same SP class, and this SP class determines a cache configuration that is different from the current one. We assume that each cache reconfiguration additionally introduces an overhead delay of 1 μs and consumes 5 μJ. The number of reconfigurations performed during the simulation experiments (f_{limit}= 4.5 GHz, A_{limit}= 1.5 mm^2, 10,000 intervals) oscillated from 0 (eon) to 846 (gzip) with an average of 278 per interval, i.e. less than 3%. These results indicate that the program patterns/phases exhibit high temporal locality, which reduces the performance overhead due to hardware reconfiguration.

Figure 3 compares the results for Real and Perfect Dynamic Adaptation. On average, the performance improvement achieved by our adaptive processor with intelligent control is 12.1% with respect to the Average-Base configuration. The maximum improvement was 28% for mgrid. In spite of the limitation on the number of configurations and the reconfiguration overhead, the Real Dynamic Adaptation processor can achieve 78% of the improvement provided by Perfect Dynamic Adaptation.

Note that, after finalizing the learning phase, the runtime selection of a cache configuration does not require previous tuning of all available microarchitectures before the selection of one with the highest performance, as proposed in [3]. Therefore, our adaptation control method requires lower overhead for the determination of the stable state of the microarchitecture than previously reported methods.

Energy, power and time-energy reductions for the benchmarks are also shown in Figure 3. These results are presented relative to the same average base machine used for performance estimations. As can be seen, real adaptation achieves a 2.7% mean reduction of energy consumption, a 14.9% mean reduction of time-energy and

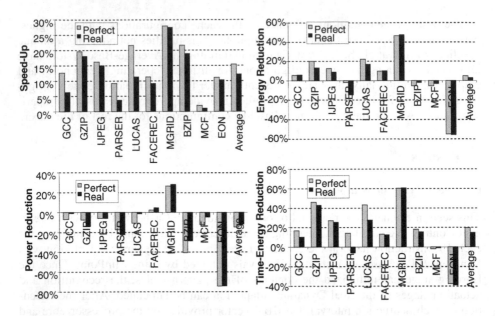

Fig. 3. Performance and power evaluation for the adaptive processor with closed-loop control (Real) and Perfect Cache Adaptation (Perfect) with respect to Average-Base for f_{limit}= 4.5 GHz and A_{limit}= 1.5 mm² (Average-Base={16KB,direct-mapping,8B/line,n=3clks})

a power increase of 12%. This phenomenon is mainly due to the frequent selection of caches with smaller sizes and lower energy cost per memory access than the average-base configurations. The dynamic adaptation enables us to increase frequency when the size is reduced, while not increasing both the number of hit cycles and CPI.

6.2 Static Adaptation

We have simulated the learning stage of static adaptation for each benchmark by selecting the cache configuration that provides the best performance during the first one billion instructions. The set of initial cache configurations depends on the pair "f_{limit}, A_{limit}" (see Table 2). In the actuation phase, cache is reconfigured with the learned configuration which is not changed until the following context switch. Fig. 4 shows the results of the analysis of static adaptation for the second one billion instruction, all limit frequencies and A_{limit}=1.5 mm². The range of performance improvements with respect to the respective average-base reference configurations is (1.1%,9%), 4.8% on average. The range of power reduction is (-11.4%,17.5%), 2.3% on average. The range of energy reduction is (-1.6%,20.7%), 7.4% on average. The range of time-energy reduction is (5.9%,23.8%), 11.9% on average.

For each f_{limit} and A_{limit}, a small set of configurations was selected in the learning stage. Contrasting the configurations selected in static adaptation, we observe that the variability of the hit latency depends on the limit frequency f_{limit} and limit area A_{limit} as in perfect dynamic adaptation; however, the range is narrower. Static

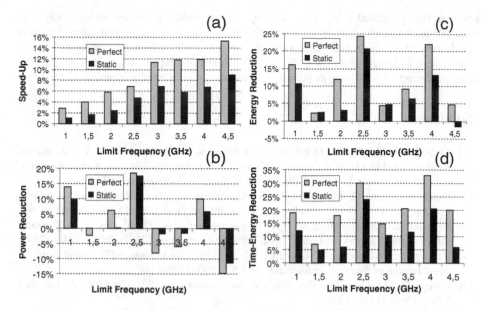

Fig. 4. (a) Execution time reduction, (b) Power reduction, (c) Energy reduction and (d) Time-Energy reduction, of the adaptive processor with Static Adaptation (Static) and Perfect Dynamic Adaptation (Perfect) for eight limit frequencies and $A_{limit}= 1.5$ mm^2

adaptation does not require more than two different hit latencies. We additionally observe that static adaptation is more efficient than the average-base configurations because f_{limit} is exploited in two ways: to perform at a higher operating frequency, although the IPC is lower; and to take advantage of the higher frequency to reduce the execution time. Both justify why static adaptation achieves more performance than the average-base configuration, and dynamic adaptation more than static adaptation.

Therefore, the potential of the static adaptation of the L1 data cache organization is as in real dynamic adaptation, the accommodation of energy per access, area, operating frequency and hit latency to program workload and ILP parallelism. The reduced cost of static adaptation with respect to dynamic adaptation is due to the absence of specialized control hardware because cache adaptation is managed by software. However, the performance of static adaptation is lower than dynamic adaptation because the phase behaviour of programs is not exploited.

7   Conclusions and Future Work

We have proposed an adaptive L1 data cache, which is managed by an intelligent control system. The main contributions are the following. The adaptive data cache: (a) improves performance and energy consumption at the same time, (b) achieves a high percentage of the improvements that a processor with perfect adaptive cache would achieve, with independence of the design-time constrains: operating frequency and chip area, (c) is superior to previous and similar approaches, (d) exploits the SimPoint mechanism, but applied to a different goal: reduce tuning overhead, (e) can be managed by two proposed control methodologies: dynamically on-line or statically off-line. We additionally discovered that high efficiency can be achieved when for each benchmark and design-time constrain, the set of preferred cache configurations is determined. In future papers, we will describe how our adaptive cache can be tuned to other preferred metrics such as power or time-energy instead of performance.

References

[1] Akiyoshi, H., Shimizu, H., Matsumoto, T., Kobayashi, K., Sambonsugi, Y.: A 320ps access, 3 GHz cycle, 144 Kb SRAM macro in 90nm CMOS technology using an all-stage reset control signal generator. In: Proc. IEEE Solid-State Circuits Conference (2003) 460–508

[2] Bahar, R., Albera, G., Manne, S.: Power and performance tradeoffs using various caching strategies. In: Proc. Symp. Low Power Electronics and Design, ACM Press (1998) 64–69

[3] Balasubramonian, R., Albonesi, D.H., Buyuktosunoglu, A., Dwarkadas, S.: A Dynamically Tunable Memory Hierarchy. IEEE Tran. on Computers, 52(10), (2003) 1243–1257

[4] Benitez, D.: Performance of Reconfigurable Architectures for Image-Processing Applications. Journal of Systems Architecture, 49(4-6), (2003) 193–210

[5] Burger, D., Austin, T.M.: The SimpleScalar Toolset, Ver. 2.0. Computer Architecture News, 25(3), (1997) 13–25

[6] Compton, K., Hauck, S.: Reconfigurable Computing: A survey of Systems and Software. ACM Computing Surveys, 34(2), (2002) 171-210

[7] Dhodapkar, A.S., Smith, J.E.: Managing Multi-Configuration Hardware via Dynamic Working Set Analysis. In: Proc. 29th Intl. Symposium on Computer Architecture. IEEE Computer Society (2002) 233–244

[8] Dropsho, S., Buyuktosunoglu, A., Balasubramonian, R., Albonesi, D.H., Dwarkadas, S., Semeraro, G., Magklis, G., Scott, M.L.: Integrating Adaptive On-Chip Storage Structures for Reduced Dynamic Power. In: Proc. Intl. Conference Parallel Architectures and Compilation Techniques. IEEE Computer Society (2002) 141–152

[9] Duda, R.O., Hart, P.E., Stork, D.G.: Pattern Classification (2nd ed.). Wiley (2000)

[10] Hu, Z., Kaxiras, S., Martonosi, M.: Timekeeping in the memory system: predicting and optimizing memory behavior. In:Proc. 29th Intl. Symposium Computer Architecture. IEEE Computer (2002) 209–220

[11] Karnik, T., Borkar, S., De, V.: Sub-90nm technologies: challenges and opportunities for CAD. In: Proc. Conference Computer-Aided Design. IEEE Computer Society (2002) 203–206

[12] Kin, J., Gupta, M., Mangione-Smith, W.: The filter cache: an energy efficient memory structure. In: Proc. 30th Intl. Symposium on Microarchitecture. IEEE Comp. Society (1997)184–193

[13] Magklis, G., Scott, M.L., Semeraro, G., Albonesi, D.H., Dropsho, S.: Profile-based Dynamic Voltage and Frequency Scaling for a Multiple Clock Domain Microprocessor. In: Proc. 30th Intl. Symposium Computer Architecture. IEEE Computer Society (2003) 14-25

[14] Moure, J.C., Rexachs, D.I., Luque, E.: The KScalar Simulator; ACM Journal of Educational Resources in Computing (JERIC), 2(1), (2002) 73–116

[15] Semeraro, G., Magklis, G., Balasubramonian, R., Albonesi, D.H., Dwarkadas, S., Scott, M.L.: Energy-Efficient Processor Design Using Multiple Clock Domains with Dynamic Voltage and Frequency Scaling. In: Proc. 8th Intl. Symposium on High-Performance Computer Architecture. IEEE Computer Society (2002) 29-42

[16] Sherwood, T., Perelman, E., Hamerly, G., Calder, B.: Automatically Characterizing Large Scale Programs. In: Proc. 10th Intl. Conference Architectural Support for Programming Languages and Operating Systems. ACM Press (2002) 45–57
[17] Sherwood, T., Sair, S., Calder, B.: Phase Tracking and Prediction. In: Proc. 30th Intl. Symposium Computer Architecture. ACM Press (2003) 336–349
[18] Shivakumar, P., Jouppi, N.P.: CACTI 3.0: An Integrated Cache Timing, Power, and Area Model. In: Compact WRL Technical Report 2001/2 (2001)

Neural Confidence Estimation
for More Accurate Value Prediction

Michael Black and Manoj Franklin

Department of Electrical and Computer Engineering,
University of Maryland, College Park, MD 20742, USA

Abstract. Data dependencies between instructions have traditionally limited the ability of processors to execute instructions in parallel. Data value predictors are used to overcome these dependencies by guessing the outcomes of instructions. Because mispredictions can result in a significant performance decrease, most data value predictors include a confidence estimator that indicates whether a prediction should be used.

This paper presents a global approach to confidence estimation in which the prediction accuracy of previous instructions is used to estimate the confidence of the current prediction. Perceptrons are used to identify which past instructions affect the accuracy of a prediction and to decide whether the prediction is likely to be correct.

Simulation studies compare this global confidence estimator to the more conventional local confidence estimator. Results show that predictors using this global confidence estimator tend to predict significantly more instructions and incur fewer mispredictions than predictors using existing local confidence estimation approaches.

1 Introduction

Much research has been done in the area of data value prediction as a means of overcoming data dependencies. The goal of data value prediction is to guess the outcome of an instruction before the instruction is actually executed, allowing future instructions that depend on its outcome to be executed sooner. Data value predictors are usually designed to look for patterns among the data produced in repeated iterations of static instructions. Accurate prediction can be attained when the repeated outcomes of a particular instruction follow easily discernable patterns.

Accuracy is a major problem with data value prediction. Even in the most advanced data predictors, as many as 30% to 60% of the predictions are incorrect [14]. If an instruction is mispredicted and the incorrect prediction is used to execute subsequent data dependent instructions, all of those dependent instructions must be executed again. For such instructions, it is typically better not to predict at all than to mispredict. For this reason, most data value predictors include a confidence estimator, which determines whether a prediction for a particular instruction is likely to be correct or not [1]. If the estimator has high confidence in a prediction, the predicted value is used by dependent instructions. Otherwise, the prediction is ignored and dependent instructions wait for the current instruction to be executed.

D.A. Bader et al. (Eds.): HiPC 2005, LNCS 3769, pp. 376–385, 2005.

A typical confidence estimator approach tries to determine the accuracy of a prediction (an instruction's *predictability*) by looking at whether the last several predictions for that instruction were correct. If they were all correct, intuitively the next prediction should also be correct. But if the instruction was recently predicted incorrectly, then the new prediction is not trusted.

This *localized* approach does not consider the effect that other surrounding instructions may have on the instruction being predicted. Correlations often exist between the predictability of different instructions, especially if one instruction is a source of data for another [14]. Hence, one instruction's prediction outcome may be correct only if a certain prior instruction's prediction outcome was correct.

However, in order to make use of other instructions' prediction accuracies, one must determine which surrounding instructions affect the current instruction. We base our estimator on the *perceptron*, a simple form of neural network. A perceptron is assigned to each instruction whose outcome needs to be predicted. Each perceptron identifies which past instructions tend to affect the instruction's prediction confidence. It then uses the prediction accuracies of those past instructions to determine a confidence value for the current prediction.

We have simulated this confidence estimator in conjunction with three different data value prediction methods that are commonly used – Last-Value, Stride, and Context – and compared it with the conventional "up-down counter" estimator [8]. Experiments show that our confidence estimator allows predictors to predict significantly more data values along with a sizeable increase in accuracy on average.

This paper is organized as follows: Section 2 discusses how perceptrons can be used to uncover predictability relationships. Section 3 describes how our perceptron-based confidence estimator works. Section 4 details our simulation setup and presents our simulation results. Section 5 summarizes the paper's conclusions.

2 Theory

2.1 Related Work

Lipasti, Wilkerson, and Shen introduced the earliest confidence estimator used in data value prediction in [8]. It is comprised of a 2-bit saturating up-down counter that chooses between three prediction states: 0 or 1 = "don't predict", 2 = "predict", and 3 = "constant" (highly predictable). If a given instruction makes a correct prediction, the counter is incremented; otherwise, it is decremented. Regardless of whether the instruction predicts correctly or mispredicts, the counter is not allowed to exceed 3 or go under 0. This approach is used in many other proposed data value predictors [1,16].

Calder, Reinman, and Tullsen proposed a confidence estimator similar to ours that uses the history of whether past value predictions were correct to index a table of saturating counters which make the confidence decision [2]. They likewise found a very large increase in speedup using this estimator, but they found their estimator to be unfeasible to implement because of the immense size of their counter table, which grows exponentially with longer histories are considered. We overcome this problem with perceptrons.

Jimenez and Lin introduced neural networks to the similar field of branch prediction in [4,5,6]. They used perceptrons to predict whether a conditional branch instruction in a program will be taken or not based on the results of prior branch instructions. Their results found that their perceptron branch predictor significantly outperforms conventional branch prediction techniques.

2.2 Predictability Relationships

Most prior work in confidence estimation for value prediction only considers local prediction history, in which only the past prediction outcomes for the current instruction are used [1,8,16]. Another source of data for confidence estimation can be the prediction outcomes of other instructions. For example, consider two instructions run sequentially:

```
1) (MUL) A = B * C
2) (ADD) D = A + A
```

If no changes are made to register A between the two instructions, a prediction outcome for the output of instruction 2 is likely to be correct only if the prediction outcome for the output of instruction 1 is correct. There is clearly a predictability relationship between the two instructions.

Of course, predictability relationships are not always so precise. An instruction's predictability could depend on the predictability of several previous instructions, and control flow changes due to branches could change which preceding instructions affect the current instruction's predictability [15].

Table 1 quantifies the commonness of predictability relationships. It shows the number of instructions whose prediction accuracy is always the same as the prediction accuracy of at least one of its previous instructions. This study is meant to indicate whether the global prediction history gives enough information to reliably estimate a prediction's confidence. If an instruction always predicts correctly when a particular instruction before it predicts correctly, and mispredicts when that past instruction mispredicts, then a predictability relationship exists between the two instructions. In this case, a confidence estimator can base its outcome on the accuracy of that past instruction's predictor. A global confidence estimator ought to perform well if this is the case for a significant percentage of instructions.

This simulation study was done using three value predictor approaches: stride, last-value, and context. All instructions with a single destination register that are executed 100 times or more are considered. For each static instruction, we keep track of the prediction correctness outcomes of the 256 instructions preceding it. Once 500 million dynamic instructions are executed, we look at each static instruction's prediction correctness results. If an instruction has a particular past instruction that *always* predicted correctly when it predicts correctly and mispredicted when it mispredicts, it is tallied under 100% dependent. If an instruction has a past instruction that predicted correctly only 99%, 95%, or 90% of the time when it predicts correctly, it is tallied under 99%, 95%, or 90% dependent, respectively.

As may be seen in the 4th row of Table 1, the stride prediction correctness of 79.12% of instructions depend entirely on some previous dynamic instruction's stride prediction correctness 90% of the time. For last-value, this is true for 77.16% of instructions, and for context 94.05% of instructions. These are significant percentages, considering that the study only considers account predictability dependence on a single previous instruction, rather than predictability dependence on the outcomes of combinations of previous instructions. It suggests that confidence estimation based on the outcomes of other instructions should be successful.

There are several possible advantages of using this global approach to confidence estimation over traditional methods that use outcomes from repeated instances of static instructions. One advantage is accuracy. If a particular instruction that was considered predictable suddenly becomes unpredictable, all of its dependent instructions may also become unpredictable. If the dependent instructions determine confidence only from their own past instances, they might choose to predict and would suffer a series of mispredictions. However, if the dependent instructions determine their confidence from the instruction that became unpredictable, they would choose not to predict, and fewer cycles would be lost due to mispredictions.

Another advantage is in warm-up. If a predictor uses a confidence estimator based on outcomes for a particular static instruction, the instruction must be executed several times before confidence can be established. However, if the confidence estimator bases its outcomes on other instructions, an instruction only needs to be executed enough times to choose which other instructions affect it. Consequently, less time elapses before accurate predictions can be made.

Table 1. Prevalence of predictability relationships in each benchmark

	Stride	Context	Last-Value
100%	39.41%	15.82%	40.57%
99%	56.49%	87.48%	57.36%
95%	70.76%	91.22%	68.99%
90%	79.12%	94.05%	77.16%

2.3 Perceptrons

To uncover predictability dependencies, we use a perceptron based confidence estimator. A perceptron is a simple neural network consisting of an adder, a threshold function, and a set of weights implemented by saturating signed integer counters. The perceptron uses these components to guess an output based on a series of inputs. Given a set of input bits, it computes the dot product of the inputs and the weights, and compares the result to a threshold value, typically 0 (an extra weight is hardwired to an input of 1 to provide a bias). If the result is greater than 0, the perceptron returns "True"; otherwise it returns "False."

The perceptron determines the values of its weights by learning. When a correct value is found, the perceptron is "trained." That is, an error value is computed by the difference between the training value and the perceptron output. This error value is multiplied by each input bit and is added to the corresponding weight. In this way, each weight is adjusted so that the desired output is realized from the particular input combination.

When applied to confidence estimation, each weight value determines the relationship between a particular past instruction and the current instruction. If a weight value is positive and large, the past instruction's predictability tends to have a direct effect on the current instruction's predictability. That is to say, the current instruction's data predictor tends to predict correctly only when the past instruction's data predictor predicted correctly. If the weight value is negative and large, the past instruction's predictability effect is inverse: the current instruction's data predictor tends to predict correctly only when the past instruction mispredicted. If the weight value's magnitude is small, the past instruction has been found to have little effect on the current instruction.

3   Confidence Estimator Organization

Figure 1 shows a block diagram of our proposed confidence estimator. It consists of two parts: a table of predictors and perceptrons, and the Global Prediction History (GPH).   The GPH contains information on whether the predictions for previous dynamic instructions were correct. It acts as a shift register, in which the prediction outcome (correct / incorrect) is shifted in when each instruction completes.

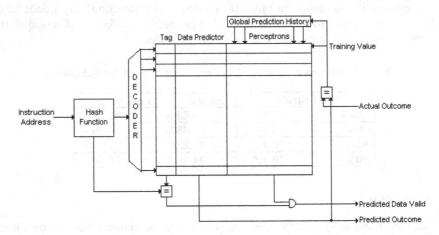

Fig. 1. Block diagram of the prediction architecture

Our prediction system works as follows: The instruction address is used to select a table entry. This table entry consists of a value predictor, which predicts a value, and a perceptron. The perceptron takes the GPH as its input and uses its weights to determine whether its output is "predict" or "don't predict". If the output is "predict", the value predictor outcome is used as a prediction; otherwise the prediction is not used. Regardless of the perceptron outcome, when the actual result of the instruction is known, it is compared against the prediction. If they match, a 1 (predicted correctly) is shifted into the GPH at the instruction's completion stage. Otherwise, a 0 (predicted incorrectly) is shifted into the GPH.   The difference between the actual result and the prediction is then used to adjust the perceptron weights and train the perceptron.

3.1 Hardware Requirements

The principal hardware cost of our perceptron estimator is storage space for the GPH and perceptron weights. The space required is proportional to the number of dynamic instructions considered and the size of the table. Each additional dynamic instruction in the history requires one extra entry in the GPH, costing 1 bit of storage, and one extra weight for each table entry. We find that typically a minimum of 6 bits but never more than 9 bits are needed for each weight.

Because an input bit is interpreted only as -1 or 1, the product between the input bit and its corresponding weight can be implemented simply by using the input bit to choose whether to invert the sign of the weight. These products are summed together, but by using a threshold of 0, only the sign of the total sum is used as the output. A threshold of 0 can be attained without any cost to the performance of the perceptron by including a bias weight: an extra input weight with a fixed input value of 1. This additional bias weight mathematically replaces the threshold [11].

4 Experimental Results

4.1 Experimentation Methodology

We performed our measurements on the PISA instruction set architecture using the SimpleScalar 2.0a tool set. The data value predictor considers every instruction that has a single destination register. Predictions are made after each instruction executes and the actual instruction output is immediately used to train the predictor. For benchmarks, we use eight programs from the SPEC2000 integer suite. Each program is run for 500 million instructions.

Our study is performed using three types of predictors: Last-Value, Stride, and Context. The Last-Value predictor simply returns the value that an instruction produced the last time it was executed. The Stride predictor computes the difference between the last two results of an instruction, and adds it to the most recent result to predict a value. The Context predictor uses the most recent four data values produced by an instruction to index a pattern table of up-down counters [16]. The counters choose one of the four data values to be the prediction. Each of the three predictors includes a table indexed by the instruction address. We use 16k table entries and a direct-map organization for the table.

The data value predictors can produce one of three outcomes: $P_{CORRECT}$ (prediction made, outcome correct), $P_{INCORRECT}$ (prediction made, outcome incorrect), and N (no prediction made). We base our metrics on the metrics used in [1]. Coverage is computed as the number of predictions ($P_{CORRECT} + P_{INCORRECT}$) divided by the total opportunities for predictions (N+ $P_{CORRECT} + P_{INCORRECT}$). Accuracy is computed as the number of correct predictions ($P_{CORRECT}$) divided by the total number of predictions ($P_{CORRECT} + P_{INCORRECT}$). Neither of these metrics alone is enough to compare predictors. Accuracy is important, since a low accuracy means that many cycles are wasted due to mispredictions. Coverage is also important in comparing predictors as an indicator of how much of the predictability potential is realized by the predictor. A good predictor must have both a high accuracy and a high coverage.

We use the saturating up-down counter as our baseline confidence estimator because of its simplicity and because it is frequently used in other proposed data value predictors. As mentioned earlier, an up-down counter estimates confidence based on the prediction history of a given static instruction. It is incremented if the predictor's outcome is correct, up to a maximum count, and decremented if the outcome is incorrect, down to 0. In our implementation, the up-down counter estimator only chooses to predict if the count is at the maximum.

An important consideration we faced in choosing an up-down counter is the size of its counting range. The range size tends to create a tradeoff between coverage and accuracy. An up-down counter estimator with a large range tends to have poor coverage because it requires many correct predictions before it becomes confident. However, because of the many correct predictions, its accuracy tends to be excellent. On the other hand, an up-down counter estimator with a small range has good coverage because it requires only a few correct predictions before it chooses to predict. By the same token, however, its accuracy is terrible.

For our baseline tests we used three up-down counters. Counter-2 counts between 0 and 1, where 1 is considered "predict" and 0 "don't predict." Counter-4 counts from 0 to 3, predicting only on 3. Counter-7 counts from 0 to 6 and predictions are only made on 6. In our benchmark comparison we use Counter-4 for comparing results because it balances coverage and accuracy. We tested our perceptron confidence estimator with three GPH sizes. Neural-4 considers the 4 previous prediction outcomes, Neural-16 the 16 previous outcomes, and Neural-256 the 256 previous outcomes. We use Neural-256 in our benchmark comparison because it best shows the potential.

4.2 Predictability of All Single Destination Register Instructions

Table 2 shows the results of Counter-4 and Neural-256 tested on Last-Value, Stride, and Context, respectively, across the eight benchmarks, where predictions are made for every instruction that has a single destination register. In Stride and Last-Value the perceptron estimator shows both improved coverage and accuracy for every benchmark. Stride's coverage improves from as low as 4.6% for bzip2 to as high as 10.6% for perlbmk, while its accuracy improves from as low as 0.05% for bzip2 to as high as 6.1% for perlbmk. Last-Value's coverage improves from as low as 4.0% for mcf to as high as 14.4% for perlbmk, while its accuracy improves from as low as 1.6% for bzip2 to as high as 10.0% for perlbmk. On average, the coverage increases by 7.8% for Stride and 9.1% for Last-Value, and the accuracy increases by 2.8% for Stride and 5.5% for Last-Value.

Context shows less improvement. The coverage is generally higher for the perceptron estimator, improving as much as 12.6% for perlbmk; however, it is 2.3% lower for twolf. On average the perceptron estimator's coverage is 5.7% higher than that of the up-down counter estimator. The accuracy, however, is actually lower for the perceptron estimator, decreasing by 2.1% on average. While the perceptron estimator shows a 2.9% accuracy improvement for perlbmk, the accuracy falls for four of the benchmarks, by as much as 10.0% for gzip. Because Context uses more complex predictability functions than Stride or Last-Value, it may be expected that the Context results can be improved when more complex neural network are used.

Table 2. Simulation results with perceptron and up-down counter

		Stride		Last-Val		Context	
		Coverage	Accuracy	Coverage	Accuracy	Coverage	Accuracy
bzip2	counter	47.30%	97.16%	36.70%	94.88%	29.68%	96.79%
	neural	51.93%	97.21%	42.66%	96.47%	33.04%	93.29%
gcc	counter	28.36%	92.20%	31.59%	87.22%	25.38%	88.43%
	neural	38.60%	95.42%	44.45%	96.20%	35.16%	89.11%
gzip	counter	28.72%	92.47%	20.23%	87.94%	10.29%	90.96%
	neural	34.61%	93.69%	27.12%	91.51%	14.78%	80.96%
mcf	counter	54.83%	97.48%	37.55%	96.12%	34.28%	96.08%
	neural	60.16%	99.65%	41.64%	99.63%	38.80%	98.47%
perlbmk	counter	20.19%	92.31%	23.00%	89.06%	19.53%	90.42%
	neural	30.82%	98.35%	37.38%	99.08%	32.06%	93.31%
twolf	counter	33.51%	91.05%	30.81%	87.77%	29.68%	96.79%
	neural	42.29%	94.63%	39.92%	93.83%	27.43%	90.00%
vortex	counter	42.96%	97.26%	43.35%	94.45%	38.49%	97.00%
	neural	52.10%	99.50%	55.17%	99.61%	47.28%	96.27%
vpr	counter	37.32%	92.66%	36.12%	91.17%	23.90%	93.06%
	neural	45.37%	96.58%	43.99%	96.16%	28.62%	91.57%

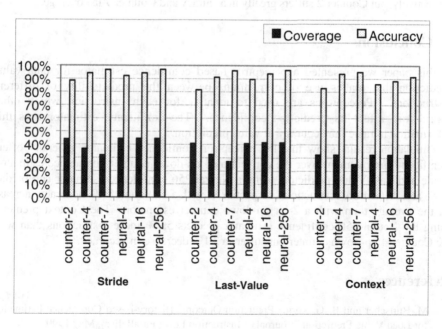

Fig. 2. Simulation results for varying estimator sizes

4.3 Sensitivity to History Size

In Figure 2 we show how the perceptron estimator is affected by its GPH size and the up-down counters by their counting range. The results are based on the average of the benchmarks.

The perceptron estimator's coverage does not change more than 0.7% when the number of previous prediction outcomes is varied. This is a result of the perceptron's bias weight balancing the output. The bias weight in a perceptron acts a threshold control for the perceptrons. If the perceptron outputs "don't predict" too often, it lowers the threshold to increase the number of "predict" outputs. Likewise, it raises the threshold if the perceptron frequently outputs "predict". This has the effect of keeping the coverage constant at the expense of accuracy.

As can be expected, the accuracy shows a marked increase when more previous instructions are considered. It grows by as much as 7.0% when 256 instead of 4 previous prediction outcomes are considered. With more previous instructions, the perceptron is likely to find a more accurate combination of previous prediction outcomes to determine the current prediction's accuracy.

We compare these to the three up-down counters, which exhibit a coverage-accuracy tradeoff. For Last-Value and Stride, the perceptron estimators have approximately the same accuracy as the largest counter and the same coverage as the smallest counter. However, their coverage is 12.7% to 14.4% better than the large counter's coverage, and their accuracy is 3.7% to 15.0% better than the small counter's accuracy. Counter-4 tends to slightly outperform the perceptron estimators for Context, as described above. Counter 2 and Counter 7 outperform the perceptron estimators in coverage and accuracy, respectively, but Counter 2 suffers greatly in accuracy and Counter 7 in coverage.

5 Conclusions

In this paper we presented a perceptron-based confidence estimator for data value prediction that makes use of correlations between the predictability of different instructions. Perceptrons are used to identify for each instruction which other instructions affect its prediction confidence. The confidence estimator uses this information to raise the accuracy of value prediction.

Simulation results show that the perceptron confidence estimator generally offers significant improvement over the conventional up-down counter confidence estimator. Stride and Last-Value predictors using a Neural-256 confidence estimator can predict 7.8% and 9.1% more instructions, respectively, with a 2.8% and 5.5% accuracy increase on the average than with a 2-bit up-down counter estimator. The Context predictor using the Neural-256 confidence estimator predicts 5.7% more instructions than with the Counter-4 up-down counter but suffers a 2.1% decrease in accuracy.

References

1. M. Burtscher and B. G. Zorn. "Prediction Outcome History-based Confidence Estimation for Load Value Prediction." Journal of Instruction Level Parallelism, May 1999.
2. B. Calder, P. Feller, A. Eustace.   "Value Profiling."   Proc 30th Intl Symp on Microarchitecture, Research Triangle Park, NC, Dec. 1997.

3. D. Jimenez and C. Lin. "Composite Confidence Estimators for Enhanced Speculation Control." Technical Report TR2002-14, Dept. of Computer Sciences, University of Texas at Austin, 2002.

4. D. Jimenez and C. Lin. "Dynamic branch prediction with perceptrons." Technical Report TR2000-08, Dept. of Computer Sciences, University of Texas at Austin, 2000.

5. D. Jimenez and C. Lin. "Neural methods for dynamic branch prediction." ACM Transactions on Computer Systems, 20(4), Nov. 2002.

6. D. Jimenez and C. Lin. "Perceptron learning for predicting the behavior of conditional branches." Proc Intl Joint Conference on Neural Networks, July 2001.

7. M. H. Lipasti and J. P. Shen. "Exceeding the Dataflow Limit via Value Prediction." Proc of the 29th Annual Intl Symposium on Microarchitecture, Dec., 1996.

8. M. H. Lipasti, C. B. Wilkerson and J. P. Shen. "Value Locality and Load Value Prediction." 7th Intl Conf on Architectural Support for Programming Languages and Operating Systems, Oct.1996.

9. T. Nakra, R. Gupta, and M. L. Soffa. "Global context-based value prediction." Proc 5th Intl Symp on High Performance Computer Architecture, Jan. 1999.

10. G. Reinman and B. Calder. "Predictive techniques for aggressive load speculation." Proc 31st International Symposium on Microarchitecture, Dec. 1998.

11. S. Russell and P. Norvig. "Artificial Intelligence: A Modern Approach." Prentice-Hall, Inc., Upper Saddle River, NJ, 1995, pp. 563-593.

12. Y. Sazeides, J. E. Smith. "Implementations of Context Based Value Predictors."Technical Report ECE-97-8, University of Wisconsin-Madison, Dec. 1997.

13. Y. Sazeides and J. E. Smith. "The Predictability of Data Values." Proc 30th Intl Symposium on Microarchitecture, Dec. 1997.

14. R. Thomas and M. Franklin, "Characterization of data value unpredictability to improve predictability." Proc Intl Conf on High Performance Computing, 2001.

15. R. Thomas and M. Franklin. "Using dataflow based context for accurate value prediction," Proc Intl Conf on Parallel Architectures and Compilation Techniques, 2001.

16. K. Wang and M. Franklin. "Highly accurate data value prediction using hybrid predictors." 30th Intl Symp on Microarchitecture, 1997.

17. H. Zhou, J. Flanagan, and T. Conte. "Detecting Global Stride Locality in Value Streams." Proc Intl Symposium on Computer Architecture, 2003.

The Potential of On-Chip Multiprocessing for QCD Machines*

Gianfranco Bilardi[1], Andrea Pietracaprina[1], Geppino Pucci[1],
Fabio Schifano[2], and Raffaele Tripiccione[2]

[1] Dipartimento di Ingegneria dell'Informazione, Università di Padova,
Via Gradenigo 6/B, 35131, Padova, Italy
{bilardi, capri, geppo}@dei.unipd.it
[2] Dipartimento di Fisica, Università di Ferrara, and INFN,
Via del Paradiso 12, 44100 Ferrara, Italy
{schifano, lele}@fe.infn.it

Abstract. We explore the opportunities offered by current and forthcoming
VLSI technologies to on-chip multiprocessing for Quantum Chromo Dynamics
(QCD), a computational grand challenge for which over half a dozen specialized
machines have been developed over the last two decades. Based on a careful study
of the information exchange requirements of QCD both across the network and
within the memory system, we derive the optimal partition of die area between
storage and functional units. We show that a scalable chip organization holds the
promise to deliver from hundreds to thousands flop per cycle as VLSI feature size
scales down from 90 nm to 20 nm, over the next dozen years.

1 Introduction

The high-end supercomputers of the near future will consist of many thousands of processing chips, each featuring billions of transistors [32]. Yet, in today's general purpose
microprocessors, only a small percentage of the available transistors goes into the functional units that actually process data, while the vast majority goes into memory and
into the circuitry that orchestrates the execution of instructions. The main reason can
be traced to a major bottleneck in computing systems stemming from the limited bandwidth available across the chip boundary. Thus, although a chip could host hundreds
of functional units, it would generally be difficult to feed them with data so to attain a
significant fraction of peak performance.

The impact of the chip I/O bottleneck varies with the application, critically depending upon its computation/communication ratio. For specialized domains, chip organizations with many functional units are receiving growing attention. Announced in 2005
are products such as the IBM/Sony/Toshiba Cell capable of 64 flop/cycle at about 4
GHz [24] and the ClearSpeed CSX600, capable of 192 flop/cycle, at 250MHz [16].

---

* This research was supported in part by MIUR of Italy under project "ALGO-NEXT: ALGO-
rithms for the NEXT generation Internet and the Web", and by the University of Padova under
Grant CPDA033838.

D.A. Bader et al. (Eds.): HiPC 2005, LNCS 3769, pp. 386–397, 2005.

A number of research projects have also been active for a few years, such as Blue-Gene/Cyclops (64 flop/cycle at 500 MHz) [5] and TRIPS (targeting 5 Tflops in a 35 nm VLSI implementation) [36].

In this paper, we explore whether aggressive on-chip multiprocessing is a viable avenue for supercomputing in the domain of Quantum Chromo Dynamics (QCD). QCD is the quantum field theory of the strong interaction explaining the structure of the hadrons, a family of particles including the proton and the neutron, as composites of "elementary" entities known as quarks, which interact by exchanging "colored" gluons. While widely believed to be the correct theory of the strong interaction, QCD is almost intractable computationally and it is regarded as a computational grand challenge [18]. Several specialized QCD computers have been designed, built, and successfully operated over the last two decades (see [28] for an extensive coverage). A sample of such machines includes the sequel of APE computers, developed in Europe [4,8,33,34,35], the CP-PACS computer, developed in Japan mainly for QCD simulations [23], and several similar efforts in the United States starting by the pioneering GF11 supercomputer [9] and continuing with the QCDSP [27] and QCDOC [15] projects.

Asymptotically, QCD computations are constrained by chip I/O bandwidth. In fact, the four dimensional nature of QCD lattices leads to communication requirements that scale with the (3/4)-th power of the computation, whereas in a (planar) chip I/O bandwidth only scales with the (1/2)-th power of the area. Suitable adaptations of arguments developed in [29,19] show that, if chip area were to grow arbitrarily large, only a vanishing fraction of it would be occupied by effectively utilized functional units[1]. In this paper, we investigate whether and when, with the evolving VLSI technology, increasing chip size (in square feature sizes) will yield diminishing returns[2]. We reach the following encouraging conclusions regarding QCD machines:

- Current (90 nm) VLSI technology is far from the asymptotic chip I/O bottleneck, which will not be severe even for a feature size of 20 nm, to become feasible around 2017 according to the International Technology Roadmap for Semiconductors (ITRS, 2004 update).
- It is currently feasible to realize QCD machines with a few thousands nodes, each containing a processing chip with 8-16 MByte of on-chip (embedded DRAM) memory and achieving 100-150 flop/cycle.
- A uniformly scalable machine organization can harness the technological potential becoming available as feature size shrinks down to 20 nm, presumably over the next 12 years. At 20 nm, a few Kflop/cycle will be achievable on one chip.

Even at modest frequencies, say 20% of state of art, the above figures appear attractive[3].

The reminder of this paper is organized as follows. In Section 2, we introduce the notion of information exchange of a computation, modeled by a function $I(n, m)$. This is the number of bits that a subsystem equipped with m bits of memory exchanges with

---

[1] Three dimensional integration would postpone, but not escape the same conclusion.

[2] Beyond this point, further increases of computational power may still be achievable, by increasing clock frequency and by assembling systems with larger numbers of chips.

[3] ITRS on-chip local clock rates are 5.2 GHz for 2005, 39.7 GHz for 2016, and 53.2 GHz for 2018.

the rest of the system when handling a subcomputaton of size n. We also consider the number $w(n)$ of operations for the same subcomputation. In terms of these quantities, we derive the optimal split of the available area between memory and functional units. In Section 3, we develop a careful study of $I(n, m)$ for the computation of the Dirac operator, which takes nearly 90% of running time for a typical QCD simulation. The main contribution of this section is the identification of schedules that reduce $I(n, m)$. The analysis precisely determines constant factors to enable finite-horizon considerations, in addition to asymptotic ones. In Section 4, we evaluate the potential of QCD machines realizable in current and future technologies, reaching the conclusions outlined above.

2   Methodology for Memory-Communication-Processing Tradeoffs

We focus on the space of multiprocessor machines realized as networks of nodes, interconnected according to a suitable topology. Each node consists of a processing chip, equipped with some on-chip memory and directly connected both to a local off-chip memory and to the processing chips of the neighbouring nodes, in the given topology.

The key result of this section is a characterization of the optimal partition of the node area between memory and functional units, in terms of the computational requirements of the target application, the number of nodes of the multiprocessor, and the area of the processing chip at a node. We begin by introducing a number of quantities related to machine, to technology, and target computation.

Machine Parameters. We characterize the machine through the following quantities: the number P of processing nodes; the number F of operations per cycle that can be executed at a node; the number m of on-chip memory bits; and the bandwidth b, in bits per cycle, between the processing chip and the rest of the system (both incoming and outgoing). This bandwidth is partitioned as $b = b_{lc} + b_{nb}$ where b_{lc} is the bandwidth with the local off-chip memory and b_{nb} is the aggregate bandwidth between the chip and its neighbors.

Technology Parameters. We consider the following parameters of the underlying VLSI technology (whose values and their scaling with VLSI feature size will be discussed in Section 4): the area A of the processing chip; the area A_F of one functional unit, pipelinable at one operation per cycle (this parameter clearly depends also on the adopted logic design); and the area A_m required for the storage of one memory bit on the processing chip.

Computation Requirements. The target computation is characterized by the following quantities, some of which assume a specific mapping onto the machine under consideration (an accurate analysis of these quantities for QCD computations will be developed in Section 3): the input size N (in QCD, the number of lattice points); the input size per node $n = N/P$; the word length L_w, in bits; the total work or number of operations $W(N)$; the work per node $w(n) = nW(N)/N$ (assuming, as reasonable, even distribution of work among the processing nodes); and the *information exchange* in bits, $I(n, m)$, between the processing chip and the rest of the system. The latter quantity can be decomposed as $I(n, m) = I_{lc}(n, m) + I_{nb}(n, m)$, where the two terms account for exchanges with the off-chip memory and near neighbors, respectively.

Under ideal conditions (i.e., all resources fully used at all times) *execution time* is

$$T^{id}(n, m) = \max\{w(n)/F, I(n, m)/b\} \ . \tag{1}$$

In order to minimize T^{id} we would like to increase both F and m, the latter because more storage on the processing chip will reduce the exchange with local off-chip memory, hence $I(n, m)$. Given the total area budget A, the question then is how to best partition it between memory and functional units. In summary, we have the following optimization problem:

$$\min \max\{w(n)/F, I(n, m)/b\} \tag{2}$$

$$s.t. \quad A_{\mathrm{F}}F + A_{\mathrm{m}}m \leq \eta A \ , \tag{3}$$

where η is the fraction of the chip area actually used for functional units and memory (as opposed to control logic, instruction caches, intra-chip communication, etc.) Estimates of η require further assumptions about chip organization; in Section 4, we provide such estimates for QCD. It is straightforward to argue that, at the optimal point for (2, 3), the two terms in the objective function (2) must be equal, and constraint (3) must be satisfied with equality, so that:

$$w(n)/F = I(n, m)/b \ , \tag{4}$$

$$A_{\mathrm{F}}F + A_{\mathrm{m}}m = \eta A \ . \tag{5}$$

Combining the two above equations yields

$$A_{\mathrm{F}}w(n)b/I(n, m) + A_{\mathrm{m}}m = \eta A \ . \tag{6}$$

If the functions $w(n)$ and $I(n, m)$ are known for the computation, given n and A, this equation determines a value $m^*(n, A)$ for m. Eq. (4) then provides the number of functional units as $F^*(n, A) = (\eta A - A_{\mathrm{m}}m^*(n, A))/A_{\mathrm{F}}$. Computation time becomes $T^{id*}(n, A) = w(n)/F^*(n, A)$. The values of n and A, which are arbitrary in the preceding analysis, could be chosen to optimize a suitable cost-performance function.

Remarks on Information Exchange. Information exchange plays a pivotal role in the methodology outlined above. The analysis of communication has often been developed by quantifying the information exchanged across a suitable partition of the system into two complementary subsystems, such as two subsets of nodes in a network (see, *e.g.*, [38,31,11]). Communication also arises within hierarchical memory systems; in this context, of particular interest is information exchanged across levels of the hierarchy, as reflected *e.g.*, in the notions of I/O complexity [21] and access complexity [10].

Our information exchange $I(n, m)$ measures simultaneously effects due to distribution of data across different nodes and across different memory levels, since the chip boundary acts as a separator for both the network and the memory system. Correspondingly, the techniques for minimizing the information exchange combine those traditionally used when optimizing the mapping of a given computation onto a fixed-topology network [25] with those used when optimizing the performance on a memory hierarchy [2,7,3]. Interactions between network and memory-hierarchy communication arise naturally in machines where speed of light is an active constraint [12]. A close relationship between network proximity and temporal locality has been exposed in [20].

3   Memory-Communication-Processing Tradeoffs for LQCD

In this section, we apply the methodology developed in Section 2 to lattice QCD (LQCD), a discretized version of QCD, currently the most amenable for non-perturbative computations. After a short introduction of the needed concepts and notations, we analyze the information-exchange of the Dirac operator, the core module within LQCD.

3.1   The LQCD Computation

The Lattice. In LQCD, space-time is discretized and represented as a a four-dimensional toroidal graph, whose arcs are the domain of the gauge field and whose vertices are the domain of the particle field. More formally, for $X \geq 0$, let $Z(X)$ denote the cyclic group $\{0, 1, \ldots, X - 1\}$ w.r.t. addition modulo X. For $\mathbf{N} = (N_1, N_2, N_3, N_4)$, let $Z(\mathbf{N}) = Z(N_1) \otimes Z(N_2) \otimes Z(N_3) \otimes Z(N_4)$ where $\otimes$ denotes the direct product operation between groups. Also, let $\mathcal{V}_d = \{\hat{\mu}_1, \hat{\mu}_2, \hat{\mu}_3, \hat{\mu}_4\}$, where $\hat{\mu}_i$ denotes the 4-tuple with the i-th component equal to 1 and the other components equal to 0. We are interested in modeling a four-dimensional toroidal space-time lattice defined as the directed graph $L(\mathbf{N}) = (Z(\mathbf{N}), E(\mathbf{N}))$, where $E(\mathbf{N}) = \{(x, x \pm \hat{\mu}) : x \in Z(\mathbf{N}), \hat{\mu} \in \mathcal{V}_4\}$. For $N = N_1 N_2 N_3 N_4$, it is $|Z(\mathbf{N})| = N$ and $|E(\mathbf{N})| = 8N$.

The Gauge Field. Recall that SU_3 denotes the *special unitary* (Lie) group of the 3×3 unitary matrices with determinant equal to one. (A matrix U is *unitary* when $U^\dagger = U^{-1}$, i.e., when its transpose conjugate is also its inverse.) A *gauge field* is a map $U : E(\mathbf{N}) \rightarrow SU_3$ which associates an SU_3 matrix with each arc of the lattice, with the property that $U(x, x + \hat{\mu}) = (U(x + \hat{\mu}, x))^{-1} = (U(x + \hat{\mu}, x))^\dagger$, for any $x \in Z(\mathbf{N})$ and $\pm\hat{\mu} \in \mathcal{V}_4$.

The Fermion Field. A *(pseudo) fermion field* is a map $\Psi : Z(\mathbf{N}) \rightarrow C^{3 \times 4}$ which associates a 3×4 complex matrix with each vertex of the toroidal lattice. In LQCD, the row index ranges over color eigenstates, while the column index ranges over spin eigenstates. It is useful to regard each 3×4 complex matrix as a 12-component complex vector.

The Dirac Operator. The Dirac operator provides the dynamical description of a given quark and depends both upon the gauge field and on a parameter K related to the mass of the quark being modeled by the operator. Mathematically, the Dirac operator is a (sparse) linear function which maps a fermion field into another fermion field.

Definition 1. *Given a real scalar quantity K and a gauge field configuration U, the Dirac operator $D_{K,U} : C^{12N} \rightarrow C^{12N}$ is the linear operator on the space of fermion fields defined by the relation*

$$\Phi(x) = [D_{K,U}\Psi](x) = K\Psi(x) + \sum_{\pm\hat{\mu}\in\mathcal{V}_4} U(x, x + \hat{\mu})\Psi(x + \hat{\mu})\Gamma_{\hat{\mu}} , \qquad (7)$$

for each $x \in Z(\mathbf{N})$, where the $\Gamma_{\hat{\mu}}$'s are the 4×4 Dirac matrices, a key feature of which is that in each row/column all entries are 0, except for one which belongs to $\{1, i, -1, -i\}$.

An LQCD computation is based on a Montecarlo-Metropolis approach that generates a random walk in the space of gauge field configurations. The key kernel of the computation is the inversion of the Dirac operator which is typically obtained through its iterated application according to Eq. (7) starting from an initial fermion field. In current simulations, this operation accounts for nearly 90% of the overall simulation time.

3.2   Computation Requirements of the Dirac Operator

In this subsection, we determine the requirements of the computation of the Dirac operator $\Phi = D_{K,U}\Psi$, defined by Eq. (7). We assume that a complex number is represented by two words, hence each 3×4 matrix $\Phi(x)$ or $\Psi(x)$ occupies 24 words, while each 3×3 matrix $U(x, x + \hat{\mu})$ occupies 18 words. To evaluate the total work $W(N)$, measured in flop (floating point operations), required by the operator, we consider the computation of $\Phi(x)$ for a lattice point x, taking into account that the multiplications by matrices $\Gamma_{\hat{\mu}}$ do not contribute any flop, since they essentially amount to permutations and sign changes. Each of the 8 $U(x, x+\hat{\mu})\Psi(x+\hat{\mu})$ products accounts for $3 \times 3 \times 4 = 36$ complex multiplications (cmul) and 24 complex additions (cadd). Further 8×12 cadd are required by the summation, which yields a total equivalent to $8 \times 36 = 288$ complex multiply and add (cmadd). Since each cmadd requires 8 flop and 24 more flop are needed to compute $K\Psi$, we conclude that the computation of $\Phi(x)$ requires $\chi = 2328$ flop, hence

$$W(N) = \chi N = 2328N . \tag{8}$$

For concreteness, we assume that the Dirac operator is implemented on a P-node 3D-torus, which is a typical topology for today's supercomputers (a similar analysis could be carried out for other topologies). We partition the lattice evenly among P processing nodes so that each node is in charge of a sublattice of size $n = k \times k \times k \times N_4$, with $k = (N/(N_4 P))^{1/3}$. For convenience, we order dimensions so that $N_4 = \min\{N_1, N_2, N_3, N_4\}$. We assume that at the beginning of the computation a node stores the Ψ and U fields restricted to lattice points and incident arcs of its assigned sublattice. Since the matrices $U(x, y)$ and $U(y, x)$ associated with the two arcs between points x and y are one the transpose conjugate of the other, we can store only one instance if both x and y are assigned to the node. At the end of the computation the node will also store the Φ fields restricted to its sublattice points. It is easy to see that all of the data residing at the node add up to $\sigma k^3 N_4 L_w$ bits, with $\sigma = 120$.

The following technical result, whose proof, omitted here for brevity, will be provided in the full version of this extended abstract, establishes the existence of efficient schedules tailored to various values of the on-chip memory size m:

Theorem 1. *With the above notation, there is a schedule for the computation of the Dirac operator which executes in time* $T(n, m) = \max\{w(n)/F, (I_{lc}(n, m) + I_{nb}(n, m))/b\}$ *where* $I_{nb}(n, m) = (\iota n/k)L_w$, *and* $I_{lc}(n, m)$ *exhibits the following dependence upon* m:

1. *for* $m = \sigma n L_w = 120 n L_w$ *(large memory),* $I_{lc}(n, m) = 0$;
2. *for* $m = (96k^3 + 432k^2 + o(k^2))L_w$ *(medium memory),* $I_{lc}(n, m) = \sigma n L_w$;
3. *for* $m = (96k^2 s + 288ks + O(1))L_w$ *with* $1 \leq s \leq k$ *(small memory),* $I_{lc}(n, m) = ((\sigma + 66/s)n + 132k^2)L_w$;

Clearly, $I_{lc}(n, m)$, hence its contribution to running time, increases as m decreases. In particular, once m goes substantially below the threshold of $\sigma n L_w$, one can see that, if the machine is balanced in the sense that the two terms in the expression for $T(n, m)$ are equal, then less than $w(n)L_w/I_{lc}(n, m) = \chi/\sigma < 20$ flop per word exchanged with the local off-chip memory can be sustained. Considerably larger values are instead achievable when $m \geq \sigma n L_w$, since in this case there is no need to transfer the fields back and forth between the processing chip and the off-chip memory, at each iterative application of the Dirac operator.

4   Performance Potential of Future QCD Machines

Based on the resource tradeoff equations derived in the previous sections, we now evaluate the performance potential of future QCD machines. We introduce a number of assumptions on technology parameters, formulated in terms of VLSI feature size λ, to allow for scaling considerations.

Technology Assumptions. We let $A = (\ell\lambda)^2$ denote the die area, assuming, for simplicity, a square of sidelength ℓ, in units of λ. Referring to an area range of 81-324 mm^2, we see that $\ell \in [1 \div 2] \cdot 10^5$ is representative of today's 90 nm scenario, while $\ell \in [4.5 \div 9] \cdot 10^5$ would represent the 20 nm scenario forecast for 2017. We express the area taken by one bit of storage as $A_m = \alpha_m \lambda^2$, estimating $\alpha_m = 50$, for embedded DRAM. We express the area of a floating-point functional unit as $A_F = \alpha_F \lambda^2$. Clearly, $\alpha_F = \alpha_F(L_w)$, that is, the area does depend upon the wordlength of the operands. For the case study below, where $L_w = 64$, we generously estimate $\alpha_F(64) = 10^8$, so that a number of registers and some auxiliary logic is also accounted for (see, *e.g.,* [17,26]). Asymptotically, $\alpha_F(L_w) = \theta(L_w^2)$, due to the $A = \theta(L_w^2/T^2)$ complexity of a VLSI multiplier (see [1,14] for lower bounds and [13] for upper bounds). However, since in our context $\alpha_F(L_w)$ is actually an average area between adder, multiplier, and register file, and since relatively small values of L_w are under consideration, we can expect the actual behavior being between linear and quadratic. Finally, we assume a chip bandwidth proportional to the perimeter, *i.e.,* $b = \beta 4\ell$. We estimate $\beta = 1/3000$, which is conservative at $\lambda = 90$ nm (for $\ell = 2 \cdot 10^5$, $b = 266$, compared to the ITRS figure of 1800 I/O signals per chip) and somewhat conservative at $\lambda = 20$ nm (for $\ell = 9 \cdot 10^5$, $b = 1200$, compared to the ITRS figure of 3000).

Input Assumptions. Consider a lattice of size $N = kP_1 \times kP_2 \times kP_3 \times N_4$ to be mapped onto a $P_1 \times P_2 \times P_3$ three-dimensional torus of $P = P_1 P_2 P_3$ nodes, each processing a $k \times k \times k \times N_4$ sublattice of $n = k^3 N_4$ points. Since (a) $N = 64^4$ is a rather large size processed on today's teraflop machines, (b) the overall computation requirements of a QCD simulation (including $O(N^{3/4})$ Dirac computations) grow approximately as $O(N^{7/4})$, and (c) a thousandfold improvement can be expected during the horizon we are investigating, it is reasonable to assume for lattice size a range $64^4 - 128 \cdot 256^3$, throughout which one can always choose to completely map within a node an entire dimension of size (approximately) $N_4 = 128$.

Machines in the Large Memory Regimen. In this regimen, where all fields are stored on chip, we have: $n = N_4 k^3$, $w(n) = \chi n$ with $\chi = 2328$ (from Eq. (8)), $m = L_w \sigma n$, with $\sigma = 120$ and $L_w = 64$, and $I(n, m) = L_w \iota n/k$, with $\iota = 288$, from Theorem 1. Then, Eq. (4) gives the number of functional units as $F = w(n)b/I(n, m) = \chi n 4\beta\ell/(L_w \iota(n/k)) = (97/576000)\ell k$. Assuming for now $\eta = 1$, Eq. (5) can be rewritten in the λ-invariant form $\alpha_F F + \alpha_m m = \ell^2$, whence, after plugging in the technology parameters and the above relation for F, we have:

$$\alpha_F F + \alpha_m m = \gamma\ell k + \delta N_4 k^3 = \ell^2 , \tag{9}$$

where $\gamma = 97 \cdot 10^5/576$ and $\delta = 384 \cdot 10^3$. Assuming $N_4 = 128$, Table 1 shows the numeric solutions of the above equations for a sample of values of chip sidelength ℓ. We can make a few observations:

- Current technology ($\ell = 10^5, 2 \cdot 10^5$) would already enable hundreds of flop/cycle.
- Projected 2017 technology ($\ell = 4 \cdot 10^5, 8 \cdot 10^5$) holds the promise of thousands of flop/cycle.
- The fraction $\alpha_F F/\ell^2$ of the die area utilized for functional units is substantial in the range being considered for ℓ, although it does decrease with ℓ. Indeed, one could derive from Eq. (9) that this fraction vanishes asymptotically as $(\gamma/(\delta N_4)^{2/3})\ell^{-1/3}$.

Machines in the Medium Memory Regimen. Consider now the case of medium memory. From Theorem 1 we have $I(n, m) = (\iota(n/k) + \sigma n)L_w$ and $m \simeq (96k^3 + 432k^2)L_w$. Hence, $F = (97/(576 + 240k))10^{-3}\ell k$ and $\ell^2 - (97/(576 + 240k))10^5\ell k + 32 \cdot 10^2(96k^3 + 432k^2)$.

As we can see from Table 1, for $\ell \leq 25000$, the medium-memory regimen achieves a better floating point performance than the large-memory regimen. The reason is that, when m is below a certain threshold, the node sublattice approaches a 1-dimensional array of N_4 points, with an unfavourable computation/communication ratio. As m and n increase with ℓ, this situation is quickly reversed, since $\alpha_F F/\ell^2$ vanishes as $(97 \cdot 10^4)/(24\ell)$. We also observe from the table that the growth rate of n as a function of ℓ is much smaller for large memory than for medium memory, so the latter regimen affords implementations with smaller numbers of nodes $P = N/n$.

Table 1. QCD chip parameters ($L_w = 64$ bits): sidelength ℓ (units of λ); b: I/O bandwidth (bits/cycle); $n = k^3 N_4$: number of sublattice points processed ($N_4 = 128$); m: on-chip memory (bits); F: flop/cycle; $\alpha_F F/\ell^2$: fraction of area devoted to FP units

$\ell/10^5$	b	Large Memory Regimen				Medium Memory Regimen			
		n	$m/10^6$	F	$\alpha_F F/\ell^2$	n	$m/10^6$	F	$\alpha_F F/\ell^2$
0.25	33	$2.57 \cdot 10^2$	1.9	5	0.84	$5.26 \cdot 10^3$	0.5	6	0.95
0.50	67	$1.54 \cdot 10^3$	11	19	0.76	$2.47 \cdot 10^5$	15	17	0.68
1.00	133	$8.48 \cdot 10^3$	62	67	0.67	$2.24 \cdot 10^6$	120	37	0.37
2.00	267	$4.36 \cdot 10^4$	319	233	0.58	$1.23 \cdot 10^7$	616	77	0.19
4.00	533	$2.12 \cdot 10^5$	1549	788	0.49	$5.68 \cdot 10^7$	2752	157	0.10
8.00	1067	$9.82 \cdot 10^5$	7194	2630	0.41	$2.45 \cdot 10^8$	11601	317	0.05
16.0	2133	$4.41 \cdot 10^6$	32298	8677	0.34	$1.02 \cdot 10^9$	47609	639	0.03

Wordlength. While Table 1 shows the parameters for $L_w = 64$, values for different wordlengths can be easily obtained from our equations, given the appropriate value for $\alpha_F(L_w)$. Simple arguments on the structure of the equations show that, for a fixed ℓ, when the wordlength is halved, then F is slightly more than doubled if α_F grows linearly with L_w, while F is slightly less than quadrupled if α_F grows quadratically with L_w. Thus, operating with the smallest wordlength that guarantees the numerical properties of the QCD algorithms, probably somewhere between 32 and 64 bits, may lead to nonnegligible savings with respect to the case for 64 bits.

4.1 Chip Organization

In the preceding analysis, by setting $\eta = 1$ in Eq. (5), we have ignored the area requirements due to control structures and intra-chip data and instruction transfers. To show how these requirements can be kept small (within 10%, corresponding to $\eta \geq 0.9$), we sketch a chip organization for machines tailored to the large-memory regimen and to suitable chip size, say $\ell \geq 10^5$.

Letting $p^2 = F/8$, we consider a chip organized as a $p \times p$ two-dimensional mesh of small processing elements (SPEs). Each SPE is endowed with m/p^2 bits of local off-chip memory and with 8 floating point units (which naturally exploit the 8 flop of the complex multiply-and-add operation, very abundant in QCD codes).

Controller. We envisage a SIMD organization with a single centralized structure in charge of flow control broadcasting control words to all SPEs. Without entering into details, we estimate its complexity to be similar to that of one functional unit and its layout to fit in an $\ell_c \times \ell_c$ region, with $\ell_c = 10^4$. A region of the same shape and size can be also budgeted for a program memory (2 Mbit of embedded DRAM). Thus, controller and program memory can be accommodated in a centrally placed (say) vertical layout strip of width $\ell_c = 10^4$.

Control Distribution. To distribute control words, and to support various reduction operations, we make provision for a binary tree rooted at the controller, with p^2 leaves at the SPEs, and with edge bandwidth $b_t = 128$ (in bit/cycle). Adopting an H-layout [37], the tree requires $(p-1)b_t < \sqrt{F/8}b_t$ bandwidth, both vertically and horizontally.

Data Transfers. For the parameter ranges we are considering, the node's sublattice can be mapped so that neighboring lattice points are assigned either to the same SPE or to two near-neighbor SPEs. Inter-node data can then be routed by p row busses and p column busses, with overall bandwidth $b/2$, both in the vertical and in the horizontal direction.

In order to translate the bandwidth requirements into area occupancy, we need to estimate the width $1/\beta_0$ (in units of λ) of a connection carrying one bit/cycle. Based on an exercise carried out on currently available 130 nm technology, we set $1/\beta_0 = 5$.

In summary, we obtain an $\ell_h \times \ell_v$ layout, where $\ell_v = \ell + (1/\beta_0)(\sqrt{F/8}b_t + b/2)$ and $\ell_h = \ell_v + \ell_c$. Considering that $F < \ell^2/\alpha_F$, $\alpha_F = 10^8$, $b = 4\beta\ell$, $1/\beta_0 = 5$, $\beta = 1/3000$, $b_t = 128$, $\ell_c = 10^4$, and $l \geq 10^5$, we can derive that $\ell_v = (1+\epsilon)\ell$, with $\epsilon = (b_t/\sqrt{8\alpha_F} + 2\beta)/\beta_0 \leq 0.026$, whence $\eta = \ell^2/\ell_h\ell_v = [(1+\epsilon^2) + (1+\epsilon)\ell_c/\ell)]^{-1} \geq 0.9$.

Table 2. Relevant parameters of current processors (see Section 4.2 for the definition of the ratio ξ). In the table, ξ_{LM} and ξ_{MM} refer, respectively, to the large and to the medium memory regimens.

	apeNEXT	BG/L	Cell	CSX600	ITANIUM2	QCDOC
frequency	200Mhz	700Mhz	3.2Ghz	250Mhz	1.6Ghz	500Mhz
λ	180nm	130nm	90nm	130nm	90nm	130nm
L_w	64	32	32/64	64	64	64
F	8	4	64/16	192	4	2
m	32kb	32Mb	20Mb	4.5Mb	72Mb	32Mb
b_{lc}	128	62.85	64	102.4	x	41.6
b_{nb}	48	24	192	256	$32 - x$	21.8
ξ_{LM}	n.a.	6.07	2.27/3.03	0.17	4.04	4.13
ξ_{MM}	0.76	9.53	0.61/1.21	0.16	2.20	6.30

4.2 Current Processors

It might be instructive to place some recent processors in the (b_{lc}, b_{nb}, m, F) space. Ideally, given b_{lc}, b_{nb} and m, the number of flop/cycle that could be sustained for the Dirac computation is $F^* = w(n)/\max(I_{lc}(n, m)/b_{lc}, I_{nb}(n, m)/b_{nb})$. Thus, the ratio $\xi = f^*/F$ can be viewed as a measure of how well the machine is balanced (for QCD). If $\xi = 1$, then the balance is perfect. If $\xi < 1$, the bandwidth and memory resources are sufficient to sustain only a fraction ξ of the available performance. If $\xi > 1$, the bandwidth and memory resources would be sufficient to sustain a multiple ξ of the available performance. Note however that ξ is only a measure of balance of one of the architectural parameters (b_{lc}, b_{nb}, m, F), once the remaining ones have been fixed, based on some criteria other than our methodology: ξ alone does not capture how suitable a given architecture is for QCD computing. Six processors [35,6,24,16,22,15] are listed in Table 2, with relevant features and the corresponding ξ metric, for both the large and the medium memory regimens. Many observations could be made, keeping in mind that it would not be appropriate to consider ξ as a figure of merit for the corresponding design, which is likely to have been optimized for different technologies (e.g., SRAM vs DRAM) and for applications different from QCD or, in the case of apeNEXT and QCDOC, for different memory regimens as well as for different codes (which do not necessarily use a schedule that minimizes the information exchange). We leave most of these observations to the interested reader, except for noting that we start to see the appearance of compute-intensive architectures and that at least in one case (the Cell processor, especially in single precision) the design parameters are remarkably close to the design space that we have identified in the previous subsections.

5 Conclusions

In this paper, we have developed an approach to analyze tradeoffs between bandwidth, memory and processing for a given computation, providing quantitative guidelines to evaluate or design a machine for such a computation. By this approach, we have shown that, broadly speaking, chips where a substantial fraction of the silicon is used for functional units could deliver from hundreds to thousands of flops per cycle on QCD computations, over the next decade.

The models of this paper only provide for a first-order analysis of the resource tradeoffs. More accurate models and analyses would obviously be needed to provide sound guidance in an actual design. In particular, wire length and corresponding delays become increasingly critical as feature size shrinks. In the chip organization sketched in Section 4.1, while near-neighbour connections should not pose a serious problem, the long wires of the tree for control distribution require further attention (pipelining the instruction stream might be sufficient for QCD codes, which have few control dependences; multiple controllers on the same chip can be another avenue). A careful analysis of these issues is a prerequisite to any credible estimate of achievable clock frequencies. Power consumption is another increasingly relevant issue, completely neglected in this preliminary study. Here, we simply observe that as static power accounts for an increasing fraction of energy consumption, as soon as area is converted into transistor a price is paid. Therefore, architectures as the one we have outlined, which maximize the percentage of area that does useful processing, become increasingly attractive from the power perspective.

At a broader level, the proposed approach and its refinements could be used to study other domains that can potentially take advantage of on-chip supercomputing.

Acknowledgments. Helpful and constructive discussions with Giacomo Marchiori, Jose Moreira, and Pratap Pattnaik are gratefully acknowledged.

References

1. H. Abelson and P. Andreae. Information transfer and area-time tradeoffs for VLSI multiplication. *Communications of the ACM*, 23(1):20–23, 1980.
2. A. Aggarwal, A.K. Chandra, and M. Snir. Hierarchical memory with block transfer. In *Proc. of the 28th IEEE Symp. on Foundations of Computer Science*, pages 204–216, 1987.
3. A. Aggarwal and J.S. Vitter. The input/output complexity of sorting and related problems. *Communications of the ACM*, 31(9):1116–1127, 1988.
4. M. Albanese et al. The APE Computer: an Array Processor Optimized for Lattice gauge Theory Simulations. *Comput. Phys. Commun.* 45:345, 1987.
5. F. Allen et al. Blue Gene: a vision for protein science using a petaflop supercomputer. IBM Systems Journal, 40(2)310-327, 2001.
6. G. Almasi et al. Design and implementation of message passing services for the Blue Gene/L supercomputer. *IBM J. Res. Develop.*, 49(2/3), 2005.
7. B. Alpern, L. Carter, E. Feig, and T. Selker. The uniform memory hierarchy model of computation. *Algorithmica*, 12(2/3):72–109, 1994.
8. C. Battista et al. The APE-100 Computer:(I) the Architecture. *Int. J. High Speed Computing* 5:637, 1993.
9. J. Beetem, M. Denneau, and D. Weingarten. The GF11 supercomputer. In *Proc.of 12th Int. Symposium on Computer Architecture*, pages 108-115, 1985.
10. G. Bilardi, A. Pietracaprina, and P. D'Alberto. On the space and access complexity of computation dags. In *Proc. of 26th Workshop on Graph-Theoretic Concepts in Computer Science*, LNCS 1928, pages 47–58, 2000.

11. G. Bilardi and F.P. Preparata. Area-time lower-bound techniques with application to sorting. *Algorithmica*, 1(1):65–91, 1986.
12. G. Bilardi and F.P. Preparata. Processor-time tradeoffs under bounded-speed message propagation: Part II, lower bounds. *Theory of Computing Systems*, 32:531–559, 1999.
13. G. Bilardi and M. Sarrafzadeh. Optimal VLSI circuits for the discrete Fourier transform. *Advances in Computing Research* 4:87–101, JAI Press, Greenwich Connecticut, 1987.
14. R.P. Brent and H.T. Kung. The chip complexity of binary arithmetic. *J. Ass. Comp. Mach.* 28(3):521–534, 1981.
15. D.Chen et al. QCDOC: A 10-teraflops scale computer for lattice QCD. In *Proc. of 18th Intl. Symposium on Lattice Field Theory (Lattice 2000)*, Bangalore, India, August 2000.
16. ClearSpeed Site: www.clearspeed.com
17. J. Clouser et al A 600-MHz superscalar floating-point processor. *IEEE Journal on Solid-State Circuits*, 34(7):1026-1029, July 1999.
18. D.E. Culler, J.P. Singh, and A. Gupta. *Parallel Computer Architecture: A Hardware/Software Approach.* Morgan Kaufmann, San Mateo, CA, 1999.
19. R. Cypher, Theoretical aspects of VLSI PIN limitations, SIAM J. Comput., Vol. 2, No.2, pp. 356-378, April 1993.
20. C. Fantozzi, A. Pietracaprina, and G. Pucci. Seamless integration of parallelism and memory hierarchy. In *Proc. of 29th Int. Colloquium on Automata, Languages and Programming*, LNCS 2380, pages 856–867, July 2002.
21. J.W. Hong and H.T. Kung. I/O complexity: The red-blue pebble game. In *Proc. of the 13th ACM Symp. on Theory of Computing*, pages 326–333, 1981.
22. Intel Itanium2 Site: www.intel.com/products/processor/itanium2/
23. Y. Iwasaki. Computers for lattice field theories. *Nuclear Physics (Proc. Suppl.)* 34:78, 1994.
24. J. Kahle, M. Suzuoki, Y. Masubuchi, Cell Microprocessor Briefing, San Francisco, February 7, 2005.
25. F.T. Leighton. *Introduction to Parallel Algorithms and Architectures: Arrays • Trees • Hypercubes.* Morgan Kaufmann, San Mateo, CA, 1992.
26. S. Mueller et al The vector floating-point unit in a synergistic processor element of a Cell processor. In *Proc. 17th IEEE Int. Symp. on Computer Arithmetic*, June 2005. To Appear.
27. R.D. Mawhinney, The 1 Teraflops QCDSP Computer, Parallel Computing 25(10–11):1281-1296, 1999.
28. *Parallel Computing*, 25(10–11), 1999. Special Issue on High Performance Computing in LQCD.
29. M. Snir, I/O Limitations on multi-chip VLSI systems, Proc. 19th Allerton Conference on Communications, Control, and Computing, Monticello, IL, 1981, pp. 224-233.
30. S.M. Sze, editor. *VLSI Technology.* McGraw-Hill, New York NY, 2nd edition, 1988.
31. C.D. Thompson. *A complexity theory for VLSI.* PhD thesis, Dept. of Computer Science, Carnegie-Mellon University, Aug. 1980. Tech. Rep. CMU-CS-80-140.
32. The Top 500 Supercomputer Sites: http://www.top500.org.
33. R. Tripiccione. APEmille. *Parallel Computing*, 25(10–11):1297–1309, 1999.
34. R. Tripiccione. LGT simulations on APEmachines. *Computer Physics Communications* 139:55, 2001.
35. R. Tripiccione. Strategies for dedicated computing for lattice gauge theories. *Computer Physics Communications* 169:442-448, 2005.
36. TRIPS: Tera-op Reliable Intelligently adaptive Processing System. www.cs.utexas.edu/users/cart/trips/.
37. J.D. Ullman. *Computational Aspects of VLSI.* Computer Science Press, Rockville MD, 1984.
38. A.C.C. Yao. Some complexity questions related to distributive computing. In *Proc. of the 11th ACM Symp. on Theory of Comp.*, pages 209–213, 1979.

Low-Power 32bit×32bit Multiplier Design with Pipelined Block-Wise Shutdown*

Yong-Ju Jang, Yoan Shin, Min-Cheol Hong,
Jae-Kyung Wee, and Seongsoo Lee

School of Electronics Engineering,
Soongsil University, Seoul 156-743, Korea
sslee@ssu.ac.kr

Abstract. This paper proposes a novel low-power 32bit×32bit multiplier with pipelined block-wise shutdown scheme. When it idles, it turns off supply voltage to reduce both dynamic and static power. It shutdowns and wakes up sequentially along with pipeline stage to avoid power line noise. In idle mode, the proposed multiplier consumes 0.013mW and 0.006mW in 0.13μm and 0.09μm technologies, respectively, and it reduces power consumption to 0.07%~0.08% of active mode. As fabrication technology becomes small, power efficiency degrades in the conventional clock gating scheme, but the proposed multiplier does not. The low-power design methodology in this paper can be easily adopted in most functional blocks with pipeline architecture.

1 Introduction

Recently, power consideration has become an important issue in VLSI design, especially for portable and battery-powered systems. In deep submicron fabrication technologies, leakage current increases exponentially according to device scaling [1]. Fig. 1 shows the trends of supply voltage, threshold voltage, dynamic power, and static power along with the progress of fabrication technology. As shown in Fig. 1, the leakage power may dominate total power in near future [2].

Shutting down some functional blocks in the chip is a promising approach to reduce both dynamic power and static power [3, 4, 5], where the supply voltage is cut off when the functional block idles. In the hardware implementation, multi-threshold voltage CMOS (MTCMOS) [2, 6, 7, 8, 9, 10] has been proposed and extensively investigated. However, severe power line noise due to the large current surge often causes system malfunction when the functional block is quite large [11], since large number of logic gates switch simultaneously during wakeup process. Recently, block-wise shutdown [11] was proposed to mitigate the power line noise. Many large-size functional blocks such as multiplier and arithmetic logic unit (ALU) employ pipeline scheme to increase the performance. When the supply voltage of the shutdown block is recovered sequentially along with the

---

* This work was supported by the Korean Research Foundation Grant. (KRF-2004-042-D00152).

D.A. Bader et al. (Eds.): HiPC 2005, LNCS 3769, pp. 398–406, 2005.
© Springer-Verlag Berlin Heidelberg 2005

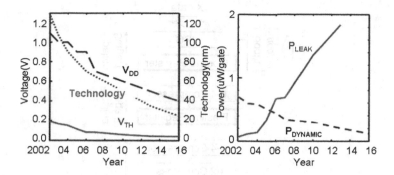

Fig. 1. Trends of supply voltage, threshold voltage, dynamic power, and static power

pipeline stages, the number of simultaneous switching gates is greatly reduced. Thus the power line noise significantly decreases [11].

In this paper, we propose a low-power 32bit×32bit multiplier with pipelined block-wise shutdown. It was designed, fabricated, and verified by chip implementation, and the power reduction was evaluated by simulation.

2   Low-Power Multiplier Design

Fig. 2 shows the proposed low-power multiplier architecture. It exploits modified Booth algorithm to reduce hardware complexity. It has three pipeline stages for high performance. To reduce the power line noise, partitioning of pipeline stage is determined by minimizing the number of maximum simultaneous switching gates. Note that the power line noise increases as the number of simultaneous switching gates increases. First stage consists of partial product blocks which generate partial product of multiplicand. Second stage consists of two 8-to-4 compressors. Third stage consists of one 8-to-4 compressor, two 4-to-2 compressors, and a final adder. The proposed multiplier was described in Verilog HDL, and it was fabricated in standard cell library. The gate counts of first, second, and third pipeline stages are about 9800, 7800, and 4800 gates, respectively.

Fig. 3 shows the supply voltage of the proposed low-power multiplier with shutdown circuitry. In the shutdown and wake-up processes, supply voltage of the whole circuit is cut off and recovered sequentially along with pipeline stage. At the same time, only one pipeline stage is shutdown and wakes up, and the number of simultaneous switching gates reduces. Thus, the power line noise is mitigated significantly.

In the conventional block-wise shutdown schemes, pipeline registers of each stage are connected to a high V_T line, and these registers are never shutdown. Thus the intermediate computation values are preserved and the block can wake up immediately. However, these pipeline registers still consume static power since their supply voltages are not cut off. In the proposed multiplier, the pipeline registers occupy about 40% of total gate count. Thus, we included all the pipeline

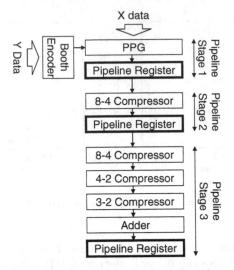

Fig. 2. Architecture of the proposed low-power 32bit×32bit multiplier

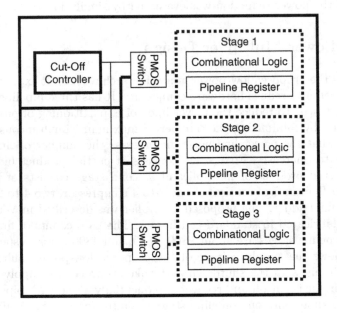

Fig. 3. Supply voltage of the proposed low-power 32bit×32bit multiplier

registers in the shutdown blocks to reduce static power, and all logic gates and pipeline registers are cut off during shutdown. Each pipeline stage has its own block power ring, and the supply voltages of all logic gates and pipeline registers are connected to it. Power is delivered in the flow as: external power supply → cut-off switch → block power ring → logic gates and pipeline registers. Cut-off

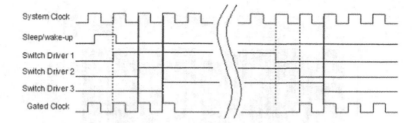

Fig. 4. Pipelined turning on and off the current switches

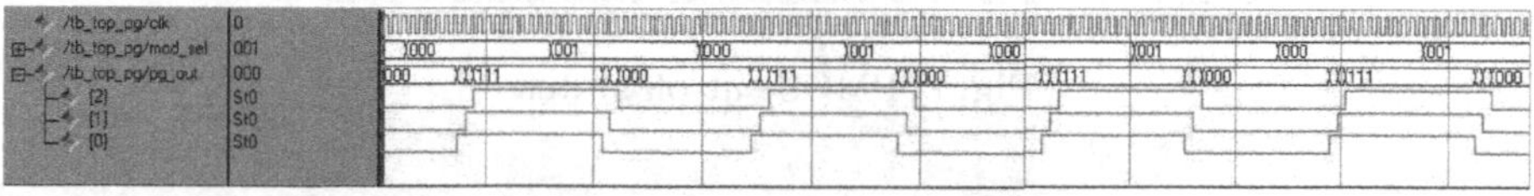

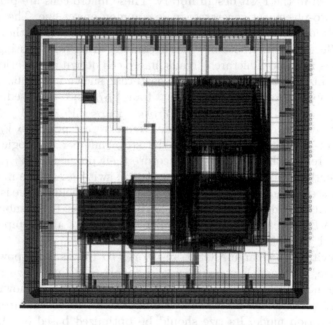

Fig. 5. Waveforms of control signals in the cut-off controller

Fig. 6. P&R result using the standard cell library

controller turns on and off cut-off switches in a pipelined manner as shown in Fig. 4 when the multiplier enters into or exits from sleep mode. When sleep signal is activated from outside of the multiplier, the cut-off controller generates pipelined control signals as illustrated in Fig. 5.

The detailed design process is as follows. First, each pipeline stage is described in Verilog HDL as separate blocks. These stages are synthesized into gate level.

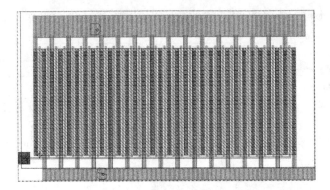

Fig. 7. PMOS cut-off switch

Each synthesized pipeline stage is placed separately, and the result is registered as a macro cell in the P&R design library. These macro cells are placed manually, since auto-placement generates single common power line. The macro cells should be placed so as to minimize interconnection length. Also they should be placed to effectively distribute heat dissipation. During auto-routing, top-level file preserving macro cell hierarchy reads macro cell design information from the P&R design library. The resulting P&R pattern is shown in Fig. 6.

In the proposed low-power multiplier, four V_{DD} lines are used, i.e. global V_{DD} line (V_{DDG}) and stage V_{DD} lines (V_{DDS1}, V_{DDS2}, V_{DDS3}). V_{DDG} supplied power to the cut-off controller. It also delivers external power to V_{DDSn} ($n = 1,2,3$) via cut-off switches, where n is the stage number. All logic gates and pipeline registers in the pipeline stage acquire their power from V_{DDSn}. V_{DDSn} is independently cut off and recovered by cut-off switches. In idle mode, cut-off switches are sequentially open, and the corresponding pipeline stage is shutdown. No cut-off switches are connected to GND lines to reduce the number of cut-off switches. GND line of the cut-off controller and those of all the pipeline stages are connected with each other.

Cut-off switches are large MOS transistors. To fully supply power to each pipeline stage, it should have very wide channels to deliver large current. To increase channel width, MOS transistors have many fingers as shown in Fig. 7. PMOS transistor is used as cut-off switch, since there is no voltage drop to deliver V_{DD} in saturation mode. Its size should be optimized based on the trade-off between area overhead and wake-up speed.

3   Simulation Results

In deep submicron fabrication technologies below 0.13μm, static power due to leakage current dominates total power consumption. Therefore, it is essential to calculate both dynamic and static power to evaluate power efficiency of the proposed multiplier. In this paper, 0.13μm and 0.09μm standard cell library

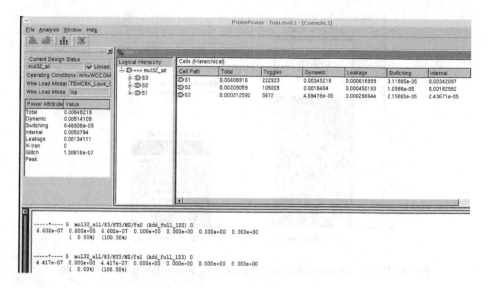

Fig. 8. Screen shot of power simulation using Synopsys Prime Power

technologies are used in the power simulation. These technologies provide both dynamic and static power simulation.

We designed the proposed multiplier with Verilog HDL description. We synthesized two versions, $0.13\mu m$ and $0.09\mu m$, from the identical Verilog HDL description. Synopsys Design Compiler supported by IC Design Education Center is used in the synthesis. It also extracts gate level netlist and delay information which are dumped into Verilog-format netlist (NETLIST) file and standard delay format (SDF) file, respectively. Mentor Graphics Modelsim generates two random 32bit input vectors as multiplicands, and performs behavioral simulation of the proposed multiplier. Then it dumps internal signal transition during behavioral simulation into value change dump (VCD) file. Power consumption of the proposed multiplier was simulated by Synopsys Prime Power. It reads power models in the standard cell library, and then calculates both dynamic and static power consumption using NETLIST, SDF, and VCD files. Fig. 8 shows a screen shot of power simulation using Synopsys Prime Power.

In the simulation, we assumed that the proposed multiplier runs at 333MHz. Supply voltage is assumed to be 1.08V in $0.13\mu m$ technology and 0.9V in $0.09\mu m$ technology, respectively. We carried out two low-power methods. The conventional clock gating [12] inserts some clock control logics in the clock driver, and disables clock transition when the block idles. Dynamic power consumption is zero in clock gating, but static power consumption cannot be reduced. On the contrary, the proposed block-wise shutdown [12] cuts off the supply voltage when the block idles, and both dynamic power and static power become zero.

Fig. 9 shows the power consumption of the proposed multiplier in active mode. In $0.13\mu m$ technology, combinational logics consume 5.156mW (29.46%), and pipeline registers consume 11.772mW (67.00%). Static power consumption

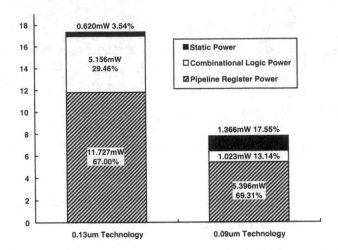

Fig. 9. Power consumption of the proposed multiplier in active mode

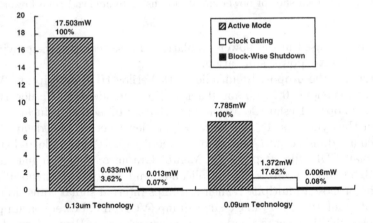

Fig. 10. Power reduction efficiency of the proposed multiplier

is 0.620mW (3.54%). Consequently, static power is negligible in 0.13μm technology. In 0.09μm technology, the situation is quite different. Combinational logics consume 1.023mW (13.14%), and pipeline registers consume 5.396mW (69.31%). Static power consumption is 1.366mW (17.55%). As a result, static power is quite comparable to dynamic power in 0.09μm technology. Note that static power increases significantly as fabrication technology becomes small.

Fig. 10 shows power reduction efficiency of the proposed multiplier. In idle mode, two low-power methods, i.e. conventional clock gating [12] and proposed block-wise shutdown [11], are compared. In 0.13μm technology, the proposed multiplier consumes 17.503mW (100%), 0.633mW (3.62%), 0.013mW (0.07%) in active mode, idle mode with clock gating, and idle mode with block-wise shutdown, respectively. In $0.09\mu m$ technology, it consumes 7.785mW (100%), 1.372mW (17.62%), 0.006mW (0.08%) in active mode, idle mode with clock gating, and idle mode with block-wise shutdown, respectively.

In active mode, the proposed multiplier consumes both dynamic and static power. In idle mode with clock gating, dynamic power is greatly reduced, while static power is never reduced. Dynamic power is not zero, since input vector signal transition induces gate capacitance charging and discharging. In idle mode with block-wise shutdown, it consumes only dynamic power of cut-off controller, static power of cut-off controller, and dynamic power for gate capacitor charging and discharging. As shown in Fig. 10, block-wise shutdown is more effective than clock gating in power reduction. Furthermore, clock gating inserts control gates into clock distribution network, which requires very careful and complicated analog circuit design. Note that power efficiency of clock gating degrades as fabrication technology becomes small, while that of block-wise shutdown does not depend on fabrication technology.

4   Conclusion

In this paper, we proposed a low-power 32bit×32 bit multiplier with pipelined block-wise shutdown scheme. Using MTCMOS technology, it shutdowns supply voltage to reduce both dynamic and static power in idle mode. It has three pipeline stages for high speed operation. Supply voltage is cut off along with pipeline stage to reduce power line noise during wake up process. Partitioning of pipeline stage is determined by equalizing the number of gates in each stage in order to reduce the number of maximum simultaneous switching gates, because power line noise is proportional to it. Supply voltage of each stage is cut off by large PMOS switch. Channel width is determined by considering the power consumption of each pipeline stage. Cut-off controller generates pipelined control signals to cut off these PMOS switches in sequential manner. The proposed multiplier is described in Verilog HDL and fabricated in standard cell library. Total gate count is about 22400 gates including cut-off controller.

Static power consumption increases significantly in deep submicron fabrication technology beyond $0.13\mu m$. We carried out power simulation in $0.13\mu m$ and $0.09\mu m$ technologies. In $0.13\mu m$ technology, the proposed multiplier consumes 17.503mW in active mode and 0.013mW in idle mode. In $0.09\mu m$ technology, it consumes 7.785mW in active mode and 0.006mW in idle mode. Thus, it reduces power consumption to 0.07%~0.08% in idle mode, while the conventional clock gating scheme reduces it to 3.62%~17.62%. Power efficiency of the conventional clock gating scheme severely degrades as fabrication technology becomes small, while that of the proposed multiplier is hardly affected. This comes from the fact that the proposed multiplier significantly reduces static power due to leakage while the conventional clock gating cannot. The low-power design methodology in the proposed multiplier can be easily adopted in most functional blocks with pipeline architecture such as arithmetic logic unit, filter and systolic array.

References

1. Sakurai, T.: Perspectives on Power-Aware Electronics, Technical Digest of International Solid-State Circuit Conference (2003) 26-29.
2. Kao, J., Narendra, S., Chandrakasan, A.: MTCMOS Hierarchical Sizing Based on Mutual Exclusive Discharge Patterns, Proceedings of Design Automation Conference (1998) 15-19.
3. Srivastava, M., Chandrakasan, A., Brodersen, R.: Predictive System Shutdown and Other Architectural Techniques for Energy Efficient Programmable Computation, IEEE Transactions on VLSI Systems **4** (1996) 42-55.
4. Benini, L., Bogliolo, A., De Micheli, G.: A Survey of Design Techniques for System-Level Dynamic Power Management, IEEE Transactions on VLSI Systems **8** (2000) 299-316.
5. Moore, G.: No Exponential is Forever: But Forever Can Be Delayed, Technical Digest of International Solid-State Circuit Conference (2003) 20-23.
6. Mutoh, S., Shigematsu, S., Gotoh, Y., Konaka, S.: Design Method of MTCMOS Power Switch for Low-Voltage High-Speed LSIs, Proceedings of Asia South Pacific Design Automation Conference (1999) 113-116.
7. Usami, K. et al.: Automated Selective Multi-Threshold Design for Ultra-Low Sleep Application, Proceedings of International Symposium on Low-Power Electronics and Design (2002) 202-206.
8. Anis, M., Mahmoud M., Elmasry, M.: Efficient Gate Clustering for MTCMOS Circuits, Proceedings of International ASIC/SOC Conference (2001) 34-38.
9. Van Der Meer, P., Staveren, A.: Effectivity of Standby-Energy Reduction Techniques for Deep Sub-micron CMOS, Proceedings of International Conference on Circuits and Systems (2001) 594-597.
10. Grochowski, E., Ayers, D., Tiwari, V.: Microarchitectural di/dt Control, IEEE Design and Test of Computers **20** (2003) 40-47.
11. Choi, J., Kim, Y., Wee, J., Lee, S.: Pipelined Wake-Up Scheme to Reduce Power line Noise for Block-Wise Shutdown of Low-Power VLSI Systems, IEICE Transactions on Electronics **87** (2004) 629 - 633.
12. Benini, L., Siegel, P., De Micheli, G.: Saving Power by Synthesizing Gated Clocks for Sequential Circuits, IEEE Design and Test of Computers **11** (1994) 32-41.

Performance Analysis of User-Level PIM Communication in the Data IntensiVe Architecture (DIVA) System

Sumit Dharampal Mediratta and Jeffrey Draper

USC Information Sciences Institute,
Marina del Rey, CA-90292, USA
{sumitm, draper}@isi.edu

Abstract: The performance of user-level messaging in PIM (Processing-In-Memory) to PIM communication is modeled and analyzed for the DIVA (Data IntensiVe Architecture) system. Six benchmarks have been used for this purpose, two from each category, namely single message transfer, parallel transfer and collective communication, as described for the PMB (Pallas MPI Benchmarks). The benchmarks used are PingPong, PingPing, SendReceive, Exchange, Barrier synchronization and AllToAll personalized exchange. The main significance of this work lies in the evaluation of an implementation of system-wide support for memory-to-memory and memory-to-host communication via a *parcel* buffer (used as a network interface). Another remarkable feature of this evaluation lies in presenting an optimal algorithm for Barrier synchronization and an optimal algorithm, with full channel utilization, for AllToAll personalized exchange for the bi-directional ring configuration of up to 8 DIVA PIMs in the memory system of a Hewlett-Packard's zx6000 server. The algorithms presented can be scaled for higher number of PIM chips with a little degradation in performance over the optimal solution. Our analysis shows that the currently employed communication mechanism can be used very efficiently for collective communication operations, and it also exposes the bottlenecks in the current design for future improvements.

1 Introduction

The user-level messaging performance of high-end architectures has always been a topic of interest to the hardware and software designers of clusters of SMP or PC nodes, and engineering and scientific communities related to the high-performance computing field. Much work has been done on evaluating performance of many commercial interconnects such as Myrinet, Quadrics and Infiniband [1], for middleware layers such as MPI (Message Passing Interface) and vendor-supplied communication primitives [2], and more recently of networks on chips [3]. Evaluating user-level messaging performance has also become crucial for PIM systems such as DIVA[1] [4], which aims to mitigate the processor-memory speed gap. DIVA targets

---

[1] This research was supported by DARPA contracts F30602-98-2-0180 and F33615-03-C-4105.

D.A. Bader et al. (Eds.): HiPC 2005, LNCS 3769, pp. 407–419, 2005.

two important classes of bandwidth-limited applications: multimedia and irregular applications, including sparse-matrix and pointer computations. DIVA accelerates both classes of applications by performing computation directly in memory, requiring novel underlying hardware structures [5][6][7].

PIM systems like DIVA have unique communication requirements because of the dense packing requirements of the memory systems and the need to provide uniform communication mechanism among heterogeneous components (PIM node processors and host processor) [4]. So, PIM systems like DIVA rely upon relatively lightweight communication protocols [8] and network [9] to provide effective memory-to-memory and memory-to-host communication. To further standardize the process of performance comparison of various high-end architectures, a set of well-defined MPI benchmarks, known as PMB (Pallas MPI Benchmarks) [10], has been developed. This paper reports an evaluation for six of these benchmarks executing on a DIVA PIM system. Instead of using MPI primitives for this purpose, native communication primitives have been developed, as increasing PIM user-level messaging performance requires judicious use of underlying system architecture and interconnection network capabilities by the software. The benchmark results obtained can be considered as the optimum performance data by the potential developers of middleware layers for DIVA, ranging from explicit message-passing to shared-memory models because of the flexibility provided by the DIVA communication mechanism. The results are also of significance to potential PIM users and other PIM system architects.

Another significance of this work is in the quantification of the performance metrics in the implementation of system-wide support for memory-to-memory communication via a *parcel* buffer for the first time. A parcel is similar to an active message [11] as it is a relatively lightweight communication mechanism containing a reference to a function to be invoked when the parcel is received. Parcels are distinguished from active messages in that the destination of a parcel is an object in memory, not a specific processor. For more DIVA architectural and communication mechanism details, please refer to [4][8][9].

Another importance of this evaluation lies in presenting optimal algorithms for collective communication routines, namely Barrier synchronization and AllToAll personalized exchange for the bi-directional ring configuration of up to 8 DIVA PIMs. The allowable network size is of 2,4 or 8 PIMs. The number of PIMs is limited by the current capacity of the zx6000's memory system. An efficient implementation of the AllToAll personalized exchange benchmark is an important achievement, as it is one of the most complicated and time-consuming communication operations in high performance computing.

In this benchmark, a personalized (or different) message is sent to every node in the network by every other node. Usually one of two types of algorithms is used to perform this operation – *Direct* or *Indirect*. Direct algorithms involve the transfer of all messages in several contention free phases directly from source node to the destination node, without involving any intermediate nodes [12]. Indirect algorithms involve the intermediate nodes in the message transfer, and a message combination step is added before the combined message is forwarded to the destination node. The indirect algorithms perform better for the cases where message length is small or communication startup time is large compared to the link transmission time [12]. This is because of the reduction in the number of phases required for indirect algorithms.

Many algorithms are available for wormhole routed torus networks [12][13][14], which can be tailored for implementation in the DIVA PIM system environment. However, the deficiencies of these algorithms due to their indirect nature [12][13] or the underlying assumptions (like use of only one virtual channel by [14]) motivated the development of the presented algorithm, which is of *Direct* type because of the reasons discussed in section 2.6.

Section 2 describes the different benchmarks analyzed, algorithms and procedure used, and gives expressions for the performance metrics; section 3 presents the plots of the performance metrics with variation in message length and system size, as applicable, and explains the results; finally, section 4 concludes this paper with some suggestions for potential improvements in the design.

2   Benchmarks – Implementation Algorithms and Analysis

The performance of six benchmarks, namely PingPong, PingPing, SendReceive, Exchange, Barrier synchronization and AllToAll personalized exchange, is analyzed for the above described PIM communication mechanism. Two benchmarks from each category, namely single message transfer, parallel transfer and collective communication (as specified in [10]) are chosen. For PingPong, PingPing, SendReceive, and Exchange, both the timing expressions and throughput calculations are given. For barrier synchronization and AlltoAll personalized exchange only timing expressions are given. These calculations are in accordance with the rules specified by [10].

2.1   PingPong

This benchmark measures the efficiency of a *single message transfer* between two processes (in this case two PIM nodes). A single message is sent from one node to another node and received back. The simplicity of bi-directional connection of two PIM nodes, for this purpose, is apparent in [9].

Send and *Receive* communication primitives have been implemented at the assembly level. The processor clock cycles ($Cycle_{proc}$) taken by each native communication primitive have been used in the performance analysis. [5][6][7] explain DIVA PIM processing logic in detail. The clock cycles taken for these primitives is fairly deterministic, as the DIVA instruction set is very generic and uses only memory operations (*Load* and *Store*) for accessing the network interface (PBuf). The parcel payload size is 32 bytes, so any message greater than 32 bytes can divided into multiple parcels and transferred using one pair of *Send* and *Receive*. This can be done in $\lceil m/32 \rceil$ *Sends* and *Receives*, where m is message length in bytes.

The bottleneck in this communication mechanism is the boundary between the PBuf and 352-bit packet serializer (send bridge). The serializer is held for 11 PiRC clock cycles ($Cycle_{PiRC}$) to convert a packet into eleven 32-bit flits, and it takes an additional 6 $Cycle_{proc}$ of handshaking to move the next packet into the serializer and start its serialization. In this implementation of DIVA, $Cycle_{PiRC}$ is twice that of $Cycle_{proc}$, and thus this bottleneck takes 28 $Cycle_{proc}$. The time taken by *Send* and *Receive* operation (25 $Cycle_{proc}$ each) is evident for one packet send, but the time taken by other *Send* commands overlaps with the above described bottleneck as another parcel can be written into the PBuf while the serializer is transferring the previous parcel. So, throughput for PingPong is dictated by the serializer bottleneck.

Below is a timing expression for the time taken (Δt) by an m-byte message transfer from one PIM node to another and back to the first node.

$$\Delta t = 2(56 + 28 \lceil m/32 \rceil) \, \text{Cycle}_{proc}$$ (1)

Based on standard throughput calculations for PingPong [10], the throughput (Γ) is given by the expression below for an m-byte message transfer in ($\Delta t/2$) seconds.

$$\Gamma = m/ (\Delta t/2) = 2m/ \Delta t$$ (2)

2.2 PingPing

This benchmark also belongs to the *single message transfer* category. The importance here is on the outgoing message being obstructed by the incoming message. A single message is sent from each node to the other node concurrently. Unlike PingPong, the message is not expected to be returned to the sender.

```
if ( ⌊ (⌈ m/32 ⌉/3) ⌋ >=1)                    if ( ⌈ m/32 ⌉% 3  >=1)
{                                              {
    for (i = 1; i = ⌊ (⌈ m/32 ⌉/3) ⌋; i = i+3 )    for (i = 1; i = ⌈ m/32 ⌉% 3 ; i = i+1 )
    {                                              {
        Send packet i;                                 Send packet i;
        Send packet i+1;                           }
        Send packet i+2;                           for (i = 1; i = ⌈ m/32 ⌉% 3 ; i = i+1 )
        Receive packet i;                          {
        Receive packet i+1;                            Receive packet i;
        Receive packet i+2;                        }
    }                                          }
}
```

Fig. 1. Algorithm for PingPing

Fig. 1 shows the algorithm for one node. The progression of two simultaneous sends at the two PIM nodes (because of bi-directional channels) is used here to overcome the above bottleneck and hide the latency of the communication network. The idea simply makes use of the idle time available or latency of the network (wasted in PingPong waiting for the packet) for other *Send* operations on each PIM node before the packet arrives from the other node. A simple analysis of the DIVA communication mechanism revealed that using 3 continuous *Send* operations, followed by 3 continuous *Receive* operations, can best overcome the serializer bottleneck and hide the latency of the network, and the resulting time taken (Δt) depends totally on the execution of *Send* and *Receive* commands. The approach here is to defer the *Receive* operation until the maximum number of packets can be received without any delay due to the serializer bottleneck (because they are buffered at the receiving node). The time is used for sending other packets otherwise.

The timing expression for PingPing is given below.

$$\Delta t = [50 \lfloor \lceil m/32 \rceil /3 \rfloor + (28 \, k + 56) \, q \,] \, Cycle_{proc}$$

$$k = \lceil m/32 \rceil \% \, 3 \text{ and,}$$

$$q = \begin{cases} 0 & \text{if } k = 0 \\ 1 & \text{if } k \neq 0 \end{cases} \qquad (3)$$

The throughput (Γ) is given below for an m-byte message transfer per Δt seconds. Please notice that throughput is not scaled by a factor of 2 (actual transfer rate is $2m$ bytes per Δt seconds) because PingPing is a single message transfer benchmark.

$$\Gamma = m/ \Delta t \qquad (4)$$

2.3 SendReceive

This benchmark belongs to the *parallel message transfer* category. The benchmarks belonging to this category aim to measure performance under global load, and performance is measured for the simultaneous execution of processes at different nodes. In a SendReceive configuration, each node n sends a message to its neighbor PIM node (n+1) mod N and receives a message from another neighbor node (n-1) mod N. The algorithm for the SendReceive is similar to algorithm provided in fig. 1, except for the fact that each PIM node n sends a message to its neighbor node (n+1) mod N and receives a message from another neighbor (n-1) mod N.

The timing expression for SendReceive remains essentially similar to that of PingPing, with one difference: the time taken (Δt) now depends on the maximum length message to be transferred among any pair of nodes.

$$\Delta t = [50 \lfloor \lceil max(m_n)/32 \rceil /3 \rfloor + (28 \, k + 56) \, q \,] \, Cycle_{proc} \qquad (5)$$

Throughput (Γ) is calculated taking into account the number of messages (N_{Msg}) injected and received by a particular node [10].

$$\Gamma = N_{Msg} \, m/ \Delta t = 2m/\Delta t \qquad (6)$$

2.4 Exchange

This benchmark also belongs to the *parallel message transfer* category. In an Exchange configuration, each PIM node n sends and receives a message to and from its neighbor nodes (n+1) mod N and (n-1) mod N. This can be viewed as two phases of SendReceive. In the first phase, each PIM node n sends a message to its neighbor node (n+1) mod N and receives a message from another neighbor (n-1) mod N. In the second phase, each node n sends a message to node (n-1) mod N and receives a message from another node (n+1) mod N. The timing expression is given below.

$$\Delta t = \{[50 \lfloor \lceil max(m_{n,0})/32 \rceil /3 \rfloor + (28 \, k_0 + 56) \, q_0 \,] \\ + [50 \lfloor \lceil max(m_{n,1})/32 \rceil /3 \rfloor + (28 \, k_1 + 56) \, q_1 \,] \} \, Cycle_{proc} \qquad (7)$$

For throughput (Γ) calculations, N_{Msg} is 4 because each node sends a message to both of its neighbors and receives a message from both.

$$\Gamma = N_{Msg} \, m / \Delta t = 4m/\Delta t \tag{8}$$

2.5 Barrier Synchronization

This benchmark belongs to the *collective communication* category, where a message is communicated among a group of processes simultaneously. Barrier Synchronization is one of the most important operations required in high performance parallel computing. Various solutions have been presented thus far to solve this problem efficiently, either by providing hardware support [15], implementing algorithms in software [16][17] or hybrid approaches [18][19][20]. The DIVA communication mechanism provides no hardware support for collective communication. Thus, an efficient algorithm is required to implement this important operation. Although one of many available algorithms [16][17] can be tailored for the DIVA platform for this purpose, a unique algorithm is desirable to make efficient use of the hardware resources available in a PIM environment.

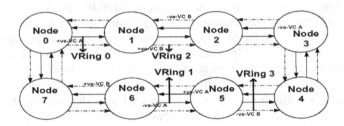

Fig. 2. 8-node network explaining the concept of Virtual Rings

DIVA PIM nodes are connected in a bi-directional ring configuration in a memory system. The concept of a Virtual Ring is introduced for collective communication algorithms. A *Virtual Ring* is defined as the complete traversal on a particular virtual channel (VC) of a physical ring (in a particular direction, i.e. +ve or −ve). Thus, there can be a maximum of four Virtual Rings in a DIVA system. VRing0, VRing1, VRing2 and VRing3 are defined as the traversal on VC A, VC A, VC B, and VC B virtual channels of +ve, -ve, +ve and −ve direction physical rings respectively (fig. 2). The nodes are labeled, as n, in an incremental manner from 0 to N-1 (where N is the total number of nodes in the ring).

The presented algorithm reduces the required barrier message transfer stages (phases) to $\log_2 N$ by forming a balanced binary tree. The number of phases (P) is optimum because a message can be distributed optimally among N nodes (without any multicast or broadcast capability) by forming a balanced binary tree. Thus,

$$P = \log_2 N \tag{9}$$

In each phase p, a node n sends barrier message *Barrier* to the destination node (or Send node), $S_{(n,p)}$, given by the expression below.

$$S_{(n, p)} = \begin{cases} [n + (N/2)(1/2^p)] \bmod N & \forall n : n \in N_1 \\ [n - (N/2)(1/2^p)] \bmod N & \forall n : n \in N_2 \end{cases}$$

(10)

Where, $N_1 = \{0, 2, \ldots, N-2\}$ & $N_2 = \{1, 3, \ldots, N-1\}$

Here, a barrier message *Barrier* consists of a single packet (256 bit payload). The barrier message received needs to be combined with the local barrier message and treated as a local barrier message for further phases. In other words, a message combination step is needed before the algorithm can begin the next phase at a particular node. The result is called the *Combined Synchronization Signature*. Further, to detect the completion of a required barrier message in a particular phase an *Expected Phase Signature* (again 256 bit wide) for each phase p is maintained at every node n.

The out of order arrival of packets at a particular node must also be addressed, since different nodes may start the barrier operation at different times. Barrier messages from nodes expected in later phases can arrive and be buffered at the PBuf and received by the current node in any order. This issue is resolved by checking whether the current phase is completed (i.e. whether expected barrier message has already been received) before entering into the *Receive* operation. Otherwise, a process will wait for the packet forever that has already been received before. These issues led to a modification of the *Receive* command for barrier synchronization. The *Send* command remains the same as for the previous operations.

Below is the expression for the virtual ring $R_{(n, p)}$ to be used by node n in phase p.

$$R_{(n, 0)} = n \bmod N_{VRings}$$
$$R_{(n, p)} = R_{(n, 0)} \qquad \forall p : p \in \{1, .., P-1\}$$

(11)

The ring $R_{(n, p)}$ used does not change with phase p. This means that contention for links is not an issue for this algorithm, even if execution of Barrier *Send* and *Receive* operations by different nodes gets out of synchronization. The channel utilization is full in the first phase. After that channels are available between two nodes, but not required by the algorithm. However, every virtual ring is used in all phases as messages are sent on minimal paths, which are disjoint in a particular phase.

The timing expression for the lower bound (when every node starts the barrier operation at the same time) on the time taken to reach complete synchronization is given below. This expression basically accumulates the single message transfer time in each phase, which is similar to the PingPong timing expression (except *Receive* takes 39 cycles). The calculations required for the destination node and ring to be used in each phase can be computed by the kernel in the initialization phase for a user program, which in turn initializes the route cache [8] so that user-level object addresses may be translated appropriately.

$$\Delta t_{lower\_bound} = [96\,P + 2N\,(1 - (1/2)^P)\,]\,Cycle_{proc}$$

(12)

The timing expression for the upper bound is also given below. Here, T_n is defined as the start time of barrier operation on node n. This worst case arises when all potential senders (in different phases) of *Barrier* to node n, which started barrier operation at min $\{T_n\}$, start the barrier operation at a time max $\{T_n\}$.

$$\Delta t_{upper\_bound} = \Delta t_{lower\_bound} + \max \{\,T_n\,\} - \min \{\,T_n\,\}$$

(13)

2.6   AllToAll Personalized Exchange

This benchmark also belongs to the *collective communication* category, as discussed above. The presented algorithm is of the *Direct* type. This type of algorithm was preferred for DIVA, since the communication startup time and latency can be *completely* hidden (for up to 8 hop distance) by using the previously presented method of 3 continuous *Sends* and *Receives* (section 2.2), with only message transfer dominating. Although DIVA tries to reduce communication by sending pointers, functions and arguments only, messages lengths can be large for applications where data use is not localized. Moreover, the message *combination and forwarding* step, used in Indirect algorithms, would have been simply overhead because the same message could have been sent to the destination node directly instead of some intermediate node.

This algorithm achieves a lower bound on the number of phases P possible for the communication of all messages with full utilization of all channels and in terms of messages required to be sent by one node. A total of $N(N-1)$ (every node sends to each node except itself) messages are required to be sent over $(N_{VC} \times N_{Directions})$ - channels, which is equivalent to N_{VRings}. Here, N_{VC} is the number of virtual channels, $N_{Directions}$ is the number of directions in the ring, and product of the former two gives, N_{VRings}, the total number of virtual rings in the network. Messages are exchanged (3 continuous *Sends* and *Receives* method), rather than only sent, in each phase. This further reduces the number of phases by a half (factor of 2 in the denominator). These messages can be exchanged on any of the N_{VRings} rings, following non-minimal paths also. The result is the expression below.

$$P = [N(N-1)] / 2\, N_{VC}\, N_{Directions} \qquad (14)$$
$$= [N(N-1)] / 2\, N_{VRings}$$

The above value of P is equal to N-1 for [N=8, N_{VRings} = 4], [N=4, N_{VRings} = 2] and [N=2, N_{VRings} = 1]. Here, N_{VRings} changes with N because required number of virtual rings decreases with N.

In each phase p, a node n sends a message the destination node, $S_{(n,p)}$.

$$S_{(n,p)} = \begin{cases} (N-n-p-1)\bmod(N-1) & \forall n : n \in N_1 \ \& \text{ if } (N-n-p-1)\bmod(N-1) \neq n \quad (15) \\ N-1 & \forall n : n \in N_1 \ \& \text{ if } (N-n-p-1)\bmod(N-1) = n \\ (3p)\bmod(N-1) & n : n = N-1 \end{cases}$$

Where, $N_1 = \{0, 1, 2, ..., N-2\}$

Below is the expression for the virtual ring $R_{(n,\,p)}$, to be used by node n in phase p.

$$R_{(n,0)} = \begin{cases} n & \forall n : n \in N_1 \\ N-n-1 & \forall n : n \in N_2, N_3 \end{cases} \qquad (16)$$

$$R_{(n,p)} = \begin{cases} R_{(n,p-1)} & \forall n : n \in N_1 \ \& \text{ if } (N-n-p-1)\bmod(N-1) \geq n \\ (R_{(n,p-1)} - 1)\bmod N_{VRings} & \forall n : n \in N_1 \ \& \text{ if } (N-n-p-1)\bmod(N-1) < n \\ (R_{(n,p-1)} - 1)\bmod N_{VRings} \ \& \ j = 0 & \forall n : n \in N_2 \ \& \text{ if } (N-n-p-1)\bmod(N-1) < n \ \& \ j = 0 \\ (R_{(n,p-1)} - 1)\bmod N_{VRings} \ \& \ j = 1 & \forall n : n \in N_2 \ \& \text{ if } (N-n-p-1)\bmod(N-1) = n \ \& \ j = 0 \\ R_{(n,p-1)} & \forall n : n \in N_2 \ \& \text{ if } j = 1 \\ R_{(n,p-1)} & \forall n : n \in N_3 \ \& \text{ if } p = \text{Even} \\ (R_{(n,p-1)} - 1)\bmod N_{VRings} & \forall n : n \in N_3 \ \& \text{ if } p = \text{Odd} \end{cases}$$

Where, $N_1 = \{0, 1, ..., N/2-1\}$, $N_2 = \{N/2, N/2+1, ..., N-2\}$ & $N_3 = \{N-1\}$

Contention for the virtual rings (or links) at the source nodes will arise for this algorithm if message lengths used by different nodes are unequal in one phase. In such a case, it would be desirable for some nodes to start their next phase early, and thus increase the traffic in the network (because of fair flow control in DIVA [9]), but this would lead to contention. This contention problem will not harm the functionality, but will increase the time taken by the algorithm because message transfer times in the allocated phases are ideally overlapped (instead of cumulative) with other message transfers. Thus, synchronization between phases is desired for AllToAll direct algorithms [12][14] for better performance. This synchronization can be achieved statically by a compiler by padding small messages with dummy messages, to make the message lengths consistent. Alternatively, a barrier command can be used after each phase. Obviously, there is a tradeoff between using a barrier operation for synchronization and simply allowing the AllToAll operation to proceed asynchronously. If the time overhead due to a barrier operation is larger than the penalty due to asynchronous treatment of differing message lengths, then the latter method is preferred.

The channel utilization is full in all phases, as messages are sent on the non-minimal disjoint paths in a particular phase. The expression for the time taken to perform a complete personalized exchange on a synchronous run of the algorithm is given below. This expression accumulates the single message transfer time in each phase, which is the PingPing time for the largest message.

$$\Delta t \quad = \sum_{p=0}^{P-1} \; [50 \lfloor \lceil \max(m_{n,p})/32 \rceil /3 \rfloor + (28 \, k_p + 56) \, q_p] \; \text{Cycle}_{proc} \qquad (17)$$

3 Performance Results and Analysis

The timing expressions for five of the benchmarks are plotted with respect to variation in the message length (fig. 3). The processor speed of 140 MHz (Cycle$_{proc}$ of 7.14ns) is chosen for this purpose, corresponding to the actual speed achievable for DIVA-II PIM chips assembled in DIMM boards and integrated in the memory system of a Hewlett-Packard Itanium2-based zx6000 server. The message length is varied from 1 byte to 4096 bytes in powers of 2, i.e. 2^i for i=0 to 12. Barrier synchronization is not shown in this graph because it is defined only for a single message length of 32 bytes. The throughput expressions are plotted for the *single* and *parallel message transfer* category benchmarks in fig. 4, as throughput is not reported for the *collective communication* category.

As seen in fig. 3, PingPong time is double that of PingPing time for message lengths up to 64 bytes, but then the difference grows with increase in the message length. This is because of the communication latency hiding technique used for PingPing, which results in reasonable time savings for large message lengths. The PingPong throughput (fig. 4) is equal to the PingPing throughput for up to two sends, and then it decreases because of the above reason. Throughput, in general, increases initially and then tends to saturate for higher message lengths. Because the impact of communication latency involved in the timing expressions of all benchmarks (for non-multiple of 3 message lengths in PingPing based expressions) becomes lesser for larger message lengths, throughput reaches the capacity limit of the underlying network. The time required for SendReceive is the same as that for PingPing, but throughput is double because total message turnover count is double (it belongs to the *parallel message transfer* category). The time taken by Exchange is twice that of SendReceive, but throughput is the same because total message turnover count is also double.

Fig. 5 shows the plots for the time needed for the Barrier operation (lower bound) and AllToAll communication with variation in the total number of PIM nodes, i.e. for

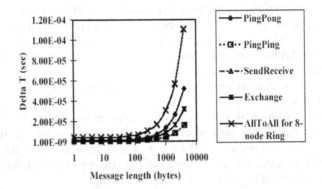

Fig. 3. Time taken for different message lengths (except for Barrier)

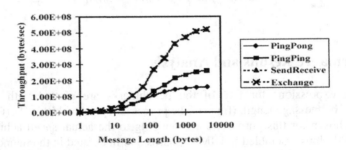

Fig. 4. Throughput achieved for different message lengths

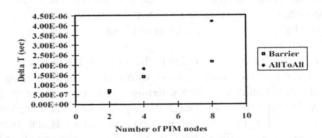

Fig. 5. Time taken by Barrier and AllToAll for different number of nodes

N=2,4, and 8, connected in a bi-directional ring configuration, as is the case for DIVA. Barrier takes more time for N=2 than AllToAll, because the Barrier *Receive* command takes more clock cycles due to synchronization signature matching for each phase. Barrier is implemented as an Indirect algorithm, but this does not imply that an Indirect algorithm will work better for AllToAll, as discussed in section 2.6.

DIVA PIMs have been assembled into DIMM boards (see fig. 8 in [8]) and inserted into the memory system of a Hewlett-Packard Itanium2-based zx6000 server[2]. Instrumented measurement experiment was conducted for PingPong benchmark for a 32-byte message transfer to validate the analysis. This experiment tested the path of parcel flow from and to the *user space* through the pbuf and PiRC. The analysis was found to be in perfect agreement with the measurement on the hardware system for this case. This experiment also verified the analyzed clock cycles taken by *Send* and *Receive* communication primitives. This indicates that analyzed performance results for other benchmarks should also be in close agreement with measurement on the hardware system.

The results cannot be fairly compared with those provided in [10] for MPI implementations on specific machines. Our analysis assumes the DIVA specific optimized implementations of communication primitives, and hence, a comparison with the results of generic MPI implementations is not appropriate. Given that DIVA is a unique architecture using PIMs, it is difficult to draw fair comparisons with general-purpose architectures, especially when considering the limited silicon area used for implementing DIVA as opposed to full-scale multiprocessors. Also, there are currently no other PIM communication results for comparison. It is simply noted that DIVA achieves sub-microsecond user-level messaging, including software overheads, in contention-free cases and also performs reasonably well for collective communication using lightweight network structures and protocols [8][9]. For more details on how other architectures perform for these communication operations, the interested reader can refer to [10] and other survey literature.

4   Conclusion and Possible Improvements

Every design, in spite of the best efforts of the designer, has some scope of improvement, and such is the case with DIVA. First, if a dedicated pbuf bus is used, pbuf load and store operations would take only one clock cycle, as they would not require arbitration through a memory controller. Secondly, RDMA capability, which has become standard in most state-of-the-art interconnection networks, is desirable, so that processor intervention is not needed in the normal case. At a minimum, a remote memory write capability would free the receiving node processor from waiting for *Receive* data (polling mode) or context switching costs (interrupt mode). Thirdly, the interaction between the send part of the pbuf and serializer, which is the bottleneck in the current design, can be greatly improved. Fourthly, providing two serializers and doubling the FIFO space in the send part of the pbuf would result in greater per-node throughput in many cases. Finally, providing some hardware support for collective communication and using a hybrid approach will definitely help systems containing larger number of PIM chips.

---

[2] We thank Tim Barrett for providing DIVA PIM testing platform on zx6000 server.

To summarize, user-level messaging performance in PIM to PIM communication is modeled and analyzed for the DIVA system in this paper. The benchmarks used for this purpose are PingPong, PingPing, SendReceive, Exchange, Barrier synchronization and AllToAll personalized exchange. A significant part of this evaluation lies in the formulation of optimal algorithms for Barrier synchronization and AllToAll personalized exchange for the bi-directional ring configuration of upto 8 DIVA PIMs in the memory system of a HP zx6000 server. The expressions for timing and throughput, as applicable, are derived for the above benchmarks, and results are thoroughly analyzed to provide insight. The results show that the currently employed communication mechanism can be used very efficiently, for collective communication operations also.

References

[1] J. Liu, B. Chandrasekara, et al. Microbenchmark performance comparison of high-speed cluster interconnects. *IEEE Micro,* Volume 24, Jan-Feb 2004, pp 42-51

[2] P.H. Worley. MPI performance evaluation and characterization using a compact application benchmark code. *MPI Developer's Conference,* July 1996, pp 170-177

[3] S.G. Pestana, et al. Cost-performance trade-offs in networks on chip: a simulation-based approach. *Design, Automation and Test in Europe Conference and Exhibition,* Volume 2, Feb 2004, pp 764-769

[4] J. Draper, et al. The Architecture of the DIVA Processing In Memory Chip. *Supercomputing,* June 2002

[5] J. Draper, et al. Implementation of a 32-bit RISC processor for the Data-Intensive Architecture processing-in-memory chip. *Application-Specific Systems, Architectures, and Processors,* 2002

[6] J. Draper, et al. Implementation of a 256-bit wideword processor for the Data-Intensive Architecture processing-in-memory chip. *28th European Solid-State Circuit Conference,* September 2002

[7] S. Mediratta, et al. A 0.18um CMOS implementation of an area efficient precise exception handling unit for processing-in-memory systems. *Midwest Symposium on Circuits and Systems,* July 2004

[8] S. Mediratta, C. Steele et al. An area efficient and protected network interface for processing-in-memory systems. *International Symposium on Circuits and Systems,* May 2005

[9] S. Mediratta, J. sondeen, J. Draper. An area efficient router for Data-Intensive Architecture (DIVA) system. *International Conference on VLSI Design,* Jan 2004.

[10] Pallas MPI Benchmarks. http://www.pallas.com/e/products/pmb/documents.htm

[11] T. V. Eicken, et al. Active messages: a mechanism for integrated communication and computation. *ISCA,* May 1992.

[12] Y.C. Tseng, et al. Bandwidth-optimal complete exchange on wormhole-routed 2D/3D torus networks: a diagonal-propagation approach. *IEEE Transactions on Parallel and Distributed Systems,* April 1997

[13] S.G. Kim, et al. Complete exchange algorithms in wormhole-routed torus networks: a divide-and-conquer strategy. *Parallel Architectures, Algorithms, and Networks,* June 1999, pp 296-301

[14] S. Hinrichs, C. Kosak, et al. An architecture for optimal all-to-all personalized communication. *ACM Symposium on Parallel Algorithms and Architectures,* 1994, pp 310-319

[15] R.E. Kessler, J.L. Schwarzmeier. Cray T3D: a new dimension for Cray research. *Compcon, Digest of Papers*, Feb 1993, pp 176-182

[16] Y. Sun, P.Y.S. Cheung. Barrier synchronization on wormhole-routed networks. *IEEE Transactions on Parallel and Distributed Systems,* June 2001, pp 583-597

[17] J.S. Yang, C.T. King. Designing tree-based barrier synchronization on 2D mesh networks. *IEEE Transactions on Parallel and Distributed Systems,* Volume 9, June 1998, pp 526-534

[18] D.K. Panda. Fast barrier synchronization in wormhole k-ary n-cube networks with multidestination worms. *HPCA,* Jan 1995, pp 200-209

[19] R. Sivaram., C.B. Stunkel, D.K. Panda. A reliable hardware barrier synchronization scheme. *Parallel Processing Symposium,* April 1997, pp 274-280

[20] J.F. Martinez, J Torrellas. Speculative synchronization: programmability and performance for parallel codes. *IEEE Micro,* Nov-Dec 2003, pp 126-134

Improved Point-to-Point and Collective Communication Performance with Output-Queued High-Radix Routers

Sameer Kumar[1], Craig Stunkel[1], and Laxmikant V. Kalé[2]

[1] IBM T.J. Watson Research Center,
Yorktown Heights, NY 10598,
{sameerk, stunkel}@us.ibm.com
[2] Department of Computer Science,
University of Illinois at Urbana-Champaign
kale@uiuc.edu

Abstract. We present an output-queued switch architecture with cross-point buffering that has improved performance for both point-to-point communication and hardware accelerated collective communication. In the past, output queuing architectures have been less popular as they require more internal speedup and buffering. However, with current technology it is possible to build output-queued switches with a relatively large number of ports. We demonstrate that our output-queued architecture performs well for point-to-point messages, specially in a fat-tree topology. We also show that output-queued architectures facilitate efficient implementations of multicasts and reductions. We present performance of multicasts and reductions on individual switches and a network of switches interconnected in a fat-tree topology. We also present simulation results based on synthetic workloads that emulate a molecular dynamics application.

1 Introduction

Communication interconnect performance is critical for several parallel applications. We present an output queued architecture with crosspoint buffering to achieve higher performance for point-to-point and collective communication operations. As output-queued switches do not suffer from head-of-line blocking, they can achieve a high utilization for point-to-point communication. In the past, output-queued architectures have been less popular because they require higher internal bandwidth and more memory. But with current ASIC technology, it is possible to build crosspoint-buffered output-queued switches. A brief intuition showing the feasibility of output-queued switches (even with a large number of ports) is presented in the next section.

Traditionally, collective operations, such as multicasts and reductions, have been optimized through processor level schemes, which send several point-to-point messages. Processor based collective optimization schemes have several problems. For short messages, completion time is dominated by CPU and network interface controller (NIC) overheads of sending messages. Large messages sent by the processors participating in the collective may contend for the same communication channels [1]. Good collective performance also requires that all intermediate processors immediately process and

D.A. Bader et al. (Eds.): HiPC 2005, LNCS 3769, pp. 420–431, 2005.

forward the incoming message. Performance is affected if one of the intermediate processors is running an operating system daemon [2], which can delay the collective operation. Moreover, with message driven execution [3] and asynchronous collectives [4] it is possible that the remote processor is busy doing other work and cannot process the message immediately, delaying broadcast completion.

For the above reasons, hardware support for collective operations is desirable. One of the approaches studied in literature implements collectives in the network interface [5]. However, performance of such NIC optimizations can be limited as several point-to-point messages are still exchanged. Instead, if multicasts and reductions are supported in the switching network, they can be finished in one network phase. On parallel systems with thousands of nodes, switch collectives will make a significant difference. Some current clustering interconnects such as Quadrics QsNet [6] and Mellanox InfiniBand [7] have multicast support in their switches. But multicast performance is restrictive as these switches have input-queued architectures [8]. Input queuing architectures require complex centralized arbitration to achieve high utilization, and are not a natural match for multicast [9,10,8]. Therefore, we use an output-queued switch architecture with cross-point buffering to improve hardware multicast performance. Popular interconnects today do not have reduction support in their switch architectures. However, we present schemes to perform network reductions efficiently and show their scalability to a large number of switch ports.

To support collectives in the network, a topology-specific spanning tree must be built over the network. In this paper, we assume a fat-tree topology [11], which is used by several popular interconnects like Quadrics QsNet [6], InfiniBand [7], and IBM SP networks. Fat-tree networks have high bisection bandwidth and can be scaled to thousands of nodes. We present schemes to build collective spanning trees on fat-tree networks, and assess the performance of collectives using those spanning trees. Our scheme conserves routing table entries, as only one tree is needed to multicast data to a group with any leaf node as the source.

We show that our output queued architecture has good performance for both multicasts and reductions through several benchmarks that simulate independent switches and networks of switches. We also present the network throughput and latency when several collectives happen simultaneously, as needed by applications like NAMD [12] and CPAIMD [13].

2   Router Architecture

Several input and output-queued architectures have been proposed for high performance interconnect switches. Input queuing (IQ) schemes allow simpler data flow but require centralized arbitration to achieve high utilization. IQ routers also suffer from head of line blocking which restricts their throughput. Using multiple virtual channels and smart buffer management improves the performance of input-queued routers [14,15]. *Virtual output queuing* [10] (VOQ), where each input queue has reserved buffer space for every output queue, can fully utilize the switch. Virtual output queuing also has a centralized arbiter and uses heuristics to select packets when there are conflicts for output ports. It requires $O(K^2)$ buffer space, where K is the number of ports.

We believe that switch design should have efficient support for multicasts and reductions (also referred to as *combines* in this paper). IQ and VOQ do not handle multicasts efficiently as they need centralized arbitration [9,8]. VOQ can achieve full utilization for multicast if every input port has $(2^K - 1)$ queues [9,8] in a KXK switch, one for every possible subset of output ports. As this requires a tremendous amount of memory, VOQ multicast scheduling algorithms use heuristics.

Two schemes have been proposed to handle multicasts in IQ routers [8,9,10]. In *no-fanout-splitting* (*fanout* is the number of multicast destination ports), a multicast packet is only sent if all destination ports are available in that arbitration cycle. Here the crossbar is used only once, but several arbitration cycles may be required to send the packet and free the input buffer for that packet. No-fanout-splitting is good for multicasts with small fanouts. In the *fanout-splitting* scheme, the packet is sent to all output ports that are available in that arbitration cycle, making the multicast packet use the crossbar for several cycles. The maximum achievable utilization for multicast is presented in [8], which is far from full utilization for many traffic patterns. IQ multicast schemes can also have deadlocks in a network of switches.

In this paper, we show the effectiveness of *output queuing* for both point-to-point communication and hardware collectives. Packets in output queuing are buffered on the output ports of a switch before being sent out. Hence, output queuing does not have the head-of-line blocking resulting in higher switch utilization. Output queuing also has distributed arbitration where each output decides which packet to send independent of other outputs. Figure 1(a) shows an output-queued router with buffers at the outputs.

Popular output-queued routers in the past have used shared buffers between output ports [14]. Such shared buffer schemes have limited scalability with respect to link bandwidth and number of ports. We use crosspoint buffering in our router architecture to make the router support high bandwidth links efficiently. Cross-point buffering guarantees that there is a reserved buffer for each pair of input and output ports. A graphic description of cross-point buffering is shown by Figure 1(b), where each input port has some reserved memory on every output port. Hence the total buffering required is $O(K^2)$. Packets arriving on input ports are immediately sent to the crosspoint determined by the destination output queue. Our architecture with cross point buffering is similar to the SAFC scheme presented in [15]. But [15] only presents the point-to-point

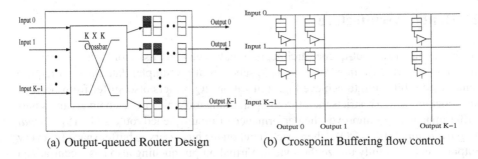

(a) Output-queued Router Design   (b) Crosspoint Buffering flow control

Fig. 1. Router Architecture

performance on one switch with limited buffer space. We are also concerned with mul-ticast and reduction performance on one switch, and on networks of switches.

Feasibility of Output-Queued Architecture: Output-queued architectures are less com-monly used as they require more memory and internal speedup to let input ports talk to several output ports simultaneously. However, with current ASIC technology it is possible to build output-queued switches. Suppose we plan to build an InfiniBand 4X switch with a bandwidth of 10Gbps (1.25 GB/s) per port. We would also like to support 20m cables or 200ns of round trip time (RTT) . Hence, we would need atleast 250 bytes of memory at each crosspoint. It is usually good to have two to four RTTs of buffer space for good switch performance. For an 8 port switch the total memory requirement is about 64KB which is easily available in modern ASICs. For a 32 port switch we need 512KB to 1MB of buffer space. With current ASIC technology 512 KB is feasible and 1MB may still be possible.

We use virtual cut through routing and credit based flow control [16] between switches. In all the schemes we present, packets are only sent out by any source switch S1 if buffer space for the entire packet is guaranteed on all the output ports of the down-stream switch S2. With crosspoint buffering, this implies that all cross-points in S2, for the input port on which S1 is connected to S2, should have buffer space available for this packet. We have considered three flow control schemes :

Scheme I : This is the simplest of all the schemes. In this scheme, each switch S1 keeps one counter for every S1 output port connected to some downstream switch S2. This counter is initially set to the number of buffers at a cross point of S2. When S1 sends a packet to S2, it decrements the counter corresponding to S2. When S2 sends this packet from S1 out, it sends credits for that packet back to S1, to receive more packets. This scheme is inefficient because packets from S1 may go to different crosspoints on S2, so the counter should not be decremented for each packet send.

Scheme II : In this scheme, each switch S1 keeps track of the buffer space available in the next switch S2 through a credit counter. As in *Scheme I*, the credit counter at each S1 output port is initialized to the size of crosspoint buffer at the corresponding downstream switch S2, and is decremented when a packet is sent out. However unlike *Scheme I*, S2 sends back a *min* of the buffer space available on all cross-points. On receiving this *min* buffer space, S1 resets the counter to *min*. This scheme has better efficiency when packets from S1 go to different output ports of S2. But, the *min* is time-warped as there is an RTT delay for it to be received by S1. So, for this scheme to work, S2 must reserve RTT additional space that would be used when the time-warped *min* results in an overflow.

Scheme III : In this scheme, each S1 output port has a copy of the routing table of the down-stream switch S2. The S1 output port also maintains K counters for each down-stream switch S2. This enables S1 to locally compute the *min* buffer space available on all crosspoints of S2 corresponding to the input port connected to S1. Each counter is incremented when packets are sent to the corresponding crosspoint in S2. Periodically, S2 sends the status of S1's packets in all its ports, which is used by S1 to update its counters corresponding to S2.

Of the three schemes, *Scheme III* would perform best as it correctly updates counters and avoids the time-warp problem of *Scheme II*. But, we do not use this scheme as it is quite complex, requires additional routing memory and may not be feasible to implement in modern ASICs. *Scheme II* leads to unusable RTT (or a packet) of buffer space, and as our experiments use a small buffer space (2 packets to 4 packets), we use *Scheme I* in all our simulations. We show that **Scheme I** performs well despite its simplicity. We also show in the next few sections that, independent of the flow control scheme chosen, multicasts and reductions are efficient and easy to implement in such an output-queued architecture.

Multicast: the credit based flow control scheme we use ensures that when any packet is sent out buffer space is available on all cross-points corresponding to this input port. For every multicast, buffer space will be available on every output port for the entire packet, thus, making each multicast **deadlock-free** [17]. On arrival, the multicast packet is immediately sent to all the ports determined by the routing table entry for the packet's destination address. The multicast packet only uses the crossbar once, resulting in better switch throughput for the multicast. With *Scheme I*, flow-control credits for this multicast packet are only sent back after all multicast packets on all the destination ports have been sent out. With *Scheme II* and *Scheme III*, the *min* is computed considering all destinations of the multicast. Hence the multicast operation can achieve full throughput and also avoid deadlocks which were possible with input queuing schemes.

Reduction: our design also supports the *Combine* operation which can be used to support reductions and barriers in hardware. We extend the barrier combine unit presented in [18] to perform reductions. The combine unit receives packets from the crossbar output and performs reductions. Every reduction has access to local state. For example, in the global sum operation the local state can store the current partial sum. For a global array sum, the local state could be an array of floating point numbers. This local state is updated by the combine unit whenever a reduction packet arrives. After all reduction packets have been processed, the combine unit sends a reduction packet back into the crossbar to be sent to the parent switch in the spanning tree.

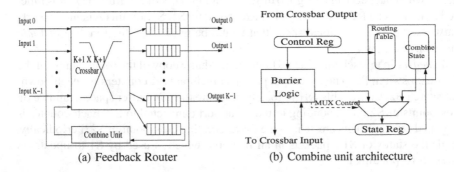

(a) Feedback Router   (b) Combine unit architecture

Fig. 2. Output-queued Router with Combine Unit

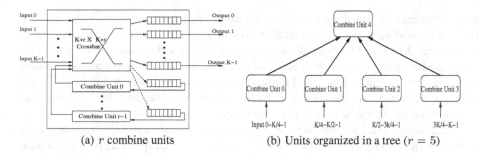

(a) r combine units

(b) Units organized in a tree ($r = 5$)

Fig. 3. Switch design with r combine units

The combine unit connects from the output port through a feedback to an input port in the switch, as demonstrated in Figure 2(a). The combine unit behaves like any other output port in the switch. Reduction packets arriving on input ports of the switch are buffered at the output port connected to the combine unit before being processed. As the combine unit behaves like any other output port in the switch, the flow control schemes *Scheme I*, *Scheme II* and *Scheme III* work here too.

The architecture of the combine unit is shown in Figure 2(b). It can take a few cycles to receive reduction packets, as the entire packet is needed to detect errors. (We do not explicitly simulate errors but we do model the delays.) The header of the packet is stored in the *control register*. The *combine logic* uses the address in the packet header to lookup the routing table for the *combine state* of the current reduction. In the following cycles, the *ALU* computes and updates the local state based on the data from the packet.

For short reductions and barriers, it may be possible to pipeline packet arrival and computation to process one packet every cycle [18]. But for larger reductions involving several data points, the combine unit may stall on each combine operation. In switches with a large number of ports, a single combine unit will become a point of contention. As ASIC speeds are much slower than custom designed CPU speeds, this may hamper the overall efficiency of the global reduction operation. Thus, affecting scalability of parallel applications on such a network.

Figure 3(a) shows the switch architecture with r combine units. The combine units are organized as a tree with $r-1$ leaves and one parent (Figure 3(b)). The leaves process the reduction packets from a subset of ports and pass their partial result to the root of the tree. Such a hierarchical design scales to more number of ports as several combines at the leaves can happen simultaneously. In Section 4, we show that even with $r \ll K$ we can achieve good performance. Hence, the reduction units are only a small additional overhead.

3   Building a Collective Spanning Tree

Spanning trees are essential to support collectives in the network hardware. These spanning trees can be directed trees where packets only travel in one direction on each hop. For example, a broadcast can be achieved by sending the packet from the source along a directed spanning tree. If the network time to do a broadcast does not depend on the root of the spanning tree, we can also build undirected spanning trees for broadcasts. Fat-tree networks can make use of such undirected trees for collectives. Here any leaf of the tree can do a broadcast with the same overhead. Our switch design has support for undirected spanning trees, as do *InfiniBand* switches. Such undirected trees save routing table memory as any leaf can send multicast messages. With directed trees [18] each sender would require a separate tree. In an InfiniBand network, the subnet manager can be requested by the application to build such trees.

The routing table has a bit vector of destination ports for each collective address, as opposed to a parent port and a list of child ports in a directed spanning tree. For a multicast operation, packets are sent to all ports except the port on which the packet arrived on. We implement combines as follows: suppose a routing table destination bit-vector has k outputs set, then the combine manager would process k-1 reduction packets and send the current partial result to the remaining port on which it did not receive a packet. Both multicasts and combines use the same routing table entries. The tag in the packet determines whether the operation is a multicast, barrier, reduction etc.

Fat-tree Networks: Fat-trees are generalizations of k-ary n-trees [11]. Figure 4 shows a 4-ary 2-tree network. Routing in a k-ary n-tree has two phases, (i) the upward phase where a packet is sent to any of the lowest common ancestors of the source and the destination, (ii) the downward phase where the packet is routed from this ancestor to the destination through a fixed path. This scheme can be extended to build collective spanning trees as follows:

```
buildTree(id, up, destlist, swlist, tlist)
id : the switch id of the current switch,  up : boolean flag that shows direction
destlist : list of processor destinations, swlist : list of previous switches
tlist : list of treeInfos (each treeInfo is list of parent and child output ports)
  begin
        swlist.insert(id);
        if(!inHighestLevel(id, destlist) && up) {        //Need to go further up
            parent = leastLoadedParent(id);
            buildTree(parent,destlist,swlist,tlist, true)
        }
        for count : 0 -> numPorts/2 - 1                  //Going down in the fat-tree
            if(child[count] routesto destlist) {
                tlist[my_pos].insert(count);
                buildTree(child[count],destlist,swlist,tlist, false);
            }
  end
```

The tree is constructed by finding a lightly loaded LCA (lowest common ancestor) between the source and every destination. The sub-routine *leastLoadedParent()* can be

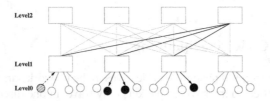

Fig. 4. Fat-tree with 16 nodes

used to find lightly loaded parents while moving up. The load of the switch is determined by the number of multicast trees passing through that switch. If the current node is an LCA for one or more destinations, the algorithm recursively finds downward paths to the destinations. This algorithm minimizes contention on the upward path of the packet and also balances the routing memory required for each collective operation.

4   Network Simulation

We simulated switches with the above architecture using POSE [19], which is a parallel event driven simulation language. We simulated 8 port and 32 port routers in a fat-tree topology with adaptive routing. Table 1 shows the parameters of our simulation. These parameters are derived from InfiniBand 4X interconnects. We first present the throughput and packet latency of point-to-point communication using the well known communication patterns [11] *transpose, bit reversal, complement and uniform.* We simulated a 256 node fat-tree network with 8 port and 32 port output queuing switches. With 8-port switches, we used 4 packet buffers as we can easily support them with current technology. For 32-port switches, we present results with both 2 packet and 4 packet buffers at the crosspoint. In the plots, **load-factor** is the ratio of the mean arrival rate and the arrival rate that saturates a link. Here, each node sends packets with a Poisson distribution that has a mean arrival rate proportional to the load-factor. Figures 5(a) and 5(b) show the throughput and response times with 8 port switches, while figures 6(a) and 6(b) show the performance of 32 port switches. Performance is good for all the above cases. In fact, with 32 ports and 4 packet buffers at each crosspoint, the network has high utilization comparable to a 256-way crossbar.

Table 1. Simulation Parameters

Speed	Packet	Channel Delay	Switch Delay	ASIC	NIC Send Ovhd.	NIC Recv. Ovhd.
10 Gbps	256 bytes	20 ns	90ns	250 Mhz	1300 ns	1300 ns

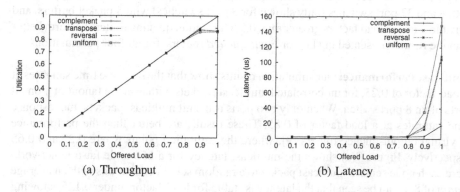

(a) Throughput   (b) Latency

Fig. 5. Throughput and Latency with 8 ports and 4 packet buffers for 256 nodes

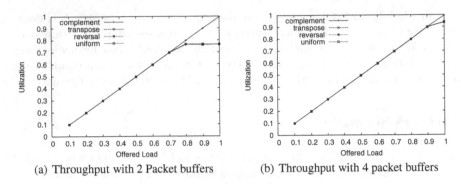

(a) Throughput with 2 Packet buffers   (b) Throughput with 4 packet buffers

Fig. 6. Results for 32 port switches

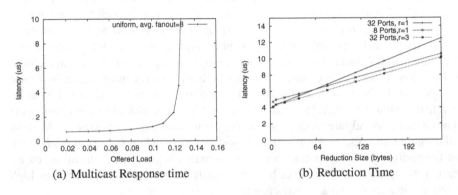

(a) Multicast Response time   (b) Reduction Time

Fig. 7. Collective communication performance on a 256 node network

With 32 port switches, the fat-tree has 32 switches organized in two levels. Since the *complement* traffic pattern is contention free [11,20], its throughput is 100% at full load. The other communication patterns *uniform, transpose* and *reversal* also have a good throughput of about 93%. This high throughput is due to *output queuing, adaptive routing* [21] in fat-trees and the fact that there are only two levels of switches or three points of contention for each packet in the entire network. Response times are also good for both 8 and 32 port switches (only shown for 8 port switches) with 4 packet buffers, and only blow up for load factors greater than 0.9. Our output queuing routers perform better than the results presented in [11] for *input-queued* routers, for all four permutations.

Multicast Performance: Our simulation results show that the response time saturates at a load-factor of 0.25, for un-correlated multicast packets with average fanout of 4 on all ports of an 8 port switch. When only two ports transmit multicast packets, the response time saturates at a load factor of 0.85. These results are better than the performance of virtual output-queued routers [9], where the saturation points where 0.22 and 0.65 respectively. Figure 7(a) shows the multicast latency for a 256 node fat-tree network. Here each node sends a multicast packet to a random set of destinations with an average *fanout* of 8. It can be seen that the latency is stable for load-factors under 0.125, showing the effectiveness of our scheme on a network of switches.

Reduction Performance: The simulated performance of global integer-array reduction is shown in Figure 7(b) for a fat-tree network. Observe that the network with 8 port switches performs better than a network with 32 port switches for message sizes greater than 64 bytes (16 integers or more). The 32 port performance is affected by the stalls in the reduction pipeline for large packets (Section 2). Multiple combine units enhance the performance of 32 port switches. We simulated the tree based hierarchical scheme presented in Section 2. Here, each leaf in the tree (Figure 3(b)) combines application data from a subset of ports in the switch. We take advantage of the fact that reductions on a fat-tree network require only one upper-level port and atmost $K/2$ lower-level ports. So, in our reduction simulation we used 3 combine units, one for the parent of the tree and two for combining application data from $K/4$ lower-level ports in each switch. Reduction completion time with 32 port switches and 3 combine units (Figure 7(b)) shows that even a small number of additional reduction units can achieve good performance for reductions with large messages.

Synthetic MD Benchmark: We present the performance of a synthetic benchmark that emulates our molecular dynamics application NAMD. Several processors in NAMD simultaneously multicast atom coordinates to a small subset of processors, which compute forces on those atoms and return results back to the source processor. The size of this subset typically varies between 13 and 40. In the synthetic benchmark, $P/16$ processors multicast data to random destinations with an average fanout of 16. So on 256 nodes, 16 nodes send multicast messages with an average fanout of 16. Here fanout represents the number of destination nodes of a multicast. The network simulator has a subnet manager component that builds multicast trees before each multicast. The application component of the simulator requests the subnet manager to build trees before each new multicast. In NAMD, each multicast is persistent between load-balancing phases and can use the same set of trees. To simulate this behavior with random multicast destinations, we made the tree build operations happen in zero time. Thus the plots effectively show the behavior of several un-correlated multicasts happening simultaneously in the network. Figures 8(a) and 8(b) show the performance of this synthetic benchmark with hardware multicast and multicast with point-to-point messages

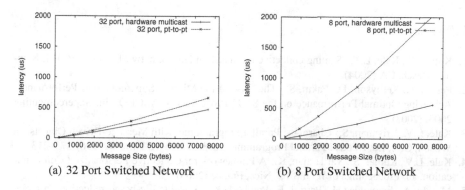

(a) 32 Port Switched Network   (b) 8 Port Switched Network

Fig. 8. Hardware multicast vs pt-to-pt messages for several simultaneous multicasts on 256 nodes with average fanout of 16. The figures clearly show the advantage of hardware multicast. As the network with 8 ports has more levels and hence more points of contention, hardware multicast has more performance gains. On parallel systems with thousands of nodes, even with 32 port switches there will be several levels and hence, more contention for switch outputs. We believe that performance gains of hardware collectives on large systems using 32 port switches is indicated by Figure 8(b).

5   Summary and Future Work

In this paper, we presented the advantages of output queuing for supporting hardware collectives in parallel system interconnects. We showed that an output-queued router with crosspoint buffering achieves better performance with multicast than some of the related work presented in literature. Multicast is quite easy to implement in our router and it avoids deadlocks and other problems faced by input queued routers.

We showed that output queuing also has impressive performance with permutations on fat-tree networks. We are able to achieve almost full network throughput for common permutations with 32 port switches and 4 packet buffers. Large radix crossbars enable us to build fat-tree networks with fewer levels, minimizing network contention. Thus, we show the need to build large radix routers. We also presented schemes to support reductions in switches. We extend a barrier scheme presented in literature to support reductions. We show that the scheme with only one barrier unit does not scale for large packet reductions. We then presented a hierarchical scheme that scales to high radix routers and large reductions. A performance comparison of the two schemes is also presented in the paper. We also described an algorithm to build spanning trees on fat-tree networks. This greedy algorithm uses heuristics that aim at minimizing contention on the upward path of packets and also the routing memory required for the collectives. We also presented the performance of several simultaneous multicasts used by NAMD.

We plan to study output queuing switch with input flow-control buffers. For InfiniBand 4X, buffer space at cross-points may not be a serious issue. However with higher bandwidth networks the RTT would become several packets, requiring input flow-control buffers. Having such an input buffer will reduce total buffer space but may increase the hardware logic, which might be an interesting trade-off to study.

References

1. Kumar, S., Kale, L.V.: Scaling collective multicast on fat-tree networks. In: ICPADS, Newport Beach, CA (2004)
2. Petrini, F., Kerbyson, D., Pakin, S.: The Case of the Missing Supercomputer Performance: Achieving Optimal Performance on the 8,192 Processors of ASCI Q. In: Supercomputing 2003. (2003)
3. Kale, L.V., Krishnan, S.: Charm++: Parallel Programming with Message-Driven Objects. In Wilson, G.V., Lu, P., eds.: Parallel Programming using C++. MIT Press (1996) 175–213
4. Kale, L.V., Kumar, S., Vardarajan, K.: A Framework for Collective Personalized Communication. In: Proceedings of IPDPS'03, Nice, France (2003)
5. Moody, A., Fernandez, J., Petrini, F., Panda, D.K.: Scalable nic-based reduction on large-scale clusters. In: Supercomputing 2003, Phoenix, AZ (2003)

6. Petrini, F., Coll, S., Frachtenberg, E., Hoisie, A.: Performance Evaluation of the Quadrics Interconnection Network. Cluster Computing **6** (2003) 125–142
7. Infiniband Trade Association: Infiniband architecture specification, release 1.0 (2000)
8. Marsan, M.A., Bianco, A., Giaccone, P., Leonardi, E., Neri, F.: On the throughput of input-queued cell-based switches with multicast traffic. In: Proceedings of IEEE Infocom. (2001)
9. Prabhakar, B., McKeown, N., Ahuja, R.: Multicast scheduling for input-queued switches. IEEE Journal of Selected Areas in Communications **15** (1997) 855–866
10. McKeown, N., Izzard, M., Mekkittikul, A., Ellersick, W., Horowitz, M.: Tiny Tera: A packet switch core. IEEE Micro **17** (1997) 26–33
11. Fabrizio Petrini and Marco Vanneschi: K-ary N-trees: High performance networks for massively parallel architectures. Technical Report TR-95-18 (1995)
12. Phillips, J.C., Zheng, G., Kumar, S., Kalé, L.V.: NAMD: Biomolecular simulation on thousands of processors. In: Proceedings of SC 2002, Baltimore, MD (2002)
13. Vadali, R., Kale, L.V., Martyna, G., Tuckerman, M.: Scalable parallelization of ab initio molecular dynamics. Technical report, UIUC, Dept. of Computer Science (2003)
14. Sivaram, R., Stunkel, C.B., Panda, D.K.: HIPIQS: A high-performance switch architecture using input queuing. IEEE Transactions on Parallel and Distributed Systems **13** (2002)
15. Tamir, Y., Frazier, G.L.: High performance multiqueue buffers for VLSI communication switches. In: Proceedings of ISCA. (1988) 343–354
16. Blackwell, T., Chang, K., Kung, H.T., Lin., D.: Credit-based flow control for ATM networks. In: Proc. of 1st Annual Conference on Telecommunications R&D in Massachusetts. (1994)
17. Sivaram, R., Stunkel, C.B., Panda, D.K.: Implementing multidestination worms in switch-based parallel systems: architectural alternatives and their impact. IEEE Transactions on Parallel and Distributed Systems **11** (2000) 794–812
18. Sivaram, R., Stunkel, C., Panda, D.: A reliable hardware barrier synchronization scheme. In: Proceedings of IPPS. (1997) 274–280
19. Wilmarth, T., Kalé, L.V.: Pose: Getting over grainsize in parallel discrete event simulation. In: 2004 International Conference on Parallel Processing. (2004) 12–19
20. Heller, S.: Congestion-free routing on the cm-5 data router. LNCS **853** (1994) 176–184
21. Aydogan, Y., Stunkel, C.B., Aykanat, C., Abali, B.: Adaptive source routing in multistage interconnection networks. In: Proceedings of ICPP. (1996) 258–267

A Clustering and Traffic-Redistribution Scheme for High-Performance IPsec VPNs*

Pan-Lung Tsai, Chun-Ying Huang, Yun-Yin Huang,
Chia-Chang Hsu, and Chin-Laung Lei

Department of Electrical Engineering,
National Taiwan University, Taipei 106, Taiwan
{charles, huangant, yunyin, oberon}@fractal.ee.ntu.edu.tw
lei@cc.ee.ntu.edu.tw

Abstract. CPE-based IPsec VPNs have been widely used to provide secure private communication across the Internet. As the bandwidth of WAN links keeps growing, the bottleneck in a typical deployment of CPE-based IPsec VPNs has moved from the last-mile connections to the customer-edge security gateways. In this paper, we propose a clustering scheme to scale the throughput as required by CPE-based IPsec VPNs. The proposed scheme groups multiple security gateways into a cluster using a transparent self-dispatching technique and allows as many gateways to be added as necessary until the resulting through-put is again limited by the bandwidth of the last-mile connections. It also includes a flow-migration mechanism to keep the load of the gateways balanced. The results of the performance evaluation confirm that the clustering technique and the traffic-redistribution mechanism together create a transparent, adaptive, and highly scalable solution for building high-performance IPsec VPNs.

1 Introduction

As the Internet becomes the most widely used transport for information delivery, it is logical for people to move away from traditional private networks consisting of dedicated circuits and turn to VPNs, or leveraging the Internet for private communication, for the reason of cost savings and convenience [1]. In order to maintain data security during the transmission of private information over the public Internet, the technologies used to implement VPNs usually make use of various encryption and authentication techniques based on cryptographic operations in such a way that the so-called tunnels are established before data transmission and the data flowing through the tunnels is protected. A commonly referred to example is the tunnel mode defined in the IPsec standard [2].

Among all approaches, Customer Premise Equipment-based (CPE-based) VPNs are one of the most popular. Fig. 1 shows a VPN between two organizations and the location of the tunnels. With CPE-based VPNs, private communication among coalition members can be achieved without requiring Internet

* This work was supported in part by the Taiwan Information Security Center, National Science Council under the grant NSC 94-3114-P-001-001-Y.

D.A. Bader et al. (Eds.): HiPC 2005, LNCS 3769, pp. 432–443, 2005.

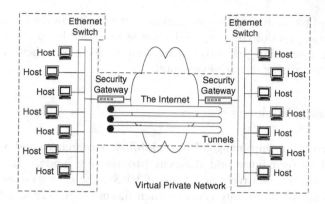

Fig. 1. Tunnels in a typical CPE-based VPN terminate on the respective security gateways residing in the both sides of the VPN

backbone (e.g., service providers) to offer special services other than network connectivity. Besides, as the IPsec protocol becomes a globally recognized standard in the industry, almost all CPE-based VPNs in the modern times run the IPsec protocol [3] and are potentially interoperable.

Despite the advantages such as interoperability and transparency (i.e., hosts behind the gateways are not required to implement the IPsec protocol), the centralized processing model limits the scalability of CPE-based IPsec VPNs. Since each security gateway serves as the endpoints of some tunnels, it has to deal with the aggregated traffic passing through the tunnels rather than just the traffic sent to and received from a single host. Worse yet, cryptographic operations are usually computation-intensive and are likely to consume a large amount of processing power even if only a small volume of traffic is being processed [4][5][6].

The drawback of poor scalability did not cause significant problems until recently in that the performance of a VPN was generally limited by the slowest link, which was one of the last-mile connections. Under such circumstances, security gateways implemented in low-cost RISC architectures with software encryption/decryption were adequate to the job. However, as broadband services like Fiber To The Home/Fiber To The Building (FTTH/FTTB) [7] become more and more popular, the bandwidth between a network and Internet backbone may grow beyond tens or even hundreds of Mbps in the near future. Since the original bottleneck is removed, the security gateways now become the new bottleneck.

In this paper, we propose a clustering scheme to scale the throughput of CPE-based IPsec VPNs. The proposed scheme aggregates the processing power of multiple security gateways by means of a transparent self-dispatching technique and hence allows as many gateways to be added as necessary until the resulting throughput is again limited by the bandwidth of the last-mile connections. We also create our own flow-migration mechanism to keep the load of the gateways balanced. The rest of the paper is organized as follows. Section 2 gives a brief introduction to some important research efforts in this area, especially those devoted to improving the performance of security gateways. Section 3 describes various parts of the proposed scheme in great details, followed by the performance evaluation presented in Section 4. Section 5 then concludes the paper by summarizing our achievements.

2 Related Work

For the purpose of improving VPN scalability, researchers and vendors have created various techniques and different provisioning models. In this section, we first introduce a relatively new type of VPNs, the network-based IP VPNs, which gains much market interest, and then discuss hardware-assisted acceleration technologies for IPsec processing.

2.1 Network-Based IP VPNs

In order to prevent customers with high-volume VPN traffic from paying for expensive security gateways, service providers and equipment vendors proposed the idea of network-based IP VPNs [8][9]. Instead of requiring each customer to acquire a separate security gateway, service providers deploy powerful provider-edge VPN devices that are capable of sustaining a large number of concurrent tunnels and begin to promote secure data transmission as a new service [10]. Thus, VPN tunnels terminate on provider-edge devices and only regular routers are needed on the customer edges. Therefore, customers may get rid of the cost incurred by the deployment and maintenance of specialized security gateways.

The major issue of network-based IP VPNs is the tradeoff between security and scalability [11]. When customers delegate the responsibility for deploying, configuring, and maintaining the devices that are endpoints of VPN tunnels, it is implied that customers must trust their service providers because data gets encrypted after it enters the networks of service providers and service providers may have a chance to inspect or modify the data to be transferred. (In fact, service providers will certainly alter the data in order to transform it into the encrypted form.) Unfortunately, service providers are not always trustworthy, and there are also customers who want better control over the data entering/leaving their networks. After all, security is the primary reason why customers are willing to pay for VPNs in the first place. The proposed scheme does not make such tradeoff. That is, it improves the scalability of CPE-based IPsec VPNs without sacrificing the security.

2.2 Hardware-Accelerated IPsec Processing

Software implementation of packet processing functions is often criticized for their low performance. Therefore, as the IPsec protocol becomes mature, it is natural for researchers and vendors to invent new hardware that accelerates the cryptographic primitives used in IPsec processing. Such performance-enhancing approach is often classified as the scale-up approach [12] because the primary focus is to increase the computation power of a single unit.

Various designs of IPsec-acceleration ASICs and architectures have been proposed and evaluated. [13] proposed a cryptographic coprocessor specialized for AES processing with a maximum throughput of 3.43 Gbps, and [14] described the design of a hardware accelerator which supported many algorithms specified in the IPsec standard, including AES, 3DES, HMAC-MD5 and HMAC-SHA1. [15] compared different hardware architectures for high-performance VPN devices and estimated the total cost of each. [16] presented the benchmarking results of a security-gateway implementation over the IXP425 network processor developed by Intel Corporation, and [17] evaluated the performance of several commercial products, some of which were also implemented on the IXP425 network processors. [18] discussed another security-gateway implementation using Intel IXP1200 network processor, a more powerful one in the family.

Although specialized hardware can be created to improve the performance of certain cryptographic primitive to a great extent, the ASIC-design approaches are also infamous for their inflexibility. If the protocol evolves or a flaw in the design is found after the production of certain ASIC, the revision of the ASIC will have to restart an entire silicon spin, which is both time-consuming and costly. Alternatively, the emerging technology of network processors [19] attempts to overcome the limitation of inflexibility by introducing highly programmable hardware and seems to have successfully drawn much attention. However, as the flexibility increases, the complexity of programming also raises. Besides, vendors have created a wide variety of incompatible programming models, resulting in long learning curves of system integration even for experienced professionals.

In contrast to the scale-up approach, our scheme follows the scale-out approach [12], in which multiple units cooperate to allow higher throughput of IPsec processing, and does not suffer the limitations mentioned above. In addition, it does not interfere with hardware acceleration and hence can be combined with the scale-up approach to achieve even higher performance.

3   The Proposed Scheme

In this section, we start by introducing the reference system architecture and then discuss two important aspects, the techniques for transparently dispatching outbound VPN traffic and the mechanisms for traffic redistribution. Subsequently, we describe the proposed scheme in detail.

3.1   System Architecture

Fig. 2 shows a part of a CPE-based VPN from the perspective of the site that interests us. It corresponds to the left half of Fig. 1 with the difference that the single security gateway on which VPN tunnels terminate is replaced by a cluster of multiple security gateways. The cluster is constructed by surrounding all member gateways with two Ethernet switches and adding a third network

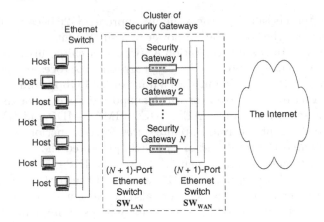

Fig. 2. The cluster is constructed by surrounding multiple security gateways with switches so that it can replace a single security gateway without changing the network topology external to the cluster

interface to each gateway for interconnection. (The third interfaces, the switch used for interconnection, and the cables are not explicitly shown in Fig. 2.) The construction of the cluster is so designed that it can be inserted into the exact location in which the original security gateway resides.

As shown in Fig. 2, there are N security gateways in the cluster. We assume that there are also N security gateways in the remote cluster (i.e., the other side of the VPN). Furthermore, for each tunnel terminating on the original gateways in Fig. 1, there is a counterpart between each pair of the ith gateway in the local cluster and the ith gateway in the remote cluster, where $1 \leq i \leq N$. That is, outbound VPN traffic sent from the ith gateway in one cluster is always destined for the ith gateway in the other cluster. The purpose of making these assumptions is merely to keep the following description away from irrelevant complications. The assumptions can be easily relaxed with minor adjustments.

3.2   Traffic-Dispatching Techniques

In a typical site-to-site scenario of CPE-based IPsec VPNs, when a host has a packet to send to the other side of the VPN, it first looks up $\mathbf{IP_G}$, the IP address of the security gateway, in its routing table. Then the host determines $\mathbf{MAC_G}$, the layer-2 MAC address corresponding to $\mathbf{IP_G}$, by excercising the ARP protocol. After the host gets $\mathbf{MAC_G}$, it encapsulates the packet with a properly constructed frame and places the frame on the wire.

Since each host generally follows the procedure described above, two techniques can be used to distribute the traffic generated by a group of hosts over multiple security gateways. First, we may assign different $\mathbf{IP_G}$'s to individual hosts by manipulating their routing-table entries so that the frames sent from the hosts will be destined for different gateways by default. Second, we may implement customized ARP processing modules on the security gateways so that individual hosts will receive different values of MAC_G for the same IP_G. Note that both techniques make the frames self-dispatched and eliminate the need for centralized dispatchers.

A major disadvantage of the first technique is the lack of transparency. This actually defeats the purpose of making a cluster. On the contrary, when the second technique is adopted, the clustered security gateways share the same IP address, IP_G, on the LAN side (i.e. on the network interfaces connected to SW_{LAN}, which is the left switch within the cluster in Fig. 2) and present the consistent image of a single but more powerful gateway to the hosts. the second technique also limits any customization within the cluster and hence does not require the hosts to change their original behavior. In addition, since most hosts implement ARP processing in a soft-state fashion, the second technique will have less trouble when the bindings between IP_G to various MAC_G's need to be altered. Therefore, the proposed scheme adopts the second traffic-dispatching technique. Section 4.2 demonstrates the effectiveness of this choice.

The two traffic-dispatching techniques mentioned above have their corresponding traffic-redistribution mechanisms. When the first technique is used, it is possible for the security gateways to indirectly update the routing tables of the hosts via ICMP Redirect messages from time to time. However, such traffic-redistribution mechanism requires cooperation from the hosts and is less feasible. As for the second traffic-dispatching technique, we can change the behavior of the customized ARP processing modules on the security gateways to always replying with the MAC address of the least-loaded gateway when responding to ARP requests. The major drawback of this mechanism is its slow response to the variation of the traffic pattern in that it acts passively and the effectiveness depends on the timeout value of the ARP-cache entries stored in individual hosts. Since both mechanisms are not good enough, we decide to create a third mechanism, which is described as follows.

3.3   Clustering with Traffic Redistribution

As shown in Fig. 2, the cluster of the security gateways also plays the role of the edge router. Therefore, in the phase of initial setup, the Ethernet interfaces connected to SW_{LAN} are configured to share IP_G as their common IP address, and those interfaces connected to SW_{WAN} are configured to have distinct public IP addresses. The routing table and the VPN software of each gateway are configured to set up VPN tunnels in the way mentioned in Section 3.1. Once properly configured, each of the security gateways is able to process VPN traffic to and from the tunnels and also route non-VPN traffic to and from the Internet. Some final steps of the initial setup are depicted in the following algorithm.

Algorithm 1. Additional preparation work in the phase of initial setup

1. Generate **V**, a set of K virtual MAC addresses, where K is a sufficiently large number, compared with the number of Ethernet interfaces in the LAN in which the cluster resides. Each address must be unique in the same segment. (Locally administered Ethernet addresses are good choices.)

2. Choose h, a hash function that takes m, a MAC address, as the input and maps it to v, an element in $\mathbf{V}$. That is,

$$h : \mathbf{M} \rightarrow \mathbf{V}, \quad \text{where } \mathbf{M} = \{m \mid m \in \mathbb{Z}, \ 0 \le m \le 2^{48} - 1\}.$$

3. Equally partition $\mathbf{V}$ into subsets $\mathbf{V}_1, \mathbf{V}_2, \mathbf{V}_3, \cdots, \mathbf{V}_N$, where N is the number of security gateways in the cluster.

Succeeding the phase of initial setup, each security gateways in the cluster runs the following four algorithms.

Algorithm 2. Frame processing

1. Define a flow, noted as $\mathbf{F}_m$, to be the set of all Ethernet frames sharing a common destination MAC address, m.
2. **In the outbound direction:** For each element v in $\mathbf{V}_i$, modify the frame-processing module running on $\mathbf{G}_i$, the ith security gateway, $1 \le i \le N$, to additionally accept and process the Ethernet frames (i.e., to pass the received frames to upper-layer VPN software) belonging to $\mathbf{F}_v$. $\mathbf{G}_i$ is referred to as the responsible owner of $\mathbf{F}_v$.
3. **In the inbound direction:** Modify the frame-processing module running on $\mathbf{G}_i$ in such a way that every frame sent to a destination MAC address m in the LAN will always have $h(m)$ as its source MAC address.

Algorithm 3. Port-learning trigger

1. For each element v in $\mathbf{V}_i$, set up a timer routine on $\mathbf{G}_i$ to periodically transmit unsolicited Ethernet frames with source MAC address being v to refresh the forwarding table of $\mathbf{SW_{LAN}}$ so that flooding can be avoided.

Algorithm 4. ARP Processing

1. Upon receiving an ARP request with respect to $\mathbf{IP_G}$, each $\mathbf{G}_i$ extracts the source MAC address m from the request and computes $x = h(m)$.
2. If $x \in \mathbf{V}_i$ (i.e., $\mathbf{G}_i$ is the responsible owner of $\mathbf{F}_x$), $\mathbf{G}_i$ sends back an ARP reply with the answer being x. Otherwise, $\mathbf{G}_i$ ignores the ARP request.

The above three algorithms together create the illusion of a super gateway device with many built-in Ethernet interfaces. With carefully selected K and h, each host in the LAN can be viewed as being directly connected to an individual Ethernet interface of the emulated super gateway device. These algorithms take advantage of the self-learning procedure implemented in every transparent bridge [20][21] and successfully realize fully decentralized traffic distribution. The details of traffic redistribution are described as follows.

Algorithm 5. Flow migration

1. Define the predecessor of $\mathbf{G}_i$, which is denoted by $p(\mathbf{G}_i)$, to be $\mathbf{G}_{i-1}$ when $2 \le i \le N$ and $\mathbf{G}_N$ when $i = 1$. Define the successor of $\mathbf{G}_i$, which is denoted by $s(\mathbf{G}_i)$, to be $\mathbf{G}_{i+1}$ when $1 \le i \le N - 1$ and $\mathbf{G}_1$ when $i = N$.

2. For every T_L seconds, $\mathbf{G}_i$ sends its load information to $p(\mathbf{G}_i)$ through the Ethernet interface designated for inter-gateway communication.
3. For every T_M seconds, $\mathbf{G}_i$ makes a flow-migration decision based on the load information it has. If the load of $\mathbf{G}_i$ is larger than the load of $s(\mathbf{G}_i)$ and the difference is greater than a threshold value L_D, then $\mathbf{G}_i$ randomly generates a subset $\mathbf{W}$ (with a fixed size) of $\mathbf{V}_i$ and notifies $s(\mathbf{G}_i)$ about its selection through the Ethernet interface designated for inter-gateway communication. Otherwise, $\mathbf{G}_i$ does nothing and waits until the next round.
4. Upon receiving a notification message from $p(\mathbf{G}_i)$, $\mathbf{G}_i$ extracts $\mathbf{W}$ from the message and, for each element w in $\mathbf{W}$, $\mathbf{G}_i$ inserts it into $\mathbf{V}_i$ (i.e., $\mathbf{G}_i$ becomes the new responsible owner for $\mathbf{F}_w$) and also sends an acknowledging Ethernet frame (with source MAC address being w) back to $p(\mathbf{G}_i)$ via $\mathbf{SW_{LAN}}$.
5. Upon receiving an acknowledging frame from $s(\mathbf{G}_i)$, $\mathbf{G}_i$ extracts the source MAC address w, which is actually one of the elements previously picked by $\mathbf{G}_i$ itself, from the frame and remove w from $\mathbf{V}_i$.

As depicted in Algorithm 5, traffic redistribution is accomplished by reassigning the flows to new responsible owners. The acknowledging frame sent in step 4 serves dual purposes. In addition to informing $p(\mathbf{G}_i)$ that $\mathbf{G}_i$ has completed the processing of the flow migration regarding w, the acknowledging frame also updates the forwarding table of $\mathbf{SW_{LAN}}$ so that $\mathbf{SW_{LAN}}$ will forward later frames destined for w through the port to which $\mathbf{G}_i$ is connected instead of the original port to which $p(\mathbf{G}_i)$ is connected. Again, Algorithm 5 relies on the self-learning behavior of $\mathbf{SW_{LAN}}$.

4   Performance Evaluation

We implement the proposed scheme as collaborating kernel modules running with Linux kernel (version 2.4.29). Openswan (version 1.0.9) is chosen to be the VPN software. The environment and the results of tests are presented as follows.

4.1   Test Environment

Fig. 3 shows the network diagram of the test environment. We use the world-recognized test equipment, SmartBits, from Spirent Communications, Inc., to generate a large volume of traffic in each test. In the lower half of Fig. 3, each unit is an x86 machine equipped with an Intel Xeon 2.4 GHz processor. The three machines on the left and the three machines on the right comprise two clusters, respectively, and the machine in the middle is configured as a router that simulates the Internet backbone. The PC in Fig. 3 runs the SmartBits control application, SmartFlow (version 3.0), on top of Microsoft Windows XP.

4.2   The Effectiveness of Clustering

In this test, we use SmartBits to simulate 224 individual hosts on each side of the VPN. Every simulated host has a distinct IP address and a unique MAC

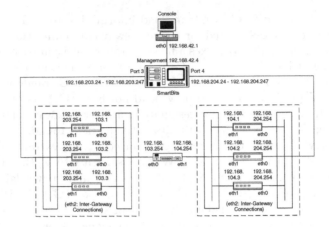

Fig. 3. SmartBits is used to generate a large volume of traffic that passes through the clusters of the security gateways in both directions

address, and will keep sending packets to its counterpart residing in the other side of the VPN once the test starts. The IP addresses and the MAC addresses of the simulated hosts on the left side of the VPN occupy the continuous ranges of 192.168.203.24–247 and 00:00:03:00:00:18–F7, respectively, and the IP addresses and the MAC addresses of the simulated hosts on the right side of the VPN occupy the continuous ranges of 192.168.204.24–247 and 00:00:04:00:00:18–F7, respectively. The endpoint pairs of the tunnels are (192.168.103.1, 192.168.104.1), (192.168.103.2, 192.168.104.2), and (192.168.103.3, 192.168.104.3), respectively.

Since the traffic pattern is fixed over time, we temporarily disable the traffic-redistribution functionality (i.e., Algorithm 5) and focus on the scalability of the chosen traffic-dispatching technique. We set $K = 256$ and make **V** consist of all MAC addresses in the range of 02:88:88:88:88:00–FF. We also choose h to be

$$h(m) = \text{0x028888888800} \mid (m \% K), \quad \text{where} \mid \text{is the bitwise-or operator and}$$
$$\% \text{ is the modulo operator.}$$

Then we measure the zero-loss throughput under the combinations of three different cluster sizes, two different frame sizes, and two popular types of IPsec ESP tunnels. Fig. 4 shows the results.

The numbers and the bars in Fig. 4 show the effectiveness of the chosen traffic-dispatching scheme. The proposed scheme consistently exhibits high scalability under all combinations. Taking the first group, 3DES-SHA1 tunnels with input frame size being 64 bytes, as an example. When the cluster size increases from 1 to 2, the zero-loss throughput is doubled. As we put three gateways into each cluster, the zero-loss throughput becomes 2.98 times, almost tripled.

4.3   The Effectiveness of Traffic Redistribution

Two parameters used in this test are different from those used in the previous test. First, the traffic-redistribution functionality is enabled. Second, the MAC

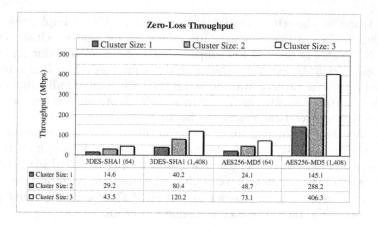

Fig. 4. Zero-loss throughput of the proposed clustering scheme with static traffic-distribution policy grows linearly as the cluster size increases

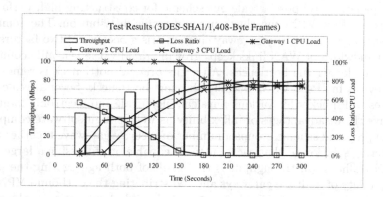

Fig. 5. Test results show that the overall throughput keeps growing to its maximum as the traffic is getting more and more evenly distributed

addresses of the simulated hosts on the two sides of the VPN are now the two arithmetic sequences 00:00:03:00:18:00, 00:00:03:00:19:00, 00:00:03:00:1A:00, $\cdots$, 00:00:03:00:F7:00 and 00:00:04:00:18:00, 00:00:04:00:19:00, 00:00:04:00:1A:00, $\cdots$, 00:00:04:00:F7:00, respectively. Using these addresses with the chosen h makes the first security gateway in each cluster become the initial responsible owner of all flows when the test starts.

This time we use 3DES-SHA1 tunnels, a value of 3 for N, a value of 1.33 for T_L, a value of 4 for T_M, a value of 10% for L_D, and a value of 4 for the size of **W**. We also take a snapshot every 30 seconds during the test. When the test starts, we instruct SmartBits to generate bidirectional 100 Mbps of traffic in the form of 1,408-byte frames for 300 seconds. The test results are shown in Fig. 5.

In Fig. 5, the bars designate the measured throughput of the bidirectional VPN traffic in the snapshots taken every 30 seconds. The values of the bars can be found on the left vertical axis in the figure. For example, the value of the rightmost bar in Fig. 5 is 100 Mbps. That means the two clusters forming the VPN are able to process 100 Mbps of left-to-right VPN traffic and another 100 Mbps of right-to-left VPN traffic at the same time. The four lines in the figure represent the ratio of frame loss and the CPU load of individual gateways in the left cluster, respectively. The values of the data points in the lines can be found on the right vertical axis in the figure.

The bars show a climbing trend of overall throughput and, around the 180th second, the throughput reaches its maximum of 100 Mbps, which is exactly the same as the offered traffic. The lines representing the CPU load provide an evident trace of the progression of traffic redistribution. As soon as a sufficient number of flows have been reassigned and the CPU load of the first gateway drops below 100%, there becomes no frame loss at all.

5   Conclusions

In this paper, we propose a scale-out scheme for constructing high-performance CPE-based IPsec VPNs via clustering and traffic distribution. The demand for higher-throughput security-gateway clusters arises because of two factors. First, as the bandwidth of WAN links keeps growing rapidly, the last-mile connections are now no longer the bottleneck. After the original bottleneck is removed, the edge devices in IPsec VPNs become the new bottleneck. Second, scale-up approaches will quickly hit various physical, electrical, and electronic limitations, and have the disadvantages of inflexibility as well as high software complexity.

The proposed scheme groups a number of security gateways together to form a clustering solution so that each cluster is capable of processing a large volume of VPN traffic. High throughput is the result of both aggregating the processing capability of multiple gateways and redistributing the outbound VPN traffic when the load of the gateways becomes unbalanced. Aggregation of the processing power is realized by a transparent self-dispatching technique that requires no dispatchers, and traffic redistribution is accomplished by flow migration. The proposed scheme is adaptive, scalable, and transparent. It responds to the variation of the traffic pattern promptly, and the throughput grows linearly as the number of security gateways in the cluster increases. All customization is limited within the clusters and no change to any other components outside the clusters is needed. In addition, the proposed scheme does not suffer the limitations of the scale-up approaches and can be combined with hardware-accelerated solutions to achieve even higher performance.

References

1. Ortiz, Jr., S.: Virtual private networks: Leveraging the Internet. IEEE Computer **30** (1997) 18–20
2. Kent, S., Atkinson, R.: Security architecture for the Internet protocol. RFC 2401 (1998)

3. Knight, P., Lewis, C.: Layer 2 and 3 virtual private networks: Taxonomy, technology, and standardization efforts. IEEE Communications Magazine **42** (2004) 124–131

4. Elkeelany, O., Matalgah, M.M., Sheikh, K.P., Thaker, M., Chaudhry, G., Medhi, D., Qaddour, J.: Performance analysis of IPSec protocol: Encryption and authentication. In: Proceedings of 2002 IEEE International Conference on Communications (ICC 2002). Volume 2. (2002) 1164–1168

5. Lin, J.C., Chang, C.T., Chung, W.T.: Design, implementation and performance evaluation of IP-VPN. In: Proceedings of 17th International Conference on Advanced Information Networking and Applications (AINA 2003). (2003) 206–209

6. Khanvilkar, S., Khokhar, A.: Virtual private networks: An overview with performance evaluation. IEEE Communications Magazine **42** (2004) 146–154

7. Kettler, D., Kafka, H., Spears, D.: Driving fiber to the home. IEEE Communications Magazine **38** (2000) 106–110

8. Metz, C.: The latest in virtual private networks: Part I. IEEE Internet Computing **7** (2003) 87–91

9. Metz, C.: The latest in virtual private networks: Part II. IEEE Internet Computing **8** (2003) 60–65

10. Carugi, M., De Clercq, J.: Virtual private network services: Scenarios, requirements and architectural constructs from a standardization perspective. IEEE Communications Magazine **42** (2004) 116–122

11. De Clercq, J., Paridaens, O.: Scalability implications of virtual private networks. IEEE Communications Magazine **40** (2002) 151–157

12. Devlin, B., Gray, J., Laing, B., Spix, G.: Scalability terminology: Farms, clones, partitions, and packs: RACS and RAPS. Technical Report MS-TR-99-85, Microsoft Research (1999)

13. Hodjat, A., Verbauwhede, I.: High-throughput programmable cryptocoprocessor. IEEE Micro **24** (2004) 34–45

14. Ha, C.S., Lee, J.H., Leem, D.S., Park, M.S., Choi, B.Y.: ASIC design of IPSec hardware accelerator for network security. In: Proceedings of 2004 IEEE Asia-Pacific Conference on Advanced System Integrated Circuits (AP-ASIC 2004). (2004) 168–171

15. Friend, R.: Making the gigabit IPsec VPN architecture secure. IEEE Computer **37** (2004) 54–60

16. Lin, Y.N., Lin, C.H., Lin, Y.D., Lai, Y.C.: VPN gateways over network processors: Implementation and evaluation. In: Proceedings of 11th IEEE Real-Time and Embedded Technology and Applications Symposium (RTAS 2005). (2005) 480–486

17. The Tolly Group, Inc.: Intel IXP425 network processors: Performance analysis of VPN devices. Document No. 204132 (2004)

18. Han, M., Kim, J., Sohn, S.: Network processor for IPSec. In: Proceedings of 6th International Conference on Advanced Communication Technology (ICACT 2004). Volume 1. (2004) 485–487

19. Comer, D.E.: Network Systems Design Using Network Processors. Pearson Prentice Hall, Inc. (2003)

20. IEEE Standards Association: IEEE standard for local and metropolitan area networks: Media access control (MAC) bridges. IEEE 802.1D-2004 (2004)

21. Seifert, R.: The Switch Book: The Complete Guide to LAN Switching Technology. John Wiley & Sons, Inc. (2000)

WDM Multistage Interconnection Networks Architectures for Enhancing Supernetworks Switching Infrastructure

Haitham S. Hamza and Jitender S. Deogun

Department of Computer Science & Engineering,
University of Nebraska-Lincoln,
Lincoln, NE 68588-0115, USA
{hhamza, deogun}@cse.unl.edu

Abstract. Multistage Interconnection Networks (MINs) provide the required switching infrastructure for many shared-memory multiprocessor systems and telecommunication networks. The concept of *Supernetworks* is evolving in response to emerging computation and communication intensive applications. Supernetworks exploit parallelism in both computing resources and communication infrastructures by interconnecting several computing clusters via high-bandwidth communication links. Wavelength Division Multiplexing (WDM) technology provides the communication infrastructure for Supernetowrks by dividing the bandwidth of a single fiber into numerous channels that can be used independently. In this paper, we investigate several architectures for WDM MINs that enhance the Supernetworks switching infrastructure. Our objective is to propose a new architecture and to evaluate its hardware complexity by comparing it to other WDM MINs architectures.

1 Introduction

Computing applications appearing on the horizon are not only computation intensive, but also *communication intensive* (e.g. genomic and visualization applications). The concept of *Supernetworks* is evolving as a solution to meet the challenges brought fourth by such applications. In Supernetworks, several distributed clusters are connected through multiple dedicated optical channels (See Figure 1). Supernetworks exploit parallelism not only in the computing resources, but also in the communication infrastructures [14].

Supernetworks can exploit parallelism in the communication infrastructure by using Wavelength Division Multiplexing (WDM) technology [14]. WDM technology exploits the tremendous bandwidth embedded in a single fiber by dividing this bandwidth into several channels (wavelengths) that can be used independently. To fully utilize the potential of WDM technology, advances in optical switch architectures are required to provide a high-performance wavelength granularity switching that is cost-effective.

Over the last few decades, considerable research efforts have been focused on developing photonic switch architectures, or Optical Cross-connects (OXCs). An

D.A. Bader et al. (Eds.): HiPC 2005, LNCS 3769, pp. 444–453, 2005.

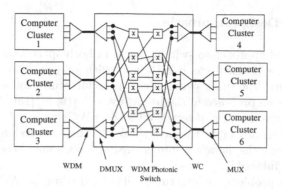

Fig. 1. A generic architecture for Supernetworks

OXC provides the basic functionality of connecting signals from input ports to the desired output ports. In WDM networks, signals may need to be switched in both *space* and *wavelength* domains in order to establish the required connection. For example, two signals on the same wavelength from two different input fibers may need to be switched to the same output fiber. In such a case, *Wavelength Converters* (WCs) are used to covert one of the signals to a different wavelength.

The main challenge in designing OXCs is to provide the required switching functionality with minimum hardware components. Unfortunately, however, as the number of wavelengths per fiber increases, larger space switches as well more WCs are needed; leading to high-cost OXC designs. *Multistage Interconnection Network* (MIN) are used to economically realize large space switches [12] [17]. In general, a MIN interconnects a set of N input ports to a set of M outputs ports using several stages of fixed-size switching modules. Most research on MINs have focused on the design and development of *pure space* MINs, where signals are only switched in the space domain [8] [21].

In this paper, we investigate WDM MINs that are capable of switching signals both in space and wavelength domains. Our objective is to propose a new architecture and to evaluate its hardware complexity by comparing it to other WDM MINs architectures. The new design transform the classical *pure space* $N-$stage Planar network [16] to a *space-wavelength* Planar network.

The reminder of the paper is organized as following. Section 2 discusses different design approaches for WDM MINs. The WDM Planar architecture and the analysis of its hardware complexity are presented in Section 3. Section 4 reviews some existing WDM MINs architectures. A comparison of architectures is given in Section 5; Conclusions are presented in Section 6.

2   Designing a WDM MIN

In this section, we briefly review some existing design approaches for WDM MINs. Then we describe the underlying design principle of the proposed WDM Planer architecture.

2.1   Existing Design Approaches

A straightforward approach to switch signals in both space and wavelength domains is to use a classical space MIN and with one or more stages of WCs before and/or after the space switch. Several architectures following this design approach have been proposed in the literature, e.g. [19] [9] [13] [20]. Most WDM architectures make use of *Full-range* WCs (or FWCs), e.g. [19]. FWCs are capable of converting any input wavelength to any output wavelength. With large number of wavelengths, however, implementing FWCs may become impractical or economically infeasible.

To avoid this problem, architectures with *Limited-range* WCs (LWCs) have received more attention over the last few years [11] [20]. Unlike FWCs, LWCs provide conversion between a sub-set of wavelengths. However, architectures with LWCs may require more complex routing algorithms; making them less attractive for packet/burst switching networks.

To reduce the cost of WDM switches, some approaches suggested the share of a pool of WCs (FWCs or LWCs) within the switch [9] [13]. However, besides their increased complexity, these architectures, in general, may have limited interconnection capabilities [22].

Another design approach to realize WDM MINs is the *wave-mixing* architectures [1][5]. Wave-mixing architectures make use of *bulk* wavelength converters, based on optical nonlinearities, to simultaneously convert several signals [1]. Wavelength conversion occurs between the switching stages. Even though wave-mixing architectures can reduce the overall number of WCs; however, up to $O(logW)$ extra stages of WCs may be added to the MIN [1]. The extra stages of WCs not only increase hardware complexity, but also increase the length of the signal path, which in turn can increase signal attenuation and accumulated cross-talk noise.

2.2   Wavelegnth Exchanging Cross-Connect — WEX

Our objective is to develop an architecture that can switch signals in space and wavelength while reducing the following factors:(1) the number of switch stages; (2) the total number of hardware components; and (3) the cost of wavelength conversion. To achieve this, the proposed architecture employs *Wavelength Exchange Optical Crossbars* (WOCs, for short) [6]. WOC integrates space and wavelength domains in a way that signals can be switched in both domains *simultaneously* and *seamlessly*. Switch architectures that use WOCs as building blocks are called *Wavelength exchanging cross-connect* (WEX) [6].

A WOC has two input ports, two output ports, and a control signal (See Figure 2). The input to an WOC is two signals S_1 at wavelength λ_1, and S_2 at wavelength λ_2. When the control signal is *OFF*; an input signal to the WOC appears at an output port with the same wavelength. This functionality is equivalent to the bar state in a traditional crossbar switch. Conversely, when the control signal is *ON*, the WOC performs both switching and conversion simultaneously.

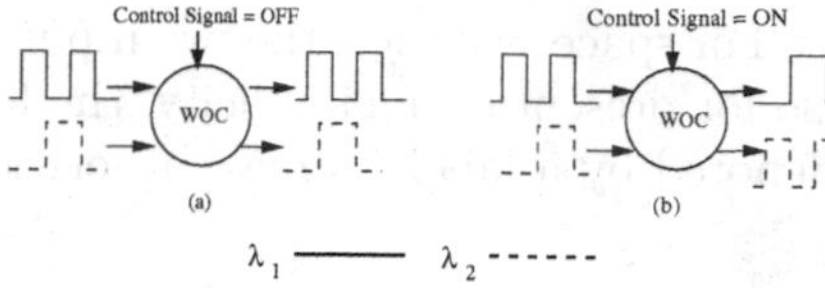

Fig. 2. The WOC and its different configurations: (a) Bar state (b) Simultaneous switching and wavelength conversion

WOCs can be realized using the *wavelength exchanging* phenomenon, where the power of two signals at different wavelengths can be simultaneously exchanged [18]. Wavelength exchanging has been theoretically and experimentally demonstrated using *Four Wave Mixing* (FWM) [10] [18], and the *2-D periodic* $\chi^{(2)}$ *nonlinear photonic crystals* [3].

The new architecture have the following main advantages:

1. Space switching and wavelength conversion are performed simultaneously, and hence, eliminating the need for separate wavelength conversion devices. This leads to a switch architecture that has a smaller total number of components as well smaller number of components in a signal path;
2. Scalability occurs in an orderly fashion, and hence, systematic methods to construct architectures with an arbitrary number of wavelengths can be developed; and
3. Wavelength conversion is performed between two *predefined* and *fixed* wavelengths. This not only can reduce the wavelength conversion cost, but also it reduces the switch configuration time and complexity.

3   The WDM $N-$Stage Planar Network

Figure 3 shows the 8-Stage Planar network [15] [16]. The $N \times N$ Planar network requires N stages of SEs. In the following, we present some notations and definitions, give a systematic method for developing a WDM $N-$Stage Planar network, and analyze its hardware complexity.

3.1   Definitions and Notations

We denote a WDM switch with F input fibers and F output fibers, where each fiber carries W wavelengths as $W^\lambda(F \times F)$. The set of fibers F are denoted as $\{f_1, f_2, ..., f_F\}$, and the set of wavelengths W are denoted as $\{\lambda_1, \lambda_2, ..., \lambda_W\}$. The dimension of $(W\lambda)F \times F$ switch is denoted by $N \times N$, where $N = FW$. Without the lose of generality, we assume both F and W are powers of 2.

A signal on wavelength λ_w in fiber j is denoted as $\lambda_w^{f_j}$. We refer to a 2×2 switching element inside the architecture as a *switch element* (SE). The *input stage* of an architecture is the first stage of the architecture, and the *output stage* is the last stage.

An *element* in the architecture refers to either a SE or a WOC. We use the notation (a, b) to denote the *label* of an element in the switch that has two inputs at wavelengths λ_a and λ_b. For space switches, the two inputs must be at the same wavelength, i.e., $a = b$, so for presentation simplicity, the label (k, k) for a 2×2 space switch is simply denoted by a label k unless otherwise is specified.

3.2   The Construction of a $W^\lambda(F \times F)$ WDM Planar Network

Figure 4 shows the $4^\lambda(2 \times 2)$ WDM Planar network. It can be noted from the structure of the Planar network (Figure 3) that stages with odd numbers (with the input stage being number 1) have larger number of elements compared to stages with even numbers. Thus, in order to reduce the number of WOCs, elements of the input stage must be SEs. This can be simply achieved by connected every two same wavelengths form input fibers to the same input switch as shown in Figure Figure 4.

We generalized the above design approach to develop the method shown in Figure 5. This method can be used to systematically develop a WDM Planar network with an arbitrary number of wavelengths and fibers. The method can be applied to construct a $2^\lambda(4 \times 4)$ Planar network as shown in Figure 6.

3.3   Hardware Complexity

The total number of components can provide a good estimation of the hardware complexity of the architecture. The following lemma gives the total number of space switches and WOCs in an $W^\lambda(F \times F)$ WDM Planar architecture.

Lemma 1. A $W^\lambda(F \times F)$ WDM Planar architecture contains $(N/2)(W - 1)$ WOCs and $(N/2)(N - W)$ space switches.

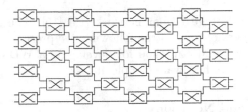

Fig. 3. The 8–Stage Planar architecture

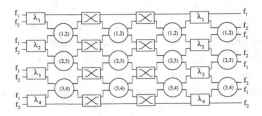

Fig. 4. The $4^\lambda(2 \times 2)$ WDM Planar architecture

$Construct - PLANAR(W, F)$:

1. **Construct** a $WF \times WF$ Planar network.
2. **Label** elements in the first two stages as follows:
   (a) **Label** every $F/2$ space switches in the input stage with label l, $l = 1, 2, ..., W$. **Repeat** labels until all input switches are labeled.
   (b) **Construct** a set $G = \{g_1, g_2, ..., g_{N/2}\}$ input stage switches' labels, in order.
   (c) **Assign** an element i in the second stage a label $L(i) = (i, i + 1)$, $i = 1, 2, ..., (N/2) - 1$.
   (d) **if** an element has a label (a, b), and $a = b$:
       Replace this element with a space switch.
   else
       Replace element with a WOC with label (a, b).
3. **Repeat** the labels of the first two stages $(WF/2) - 1$ times.

Fig. 5. The $Construct - Planar(W, F)$ method

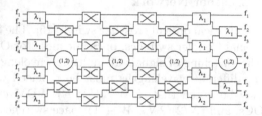

Fig. 6. The $(2\lambda)4 \times 4$ WDM Planar architecture

Proof. From the construction of the WDM Planar network, similar wavelengths from each pair of input fibers are connected to the same switch in the input stage. Consider a single wavelength λ_i; there are $F/2$ consecutive switches in the input stage that have input signals at λ_i. Therefore, in the second stage, there should be $(F/2) - 1$ space switches with input signals at λ_i. For W wavelengths, the second stage should contain a total of $W[(F/2) - 1]$ space switches. Therefore, the total number of WOCs in the second stage is $(WF/2) - 1 - W[(F/2) - 1] = W - 1$ WOCs. From the recursive structure of the network, there are $N/2$ stages with the same structure of the second state, thus, there are $(N/2)(W - 1)$ WOCs in the architecture . Since the Planar network has $(N/2)(N - 1)$ elements; thus, there are $(N/2)(N - 1) - (N/2)(W - 1) = (N/2)(N - W)$ space switches. $\square$

4   Architectures for Comparison

In this section, we review some existing WEX architectures that we use in Section 6 for evaluating the hardware complexity of the proposed Planer WDM network.

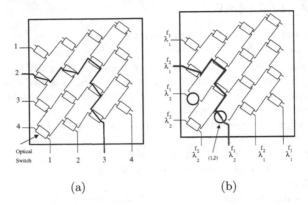

(a)   (b)

Fig. 7. (a) The 4×4 single crossbar, and (b) The $2^\lambda (2 \times 2)$ WDM crossbar [7]

4.1   The WDM Crossbar Network

Figure 7- a shows the classical optical crossbar switch [15]. The SEs of the switch can be 2×2 Directional Coupler switches, or Semiconductor Optical Amplifiers (SOAs). To allow the switching of signals at multiple wavelengths, any SE with two input signals at different wavelengths must be replaced with a WOC (see Figure 7- b). It may be noted that, a $W^\lambda (F \times F)$ WDM crossbar, contains $(N/2)(W-1)$ WOCs, and $(N/2)(2N - W + 1)$ space switches [7].

4.2   The WDM Benes Network

Benes network [2] is known as one of the most efficient MINs due to its small number of switching elements and its low insertion loss. One possible design to realize a $2^\lambda (2 \times 2)$ WDM Benes network is the one shown in Figure 8-a [6]. To construct the $2^\lambda (2 \times 2)$ WDM Benes, we start with a 4×4 space Benes network, and replace the two SEs in the middle stage with a pair of WOCs. From the connectivity pattern between the input and middle stage, it can be seen that both WOCs have the label $(1, 2)$. It is worth noting that, other constrcutions for the $2^\lambda (2 \times 2)$

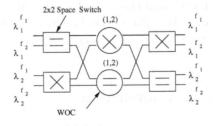

Fig. 8. The $2^\lambda (2 \times 2)$ WDM Benes architecture

WDM Benes can be obtained by changing the connectivity pattern of the wavelengths to the input stage switches [6]. It can be shown that a $W^\lambda(F \times F)$ Benes architecture contains WF space switches and $(WF/2)(2log(WF) - 3)$ WOCs.

4.3 The WDM Clos Architecture

Let $N = rn$, a 3-stage $N \times N$ Clos network [4], denoted by Clos(m, n, r), has r switches of size $n \times m$ at the first stage; m switches of size $r \times r$ in the second stage; and r switches of size $m \times n$ in the third stage. Clos network is *strictly nonblocking* if $m \geq 2n - 1$ and it is *rearrangeable nonblocking* if $m \geq n$.

One possible design to develop a WDM Clos network is to connect similar wavelengths from all input fibers to the same switch in the input stage [7]. In such a design, the first and third stages of the architecture contain pure space switches. The middle stage, however, contains m WDM crossbar switches (see Section 5.1). The total number of WOCs is determined by the middle stage of the switch. It can be seen that a $W^\lambda(F \times F)$ WDM Clos with $m = n$ contains $(mW/2)(4F + W + 1)$ space switches and $(mW/2)(W - 1)$ WOCs [7].

5 Comparison of Architectures

In this section, we compare the total number of SEs and WOCs for different WDM MIN networks discussed in Sections 3 and 4. Figures 9-a and 9-b, respectively, compare the total number of SEs and WOCs of different architectures for $F = 8$ and $W = 2, 4, 8$, and 32. For the Clos network, we chosen $m = F = 8$.

As shown in the figures, for small number of wavelengths ($W \leq 8$), Benes network requires more WOCs compared to all other architectures. Moreover, the cost of the Planar network seems to be very close to that of the Clos network

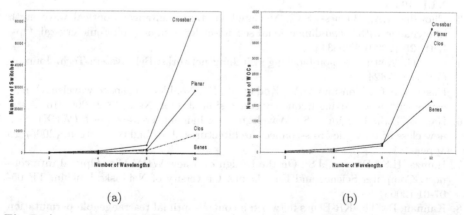

(a)   (b)

Fig. 9. A comparsion of different architectures with $F = 8$ (a) Total number of SEs, and (b) The total number of WOCs

in terms of the number of SEs and WOCs. When the number of wavelengths increases($W > 8$), Benes network requires the least number of components compared to all other architectures.

In general, it can be concluded that, for networks with large number of wavelengths, Benes network can provide a cost-effective design in terms of the number of SEs and WOCs. It is worth noting that, although Benes network requires less number of elements compared to Planar network, however, Benes network contains large number of waveguide crossovers as opposed to the Planar network which contains no waveguide crossovers. Crossover can potentially complicate the implementation of the switch in integrated optics. Moreover, the repetitive multistage interconnection design of the Planar network can reduce its implementation cost compared to other architectures.

6   Conclusions

In this paper, we developed a new WDM MIN that transforms the pure space N−stage Planar network to a WDM Planar network that is capable of switching signals in both space and wavelength domains. The new design requires the same number of stages as in the pure space Planar network. In addition, wavelength conversion in the proposed design is performed between two *fixed* and *predefined* wavelengths, which eliminates the need for the expensive FWCs and simplifies the configuration of the switch.

References

1. Antoniades, N., Yoo, S.J.B., Bala, K., Ellinas, G., Stern, T.E.: An architecture for a wavelength-Interchanging cross-connect utilizing parametric wavelength converters. J. of Lightwave Tech.**17** (1999) 1113-1125
2. Benes, V.E.: On rearrangeable three-stage connecting networks. Bell Syst. Tech. J. **XLI** (1962)
3. Chowdhury,A., Hagness, S.C., McCaughan, L.: Simultaneous optical wavelength interchange with a two-dimensional second-order nonlinear photonic crystal. Opt. Lett. **25** (2000) 832-834
4. Clos, C.: A study of non-blocking switching networks. Bell System Tech. Journal. (1958) 407-424
5. Dasylva, A.C., Montuno, D.Y., Kodaypak, P. : Nonblocking space-wavelength networks with wave-mixing frequency conversion. J. Opt. Net. (2002) 206-216
6. Hamza, H.S., Deogun, J.S.: Wavelegnth exchanging cross-connect (WEX) — a new class of photonic cross-connect architectures. J. of Lightwave Tech. (2005) To appear.
7. Hamza, H.S., Deogun, J.S.: On the Design of a new WDM Clos Optical Interconnect. Computer Science and Eng. Dept., University of Nebraska-Lincoln. TR-05-07-01 (2005)
8. Kannan, R.: The KR-Benes network: a control-optimal rearrangeable permutation network. IEEE Transactions on Computers **54** (2005)
9. Lee K.-C., Li, V.O.K.: A wavelength-convertible optical network. IEEE/OSA J. of Lightwave Tech. **11** (1993) 962-970

10. Moei, K., Takara H., Saruwatari, M.: Wavelength interchange with an optical parametric loop mirror. Electronics Lett. **33** (1997) 520 -522
11. Ngo,H.Q., Pan, D., Yang, Y.: Optical switching networks with minimum number of limited range wavelength converters. Proc. 24th IEEE International Conf. on Computer Communications (Infocom 2005) (2005)
12. Tewksbury, S.K., Hornak,L. A.: Communication network issues and high density interconnects in large-scale distributed computing systems. IEEE J. Select. Areas Commun. **6** (1988) 587-609
13. Torrington-Smith, N.P. , Mouftah, H.T., Rahman, M.H.: An evaluation of optical switch architectures utilizing wavelength converters. Electrical and Computer Eng., 2000 Canadian Conference **2** (2000) 1008 -1013
14. Smarr, L.L. et al.: The OptIPuter. Communication of the ACM **46** (2003) 59-66
15. Spanke, R.A.: Architectures for guided-wave optical space switching systems. IEEE Communication Magazine, 25(1987) 42 -48
16. Spanke, R.A., Benes, V.E.: An N-stage planar optical permutation netowrk. Applied Optics **26** (1987)
17. Stern, T.E., Bala, K.: Multiwavelength optical networks: a layered approach. Addison Wesley (1999)
18. Uesaka, K., Wong, K. K-Y., Marhic, M.E., Kazovsky, L.G.: Wavelength Exchange in a Highly Nonlinear Dispersion-Shifted Fiber: Theory and Experiments. IEEE J. of Selected Topics in Quantum Elect. **8** (2002) 560-568
19. Wilfong,G., Mikkelsen,B., Doerr, C., Zirngibl, M.: WDM cross-connect architectures with reduced complexity. J. of Lightwave Tech. **17** (1999) 1732-1741
20. Yang, Y., Wang, J.: Designing WDM optical interconnection with full connectivity by using limited wavelength conversion. IEEE Trans. on Computers **53** (2004)
21. Yang, Y., Wang, J.: A fault-tolerant rearrangeable permutation network. IEEE Trans. on Computers **53** (2004) 414-426
22. Yang, Y., Wang, J.: Cost-effective designs of WDM optical interconnects. IEEE Transactions on Parallel and Distributed Sys. **16** (2005) 51-66

Learning-TCP: A Novel Learning Automata Based Congestion Window Updating Mechanism for Ad hoc Wireless Networks

B. Venkata Ramana and C. Siva Ram Murthy

Department of Computer Science and Engineering,
Indian Institute of Technology, Madras, India
vramana@cs.iitm.ernet.in, murthy@iitm.ac.in

Abstract. The use of traditional TCP, in its present form, for reliable transport over Ad hoc Wireless Networks (AWNs) leads to a significant degradation in the network performance. This is primarily due to the congestion window (*cwnd*) updation and congestion control mechanisms employed by TCP and its inability to distinguish congestion losses from wireless losses. In order to provide an efficient reliable transport over AWNs, we propose Learning-TCP, a novel learning automata based reliable transport protocol, which efficiently adjusts the *cwnd* size and thus reduces the packet losses. The key idea behind Learning-TCP is that, it dynamically adapts to the changing network conditions and appropriately updates the *cwnd* size by observing the arrival of acknowledgment (ACK) and duplicate ACK (DUPACK) packets. Learning-TCP, unlike other existing proposals for reliable transport over AWNs, does not require any explicit feedback, such as congestion and link failure notifications, from the network. We provide extensive simulation studies of Learning-TCP under varying network conditions, that show increased throughput (9-18%) and reduced packet loss (42-55%) compared to that of TCP.

1 Introduction

Ad hoc Wireless Networks (AWNs) are formed dynamically by a collection of mobile wireless nodes in the absence of any fixed infrastructure. Communication between any two nodes that are not within the radio range of each other takes place in a multi-hop fashion, with other nodes acting as routers. The AWNs are typically characterized by unpredictable and unrestricted mobility of the nodes and the absence of a centralized administration. The AWNs are considered as resource constrained networks because of the limited bandwidth on the network and limited processing and battery powers at the mobile nodes. These networks are very useful in military, emergency rescue operations, where the existing infrastructure may be unreliable or even unavailable, and also in commercial applications, such as on-the-fly conferences and electronic classrooms. Extensive research work on ad hoc wireless networking has been carried out on

D.A. Bader et al. (Eds.): HiPC 2005, LNCS 3769, pp. 454–464, 2005.

issues, such as medium access and routing [1] and [2]. In our paper, we focus on the issues related to reliable and adaptive data transport over AWNs.

Transmission Control Protocol (TCP) is the de-facto protocol used for reliable transport over the Internet. However, using TCP in its present form for the AWNs, degrades the performance of the networks, in terms of increase in the packet loss and reduction in the average network throughput. The reasons are as follows. TCP uses a deterministic *cwnd* updating mechanism that increases *cwnd* on receipt of an ACK packet and decreases the *cwnd* on receiving three successive DUPACK packets or upon the occurrence of an RTO. This deterministic approach may often lead to the occurrence of congestion situations in the network which further results in a high packet loss. The high packet loss affects AWNs severely as they are highly constrained by the bandwidth and the battery power. Another reason is the TCP's inability to distinguish congestion losses from the wireless losses, as a result of which TCP invokes the congestion control mechanism for both types of the losses. Hence, in order to use TCP in AWNs, an efficient mechanism must be provided with TCP for updating the *cwnd* based on the network conditions and distinguishing the congestion losses from wireless losses in order to take the appropriate action for each type of loss.

In order to provide an efficient reliable transport over AWNs, in this paper, we propose Learning-TCP, a novel *cwnd* updating mechanism that aims at minimizing the losses due to congestion. Learning-TCP works without the explicit feedback from the network and uses learning automata [8] to learn the network behavior. Our work is motivated by the advantages of learning automata over that of conventional mechanisms, such as machine learning and fuzzy logic techniques. Learning automata based solutions learn the network state better and faster and do not require modeling of the network. Unlike other techniques, learning automata require simple algorithms for updating probabilities of various actions, and the amount of information to be maintained and the number of computations to be done are significantly low.

The learning automata consists of a learning automaton which provides a simple model for adaptive decision making with unknown random environments. The learning automaton interacts with the environment by selecting an action from a set of actions. When a specific action is performed, the environment provides either a favorable or an unfavorable response. The selection of action could be either deterministic or stochastic. In the latter case, probabilities are maintained for each possible action to be taken, which are updated with the reception of each response from the environment. The objective in the design of the learning automaton is to determine how the previous actions and responses should affect the choice of the current action to be taken, and to improve or optimize some predefined objective function.

The models of learning automata can be classified based on the number of actions in the action set. The Finite Action-set Learning Automata (FALA) is one such model which contains the finite number of actions and each action corresponds to a range of responses provided by the environment. In FALA, the number of actions need to be finite. However, such discretization may not be possible in all the situations as the discretization may be too coarse for the problem, and a finer discretization may result in a large number of actions; these large number of actions may increase the time to update the action probabilities and also complicate the process of decision making.

A natural choice in such a case, would be to use the Continuous Action-set Learning Automaton (CALA) [9] which has an infinite number of actions. The basic idea is that, the CALA selects an action x at each time instant n, where x is a real number chosen from a normal distribution with mean $\mu(n)$ and standard deviation $\sigma(n)$. For this purpose, the CALA maintains $\mu(n)$ and $\sigma(n)$ of the probability distribution and it updates these values at each time instant based on the reinforcement received from the environment. The details about the updating function used in CALA are provided in Section 3. For a detailed description of learning automata, readers are referred to [8] and [9].

The rest of the paper is organized as follows. Section 2 discusses the related work. Section 3 provides the design and functional details of Learning-TCP. Section 4 presents the simulation results and analysis of the results. Finally, Section 5 summarizes our work.

2   Related Work

Several proposals ([3]-[7]) have been made in order to address problems related to the reliable transport over AWNs. These proposals, follow different approaches for providing reliable transport over AWNs, and can be classified broadly into two categories – network dependent and network independent approaches. Fig. 1 gives a detailed classification of these proposals. The proposals [3], [4], [5], and [6] are network dependent as they rely on explicit feedback information from the network, such as congestion, link failure, and available bandwidth notifications, in order to work efficiently. However, network dependent proposals may perform poorly when the feedback information from network is unavailable or unreliable, which is more common in the AWNs. Hence, network independent approaches are gaining research focus. The proposal [7], which belongs to this category, does not require any feedback from the network. Instead, it performs loss classification and takes appropriate actions, using fuzzy logic based approach. Here, the fuzzy engine monitors the rate of change of RTT and the number of hops (obtained from the routing protocol) in the TCP session and detects congestion when RTT increases n times successively, in which every increment is larger than a given α (where n and α are predefined). The channel error losses are detected when the mean RTT is small. The path break errors are determined in a more deterministic way - on the occurrence of an RTO.

There are several TCP variants, such as TCP Vegas and TCP Westwood that focus on the updating the $cwnd$ in wired/wireless environment. These proposals follow a deterministic approach for updating the $cwnd$. However, to the best of our knowledge, there exists no proposal for AWNs that focuses on the updation of the $cwnd$ by following a probabilistic approach and without relying on the explicit network feedback.

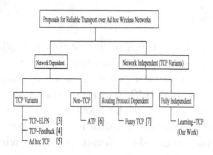

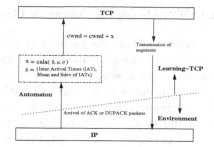

Fig. 1. Classification of the proposals

Fig. 2. Relationship between the learning automaton and the environment

3   Design and Implementation of Learning-TCP

Learning-TCP uses a learning automata based approach to adapt to the network conditions and updates the *cwnd* size, accordingly. Almost all the proposals made for reliable transport over AWNs are dependent on the explicit network feedback and moreover, they focus on distinguishing congestion and wireless losses in order to take an appropriate action for each type of loss. However, Learning-TCP is different from these proposals in three ways. Firstly, it does not rely on explicit network feedback and secondly, the focus is to learn the network conditions and accordingly update the *cwnd* size. Finally, the learning and controlling methods used by Learning-TCP are probabilistic rather than deterministic.

Fig. 2 shows the relationship between various components of Learning-TCP and the interactions of Learning-TCP with the environment. Learning-TCP is compatible with TCP as it does not modify the semantics and header format of TCP, but changes the *cwnd* updating mechanism. In order to learn the network state and appropriately adjust *cwnd*, a learning automaton that implements CALA is placed at each node that works along with TCP.

As shown in the figure, the learning automaton learns the network conditions by observing the inter arrival times (IATs), where IAT denotes the time difference between the arrival times of any two successive TCP *acks*[1]. Henceforth, in this paper, we use *network response* to represent the IATs of the TCP packets. This network response is transformed into a performance index that ranges between [0,1]. The notation $\beta(n)$ is used to represent the performance index at any time step n. $\beta(n) = 1$ indicates the highest reward, and $\beta(n) = 0$ indicates a penalty. Using the $\beta(n)$, the probability updating algorithm, which is discussed in Section 3.2, updates the action probability distribution. Then, the learning automaton selects an action that is essentially an amount of increment or decrement in the *cwnd* size, using its current action probability distribution. Learning-TCP handles DUPACKs in the same way as they are handled by TCP. However, in all the cases, the *cwnd* updations are done only through the learning mechanism.

---

[1] Throughout this paper, the term *ack* refers to both ACK and DUPACK packets.

3.1 Obtaining the Network Conditions

In order to obtain the network conditions, the automaton maintains a window called *time_window*, which contains the N most recent IATs of the TCP *acks* and calculates the average (*mean*) and standard deviation (*sdev*) of these N values. It is often possible that an *ack* with a very high IAT causes a large (or sudden) increase in the *mean*, which may adversely affect the learning mechanism. These fluctuations may be due to the poor channel conditions in the reverse path (from receiver to sender, the case when *acks* take different path), in which case Learning-TCP may wrongly interpret these conditions as congestion on the forward path. In order to prevent these fluctuations, an IAT is added directly to the *time_window* and further considered in the *mean* calculation, only if the IAT is in range [*mean-sdev, mean+sdev*]; otherwise the IAT is appropriately reset to the nearest bound and added to the *time_window*. This updated value is used in future *mean* calculations. As the IATs are always bounded, the effect of a very low (or high) IAT on the learning mechanism is minimized.

Upon receiving a TCP *ack* from IP at any time step n, using *mean* and current IAT, as shown in Eq. 1, the learning automaton captures the network state into a parameter called *time_ratio*. The *time_ratio* has an upper bound of 1 and a lower bound of δ which takes the value -2. We assume that the maximum time interval between any two consecutive events to be three times the *mean*, and hence the value for δ is -2. Any *ack* that arrives later than three times that of the current *mean*, is considered as a *late ack* and the *time_ratio* is taken as -2. The *time_ratio* obtains a value close to 1 when the current IAT is negligible compared to the *mean*, in which case, the *ack* is called an *early ack*.

$$time\_ratio = \frac{mean - current\ IAT}{mean} \tag{1}$$

$$\gamma = \frac{time\_ratio - \delta}{1 - \delta} \tag{2}$$

As shown in Eq. 2, the normalized network response from *time_ratio* is stored as the parameter γ. Depending on the state of automaton, Increase-state or Decrease-state (as described in Section 3.3), the automaton takes the performance index $\beta(n)$ as γ or $1 - \gamma$. Using the performance index $\beta(n)$ and the continuous action updating algorithm, the automaton updates the $\mu(n)$ and $\sigma(n)$ of the action probability distribution. Section 3.3 provides more details about the interpretation of the network response, the states of automaton, and the expected behavior of the automaton for varying network conditions.

3.2 The Continuous Action Updating Algorithm

In this section, we provide an account of the updating algorithm used in the Continuous Action Learning Automata (CALA). In CALA, the number of actions are infinite. The action probability distribution is assumed to follow a normal distribution. The CALA is becoming popular as the functions for updating the action probability distribution are simple and it does not require the discretization of the action set. Eqs. 3 and 4 correspond to the updating functions for $\mu(n)$ and $\sigma(n)$ of the action probability distribution at any time step n, respectively. In our problem, $\mu(n)$ and $\sigma(n)$ represent the mean and deviation of an effective amount of update in the *cwnd* size.

$$\mu(n+1) = \mu(n) + \lambda \, \frac{\beta_{x(n)} \times MSS}{\phi(\sigma(n))} \, \frac{x(n) - \mu(n)}{\phi(\sigma(n))} \tag{3}$$

$$\sigma(n+1) = \sigma(n) + \lambda \, \frac{\beta_{x(n)} \times MSS}{\phi(\sigma(n))} \left[\left(\frac{x(n) - \mu(n)}{\phi(\sigma(n))} \right)^2 - 1 \right] - \frac{\lambda \times K \times (\sigma(n) - \sigma_L)}{MSS} \tag{4}$$

where, $\phi(\sigma) = \sigma_L$ for $\sigma \leq \sigma_L$ (or) $\phi(\sigma) = \sigma$ for $\sigma > \sigma_L > 0$, $x(n)$ is the action taken at any time step n, λ is the learning parameter controlling the step size ($0 < \lambda < 1$), K is a large positive constant, σ_L is the lower bound on σ, and MSS represents the maximum size of a TCP segment in bytes.

The idea behind the updating function is as follows. It essentially shifts $\mu(n)$ towards $x(n)$ for an action $x(n)$ (amount of increment or decrement at time step n). When the feedback is better ($\beta(n)$ close to 1), the change in $\mu(n)$ is higher, compared to when the feedback is poor ($\beta(n)$ close to 0). The updating function increases $\sigma(n)$ in the following two cases: when $|x(n) - \mu(n)| > \sigma(n)$ and obtains a better response from the environment or $|x(n) - \mu(n)| < \sigma(n)$ and obtains a poor response from the environment; in all other cases $\sigma(n)$ is decreased. There is a reduction term in Eq. 4, the purpose of which is to gradually decrease $\sigma(n)$ towards σ_L, so that the automaton slowly moves towards a stable state.

3.3   Discussion About Learning-TCP

Here, we provide the details on the interpretation of the network response, the states of automaton, and expected behavior for automaton in different network conditions.

When the network is lightly loaded, the *ack* packets sent by the TCP receiver arrive at the TCP sender at faster rate (i.e. with low IAT) than when the load in the network is high. These packets are called *early acks*. The *early acks*, which result in a value close to one for γ computed from Eq. 2, give an indication that the network is currently lightly loaded and the *cwnd* should be increased at a high rate, to make use of the network resources efficiently and to improve the throughput. The high value for γ shifts the μ value towards a higher value. We choose initial values for $\mu(0)$ to be 1 Byte and $\sigma(0)$ to be 1 *MSS number of Bytes*. When μ is reaching higher values, the probability for selecting actions that increase the *cwnd* size increases. When the automaton continuously receives *late acks*, the μ value is reduced to a low value (zero or even lesser). This causes the automaton to select the actions that reduce the current *cwnd* size.

In all the cases, the rate of increase or decrease in the μ value depends on the *learning parameter* (λ) and the degree of favorable response (γ) obtained from the network. In all our simulations, we have fixed the λ value as 0.1. The selection of the learning parameter is a trade-off between the accuracy and speed of adaptability. The lower values of λ improve the accuracy of learning. They avoid the unnecessary reduction in μ because of the accidental losses occurring on the lightly network. Higher values of λ cause the automaton to adapt to the changes in the network rapidly. However, in this case the accuracy is low.

We assume two states, *increase* and *decrease*, for the automaton. When $x(n) - \mu(n) > 0$, the automaton is said to be in the *increase* state. Otherwise it is in the *decrease* state. The network response, γ is treated differently depending on the state of the automaton. When the automaton is in *increase* state, an *early ack* corresponds to a positive feedback (γ close to 1). This is because when an *early ack* is received, an action which increases the *cwnd* should be preferred. Hence, we directly take the network response as performance index ($\beta = \gamma$) when the automaton is in *increase* state. When the automaton is in the *decrease* state, a *late ack* (γ close to 0) indicates a positive feedback and we use $\beta = (1 - \gamma)$. This is because when a *late ack* is received, an action which decreases the *cwnd* should be preferred.

4   Simulation Results

We have carried out extensive simulations using GloMoSim for measuring the performance of our protocol and compared our results with that of TCP-reno. The various parameters used in our simulation are listed in Table 1. The metrics used for comparison are *average packet loss*, *network throughput*, and *average cwnd size* (average over number of *cwnd* updations).

Figs. 3 and 4 present the packet losses of TCP and Learning-TCP for 5 and 15 simultaneous flows, for different mobility values. We observe a significant and consistent reduction in the packet loss of Learning-TCP over that of TCP. We observe that both TCP and Learning-TCP show an increase in the packet loss with increasing mobility. However, Learning-TCP shows a significantly lower packet loss for different mobility values compared to that of TCP. The reduction in the packet loss ranges between 45% and 55% for 5 flows, and between 42% and 48% for 15 flows. The reason for the poor performance of TCP is the deterministic approach used for updating the *cwnd*. TCP strictly increases the *cwnd* even when the *ack* packets are received with high delays (*late acks*). Moreover, in the *slow-start* phase, TCP increases the *cwnd* by 1 MSS for every ACK packet it receives. Though this aggressive (or exponential) increase helps TCP to probe

Table 1. Simulation Parameters Used in GloMoSim

Description	Value
Simulation area/Node placement	1000m × 1000m/Random
Transmission power/range	5dBm/195m
Application/Routing/MAC protocols	FTP/AODV/802.11b with 2Mbps
Mobility model/Pause time	Random way-point/zero seconds
Number of nodes/Simultaneous flows	100/5,10, and 15
Per flow data transfer/Mobility Range	1.1Mbytes/[0-16]m/s
Number of different seeds	25 and confidence interval 95%

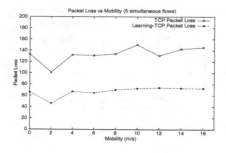

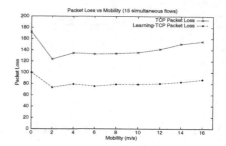

Fig. 3. 5 flows: Packet Loss vs Mobility   **Fig. 4.** 15 flows: Packet Loss vs Mobility

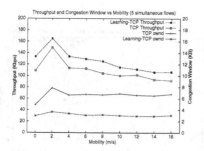

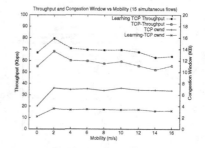

Fig. 5. 5 flows: Throughput & cwnd vs Mobility

Fig. 6. 15 flows: Throughput & cwnd vs Mobility

and fill to the maximum network capacity at a fast rate, it leads to heavy loss, when the network reaches a highly congested state. In contrast, the updating function of Learning-TCP follows a proactive approach to counter congestion in the network. Upon receipt of a *late ack*, the learning automaton favors the actions that decrease the *cwnd*. As a result, Learning-TCP experiences lower packet loss when the network is congested. This directly reduces the possibility of congestion, and in turn minimizes the number of retransmissions. Due to lower congestion in the network, the control overhead generated by the *false* route error messages (RERR) may also be minimized. Note that on detecting link break with down-stream node, an intermediate node generates RERR, which leads to route re-establishment. However, some of these may be *false* as they are generated even during the congestion. This is due to the fact that the intermediate nodes are incapable of distinguishing congestion from link breaks. Learning-TCP helps the network to recover quickly from congestion, since all the nodes which experience higher delays are most likely to perform similar decrease actions. Hence, the congestion recovery can be done faster with fewer congestion losses. Thus, the channel utilization and energy consumption at the nodes improve.

Figs. 5 and 6 show corresponding throughput and average *cwnd* size for the same setup. For both 5 and 15 flow scenarios, we observe a consistent improvement in the throughput of Learning-TCP over TCP and a higher *cwnd* for

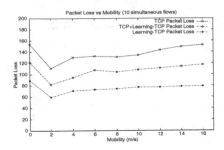

Fig. 7. Packet Loss vs Mobility (Only TCP, TCP+Learning-TCP, and Learning-TCP flows)

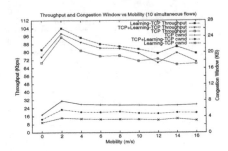

Fig. 8. The corresponding Throughput & cwnd vs Mobility

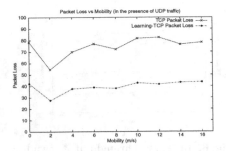

Fig. 9. Packet Loss vs Mobility in the presence of UDP traffic

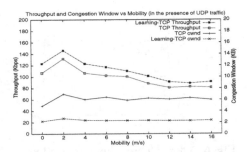

Fig. 10. The corresponding Throughput & cwnd vs Mobility

TCP across all the mobilities than for Learning-TCP. The improvement in the throughput of Learning-TCP over that of TCP ranges between 9% and 18% for 5 flows, and between 13% and 18% for 15 flows. Unlike in TCP, the increments and decrements in *cwnd* are not restricted in Learning-TCP. Learning-TCP increases the *cwnd* by higher amounts (>1 MSS) when it perceives the network to be lightly loaded. That is, when the load in the network is relatively low, μ reaches a high value and the *cwnd* is increased by large amounts. However, when the inter arrival times of *acks* are high, the updating algorithm decreases μ.

Figs. 7 and 8 represent the performance of Learning-TCP in the presence of TCP traffic. For 10 TCP flows alone, 10 Learning-TCP flows alone, and a mixture of 5 TCP and 5 Learning-TCP flows, we compared the overall packet loss, average size of *cwnd*, and throughput. The increase in packet loss with mobility for Learning-TCP only, is significantly lower. The packet loss for Learning-TCP is lower than that of a composition of Learning-TCP and TCP flows, which in turn has lower packet loss compared to TCP flows alone. This implies that Learning-TCP works well even in the presence of TCP flows.

We also tested Learning-TCP in the presence of UDP (User Datagram Protocol) traffic. Figs. 9 and 10 correspond to the results in the presence of CBR (Constant Bit Rate) flows that generate UDP traffic. Each CBR flow generates packets of size 512 bytes with packet inter departure time of 100ms. Fig. 9 shows packet loss versus mobility, and Fig. 10 shows throughput and average *cwnd* size versus mobility for both Learning-TCP and TCP. There are 5 UDP flows running in parallel with the 5 TCP or Learning-TCP flows. Learning-TCP outperforms TCP by showing significant reduction in the packet loss and consistent improvement in the throughput with increasing mobility. The reduction in packet loss and improvement in throughput for Learning-TCP over TCP, are in range 45%-49% and 7%-14%, respectively, which confirms that the learning mechanism works efficiently even in the presence of non-TCP traffic.

In all the simulation studies, we observe a higher *cwnd* for TCP than for Learning-TCP. This is mainly due to the aggressive (or exponential) increase in the *cwnd* size during the slow-start phase of TCP. Though the exponential increase results in a high *cwnd*, it leads to a high packet loss for TCP, which further results in a long loss recovery period. Hence, TCP shows a lower throughput even with the higher *cwnd*, compared to that of Learning-TCP.

5   Conclusions

In this paper, we proposed a novel Learning-TCP for AWNs, which does not depend on explicit feedback from the network. Learning-TCP adapts to the network conditions by observing the inter arrival times of *ack* packets. We used learning automata because of its unique feature of learning the network state better and faster by maintaining considerably lesser amount of information and with negligible computational requirements. Learning-TCP was simulated using GloMoSim and the simulation results have clearly indicated an efficient *cwnd* updation that resulted in fewer packet losses and improved throughput. We showed through extensive simulation studies that Learning-TCP results in the reduction of packet loss by about 42-55% and an increase in the throughput by about 9-18% compared to that of TCP, resulting in better network utilization and lower energy consumption by the nodes, which is vital for resource constrained AWNs. Moreover, Learning-TCP is compatible with the traditional TCP as it does not modify the semantics and header format of TCP.

References

1. I. Chlamtac, M. Conti, and J. J. N. Liu, "Mobile Ad Hoc Networking: Imperatives and Challenges," *Journal of Ad Hoc Networks*, vol. 1, no. 1, pp. 13-64, July 2003.
2. C. Siva Ram Murthy and B. S. Manoj, *Ad Hoc Wireless Networks: Architectures and Protocols,* Prentice Hall, New Jersey, 2004.
3. G. Holland and N. Vaidya, "Analysis of TCP over Mobile Ad Hoc Networks," *in Proc. ACM MobiCom*, pp. 219-230, August 1999.
4. K. Chandran, S. Raghunathan, S. Venkatesan, and R. Prakash, "A Feedback Based Scheme for Improving TCP Performance in Ad Hoc Wireless Networks," *IEEE Personal Communications Magazine*, vol. 8, no. 1, pp. 34-39, February 2001.
5. J. Liu and S. Singh, "ATCP: TCP for Mobile Ad Hoc Networks," *IEEE Journal on Selected Area in Communications*, vol. 19, no. 7, pp. 1300-1315, July 2001.

6. K. Sundaresan, V. Anantharaman, H. Hsieh, and R. Sivakumar, "ATP: A Reliable Transport Protocol for Ad-hoc Networks," *in Proc. ACM MobiHoc*, June 2003.
7. R. Oliveira and T. Braun, "A Delay-based Approach Using Fuzzy Logic to Improve TCP Error Detection in Ad Hoc Networks," *in Proc. IEEE WCNC*, March 2004.
8. K. S. Narendra and M. A. L. Thathachar, *Learning Automata: An Introduction,* Prentice Hall, New Jersey, 1989.
9. M. A. L. Thathachar and P. S. Sastry, *Networks of Learning Automata: Techniques for Online Stochastic Optimization,* Kluwer Academic, New Jersey, 2004.

Design and Implementation
of the HPCS Graph Analysis Benchmark
on Symmetric Multiprocessors

David A. Bader* and Kamesh Madduri

College of Computing,
Georgia Institute of Technology, Atlanta, GA 30332, USA
{bader, kamesh}@cc.gatech.edu

Abstract. Graph theoretic problems are representative of fundamental computations in traditional and emerging scientific disciplines like scientific computing and computational biology, as well as applications in national security. We present our design and implementation of a graph theory application that supports the kernels from the Scalable Synthetic Compact Applications (SSCA) benchmark suite, developed under the DARPA High Productivity Computing Systems (HPCS) program. This synthetic benchmark consists of four kernels that require irregular access to a large, directed, weighted multi-graph. We have developed a parallel implementation of this benchmark in C using the POSIX thread library for commodity symmetric multiprocessors (SMPs). In this paper, we primarily discuss the data layout choices and algorithmic design issues for each kernel, and also present execution time and benchmark validation results.

1 Introduction

One of the main objectives of the DARPA High Productivity Computing Systems (HPCS) program [1] is to reassess the way we define and measure performance, programmability, portability, robustness and ultimately *productivity* in the High Performance Computing (HPC) domain. An initiative in this direction is the formulation of the Scalable Synthetic Compact Applications (SSCA) [2] benchmark suite. These synthetic benchmarks are envisioned to emerge as complements to current scalable micro-benchmarks and complex real applications to measure high-end productivity and system performance. Each SSCA benchmark is composed of multiple related kernels which are chosen to represent workloads within real HPC applications and is used to evaluate and analyze the ease of use of the system, memory access patterns, communication and I/O characteristics. The benchmarks are relatively small to permit productivity testing and

---
* This work was supported in part by DARPA Contract NBCH30390004; and NSF Grants CAREER ACI-00-93039, NSF DBI-0420513, ITR ACI-00-81404, DEB-99-10123, ITR EIA-01-21377, Biocomplexity DEB-01-20709, and ITR EF/BIO 03-31654.

D.A. Bader et al. (Eds.): HiPC 2005, LNCS 3769, pp. 465–476, 2005.

programming in reasonable time; and scalable in problem representation and size to allow simulating a run at small scale or executing on a large system at large scale. They are also described in sufficient detail to drive novel HPC programming paradigms, as well as architecture development and testing.

SSCA#2 [3] is a graph theoretic problem which is representative of computations in the fields of national security, scientific computing, and computational biology. The HPC community currently relies excessively on single-parameter microbenchmarks like LINPACK [4], which look solely at the floating-point performance of the system, given a problem with high degrees of spatial and temporal locality. Graph theoretic problems tend to exhibit irregular memory accesses, which leads to difficulty in partitioning data to processors and in poor cache performance. The growing gap in performance between processor and memory speeds, the memory wall, makes it challenging for the application programmer to attain high performance on these codes. The onus is now on the programmer and the system architect to come up with innovative designs.

Symmetric Multiprocessors (SMPs) with modest shared memory have emerged as a popular platform for the design of scientific and engineering applications. SMP clusters are now ubiquitous in high-performance computing, consisting of clusters of multiprocessors nodes (e.g., IBM pSeries, Sun Fire, Compaq AlphaServer, and SGI Altix) inter-connected with high-speed networks (e.g., vendor-supplied, or third party such as Myricom, Quadrics, and InfiniBand). Current research has shown that it is possible to design algorithms for irregular and discrete computations [5,6,7] that provide efficient and scalable performance on SMPs. To analyze SMP performance, we use a complexity model similar to that of Helman and JáJá [8] which has been shown to provide a good cost model for shared memory algorithms on current symmetric multiprocessors [5,8,9]. The model uses two parameters: the problems input size n, and the number p of processors. There are two parts to an algorithm's complexity in this model: M_E, the maximum number of non-contiguous memory accesses required by any processor, and T_C, the computation complexity. This model, unlike the idealistic PRAM, is more realistic in that it penalizes algorithms with non-contiguous memory accesses that often result in cache misses.

This paper is organized as follows. Sections 3-7 discuss the scalable data generation stage and each of the four kernels in detail: we present the kernel specification, the design trade-offs involved in implementation, illustrations of our data layouts, and relevant algorithms. Section 8 summarizes the execution time and memory usage results, primarily on the Sun E4500 shared memory SMP. In the final section, we present our conclusions and plans for future work.

2   Preliminaries

2.1   Definitions

Let $G = (V, E)$ be a directed, weighted multi-graph, where $V = \{v_1, v_2, ..., v_n\}$ is the set of vertices, and $E = \{e_1, e_2, ..., e_m\}$ is the set of weighted, directed edges. An edge $e_i \in E$ is represented by the tuple $\langle u, v, w_i \rangle$, where $u, v \in V$,

w_i is either a positive integer from a bounded universe or a character string of fixed length, and the edge e_i is directed from u to v. There are no self loops in the SSCA#2 graph, i.e., for any edge $e_i = \langle u, v, w_i \rangle \in E$, we have $u \neq v$. Two vertices u, v are said to be *linked* if there exists at least one directed edge from u to v or v to u. We define a set of vertices $C \subseteq V$ to be a *clique*, if each pair of vertices $\{u, v\} \in C$ is *linked*. This means that a clique has edges between each pair of vertices, but not necessarily in both directions. A *cluster* $S \subseteq C \subseteq V$ is loosely described as a maximal set of *highly inter-connected* vertices.

2.2 Benchmark Input Parameters

Some user-defined constants are used for the data generation step and subsequent kernels.

1. *totVertices* : the number of vertices in the graph. We also use n to represent the number of vertices, and m the number of directed edges in sections of the paper.
2. *maxCliqueSize* : the maximum size of a clique in the graph. Clique sizes are uniformly distributed in the interval $[1, maxCliqueSize]$.
3. *maxParalEdges* : the maximum number of parallel edges between two vertices. The number of edges between any two vertices are uniformly distributed in the interval $[1, maxParalEdges]$
4. *probUnidirectional* : probability that the connections between two vertices will be unidirectional as opposed to bidirectional
5. *probInterClEdges* : the probability of inter-clique edges
6. *percIntWeights* : percentage of edges assigned integer weights
7. *maxIntWeight* : the maximum integer weight
8. *maxStrLen* : maximum number of characters in the string weight
9. *subGrEdgeLength* : maximum edge length in graphs generated by Kernel 3
10. *maxClusterSize* : maximum cluster size generated by the cuts in Kernel 4

3 Scalable Data Generation

The Scalable Data Generation stage takes user parameters as input and generates the graph as tuples of vertex pairs and their corresponding weights. The intended graph has a hierarchical nature, with random-sized *cliques*, and inter-clique edges assigned using a random distribution. The edge weights can be integer values or randomly generated character strings. The scalable data generator need not be parallelized, and is not timed.

3.1 Implementation

This step's output should be an edge list with each element of the form $\langle u, v, w \rangle$, where the edge is directed from u to v, and w is a positive integer weight or a character string. Our implementation returns four one-dimensional array constructs: two arrays corresponding to the start and end vertices, and the two other arrays representing the integer and string weights. Although this stage is not timed, we parallelize the main steps for practical considerations.

Note that the SSCA#2 graph has some very specific properties. It is essentially a collection of cliques (defined in the earlier section), with the *inter-clique edges* assigned using a hierarchical distribution, based on the distance between the cliques. The fourth kernel deals with extraction of highly inter-connected *clusters* from the graph, and we would like the extracted clusters to be as close as possible to the original cliques. The implementation details of the data generation stage are discussed in an extended version of this paper [10].

4   Kernel 1: Graph Generation

This kernel constructs the graph from the data generator output tuple list. The graph can be represented in any manner, but cannot be modified by subsequent kernels. The number of vertices in the graph is not provided and needs to be determined in this kernel. It is also suggested that statistics be collected on the graph to aid verification of subsequent kernels.

4.1   Details

There are many figures of merit for each kernel, including but not limited to memory use, running time, ease of programming, ease of incrementally improving, and so forth. Thus, a figure of merit for any implementation would be the total space usage of the graph data structure. Also, the graph data structure (or parts of it) *cannot be modified or deleted* by subsequent kernels. So we need to choose a data layout which can be created quickly and easily (since Kernel 1 is timed), is space efficient, and is optimized for efficient implementations of Kernels 2, 3 and 4.

Kernels 2 and 3 operate on the directed graph, but for Kernel 4, the specification states that multiple edges, edge directions, and edges weights, are to be ignored. This complicates the design and implementation – if we plan to use a separate graph layout for Kernel 4, we need to construct it in Kernel 1, and it cannot be modified in Kernels 2 and 3. The developer now must design a data structure and layout which considers all these competing optimization criteria, and this is the core challenge in the benchmark.

An adjacency matrix representation is easy to implement and well-suited for dense graphs. In this case, however, the generated graph is sparse and a matrix representation would be very inefficient in memory usage. Another common method of representing directed and weighted graphs is the adjacency list representation. This is easy to implement and also space efficient. However, repeated memory allocation calls while constructing large graphs, and irregular memory accesses in the subsequent kernels will hurt performance. For our current implementation, we follow an adjacency list representation, but using the more cache-friendly *adjacency arrays* [11] with auxiliary arrays.

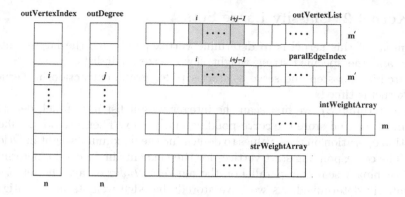

Fig. 1. The data layout for representing the directed graph – Kernel 1

Since multiple edges between two vertices can be ignored for Kernels 3 and 4, we do not store them explicitly, but have another array to keep track of these edges and to map an edge to its corresponding weight. We first construct the part of the data structure to store the directed graph information. We use two arrays of size *totVertices* to index and access the adjacencies corresponding to each vertex. The adjacency list (without multiple edges) is stored in a contiguous memory location, and so is the array storing the multiple edge information. The data layout used is illustrated in Fig. 1.

Graph construction (for our adjacency array representation) is inherently sequential, but since we have a sorted edge tuple list, we can extract some parallelism. First, the size of the graph can be easily determined by finding the maximum vertex number in the start vertex or the end vertex list. Assuming the tuple list is sorted by start vertex, the value can be determined in constant time by reading off the last element in the startVertex array. Otherwise we can determine the maximum value in parallel in $T_C = O(m/p + \log p)$ time. Processors then scan independent sections of the tuple list to determine the outdegree of each vertex. We have a parallel time overhead of $O(p)$ for bookkeeping purposes. In the next pass, we allocate memory for the *outVertexList* and *paralEdgeList* arrays and fill in entries in parallel in $O(m'/p + \log p)$ time, where m' is the number of unique directed edges (removing the parallel edges).

We construct the implied edge list by scanning the *outVertexList* in parallel. For each edge $\langle u, v \rangle$, we check if the *outVertexList* has the edge $\langle v, u \rangle$. If not, we add u to the implied edge list of v. This step has an asymptotic time complexity of $T_C = O(m'/p + \log p)$ and involves $m' + m/p$ non-contiguous memory accesses. We also need to use mutex locks to prevent race conditions, which affects performance. The integer and string weight arrays can be trivially constructed in constant time, since we retain the vertex ordering in the edge tuples. In sum, the computational complexity for Kernel 1 is given by $T_C = O(m/p + \log p)$, and $M_E = m' + 2m/p$. The asymptotic space requirements for the storing the tuple list and the graph data structure are both $O(m)$. The memory requirements in both these cases are further compared in Section 9.

5   Kernel 2: Classify Large Sets

The intent of this kernel is to determine vertex pairs with the largest integer weight and the specified string weight. Two vertex pair lists, S_I and S_C, are generated in this step and serve as start sets for graph extraction in Kernel 3. This kernel is timed.

To determine S_I, we first scan the integer weight list in parallel, determine local maxima, and store the corresponding end vertex. Then, we do an efficient reduction operation on the p values to determine the maximum weight in $O(\log p)$ time. The corresponding start vertices for the elements in S_I can be determined by a fast binary search in parallel on the *outVertexIndex* array. The set S_C can be similarly determined. As we have stored the edge weights in a contiguous block, we have the work equally distributed among all processors. Finding the maximum weighted edge is the dominant step in this stage and $T_C = O(m/p + \log p)$ for this kernel.

6   Kernel 3: Extracting Sub-graphs

Starting from each of the vertex pairs in the sets S_I and S_C, this kernel produces sub-graphs which consist of the vertices and edges along all paths of length less than *subGrEdgeLength*. The recommended algorithm for graph extraction in the specification is Breadth First Search.

6.1   Implementation

We use a Breadth First Search (BFS) algorithm starting from the *endVertex* of each element in S_I and S_C, up to a depth of *subGrEdgeLength*. Now *subGrEdge-Length* is typically chosen to be a small number, a constant value in comparison to the number of graph vertices. We also know that this graph is essentially a collection of cliques (whose maximum size is bounded), and so a BFS up to a constant depth would yield a subgraph $G' = (V', E')$ such that $|V'| \ll |V|$. Even though the BFS computational complexity is of the same order as the previous kernels ($T_C = O(m')$), we can expect this kernel to finish much faster. We have not implemented a fine-grained parallel BFS yet. Currently, we just distribute the vertices in S_I to the available processors and run BFS in parallel on each of these, which limits the concurrency to $|S_I| + |S_C|$. The queue ADT we use in this algorithm is implemented using a dynamic array, a linked list and a simple one-dimensional array. Since the extracted graph is quite small, we find that all three representations give similar results. Note that linked lists are easy to implement, space-efficient and could be used for small problem sizes, since we will not be performing any further operations with the extracted graph.

7   Kernel 4: Graph Clustering

The intent of this kernel is to partition the graph into highly inter-connected *clusters* and minimize the number of links between these clusters. Multiple edges, edge directions and weights can be ignored. Since exact solutions to this problem are NP-hard, heuristics are allowed, provided they satisfy the kernel validation criterion. This kernel should not utilize any auxiliary information collected in the previous kernels or in the graph generation process.

7.1   Details

This kernel is based on the partitioning problem formulated by Kernighan and Lin [12], with all the edge costs considered equal. Sangiovanni-Vincentelli, Chert, and Chua [13,14] have earlier applied this work for solving circuit problems. The maximal clique problem [15] is a well-studied NP-complete problem, and several heuristics have been proposed to solve this [16]. Our problem is not as difficult as the maximal clique problem, because of the manner in which the graph is generated, and also due to the restriction on the maximum clique size.

We cannot apply popular multi-level graph partitioning tools like Chaco [17] and METIS [18] to solve this kernel. These tools use a variety of heuristics and are highly refined, but they are primarily used to partition nearly-regular graphs into *equal-sized blocks*, while minimizing edge cut. Graph partitioning results using Chaco are presented in [10]. The required partitioning in this problem, however, is highly irregular and cannot be found accurately using these tools.

The specification suggests an algorithm for solving this kernel, which is a variant of a graph clustering algorithm given by Koester [19]. This sequential algorithm iteratively forms a sequence of disjoint clusters, which are subgraphs no larger than *maxClusterSize* vertices. As each cluster is selected, its vertices are removed from further consideration. To select the vertices in a cluster, the algorithm starts with some remaining vertex (which forms the initial one-element cluster), and its links to any remaining vertices (which form the initial adjacent set). It then expands the cluster by repeatedly moving an adjacent set vertex to the cluster, and adding that vertex's non-cluster links to the adjacent set. The new vertex is chosen depending on how tightly it and its links are connected to the existing cluster, and how many links it adds to the adjacent set. The cluster is complete if the adjacent set is empty. Otherwise when the cluster reaches *maxClusterSize* vertices in size, the cluster elements are marked used, the cluster is added to the cluster list, and size of the adjacent set is added to the count of interclique links.

The reference implementation uses this algorithm for solving Kernel 4 and reports good results. The specification suggests statistical validation for assessing the quality of the clustering algorithm. One recommended empirical measure is to check if $interClusterLinkNum < refcutLinksNum$, where $refcutLinksNum$ is given by $\dfrac{intercliqueLinkNum}{\sqrt{(maxClusterSize/maxCliqueSize)}}$ and $interCliqueLinkNum$ refers to the number of inter-clique vertex pairs connected by at least one directed edge. Algorithms with $interClusterLinkNum$ within 5% of the value $refCutLinksNum$ are acceptable. It is also suggested that for small problem sizes, the algorithm correctness be checked rigorously, and parallel results be verified against serial results.

This algorithm is however inherently sequential. Cliques of size less than *max-ClusterSize* with inter-clique edges may not be extracted correctly. We propose a new parallel greedy algorithm (pseudo-code is given in [10]) to extract clusters. The quality of results is comparable to the reference algorithm, and some results are presented in the next section.

Our parallel algorithm works as follows. We first sort the vertices in parallel in the decreasing order of their degree. The parallel radix sort uses a linear-time counting sort for a constant number of iterations. A shared array *vStatus* of size n is maintained to keep track of the status of each vertex – whether it is unassigned yet, or assigned to a unique cluster. Each processor chooses a vertex from the top of the queue, colors the vertex and its adjacencies (both the out-vertices and the implied edges) with a unique number, given by $i \times$ *current iteration number*, where i is the processor index. The adjacencies of each vertex in the cluster are inspected, and if more than a certain threshold of them are similarly colored, it is accepted. Otherwise it is rejected and the vertex is unmarked. We also update the *edgeCut* simultaneously — if we decide that an originally colored vertex does not belong to the cluster, we add all the inter-clique edges to the cut-set. The vertex degree is bounded by $O(maxClusterSize)$. The clustering algorithm runs in linear time in the worst case (a single clique of size $O(n)$), with M_E given by $O(n/p)$. If *maxClusterSize* is chosen to be a constant value, $T_C = M_E = O(n/p)$.

The heuristic correctly extracts nearly all cliques, except for those of very small sizes (with 3-4 elements), as it is tough to define acceptance thresholds. We have two choices in such cases: either classify these vertices as clusters of smaller sizes (say 1 or 2), or add these vertices to existing clusters. The former approach is a more conservative method of forming clusters and *false positives* (vertices wrongly assigned to a cluster) are avoided, but it would also lead to an inflated number of extracted clusters and inter-cluster edges. We thus have a trade-off between graph clustering *specificity* (corresponds to exact clique extraction) and *sensitivity* (correlates to minimization of intra-cluster links) in this case. We can define the threshold values for accepting a vertex into a cluster according to what our primary optimization criterion is — retaining specificity, or minimizing inter-clique edges and increasing sensitivity. The suggested validation scheme for this kernel is to compare the inter-clique links with the inter-cluster links, and so we optimize for the inter-cluster edges when reporting the results in Section 9.

8   Experimental Results

This section summarizes the experimental results of our SSCA#2 implementation, tested on the Sun E4500, a uniform-memory-access (UMA) shared memory parallel machine with 14 UltraSPARC II 400MHz processors and 14 GB of memory. Each processor has 16 Kbytes of direct-mapped data (L1) cache and 4 Mbytes of external (L2) cache.

We use a binary scaling heuristic *SCALE* to uniformly express the input parameter values. The following values have been used for reporting results in this section: $totVertices = 2^{SCALE}$, $maxCliqueSize = 2^{(SCALE/3)}$, $maxParalEdges$

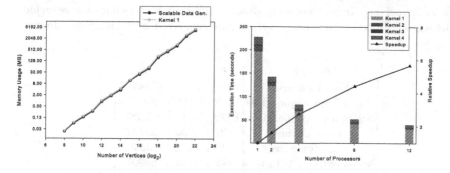

Fig. 2. Memory Usage (left) and Execution Time (right)

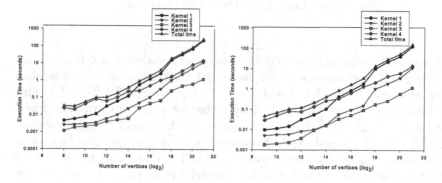

Fig. 3. Execution time of Kernels 1, 2, 3, and 4, on four and eight processors, in the left and right plots, respectively

$= 3$, $probUnidirectional = 0.3$, $probInterClEdges = 0.5$, $percIntWeights = 70$, $maxIntWeight = 2^{SCALE}$, $maxStrLen = SCALE$, $subGrEdgeLength = SCALE$, and $maxClusterSize = 2^{(SCALE/3)}$.

Fig. 2 compares memory utilization of the data generator and our graph layout (described in Section 5). Note that we explicitly store implied edge information in Kernel 1, causing the graph data structure to use slightly more memory than the data generator output. One of the figures of merit of the implementation is the largest problem size that can be solved on a given architecture. On the Sun E4500, memory proves to be the bottleneck to scaling. The largest problem size that can be handled with these parameters is 2^{21} vertices, which generates 156M edges for the above input parameters. We could further solve a problem size of 2^{22} vertices, by writing the data generator output to disk.

The running times for multi-processor runs are also given in Fig. 2. The execution time is dominated by graph generation, which scales reasonably with the number of processors for various problem sizes. We use a locking scheme to construct the implied edge list in parallel, which leads to a moderate slowdown of Kernel 1. There is also limited parallelism in Kernel 3 dependent on the size of the Kernel 2 start sets.

Table 1. Kernel 4 – Graph Clustering Results. (intra and inter-clique edges include parallel edges; a link is defined as a vertex pair connected by at least one directed edge).

SCALE	12	16	20
No. of Vertices	4096	65536	1048576
No. of intra-clique edges	40850	361114	39511513
No. of inter-clique edges	8472	72365	645787
No. of cliques	486	3990	32167
Avg. clique size	8.42	16.42	32.6
No. of extracted clusters	383	3142	25201
Avg. cluster size	10.69	20.85	41.6
No. of inter-clique links	5230	49907	422292
No. of inter-cluster links	1968	18892	185250

Fig. 3 gives the running times of the four kernels for various problem scales, on four and eight processors respectively. Note that the number of non-contiguous memory accesses $M_E = O(m')$ and $T_C = O(n/p + \log p)$ for Kernel 1, and so the benchmark execution time is dominanted by graph construction. Since $maxClusterSize = 2^{SCALE/3}$, we find a sharp rise in Kernel 1 execution time for $SCALE = 9, 12, 15$, and 18, as the number of edges generated in these cases is comparatively higher than the previous value. The dominant step in Kernel 1 is construction of the implied edge list. Kernel 3 takes the least time, as the search depth value is very small.

Rigorous verification of full-scale runs is prohibitive, and so the benchmark specification suggests a statistical validation scheme. Table 1 summarizes validation results for Kernel 4. The number of clusters extracted and the number of inter-cluster links are reported for three different problem sizes (for a four-processor run). The quality of the results is chiefly dependent on two input parameters: *probUnidirectional* and *probInterClEdges*. We have tested the correctness of our implementation on small graph sizes. We also find the clustering results to be consistent across multi-processor runs, as we do not use locking in this kernel. Note that in cases when the graph has a high percentage of inter-clique edges, we have a trade-off between exact clique extraction and minimization of inter-cluster edges, as discussed in the previous section.

9   Conclusions

In this paper, we present the design and implementation of the SSCA#2 graph theory benchmark. This benchmark consists of four kernels with irregular memory access patterns that chiefly test a system's memory bandwidth and latency. Our parallel implementation uses C and POSIX threads and has been tested on the Sun Enterprise E4500 SMP system. The dominant step in the benchmark is the construction of the graph data structure, which limits scaling on the

Sun E4500. We are currently working on implementations of SSCA#2 on other shared-memory systems such as the Cray MTA-2 and the Cray XD1.

Acknowledgments

We thank Bill Mann, Jeremy Kepner, John Feo, David Koester, John Gilbert, Ram Rajamony, and other members of the HPCS working group for trying out early versions of our implementation, discussions of the benchmark specifications, and their valuable suggestions.

References

1. DARPA Information Processing Technology Office: High productivity computing systems project (2004) http://www.darpa.mil/ipto/programs/hpcs/.
2. Kepner, J., Koester, D.P., et al.: HPCS Scalable Synthetic Compact Application (SSCA) Benchmarks (2004) http://www.highproductivity.org/SSCABmks.htm.
3. Kepner, J., Koester, D.P., et al.: HPCS SSCA#2 Graph Analysis Benchmark Specifications v1.0. (2005)
4. Dongarra, J., Bunch, J., Moler, C., Stewart, G.: LINPACK Users' Guide. SIAM, Philadelphia, PA. (1979)
5. Bader, D., Sreshta, S., Weisse-Bernstein, N.: Evaluating arithmetic expressions using tree contraction: A fast and scalable parallel implementation for symmetric multiprocessors (SMPs). In Sahni, S., Prasanna, V., Shukla, U., eds.: Proc. 9th Int'l Conf. on High Performance Computing (HiPC 2002). Volume 2552 of Lecture Notes in Computer Science., Bangalore, India, Springer-Verlag (2002) 63–75
6. Bader, D.A., Cong, G.: A fast, parallel spanning tree algorithm for symmetric multiprocessors (SMPs). In: Proc. Int'l Parallel and Distributed Processing Symp. (IPDPS 2004), Santa Fe, NM (2004)
7. Bader, D.A., Cong, G.: Fast shared-memory algorithms for computing the minimum spanning forest of sparse graphs. In: Proc. Int'l Parallel and Distributed Processing Symp. (IPDPS 2004), Santa Fe, NM (2004)
8. Helman, D.R., JáJá, J.: Designing practical efficient algorithms for symmetric multiprocessors. In: Algorithm Engineering and Experimentation (ALENEX'99). Volume 1619 of Lecture Notes in Computer Science., Baltimore, MD, Springer-Verlag (1999) 37–56
9. Helman, D.R., JáJá, J.: Prefix computations on symmetric multiprocessors. Journal of Parallel and Distributed Computing **61** (2001) 265–278
10. Bader, D.A., Madduri, K.: Design and implementation of the HPCS graph analysis benchmark on symmetric multiprocessors. Technical report, Georgia Instutite of Technology (2005)
11. Park, J., Penner, M., Prasanna, V.: Optimizing graph algorithms for improved cache performance. In: Proc. Int'l Parallel and Distributed Processing Symp. (IPDPS 2002), Fort Lauderdale, FL (2002)
12. Kernighan, B., Lin, S.: An efficient heuristic procedure for partitioning graphs. The Bell System Technical Journal **49** (1970) 291–307
13. Sangiovanni-Vincentelli, A., Chert, L., Chua, L.: A new tearing approach: Node tearing nodal analysis. In: Proc. IEEE Int'l Symp. on Circ. and Syst., Phoenix, AZ (1975) 143–147

14. Sangiovanni-Vincentelli, A., Chert, L., Chua, L.: An efficient heuristic cluster algorithm for tearing large-scale networks. IEEE Trans. Circuits and Systems (1977) 709–717

15. Bomze, I., Budinich, M., Pardalos, P., Pelillo, M.: The maximum clique problem. In Du, D.Z., Pardalos, P.M., eds.: Handbook of Combinatorial Optimization. Volume 4. Kluwer Academic Publishers, Boston, MA (1999)

16. Johnson, D., Trick, M., eds.: Cliques, Coloring, and Satisfiability: Second DIMACS Implementation Challenge, October 11-13, 1993. Volume 26 of DIMACS Series in Discrete Mathematics and Theoretical Computer Science. American Mathematical Society (1996)

17. Hendrickson, B., Leland, R.: A multilevel algorithm for partitioning graphs. In: Proc. Supercomputing '95, San Diego, CA (1995)

18. Karypis, G., Kumar, V.: MeTiS: A Software Package for Partitioning Unstructured Graphs, Partitioning Meshes, and Computing Fill-Reducing Orderings of Sparse Matrices. Department of Computer Science, University of Minnesota. Version 4.0 edn. (1998)

19. Koester, D.P.: Parallel Block-Diagonal-Bordered Sparse Linear Solvers for Power Systems Applications. PhD thesis, Syracuse University, Syracuse, NY (1995)

Scheduling Multiple Flows on Parallel Disks*

Ajay Gulati[1] and Peter Varman[2]

[1] Department of Computer Science,
[2] Department of Electrical Engineering and Computer Science,
Rice University, Houston TX 77005, USA
{gulati, pjv}@rice.edu

Abstract. We examine the problem of scheduling concurrent independent flows on multiple-disk I/O storage systems. Two models are considered: in the shared buffer model the memory buffer is shared among all the flows, while in the partitioned buffer model each flow has a private buffer. For the parallel disk model with $d > 1$ disks it is shown that the problem of minimizing the schedule length of $n > 2$ concurrent flows is NP-*complete* for both buffer models. A randomized scheduling algorithm for the partitioned buffer model is analyzed and probabilistic bounds on the schedule length are presented. Finally a heuristic based on static buffer allocation for the shared buffer model is discussed.

1   Introduction

Advances in disk drive and networking technologies and a sharp increase in data-intensive applications have revolutionized the architecture and usage paradigms of modern storage systems. Resource management issues have become increasingly important in data centers that must coordinate the operation of large numbers of concurrent devices and serve hundreds of gigabytes of data per second. Both commercial and scientific workloads require high-bandwidth access to large data sets that reside on shared storage facilities and are accessed by multiple applications. Shared storage servers are being increasingly proposed as a cost-effective solution for maintaining data repositories, which can take advantage of economies of scale and consolidated management. Sharing a storage server raises two issues: effective use of server resources and fair scheduling of individual clients. We examined the issue of performance isolation and providing QoS guarantees to individual clients in [1]. In this paper we address the problem of efficiently utilizing the resources of a shared storage system when servicing multiple concurrent flows.

Previous work on reducing latency in parallel I/O systems has dealt with the scheduling of a single flow to effectively exploit prefetching and caching from multiple disks; efficient algorithms are now known that maximize disk system throughput for a single flow (see [2, 3, 4, 5, 6, 7, 8] for example). Our work in this paper is a generalization of the parallel disk model [9] to the case when we have more than one concurrent flow simultaneously sharing the I/O system. The problems of sharing

---

* Support by the National Science Foundation under Grant CCR-0105565 and the IR/D program is gratefully acknowledged.

D.A. Bader et al. (Eds.): HiPC 2005, LNCS 3769, pp. 477–487, 2005.

the parallel disks among multiple concurrent flows have only been recently considered [10, 11, 12, 13, 14, 1]. All of these works are mostly concerned with different models of storage virtualization and QoS-based resource allocation.

We consider two models of shared servers. In both models the disk subsystem is shared by all the flows. The models differ in how the memory buffer is allocated to the flows: in the shared buffer model all flows share a common buffer, while in the partitioned buffer each flow has its own private buffer. We show that in both models the problem of scheduling a set of requests from each flow to minimize the number of parallel I/O steps is NP-*Complete*. In contrast the case of a single flow is known to have efficient polynomial-time scheduling algorithms [2, 4, 5, 6, 7]. We also show that the congestion-removal techniques of Leighton, Maggs and Rao [15] can be extended to obtain a randomized scheduling algorithm for the partitioned buffer model with probabilistically bounded schedule length. Finally for the shared buffer model, we present a heuristic that combines a novel static buffer allocation strategy with the randomized approach used in the partitioned buffer model.

The off-line scheduling of a single flow to obtain the minimum-length schedule has been extensively studied [16, 2, 7, 6, 17, 4, 5]. Polynomial time algorithms in the parallel disk model [9] and stall model [3] have been obtained in [6, 17, 4, 5] and [7, 2] respectively. However, the problem of scheduling multiple flows has not been formally considered previously in either parallel I/O model.

2 I/O Models

When multiple applications share the disk system we look at two related models, one where the buffer is shared among all the flows, and the other in which the buffer is statically partitioned among them. In the Shared Disk Partitioned Buffer (SDPB) model, the disks are shared by all the flows but the buffer is partitioned among them. The most common scenario for such an organization is where the buffering is done in individual client nodes and the disks are accessed over a SAN. In the Shared Disk Shared Buffer (SDSB) model, the buffer is logically shared among all workloads. This may be either in the form of a centralized storage buffer or made up of distributed shared memory.

There are n independent *flows* (applications) that are simultaneously accessing the storage system. Each flow is abstracted by a *reference string* consisting of the ordered sequence of *blocks* that it accesses. Blocks need to be delivered to the application in the order in which they appear in the reference string. We focus on read-once reference strings where each block is unique, which model workloads like multimedia streaming. There is a fixed amount of buffer memory that the system uses for *prefetching*. A buffered block is consumed as soon as it becomes the first unconsumed block in its reference string. Each flow knows a subsequence of the reference string beyond its last access; this subsequence is the *lookahead window* for the flow. If the lookahead window includes the entire reference string then the schedule is said to be off-line. We assume that requests in one lookahead window are serviced before newly arriving requests are scheduled.

A model of the SDSB configuration is shown in Figure 1. Flow queues hold the requests in the currently visible portion of the reference strings. The high-level sched-

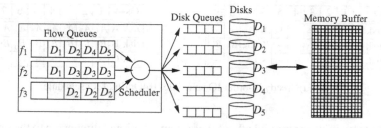

Fig. 1. Shared disk and Shared buffer configuration

uler dispatches requests from the flow queues to the disk queues in accordance with the schedule that it constructs. In every parallel I/O step a number of requests, at most one request for each disk, are dispatched by to the disk queues. Individual disks service the requests in their disk queues one at-a-time. The blocks read from the disks are placed in the memory buffer. In the SDPB model a block can only be placed in the partition belonging to the flow; in the SDSB model a common buffer of capacity M blocks is shared by all the flows. A flow will consume blocks in the order of the reference string from this buffer. Blocks that are fetched out-of-order of the reference string will be buffered until required by the flow. Each disk is free to reorder the requests in its disk queue to optimize physical access times by exploiting spatial locality in the data placement. The high-level scheduler must ensure that the requests it has dispatched to the disk queues at any time do not overflow the capacity of the buffer.

If the memory buffer can hold data from all outstanding requests in a lookahead window, then the scheduling is trivial. The interesting case arises when the number of requests exceeds the buffer capacity. The scheduler must then determine which of the requests to dispatch at each step, so as to minimize the total number of I/O steps required to service all the requests.

We illustrate the scheduling problem by considering a single flow. Even for this special case the optimal schedule is not straightforward to construct. As an example we consider a single flow with reference string $\mathscr{R} = A_1 A_2 A_3 A_4 A_5 B_1 B_2 A_6 B_3 B_4 B_5 A_7 C_1 C_2 B_6 B_7$. Here, A_i (B_i, C_i) stand for the ith block from disk A (respectively disk B and disk C). Figure 2(a) shows the schedule created by an intuitive, greedy in-order prefetching strategy. A greedy prefetching algorithm tries to keep as many disks as possible busy at each I/O step; in-order fetching means it always chooses a block earlier in the reference string in preference to one that occurs later. The schedule assumes a shared buffer of size $M = 6$ blocks. In step 1 of Figure 2(a), blocks A_1, B_1 and C_1 are fetched from the three disks respectively. A_1 is consumed, and a request for the next block A_2 is made. Since there are four free blocks in the buffer at this time, the system will prefetch the next block from disks B and C along with A_2 in the next I/O step. On making the reference to A_3, the buffer holds 4 blocks; hence it cannot prefetch blocks from both B and C, but must choose to prefetch from one or the other of the disks. A greedy in-order prefetching scheduling algorithm will fetch B_3 in preference to C_3 since it occurs earlier in the reference string. Continuing in this manner, we obtain the schedule of length 9 to service the entire reference string. In contrast, Figure 2(b) shows the optimal-length schedule for this reference string consisting of 7 parallel I/O steps.

IOs→	1	2	3	4	5	6	7	8	9
Disk A	A_1	A_2	A_3	A_4	A_5	A_6	A_7	-	-
Disk B	B_1	B_2	B_3	-	-	B_4	B_5	B_6	B_7
Disk C	C_1	C_2	-	-	-	-	-	-	-

IOs→	1	2	3	4	5	6	7
Disk A	A_1	A_2	A_3	A_4	A_5	A_6	A_7
Disk B	B_1	B_2	B_3	B_4	B_5	B_6	B_7
Disk C	-	-	-	-	-	C_1	C_2

Fig. 2. (a)Greedy In-order IO schedule. (b)Optimal IO schedule.

When the input includes multiple flows there is an additional dimension to the problem. Not only does the scheduler need to determine which disks to fetch from in an I/O step, it also needs to decide which flow should be allocated a disk at each I/O step.

2.1 Scheduling Multiple Flows

The problem of constructing a minimum-length schedule for two or more reference strings in the SDPB and SDSB models is formally defined below. In the following, the number of flows is denoted by n, the number of disks is d and the set of disks is denoted by $\mathcal{D} = \{D_1, D_2, \cdots D_d\}$.

Partitioned Buffer: The problem will be denoted by SDPB(n, d, Σ, L). $\Sigma = [m_1, m_2, \cdots m_n]$ is a vector specifying buffer allocations where m_i is the buffer size of flow f_i, and L is a desired bound on the schedule length. The decision problem is to determine if there exists a schedule that requires less than or equal to L I/O steps. Note that in a valid schedule, each reference string consumes blocks in the order specified, but there is no restriction on the ordering across strings. Furthermore, flow f_i can hold at most m_i blocks in the buffer at any time.

Shared Buffer: The problem will be denoted by SDSB(n, d, M, L), where the memory buffer of size M blocks is shared among all flows, and L is a desired bound on the schedule length. The decision problem is to determine if there exists a schedule that requires less than or equal to L I/O steps. Note that in a valid schedule the buffer can hold at most M blocks at any time.

3   NP-Completeness

In this section, we present the proofs for NP-completeness of the scheduling problems for the two models. The input length is the sum of lengths of all input reference strings. Each reference string will be assumed to be fully enumerated by the sequence of blocks that it accesses, so that a reference string with l requests is assumed to require $\Omega(l)$ bits to represent. That is we do not assume any form of compression of the input reference string. This assumption does not limit the generality of our model, and reflects the natural encoding in the application domain.

We will use the following known NP-complete problem, 3-Partition, to show NP-completeness of our problems.

Definition 1. 3-Partition: *Given a multi-set $\mathcal{A} = \{a_1, a_2, ..., a_{3w}\}$ and a positive integer B, such that $\forall i, 1 \leq i \leq 3w, B/4 < a_i < B/2$, and $\sum_{i=1}^{3w} a_i = wB$. Does there exists a* partition of $\mathcal{A}$ into w subsets $\{A_1, A_2, ..., A_w\}$, such that each A_s has exactly 3 elements and $\sum_{a_i \in A_s} a_i = B, \forall i, 1 \leq s \leq w$.

Lemma 1. *3-Partition is NP-Complete even when the input is assumed to be of size* wB *[18].*

Our first result is that the SDPB problem is NP-complete. This result actually follows by showing the correspondence between the special case of SDPB when each $m_i = 1$ and the job shop scheduling problem that has been extensively studied in the literature (see [15, 19, 20]). The proof is omitted due to lack of space. The complexity of the SDSB problem however does not follow directly from job-shop scheduling or its variants, and we prove its NP-*Completeness* from first principles. To specify a reference string we will use the following notation: $r \times D_j$ will mean r distinct consecutive requests to disk D_j. The concatenation of two reference strings α and β will be denoted by $\alpha * \beta$, and the concatenation of α to itself s times will be denoted by α^s.

Theorem 1. SDPB(n, d, Σ, L), n arbitrary, is NP-complete.

Theorem 2. SDSB(n, d, M, L), n arbitrary, is NP-complete.

Proof. It is easy to verify that the problem is in NP.

We reduce 3-Partition to SDSB(n, d, M, L). Given w, B and a multi-set $A = \{a_1, a_2, ..., a_{3w}\}$, we construct an instance of SDSB(n, d, M, L) with $n = 3w + 1$, $M = cB$ for some constant c, and $L = 2w\alpha$, where $\alpha = B + M - 1$. The reference strings $R_1, R_2, \cdots, R_{3w}, R_{3w+1}$ are defined as follows.

$$R_i = (a_i \times D_1) * (a_i \times D_2), 1 \leq i \leq 3w$$
$$R_{3w+1} = (\alpha \times D_2) * [((\alpha + M - 1) \times D_1) * ((\alpha + M - 1) \times D_2)]^{(w-1)}$$
$$*(\alpha + M - 1 \times D_1)$$

For all i, $1 \leq i \leq 3w$, the schedule for R_i consists of a_i fetches from D_1 followed by a_i fetches from D_2 with some overlap possible between the fetches from the two disks. The amount of overlap can be no more than $M - 1$. R_{3w+1} consists of α blocks from D_2 followed by a repeating pattern consisting of $\alpha + M - 1$ blocks from D_1 followed by $\alpha + M - 1$ blocks from D_2; this pattern is repeated $w - 1$ times, and is finally followed by $\alpha + M - 1$ blocks from D_1. Individual schedules for R_i are shown in Figure 3. There can be at most $M - 1$ blocks prefetched at any time. Hence, in order for R_{3w+1} to complete within the schedule $L = 2\alpha w$, the following is necessary: in the interval $[(k - 1)\alpha, k\alpha]$, $k = 1, 2, \cdots 2w$, α blocks of R_{3w+1} must be fetched from D_2 if k is odd, and from D_1 if k is even. In order to meet this schedule $M - 1$ blocks of R_{3w+1} must be prefetched from the other disk in every interval $[(k - 1)\alpha, k\alpha]$; these prefetched blocks are shown by the vertical stripes in the unshaded regions. This implies that there are no free buffers available for prefetching blocks of R_i, $1 \leq i \leq 3w$, which must be fetched one block at a time.

Now we will show that there exists a 3-Partition if and only if a schedule of length L is possible.

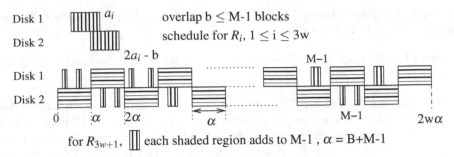

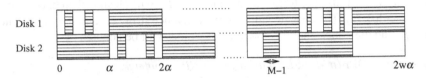

Fig. 3. Individual schedules for SDSB model

Fig. 4. A feasible schedule with length $2w\alpha$

Case 1: Suppose a partition of A into w subsets $\{A_1, A_2, A_3, \cdots, A_w\}$ exists, such that $A_k = \{A_{k_1}, A_{k_2}, A_{k_3}\}$, $\sum_{i=1}^{3} a_{k_i} = B$, then the following schedule of length L is certainly possible.

In Figure 4, for each unshaded region $[2(k-1)\alpha, (2k-1)\alpha]$, $1 \le k \le w$, requests corresponding to A_k, in particular reference strings R_{k_1}, R_{k_2} and R_{k_3}, are scheduled on D_1. The corresponding requests from D_2 are scheduled in the unshaded interval $[(2k-1)\alpha, 2k\alpha]$. In the unshaded interval $[2(k-1)\alpha, (2k-1)\alpha]$, we fetch the B blocks belonging to R_{k_1}, R_{k_2}, and R_{k_3} on D_1, and prefetch the next set of $M-1$ blocks from R_{3w+1} on D_2; their relative ordering within the region is unimportant. Similarly the unshaded region, $[(2k-1)\alpha, 2k\alpha]$ is used to schedule the B blocks on D_1 belonging R_{k_1}, R_{k_2} and R_{k_3}, and on D_2 (except for the last unshaded region) prefetch the next $M-1$ blocks from R_{3w+1}, as shown in figure 4. Hence if a 3-Partition exists, so does a schedule with length L.

Case 2: If a schedule of length $L = 2w\alpha$ exists, then we show that set $\mathcal{A}$ can be partitioned into w desired subsets $A_1, A_2, \cdots, A_w$. First observe that in a valid schedule of length $2w\alpha$, R_{3w+1} needs all the shaded regions in Figure 4, and $M-1$ of each unshaded interval (except the last one of D_2) for fetching its blocks. Consequently the remaining reference strings, R_i, $1 \le i \le 3w$, must be scheduled within the unshaded regions; since in any of these intervals, $M-1$ of the time is used to prefetch blocks of R_{3w+1}, the blocks of R_i, $1 \le i \le 3w$, must be fetched in time $\alpha - (M-1) = B$. The conditions are exactly the same as in Theorem 1, where the short reference strings needed to be scheduled in intervals of length B on each disk. Using identical reasoning, we can conclude that the scheduling of the R_i, $1 \le i \le 3w$, in the unshaded intervals induces a 3-Partition of $\mathcal{A}$.

Note that the sum of lengths of all reference strings is $2wB + 2w\alpha + (2w-2)(M-1) = 4wB + (4w-2)(M-1)$ which is polynomially related to wB when $M = \Theta(B)$.

4   Randomized Scheduling

In the previous section we showed that the general problem of scheduling multiple reference strings with the minimum number of I/O's is NP-*complete* for both the SDPB and SDSB models. In this section we look at the complexity of the optimal scheduling problem in more favorable situations. In the first situation we show how the fundamental result of Leighton, Maggs and Rao [15] for congestion removal in networks using randomization can be applied to the SDPB scheduling problem. This results in an algorithm with probabilistically bounded schedule length. For the SDSB problem we describe a heuristic scheme based on quasi-static memory allocation followed by congestion removal via randomization.

4.1   SDPB Model

We first consider the SDPB model where each flow has its own independent buffer. For each flow f_i with buffer size m_i we use the optimal scheduling algorithm [4,5] to find a schedule that minimizes the number of I/O steps required to fetch all the blocks of f_i, when executed in isolation. Let T_i be the number of steps required by the optimal schedule of f_i using buffer m_i. Let $T = max_{1 \leq i \leq n} T_i$ be the maximum schedule length of any of the flows. Note that T is a *lower bound* on the length of any schedule for the SDPB problem instance.

At any time step t between 1 and T_i, the schedule for f_i indicates which disks will be active fetching blocks of f_i at that time step. If we overlay the schedules of each f_i, $1 \leq i \leq n$, then *disk conflicts* may occur. That is at time step t, the schedules of two or more flows may access the same disk D_k. A simple way to handle the congestion is to simulate each step of the overlay schedule by a number of sub-steps, where only one block is fetched from any disk in each sub-step. Let $c(i,t)$ indicate the congestion (number of contending flows) for disk D_i at time step t. Let $c^*(t) = max_{1 \leq i \leq d} c(i,t)$ be the maximum congestion on any disk at time step t. Then the accesses at step t can be simulated in a conflict-free manner in $c^*(t)$ steps. Hence the entire schedule can be simulated in $\Sigma_{1 \leq t \leq T} c^*(t)$ steps. Since in the worst-case $c^*(t)$ can equal d, this gives a worst-case bound of Td for the length of the schedule using this strategy.

We now present a randomized scheduling method for SDPB with bounded performance. The strategy uses the fundamental idea of Leighton, Maggs and Rao [15] who showed how to remove congestion in a routing network using randomization. The relation of the network routing problem to *shop scheduling* problems was shown in [20]. In the *job shop scheduling* problem there are d machines and n jobs; each job is made up of an ordered sequence of operations which must be serviced in order. Each operation must be assigned to a particular machine depending on the operation. A machine may work on only one operation at any time step, and each job may be processed by only one machine at a time. A special case of job shop where each job is processed exactly once on each machine is known as the *flow shop* problem.

The routing problem in [15] is a special case of flow-shop scheduling where each operation requires unit time. In [20] the technique of [15] was generalized to allow operations to have different and arbitrary lengths, and to allow a job to use the same machine several times. However, each job could only be processed by one machine at any time. It is possible to visualize the SDPB problem considered in this paper within this framework. Associate flows with jobs, disks with the machines, and each block of a flow as a unit-time operation of the job. Then the special case when the memory assigned to each flow is one ($m_i = 1$), corresponds exactly to the job shop scheduling problem. When $m_i > 1$, SDPB differs from the standard job shop problem in that several machines (disks) may service the operations (fetch blocks) of the same job (flow) at the same time step. That is, there is parallelism among the operations of a single job which is disallowed in the standard job shop scheduling model.

We show below that the techniques of [15, 20] can be extended to apply to this more general model of SDPB as well. Let N be the total number of blocks in all flows combined. Recall that the length of the optimal schedule for flow f_i with memory allocation m_i is denoted by T_i, and that $T = max_{1 \leq i \leq n} T_i$ is the maximum length of the schedule for any flow. For any buffer size m_i the optimal schedule for f_i can be constructed using the algorithm in [4, 5]. Let the total number of blocks fetched from disk D_i be B_i, and let $B = max_{1 \leq i \leq d} B_i$ denote the maximum number of blocks fetched from any disk. Finally let the length of the optimal schedule for SDPB be T^*. Then $T^* \geq T$ and $T^* \geq B$; so $T + B \leq 2T^*$. We show that with high probability all accesses can be scheduled to complete in $cT^* \log N$ steps, for a constant c.

The scheme parallels that in [15, 20]. The schedule for flow f_1 begins at time step 1. The start time for the schedule for each flow f_i, $i > 1$ is staggered by an integer number of steps randomly chosen between 1 and B. The n schedules with start times between 1 and B when overlaid result in a schedule of length at most $T + B$. However, there may be contention for disks at different time steps, depending on how the rectangles representing the schedules line up. We show below that the worst-case disk conflict is small with high probability.

Theorem 3. *Let N be the total number of blocks in the reference strings. It is possible to find a schedule for SDPB that with high probability is within a factor $c \log N$ of the optimal length, for a constant c.*

Proof. The overlapped schedules obtained by staggering the flows is of length no more than $T + B$. We show that with high probability the maximum load on any disk is upper bounded by $O(\log N)$.

Consider an arbitrary time step u and an arbitrary disk s. At most B blocks are candidates for scheduling on disk s at time step u. The probability of any of the candidate blocks being scheduled in that time step is $1/B$, since the random offset of each flow gives the block a probability $1/B$ of being scheduled in that slot. Hence the probability that k or more blocks in are scheduled in this slot is upper bounded by $\binom{B}{k}(1/B)^k \leq (eB/k)^k(1/B)^k = (e/k)^k$. The probability that k or more blocks are scheduled in any of the disks at time step u is therefore less than $d(e/k)^k \leq N(e/k)^k$. If we choose $k = 2e \log N$, then this probability is upper bounded by $(1/N^{2e-1})$. Now $T \leq N$ and $B \leq N$, and so we have $T + B \leq 2N$. Since there are at most time $T + B$ steps in the schedule, the probability that the maximum load on a disk at any time step is greater that $k = 2e \log N$ is upper bounded by $2N * (1/N^{2e-1}) < 1/N^3$. Hence with high probability each time step in the schedule of length $T + B$ can be simulated contention free in $k = O(\log N)$ steps. Since $T + B \leq 2T^*$, the overall schedule is of length $cT^* \log N$, for a constant c, with high probability.

4.2   SDSB Model

For the SDSB model we first statically partition the buffer among the flows so that in the absence of any disk contention, it minimizes the maximum schedule length of any flow. We then proceed exactly as in the SDPB model. That is first construct the shortest schedule for each flow individually based on the storage it has been allocated, and then stagger their starting times randomly to reduce disk contention. To capture phases of the workload the buffer can be reparititoned for small disjoint sections of the reference string.

Let $T_i(k)$, $1 \leq i \leq n$, denote the length of the optimal schedule of flow f_i with a buffer allocation of k blocks. The buffer allocation is denoted by a vector $\mathcal{M} = [m_1, m_2, \cdots, m_n]$, $\sum_{i=1}^{n} m_i = M$, where m_i is the amount of buffer allocated to f_i. The aim is to find the vector $\mathcal{M}$ that minimizes the completion time of all flows; *i.e.* minimizes $max\{T_i(m_i) \mid \sum_i m_i = M, 1 \leq i \leq n, \}$. A straightforward algorithm will evaluate $T_i(k)$ for all f_i, $1 \leq i \leq n$, and for all values of k, $1 \leq k \leq M$. This can be done in $\Theta(nML)$ time, where L is the length of the longest reference string. It then considers all n-partitions of M, and chooses the partition that minimizes the maximum length schedule. Since there are $\Theta(M^{n-1})$ such partitions, the total time required is $\Theta(nML + nM^{n-1})$; this straightforward algorithm is clearly infeasible in practice. We present a more efficient algorithm that runs in time $\Theta(Ln \log^{\log n} M)$.

The algorithm for the case $n = 2$ is shown in Figure 5. The main idea is as follows. We compute $T_1(m)$ and $T_2(M - m)$. If $T_1(m)$ is the larger of the two, we allocate more memory to f_1 and less to f_2, or vice versa if $T_2(M - m)$ is larger. By reallocating the excess buffer between the two strings, using a binary-search like pattern, we can converge on the optimal in a logarithmic number of probes.

For n reference strings, we use a recursive version of the above method. Divide the strings into two sets with $n/2$ strings in each. Assign m_1 buffer blocks to the first set and m_2 buffer blocks to the other partition. Initially, $m_1 = m_2 = M/2$. Recursively, find the best schedule length for each partition using m_1 and m_2 buffers respectively. If the first set has the longer schedule length, then move buffers from set 2 to set 1, or vice versa as necessary, exactly as if there were two strings. Continue the process till one converges on the best allocation. We will require to do $\Theta(\lg M)$ probes to each of the subproblems to find the best allocation. Formally this recursion can be represented as:

```
l = 1; h = M - 1
while (l < h)
Note: Lⁿₘ is the minimum schedule length of flow n using m buffers
    m = (l + h)/2; /* Probe mid point of memory range */
    d₁ = L₁ˡ - L₂ᴹ⁻ˡ;   d₂ = L₁ᵐ - L₂ᴹ⁻ᵐ;
    if (d₁ × d₂ < 0) h = (m - 1)
    else if (d₁ × d₂ > 0) l = (m + 1);
    else return m
return l;
```

Fig. 5. $O(L \log M)$ algorithm to optimally allocate memory to two flows

$$T(n,M) \le c_1 \lg M \left(2\, T(n/2,M)\right) \text{ and } T(1,M) \le c_2 L \tag{1}$$

Solving the recurrence, we get:

$$T(n,M) \le c_1^k \cdot \log^k M \cdot 2^k \cdot T(n/2^k,M) = \Theta(n\, L\, \lg^{\lg n} M) \tag{2}$$

5 Conclusions

In this paper, we present an analytical model to study scheduling of multiple flows in a parallel I/O system. We show that obtaining a minimum length schedule in this case is NP-*complete* for both the partitioned buffer and shared buffer models. A randomized algorithm based on the congestion removal technique of [15] was analyzed and the dilation in the schedule length for the SDPB model was shown to be bounded by a factor logarithmic in the number of blocks, with high probability. A heuristic for the SDSB model was presented based on quasi-static partitioning of the buffer among the flows, followed by randomized congestion removal. An interesting open issue is the complexity of the problem with a fixed number of reference strings, and a variable number of disks or buffer size. Another interesting problem is to come up with approximation algorithms for the SDSB model with guarantees on performance.

References

1. Gulati, A., Varman, P.: Lexicographic QoS scheduling for parallel I/O. In: Proceedings of the ACM Symposium on Parallelism in Algorithms and Architectures, Las Vegas, Nevada, United States, ACM Press (2005) 29–38
2. Albers, S., Garg, N., Leonardi, S.: Minimizing stall time in single and parallel disk systems. J. ACM **47** (2000) 969–986
3. Cao, P., Felten, E.W., Karlin, A.R., Li, K.: A study of integrated prefetching and caching strategies. In: Proc. of the Joint Intl. Conf. on Measurement and Modeling of Computer Systems, ACM Press (1995) 188–197
4. Kallahalla, M., Varman, P.J.: PC-OPT: optimal offline prefetching and caching for parallel I/O systems. IEEE Transactions on Computers **51** (2002) 1333–1344
5. Kallahalla, M., Varman, P.: Optimal read-once parallel disk scheduling. Algorithmica, (A preliminary version appeared in the 6th ACM IOPADS, 1999) (2005)
6. Hutchinson, D.A., Sanders, P., Vitter, J.S.: Duality between prefetching and queued writing with parallel disks. In: Proceedings of the 9th Annual European Symposium on Algorithms, Århus, Denmark, Springer-Verlag (2001) 62–73
7. Kimbrel, T., Karlin, A.R.: Near-optimal parallel prefetching and caching. SIAM J. Comput. **29** (2000) 1051–1082
8. Patterson, R.H., Gibson, G., Ginting, E., Stodolsky, D., Zelenka, J.: Informed prefetching and caching. In: Proc. of the 15th ACM Symp. on Operating Systems Principles. (1995)
9. Vitter, J.S., Shriver, E.A.M.: Optimal disk I/O with parallel block transfer. In: Proceedings of the twenty-second annual ACM symposium on Theory of computing, ACM Press (1990)
10. Jin, W., Chase, J.S., Kaur, J.: Interposed proportional sharing for a storage service utility. SIGMETRICS Perform. Eval. Rev. **32** (2004) 37–48
11. Huang, L., Peng, G., Chiueh, T.C.: Multi-dimensional storage virtualization. SIGMETRICS Perform. Eval. Rev. **32** (2004) 14–24

12. Lumb, C., Merchant, A., Alvarez, G.: Façade: Virtual storage devices with performance guarantees. File and Storage technologies (FAST'03) (2003) 131–144
13. Gulati, A.: Scheduling with QoS in parallel I/O systems. Master's thesis, Rice University, Department of Computer Science (2004)
14. Gulati, A., Varman, P.: Scheduling with QoS in parallel I/O systems. In: International Workshop on Storage Network Architecture and Parallel I/Os, (held in conjunction with PACT), Antibes Juan-les-pins, France (2004)
15. Leighton, F.T., Maggs, B.M., Rao, S.B.:   Packet routing and job-shop scheduling in O(congestion + dilation) steps. In: Combinatorica. Volume 14. (1994) 167–186
16. Barve, R.D., Kallahalla, M., Varman, P.J., Vitter, J.S.: Competitive parallel disk prefetching and buffer management. Journal of Algorithms **36** (2000) 152–181
17. Kallahalla, M., Varman, P.J.: Optimal prefetching and caching for parallel I/O systems. In: Proc. of 13th ACM Symp. on Parallel Algorithms and Architectures, ACM press (2001)
18. Garey, M.R., Johnson, D.S.: Computers and intractability; a guide to the theory of NP-completeness. W. H. Freeman (1979)
19. Gonzalez, T., Sahni, S.: Flowship and jobshop schedules: Complexity and approximation. Operations research **26** (1978) 36–52
20. Shmoys, D.B., Stein, C., Wein, J.: Improved approximation algorithms for shop scheduling problems. In: SIAM Journal on Computing. Volume 23. (1994) 617–632

Snap-Stabilizing Detection of Cutsets

Alain Cournier, Stéphane Devismes, and Vincent Villain

LaRIA, CNRS FRE 2733, Université de Picardie Jules Verne, Amiens, France
{cournier, devismes, villain}@laria.u-picardie.fr

Abstract. A *snap-stabilizing protocol*, starting from any configuration, always behaves according to its specification. Here, we present the first snap-stabilizing protocol for arbitrary rooted networks which detects if a set of nodes is a cutset. This protocol is based on the depth-first search (DFS) traversal and its properties. One of the most interesting properties of our protocol is that, despite the initial configuration, as soon as the protocol is initiated by the root, the result obtained from the computations will be right. So, after the first execution of the protocol, the root is able to take a decision: "the input set is a cutset or not", and this decision is right.

1 Introduction

In this paper, we present the first snap-stabilizing protocol for detecting if a set of processors is a cutset of an arbitrary rooted network. Consider a connected undirected graph $G = (V, E)$, where V is the set of N nodes and E the set of edges. $CS \subseteq V$ is a *cutset* (or a *separator*) of G if and only if the removal of all nodes of CS disconnects G. The detection of cutsets is an important issue in many applications such as evaluating the reliability of networks. Thus, from the fault tolerance point of view, detecting if a set of processors is a cutset of a network is essential. The concept of *self-stabilization* [1] is the most general technique to design a system tolerating arbitrary transient faults. A self-stabilizing system, regardless of the initial states of the processors and messages initialy in the links, is guaranteed to converge to the intended behavior in a finite time. *Snap-stabilization* was introduced in [2]. A *snap-stabilizing* protocol guaranteed that it always behaves according to its specification. In other words, a snap-stabilizing protocol is also a self-stabilizing protocol which stabilizes in 0 time unit. Obviously, a *snap-stabilizing* protocol is optimal in stabilization time.

Related Works. In the graph theory area, researchers are interested to scan all minimal cutsets of a graph. But, Provan and Ball proved that scanning all cutsets of a given graph in an NP-hard problem [3]. Thus, some heuristics have been designed for arbitrary graphs [4] and polynomial complete methods has developped for some particular class of graphs [5,6]. Several works have been also proposed in distributed (non self-stabilizing) systems [7,8]. To our best knowledge, nothing about cutsets has been proposed in self-stabilizing systems until now (so, neither in snap-stabilizing systems).

Contribution. In this paper, we present the first snap-stabilizing protocol for detecting if a set of processors is a cutset of an arbitrary rooted network. One of the most interesting

D.A. Bader et al. (Eds.): HiPC 2005, LNCS 3769, pp. 488–497, 2005.

properties of our protocol is that, despite the initial configuration, as soon as the protocol is initiated by the root, the result obtained from the computations will be right. So, after the first execution of the protocol, the root is able to take a decision: "the input set is a cutset or not", and this decision is right. The presented protocol is the composition of a distributed cutset test algorithm with a previous snap-stabilizing DFS wave protocol [9]. The drawback of our solution is high cost memory requirement due to the snap-stabilizing DFS wave protocol. But, our cutset test algorithm may be composed with any self-stabilizing DFS wave protocol in order to improve the memory requirement. However, in this case, the resulting protocol will be self-stabilizing only.

The rest of the paper is organized as follows: in Section 2, we describe the model in which our protocol is written. In Section 3, we present some useful properties about cutsets. We describe our protocol in Sections 4. In Section 5, we give a sketch of the proof of snap-stabilization of our protocol[1]. Finally, after presenting some complexity results (Section 6), we make concluding remarks (Section 7).

2   Preliminaries

Network. We consider a *network* as an undirected connected graph $G = (V, E)$ where V is a set of *processors* ($|V| = N$) and E is the set of *bidirectional communication links*. We consider networks which are *asynchronous* and *rooted*, i.e., among the processors, we distinguish a particular processor called *root*. We denote the root processor by r. A communication link (p, q) exists if and only if p and q are neighbors. Every processor p can distinguish all its links. To simplify the presentation, we refer to a link (p, q) of p as the *label* q. We assume that the labels of p, stored in the set $Neig_p$, are locally ordered by $\prec_p$. We assume that $Neig_p$ is a constant and is an input from the system.

Computational Model. In the computation model we use, each processor executes the same program except r. We consider the local shared memory model of communication. The program of every processor consists in a set of *shared variables* (henceforth, referred to as variables) and a finite set of actions. A processor can only write to its own variables, and read its own variables and variables owned by the neighboring processors. Each action is constituted as follows: $< label > :: < guard > \rightarrow < statement >$. The guard of an action in the program of p is a boolean expression involving the variables of p and its neighbors. The statement of an action of p updates one or more variables of p. An action can be executed only if its guard is satisfied. We assume that the actions are atomically executed, meaning, the evaluation of a guard and the execution of the corresponding statement of an action, if executed, are done in one atomic step. The *state* of a processor is defined by the value of its variables. The *state* of a system is the product of the states of all processors ($\in V$). We will refer to the state of a processor and system as a (*local*) *state* and (*global*) *configuration*, respectively. Let $\mathcal{C}$ be the set of all possible configurations of the system. An action A is said to be enabled in $\gamma \in \mathcal{C}$ at p if the guard of A is true at p in γ. A processor p is said to be *enabled* in γ ($\gamma \in \mathcal{C}$) if there exists an enabled action in the program of p in γ. Let a distributed protocol $\mathcal{P}$ be a collection of binary

---

[1] See http://www.laria.u-picardie.fr/$\sim$devismes/tr2005-04.pdf for a complete proof.

transition relations denoted by $\mapsto$, on $\mathcal{C}$. A *computation* of a protocol $\mathcal{P}$ is a *maximal* sequence of configurations $e = (\gamma_0, \gamma_1, ..., \gamma_i, \gamma_{i+1}, ...)$, such that for $i \geq 0, \gamma_i \mapsto \gamma_{i+1}$ (called a *single computation step* or *move*) if γ_{i+1} exists, else γ_i is a terminal configuration. *Maximality* means that the sequence is either finite (and no action of $\mathcal{P}$ is enabled in the terminal configuration) or infinite. All computations considered in this paper are assumed to be maximal. In a step of computation, first, all processors check the guards of their actions. Then, some *enabled* processors are chosen by a *daemon*. Finally, the "elected" processors execute one or more of their *enabled* actions. There exists several kinds of *daemon*. Here, we assume an *unfair distributed daemon*. The *unfairness* means that the daemon can forever prevent a processor to execute an action except if it is the only enabled processor. The *distributed* daemon implies that, during a computation step, if one or more processors are enabled, the daemon chooses at least one (possibly more) of these enabled processors to execute an action. We consider that any processor p executed a *disabling action* in the computation step $\gamma_i \mapsto \gamma_{i+1}$ if p was *enabled* in γ_i and not enabled in γ_{i+1}, but did not execute any action between these two configurations. (The disabling action represents the following situation: at least one neighbor of p changes its state between γ_i and γ_{i+1}, and this change effectively made the guard of all actions of p false.) In order to compute the time complexity, we use the definition of *round* [10]. This definition captures the execution rate of the slowest processor. Given a computation e, the *first round* of e (let us call it e') is the minimal prefix of e containing the execution of one action (an action of the protocol or the disabling action) of every enabled processor from the first configuration. Let e'' be the suffix of e such that $e = e'e''$. The *second round* of e is the first round of e'', and so on.

Snap-Stabilizing Systems. The concept of *Snap-stabilization* was first introduced in [2] as follows: a snap-stabilizing protocol guarantees that it always behaves according to its specification. In [11], authors discuss and formalize the definition to clarify the concept. In particular, they recall that snap-stabilization does not guarantee that all components of the system never work in a fuzzy manner. Snap-stabilization just ensures that if an execution of the protocol is initiated by some processor, then the protocol behaves as expected. The protocol we present is a *wave protocol* as defined by Tel in [12]. By definition, any execution of a wave protocol contains at least one initialization action. So, following [11], we propose a more simple definition of snap-stabilization holding for wave protocols.

Definition 1 (Snap-stabilization for Wave Protocols). *Let $\mathcal{T}$ be a task, and $S\mathcal{P}_{\mathcal{T}}$ a specification of $\mathcal{T}$. A wave protocol $\mathcal{P}$ is snap-stabilizing for $S\mathcal{P}_{\mathcal{T}}$ if and only if (i) at least one processor eventually executes a particular action of $\mathcal{P}$, and (ii) the result obtained with $\mathcal{P}$ from this particular action always satisfies $S\mathcal{P}_{\mathcal{T}}$.*

3 Basis of the Algorithm

3.1 Definitions

We call *path* of $G = (V, E)$ any sequence of processors $P = p_0, p_1, ..., p_k$ such that $\forall i, 1 \leq i \leq k, (p_{i-1}, p_i) \in E$. P is said *elementary* if $\forall i, j, 0 \leq i < j \leq k, p_i \neq p_j$. If

$p_0,....,p_{k-1}$ is elementary and $p_0 = p_k$, then P is called a *cycle*. The processors p_0 and p_k are termed as the *extremities* of the path. The *length* of P, noted $|P|$, is the number of edges which compose P. $G_S = (V_S, E_S)$ is the subgraph of $G = (V, E)$ induced by V_S if and only if $V_S \subseteq V$ and $E_S = E \cap (V_S)^2$. $G = (V, E)$ is said connected if and only if $\forall p, q \in V$ there exists a path between p and q in G. A connected componante of G is any connected subgraph of G maximal by inclusion. A connected undirected graph without any cycle is called a tree. The graph $T = (V_T, E_T)$ is a spanning tree of $G = (V, E)$ if and only if T is a tree, $V_T = V$, and $E_T \subseteq E$. Let $Tree(r) = (V, E_T)$ be a spanning tree of G rooted at r. The *height* of a node p in $Tree(r)$, noted $h(p)$, is the length of the elementary path from r to p in $Tree(r)$. $H = \max_{p \in Tree(r)}\{h(p)\}$ represents the height of $Tree(r)$. For a node $p \neq r$, a node $q \in V$ is said to be the *parent* of p in $Tree(r)$ if and only if q is the neighbor of p (in $Tree(r)$) such that $h(p) = h(q) + 1$. Conversely, p is said to be the *child* of q in $Tree(r)$. A node p_0 is said to be an ancestor of another node p_k in $Tree(r)$ (with $k > 0$) if there exists a sequence of nodes $p_0,...,p_k$ such that $\forall p_i$, with $0 \leq i < k$, p_i is the parent of p_{i+1} in $Tree(r)$. Conversely p_k is said to be a *descendant* of p_0. We note $Tree(p)$ the subtree of $Tree(r)$ rooted at p ($\in V$), i.e., the subgraph of $Tree(r)$ induced by p and its descendants in $Tree(r)$. We call *tree edges* the edges of E_T and *non-tree edges* the edges of $E \setminus E_T$. We call *non-tree neighbors* of p, nodes linked to p by a non-tree edge. $Tree(r)$ is a DFS spanning tree of $G = (V, E)$ if and only if $\forall (p,q) \in E$, $p \in Tree(q)$ or $q \in Tree(p)$.

3.2 Approach

Let $CS \subseteq V$. Let $G' = (V', E')$ be the subgraph of G induced by $V' = V \setminus CS$. Let $Tree(r) = (V, E_T)$ be a DFS spanning tree of G rooted at r. By definition, CS is a cutset of G if and only if there exists at least two connected componantes in G'. So, in the following, we particularize a node, called $CCRoot$, for each connected componantes in G'. Then, we deduce some results, the last one is a technical lemma which provide a way to locally detect if a node is a $CCRoot$.

Definition 2 (CCRoot). *We call $CCRoot$ of a connected componante C of G', a node $p \in C$ satisfying $h(p) \leq h(p')$, $\forall p' \in C$ (i.e., p is a node of C with the minimal height in $Tree(r)$). In particular, by definition, r is a CCRoot if $r \notin CS$.*

Lemma 1. *Let C be a connected componante of G' and p be a $CCRoot$ of C. $Tree(p)$ contains (at least) every node of C.*

Corollary 1. *There only exists one $CCRoot$ in each connected componante of G'.*

Theorem 1. *CS is a cutset if and only if there exists at least two $CCRoot$ in G'.*

Lemma 2. *Let C be a connected componante of G'. A node p is the $CCRoot$ of C if and only if p satisfies the two following conditions: (i) $p \in C$, (ii) $\forall x \in Tree(p)$ such that $x \in C$, $\forall y \in Neig_x$: $y \notin CS \Rightarrow h(y) \geq h(p)$.*

4   Algorithm

In this section, we propose a snap-stabilizing protocol for detecting if a set of processors is a cutset of the network. Our protocol is the *conditional composition* of two other protocols: Algorithm $\mathcal{DFS}$ and Algorithm $\mathcal{CCRC}$ (the CCRoots Counting Algorithm). Algorithm $\mathcal{DFS}$ refers to the snap-stabilizing depth-first search (DFS) protocol of [9]. Algorithm $\mathcal{CCRC}$ uses the DFS properties in order to count the $CCRoot$ of the network as explained in the previous section. So, after recalling the definition of the *conditional composition*, we present Algorithm $\mathcal{DFS}$. We then introduce the data structures used by Algorithm $\mathcal{CCRC}$. Finally, we explain the behavior of the conditional composite algorithm $\mathcal{CCRCDFS}$, i.e., the conditional composition of Algorithm $\mathcal{CCRC}$ and Algorithm $\mathcal{DFS}$.

4.1   Conditional Composition

The *conditional composition* is a protocol composition technique which has been introduced by Datta et al in [13]. This general technique allows to simplify the design and proofs of Algorithm $\mathcal{CCRCDFS}$.

Definition 3 (Conditional Composition). *Let S_1 and S_2 be protocols such that variables written by S_2 are not referred by S_1. The conditional composition of S_1 and S_2, denoted by $S_2 \circ_{|\mathcal{G}} S_1$, is a protocol that satisfies the following conditions:*

1. *It contains all the variables and actions of S_1 and S_2.*
2. *$\mathcal{G}$ is a set of predicates and is a subset of the guards of S_1.*
3. *Every guard of S_2 has the form $g \wedge h$ or $\neg g \wedge h$ where g is a logical expression using the guards $\in \mathcal{G}$.*
4. *Since some actions of S_2 may also be enabled when an action of S_1 is enabled, the order of execution is the following: the action of S_2 followed by the action of S_1 (in the same step).*

4.2   Algorithm $\mathcal{DFS}$

We now roughly present Algorithm $\mathcal{DFS}$ (see [9] for more details). In Algorithm $\mathcal{DFS}$, the root processor (r) eventually initiates a traversal of the network. During the traversal, all the processors are sequentially visited in DFS order. Algorithm $\mathcal{DFS}$ is snap-stabilizing. The snap-stabilizing property guarantees that, since r initiates the protocol, the traversal is performed as expected. In particular, the traversal cannot be corrupted by any abnormal behavior. The traversal performed by Algorithm $\mathcal{DFS}$ progresses in the network as a token circulation:

- The traversal begins when r creates a token by Action F.
- Each non-root processor p executes Action F when it receives the token for the first time.
- A processor p executes Action B each time the token is backtracked to it: If p has sent the token to q, then, since the traversal ends at q (i.e., q holds the token and the token has visited all its neighbors), q backtracks the token to p.

Obviously, the traversal performed by Algorithm $\mathcal{DFS}$ follows a DFS spanning tree of the network. Frow now on, we note $Tree(r) = (V, E_T)$ this tree. Also, we note $h(p)$ the height of the node p in $Tree(r)$ and H the height of $Tree(r)$.

4.3 Algorithm $\mathcal{CCRC}$

Algorithm $\mathcal{CCRC}$ is just an application of the properties shown in Section 3. We now describe the inputs, variables, and actions of Algorithm $\mathcal{CCRC}$.

Algorithm 1 Algorithm ($\mathcal{CCRC}$) CCRoots Counting for $p = r$

Input:
$Neig_p$: set of neighbors (locally ordered);
$S_p \in Neig_p \cup \{idle, done\}$: variable from Algorithm $\mathcal{DFS}$;
$Forward(p), Backward(p), LockedF(p), LockedB(p)$: predicates from Algorithm $\mathcal{DFS}$;
$Next_p$: macro from Algorithm $\mathcal{DFS}$;
$InCS_p$: boolean;
Constant: $Level_p = 0$;

Variables: $IsCutset_p$: boolean; Cnt_p: integer;

Macros:
$InitCnt_p \quad\quad = \textbf{if}\,(InCS)\ \textbf{then}\ Cnt_p := 0;\ \textbf{else}\ Cnt_p := 1;$
$UpdIsCutset_p = \textbf{if}\,(Next_p = done)\ \textbf{then}\ IsCutset_p := (Cnt_p \geq 2);$
Actions:
$Forward(p) \wedge \neg LockedF(p) \quad \rightarrow InitCnt_p;\ UpdIsCutset_p;$
$Backward(p) \wedge \neg LockedB(p) \rightarrow Cnt_p := Cnt_{S_p};\ UpdIsCutset_p;$

Inputs. Algorithm $\mathcal{CCRC}$ reads two inputs from Algorithm $\mathcal{DFS}$: S_p and $Next_p$. The current successor (resp. predecessor) of a processor p in the traversal is maintained in S_p (resp. P_p). Note that $S_p \in Neig_p \cup \{idle, done\}$ meaning that p is ready to receive the token ($S_p = idle$), the traversal from p is done ($S_p = done$), or the traversal from p is in progress (and S_p designates its current successor in the traversal). Moreover, using the S variables, p can dynamically evaluate its parent P_p in $Tree(r)$ as follows: $P_p = q$ where $S_q = p$ (see Macro P_p). Finally, Macro $Next_p$ allows to compute a new value for S_p. In Algorithm $\mathcal{CCRC}$, we only use this macro to know when the traversal from p is done, i.e., when $Next_p = done$. To simplify the design of the algorithm, we assume that every processor p knows if it belongs to the set to test (noted CS) thanks to the boolean $inCS_p$. In fact, we show $inCS_p$ as an input of the system but we could provided CS (using a set of Ids) in the input of r only and, after, propagated it to all other processors using Algorithm $\mathcal{DFS}$.

Variables. In Algorithm $\mathcal{CCRC}$, each processor p maintains the following datas: (i) $Level_p$, Cnt_p, and $IsCutset_p$ for $p = r$; (ii) $Level_p$, $Back_p$, and Cnt_p for $p \neq r$. $Level_p$ refers to as the height of p in $Tree(r)$. In $Back_p$, we compute the value $UNNTC(p)$ (i.e., the Uppermost Non-Tree Neighbor of $Tree(p)$ in C_p) as follows: **If** $p \in CS$, $UNNTC(p) = -1$. **Otherwise**, p belongs to a connected componante of G', noted C_p, and $UNNTC(p)$ is equal to the minimal value among the height of each node of $Tree(p) \cap C_p$ and the height of their non-tree neighbors q such that $q \in C_p$. From the definition of $UNNTC$ and Lemma 2, the following theorem shows that if $Level_p$ and $Back_p$ are correctly evaluated (i.e., if $Level_p = h(p)$ and $Back_p = UNNTC(p)$), then we can locally detect if p is a $CCRoot$ or not.

Theorem 2. $\forall p \in V \backslash \{r\}$, p is a $CCRoot$ if and only if $p \notin CS$ and $h(p){=}UNNTC(p)$.

Algorithm 2 Algorithm ($\mathcal{CCRC}$) CCRoots Counting for $p \neq r$

Input:
$Neig_p$: set of neighbors (locally ordered);
$S_p \in Neig_p \cup \{idle, done\}$: variable from Algorithm $\mathcal{DFS}$;
$Forward(p)$, $Backward(p)$, $LockedF(p)$, $LockedB(p)$: predicates from Algorithm $\mathcal{DFS}$;
$Next_p$: macro from Algorithm $\mathcal{DFS}$;
$InCS_p$: boolean;
Variables: Cnt_p, $Level_p$, $Back_p$: integers;

Predicate:
$IsCCRoot(p) \equiv (Back_p = Level_p)$

Macros:
$$P_p \qquad\qquad\qquad = (q \in Neig_p :: S_q = p);$$
$$NonCSAncLevel_p = \{x \in \mathbb{N} :: (\exists q \in Neig_p :: Level_q = x \wedge Level_q < Level_p \wedge \neg inCS_q)\};$$
$$NonCSDescBack_p = \{x \in \mathbb{N} :: (\exists q \in Neig_p :: Back_q = x \wedge Level_q > Level_p \wedge \neg inCS_q)\};$$
$$UpdBack_p \qquad = \text{if } (InCS_p) \text{ then } Back_p := -1;$$
$$\qquad\qquad\qquad\quad \text{else } Back_p := \min(\{Level_p\} \cup NonCSAncLevel_p \cup NonCSDescBack_p);$$
$$UpdCnt_p \qquad\quad = \text{if } (IsCCRoot(p)) \text{ then } Cnt_p := Cnt_p + 1;$$
$$Update_p \qquad\qquad = \text{if } (Next_p = done) \text{ then } UpdBack_p; UpdCnt_p;$$
Actions:
$$Forward(p) \wedge \neg LockedF(p) \quad \rightarrow \quad Level_p := Level_{P_p} + 1; Cnt_p := Cnt_{P_p}; Update_p;$$
$$Backward(p) \wedge \neg LockedB(p) \rightarrow \quad Cnt_p := Cnt_{S_p}; Update_p;$$

Thus, thanks to the $Level$ and $Back$ variables, we can locally detect the $CCRoots$. So, in addition, we use the Cnt variables to count the $CCRoots$ of the network. Finally, the boolean $IsCutset_r$ is used as a flag to mark if CS is a cutset or not.

Actions. Using the conditionnal composition, the actions of Algorithm $\mathcal{CCRC}$ are executed in the same step of Actions F and B of Algorithm $\mathcal{DFS}$ (see Definition 3). Action F is enabled at p when p satisfies $Forward(p) \wedge \neg LockedF(p)$. Respectively, Action B is enabled at p when p satisfies $Backward(p) \wedge \neg LockedB(p)$.

During a traversal, when Processor p receives the token for the first time (Action F), p can compute a value depending on it and its parents: a *prefix action*. In Algorithm $\mathcal{CCRC}$, the prefix action allows to compute $Level_p$ for non-root processors and to initialise Cnt_p for the root (Definition 2 allows to determine if r is a $CCRoot$ or not). Then, when the traversal locally ends at p (p executes Actions F or B while $Next_p = done$), p can calculate a result depending on it, its neighbors and/or its descendants: a *postfix action*. Indeed, in this case, $Tree(p)$ is entirely computed and the token has visited all neighbors of p. In Algorithm $\mathcal{CCRC}$, the postfix action allows to:

- Compute $Back_p$ for $p \neq r$. Indeed, when the traversal ends at p, its neighbors have computed their height and its descendants have evaluated their $Back$ Variable.
- Update Cnt_p for $p \neq r$. As $Back_p$ and $Level_p$ are evaluated, by Theorem 2, p knows if it is a $CCRoot$ and, if necessary, it increments Cnt_p.
- Update $IsCutset_p$ for $p = r$. When the traversal ends at r, the traversal is entirely done. So, r knows the number of $CCRoots$ of the network and, using Theorem 1, Macro $UpdIsCutset_p$ updates $IsCutset_p$ as well.

Finally, some actions of Algorithm $\mathcal{CCRC}$ have to be executed at each step of Algorithm $\mathcal{DFS}$ (when Actions F or B are executed). These actions allow to maintain in the Cnt variables the number of $CCRoots$ currently discovered.

4.4 Algorithm $\mathcal{CCRCDFS}$

Algorithm $\mathcal{CCRCDFS}$ is shown as Algorithm 3. Informally, Algorithm $\mathcal{CCRCDFS}$ works as follows. The root, r, begins the traversal by creating a token and initialises Cnt_r to 0 or 1 according to Definition 2. Then, each time a processor $p \neq r$ receives the token for the first time, it initialises Cnt_p ($Cnt_p := Cnt_{S_p}$) and computes its height in $Level_p$. Each time the token is backtracked to a processor q, q updates Cnt_q. When the traversal ends at q, q computes $Back_q$. Indeed, all its neighbors have computed their $Level$ variables and all its descendants have already computed their $Back$ variables. Thus, by Theorem 2, q can decide if it is a $CCRoot$ or not and updates Cnt_q as well. Finally, when the traversal is completely done (i.e., the token is backtracked to r and the token has visited all its neighbors), r can decide if CS (the set of nodes to test) is a cutset (according to Theorem 1) and updates $IsCutset_r$ as well. Thus, from any initial configuration, after the end of a DFS traversal initiated by r, we obtain a configuration similar to the one shown in Figure 1. In this exemple, $CS = \{1, 6, 8\}$ and r, 2 are $CCRoots$. The root processor r is a $CCRoot$ because $r \notin CS$ (Definition 2). Processor 2 is a $CCRoot$ because $2 \neq r$, $2 \notin CS$, and $Level_2 = Back_2$. During the traversal, the Cnt variables count the number of $CCRoots$ (here, equal to 2) and $IsCutset_r$ is set to true at the end of the traversal according to Theorem 1.

Algorithm 3 Algorithm ($\mathcal{CCRCDFS}$) CCRoots Counting and Depth-First Search

$\mathcal{CCRC} \circ |_{\{Forward, LockedF, Backward, LockedB\}} \mathcal{DFS}$

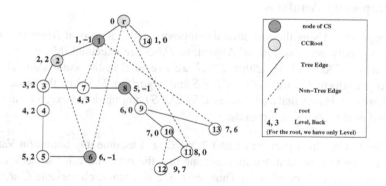

Fig. 1. State of the network after the end of a DFS traversal initiated by r

5 Sketch of Proof

In this section, we show that Algorithm $\mathcal{CCRCDFS}$ (i.e., the conditional composition of Algorithm $\mathcal{DFS}$ and Algorithm $\mathcal{CCRC}$) is snap-stabilizing under an unfair daemon. First, we can remark that Algorithm $\mathcal{CCRC}$ does not change the variables used by Algorithm $\mathcal{DFS}$. Moreover, no action of Algorithm $\mathcal{CCRC}$ can prevent any action of Algorithm $\mathcal{DFS}$ since, when an action of Algorithm $\mathcal{CCRC}$ is executed at p, it is done in the same step of an action of Algorithm DFS at p (because of the conditional composition). So, Algorithm $CCRC$ has no impact on the behavior of Algorithm DFS. From [9], we know that Algorithm DFS is snap-stabilizing, i.e., r eventually initiates the protocol and since r initiates the protocol, Algorithm DFS satisfies its specification. More precisely, starting from any initial configuration, r eventually initiates a traversal of the network. During this traversal, all the processor are sequentially visited in DFS order. In particular, the snap-stabilizing property guarantees that the traversal performed by Algorithm DFS cannot be corrupted by any abnormal behavior. Since Algorithm $CCRC$ cannot prevent Algorithm DFS to work as expected, we will observe the system from the moment when r initiates the protocol and we focus on the traversal performed from r only (we do not take care of any abnormal behavior related to Algorithm DFS). So, if we focus on the traversal performed from r, if easy to verify that, after receiving the token for the first time, any $p \in V$ satisfies $Level_p = h(p)$ until the end of the traversal. Then, when the traversal ends at p, $Back_p = UNNTC(p)$ and, by Theorem 2, p is able to decide if it is a $CCRoot$ or not as explained in Section 4. Hence, at the end of a traversal initiated by r, r knows the number of $CCRoots$ and takes the right decision, i.e., $IsCutset_r = true$ if and only if CS is a cutset. Finally, in [9], Algorithm DFS is proven assuming an unfair daemon. Now, by Definition 3, Algorithm $CCRCDFS$ works with the same number of steps than Algorithm DFS and it is snap-stabilizing under the unfair daemon.

Theorem 3. *Under an unfair daemon, Algorithm $CCRCDFS$ is snap-stabilizing and detects if CS is a cutset.*

6 Complexity Analysis

Time Complexity. Using the conditional composition, the actions of Algorithm $CCRC$ are executed only when actions of Algorithm DFS are executed. Moreover, actions of Algorithm $CCRC$ and Algorithm DFS are executed in the same step. Thus, the complexity results of Algorithm $CCRCDFS$ and Algorithm DFS are the same. Hence, from [9], we can deduce that a complete $CCRCDFS$ computation is executed in $O(N^2)$ moves and in at most $6N - 1$ rounds.

Space Complexity. In Algorithms 1 and 2, we do not assume any bound on Variables Cnt, $Level$, and $Back$. But, we may assume that the maximal value of each of these variables is any upper bound of N. Thus, we can claim that each variable Cnt, $Level$, or $Back$ can be stored in $O(\log N)$ bits and, by taking account of the other variables, we can deduce that the space requirement of Algorithm $CCRC$ is $O(\log(N))$ bits per processor. From [9], we can conclude that the space requirement of Algorithm $CCRCDFS$ is $O(N \times \log(N) + \log(\Delta))$ bits per processor (where Δ is an upper bound on the degree of the processors).

7 Conclusion

In this paper, we have presented the first snap-stabilizing protocol for detecting if a set of processors is a cutset of an arbitrary rooted network called Algorithm $CCRCDFS$.

This protocol, which is a conditionnal composition of Algorithms $CCRC$ and DFS, works assuming an unfair daemon, i.e., the weakest scheduling assumption. The snap-stabilizing property guarantees that despite the initial configuration, as soon as our protocol is initiated by the root, the result obtained from the computations will be right. Moreover, as our protocol is snap-stabilizing, our protocol is optimal in stabilization time. In addition, note that a complete computation of Algorithm $CCRCDFS$ is executed in $O(N)$ rounds and $O(N^2)$ moves. Finally, the space requirement of our solution is $O(N \times \log(N) + \log(\Delta))$ bits per processor. Algorithm $CCRC$ can be combined with any self-stabilizing DFS wave protocol (e.g. [14,15]) in order to improve the memory requirement. Of course, in this case, the resulting protocol will be self-stabilizing only.

References

1. Dijkstra, E.: Self stabilizing systems in spite of distributed control. Communications of the Association of the Computing Machinery **17** (1974) 643–644
2. Bui, A., Datta, A., Petit, F., Villain, V.: State-optimal snap-stabilizing PIF in tree networks. In: Proceedings of the Fourth Workshop on Self-Stabilizing Systems, IEEE Computer Society Press (1999) 78–85
3. Provan, J.S., Ball, M.O.: The complexity of counting cuts and of computing the probability that a graph is connected. SIAM Journal of Computing **12** (1983) 777–788
4. Karger, D.R.: Minimum cuts in near-linear time. Journal of the ACM **47** (2000) 46–76
5. Ahmad, S.H.: Simple enumeration of minimal cutsets of acyclic directed graph. IEEE Transactions on Reliability **37** (1988) 484–487
6. Whited, D.E., Shier, D.R., Jarvis, J.P.: Reliability computations for planar networks. OSRA Journal of Computing **2(1)** (1990) 46–60
7. Fard, N.S., Lee, T.H.: Cutset enumeration of network systems with link and node failure. Reliability Engineering and System Safety **65** (1999) 141–146
8. Rai, S.: A cutset approach to reliability evaluation in communication networks. IEEE Transactions on Reliability **31** (1982) 428–431
9. Cournier, A., Devismes, S., Petit, F., Villain, V.: Snap-stabilizing depth-first search on arbitrary networks. In: OPODIS'04, International Conference On Principles Of Distributed Systems Proceedings. (2005) 267–282
10. Dolev, S., Israeli, A., Moran, S.: Uniform dynamic self-stabilizing leader election. IEEE Transactions on Parallel and Distributed Systems **8** (1997) 424–440
11. Cournier, A., Datta, A., Petit, F., Villain, V.: Enabling snap-stabilization. In: 23th International Conference on Distributed Computing Systems (ICDCS 2003). (2003) 12–19
12. Tel, G.: Introduction to distributed algorithms. Cambridge University Press (Second edition 2001)
13. Datta, A.K., Gurumurthy, S., Petit, F., Villain, V.: Self-stabilizing network orientation algorithms in arbitrary rooted networks. In: International Conference on Distributed Computing Systems. (2000) 576–583
14. Huang, S., Chen, N.: Self-stabilizing depth-first token circulation on networks. Distributed Computing **7** (1993) 61–66
15. Datta, A., Johnen, C., Petit, F., Villain, V.: Self-stabilizing depth-first token circulation in arbitrary rooted networks. Distributed Computing **13(4)** (2000) 207–218

Scheduling Divisible Loads with Return Messages on Heterogeneous Master-Worker Platforms

Olivier Beaumont[1], Loris Marchal[2], and Yves Robert[2]

[1] LaBRI, UMR CNRS 5800, Bordeaux, France
Olivier.Beaumont@labri.fr
[2] LIP, UMR CNRS-INRIA-UCBL 5668, ENS Lyon, France
{Loris.Marchal, Yves.Robert}@ens-lyon.fr

Abstract. In this paper, we consider the problem of scheduling divisible loads onto an heterogeneous star platform, with both heterogeneous computing and communication resources. We consider the case where the workers, after processing the tasks, send back some results to the master processor. This corresponds to a more general framework than the one used in many divisible load papers, where only forward communications are taken into account. To the best of our knowledge, this paper constitutes the first attempt to derive optimality results under this general framework (forward and backward communications, heterogeneous processing and communication resources). We prove that it is possible to derive the optimal solution both for LIFO and FIFO distribution schemes. Nevertheless, the complexity of the general problem remains open: we also show in the paper that the optimal distribution scheme may be neither LIFO nor FIFO.

1 Introduction

This paper deals with scheduling divisible load applications on heterogeneous platforms. As their name suggests, divisible load applications can be divided among worker processors arbitrarily, i.e. into any number of independent pieces. This corresponds to a perfectly parallel job: any sub-task can itself be processed in parallel, and on any number of workers. In practice, the *Divisible Load Scheduling* model, or DLS model, is an approximation of applications that consist of large numbers of identical, low-granularity computations.

Quite naturally, we target a master-worker implementation where the master initially holds (or generates data for) a large amount of work that will be executed by the workers. In the end, results will be returned by the workers to the master. Each worker has a different computational speed, and each master-worker link has a different bandwidth, thereby making the platform fully heterogeneous. The scheduling problem is first to decide how many load units the master sends to each worker, and in which order. After receiving its share of the data, each worker executes the corresponding work and returns the results to the master. Again, the ordering of the return messages must be decided by the scheduler.

The DLS model has been widely studied in the last several years, after having been popularized by the landmark book [7]. The DLS model provides a practical

D.A. Bader et al. (Eds.): HiPC 2005, LNCS 3769, pp. 498–507, 2005.
© Springer-Verlag Berlin Heidelberg 2005

framework for the mapping of independent tasks onto heterogeneous platforms, and has been applied to a large spectrum of scientific problems. From a theoretical standpoint, the success of the DLS model is mostly due to its analytical tractability. Optimal algorithms and closed-form formulas exist for important instances of the divisible load problem. A famous example is the closed-form formula given in [4,7] for a bus network. The hypotheses are the following: (i) the master distributes the load to the workers, but no results are returned to the master; (ii) a linear cost model is assumed both for computations and for communications (see Section 2.1); and (iii) all master-worker communication links have same bandwidth (but the workers have different processing speeds). The proof to derive the closed-form formula proceeds in several steps: it is shown that in an optimal solution: (i) all workers participate in the computation, then that (ii) they never stop working after having received their data from the master, and finally that (iii) they all terminate the execution of their load simultaneously. These conditions give rise to a set of equations from which the optimal load assignment α_i can be computed for each worker P_i.

Extending this result to a star network (with different master-worker link bandwidths), but still (1) without return messages and (2) with a linear cost model, has been achieved only recently [5]. The proof basically goes along the same steps as for a bus network, but the main additional difficulty was to find the optimal ordering of the messages from the master to the workers. It turns out that the best strategy is to serve workers with larger bandwidth first, independently of their computing power.

The next natural step is to include return messages in the picture. This is very important in practice, because in most applications the workers are expected to return some results to the master. When no return messages are assumed, it is implicitly assumed that the size of the results to be transmitted to the master after the computation is negligible, and hence has no (or very little) impact on the whole DLS problem. This may be realistic for some particular DLS applications, but not for all of them. For example suppose that the master is distributing files to the workers. After processing a file, the worker will typically return results in the form of another file, possibly of shorter size, but still non-negligible. In some situations, the size of the return message may even be larger than the size of the original message: for instance the master initially scatters instructions on some large computations to be performed by each worker, such as the generation of several cryptographic keys; in this case each worker would receive a few bytes of control instructions and would return longer files containing the keys.

Because it is very natural and important in practice, several authors have investigated the problem with return messages: see the papers [3,8,9,2,1]. However, all the results obtained so far are very partial. Intuitively, there are hints that suggest that the problem with return results is much more complicated. The first hint lies in the combinatorial space that is open for searching the best solution. There is no reason for the ordering of the initial messages sent by the master to be the same as the ordering for the messages returned to the master by the workers after the execution. In some situations a FIFO strategy (the worker first served by the master is the first to return results, and so on) may be preferred, because it provides a smooth and well-structured pipelining scheme. In other situations, a LIFO strategy (the other way round, first served workers are the last to return results) may provide better results, because faster workers would work a longer period if we serve them first and they send back their results last. True, but what if these fast workers have slow communication links? In fact, and here comes the second hint, it is not even clear whether all workers should be enrolled in the computation by the master. This is in sharp contrast to the case without return messages, where it is obvious that all workers should participate. To the best of our knowledge, the complexity of the problem remains open, despite the simplicity of the linear cost model. In [1], Adler, Gong and Rosenberg show that all FIFO strategies are equally performing on a bus network, but even the analysis of FIFO strategies is an open problem on a star network.

The main contributions of this paper are the characterization of the best FIFO and LIFO strategies on a star network, together with an experimental comparison of them. While the study of LIFO strategies nicely reduces to the original problem without return messages, the analysis of FIFO strategies turns out to be more involved; in fact, the optimal FIFO solution may well not enroll all workers in the computations. Admittedly, the complexity of the DLS problem with return messages remains open: there is no a priori reason that either FIFO or LIFO strategies would be superior to solutions where the ordering of the initial messages and that of return messages are totally uncorrelated (and we give an example of such a situation in Section 2). Still, we believe that our results provide an important step in the understanding of this difficult problem, both from a theoretical and practical perspective. Indeed, we have succeeded in characterizing the best FIFO and LIFO solutions, which are the most natural and easy-to-implement strategies. Due to space limitations, the overview of related work is not included in this paper, please refer to the extended version [6]. Similarly, all proofs are omitted, but they are all detailed in [6].

2   Framework

2.1   Problem Parameters

A *star network* $\mathcal{S} = \{P_0, P_1, P_2, \ldots, P_p\}$ is composed of a master P_0 and of p workers P_i, $1 \le i \le p$. There is a communication link from the master P_0 to each worker P_i. In the linear cost model, each worker P_i has a (relative) computing power w_i: it takes $X.w_i$ time units to execute X units of load on worker P_i. Similarly, it takes $X.c_i$ time units to send the initial data needed for computing X units of load from P_0 to P_i, and $X.d_i$ time units to return the corresponding results from P_i to P_0. Without loss of generality we assume that the master has no processing capability (otherwise, add a fictitious extra worker paying no communication cost to simulate computation at the master). Note that a *bus network* is a star network such that all communication links have the same characteristics: $c_i = c$ and $d_i = d$ for each worker P_i, $1 \le i \le p$.

It is natural to assume that the quantity $\frac{d_i}{c_i}$ is a constant z that depends on the application but not on the selected worker. In other words, workers who communicate faster with the master for the initial message will also communicate faster for the return message. In the following, we keep using both values d_i and c_i, because many results are valid even without the relation $d_i = zc_i$, and we explicitly mention when we use this relation.

Finally, we use the standard model in DLS problem for communications: the master can only send data to, and receive data from, a single worker at a given time-step. A given worker cannot start execution before it has terminated the reception of the message from the master; similarly, it cannot start sending the results back to the master before finishing the computation. However, there is another classic hypothesis in DLS papers which we do not enforce, namely that there is no idle time in the operation of each worker. Under this assumption, a worker starts computing immediately after having received its initial message, which is no problem, but also starts returning the results immediately after having finished its computation: this last constraint does reduce the solution space arbitrarily. Indeed, it may well prove useful for a worker P_i to stay idle a few steps before returning the results, waiting for the master to receive the return message of another worker $P_{i'}$. Of course we could have given more load to P_i to prevent it from begin idle, but this would have implied a longer initial message, at the risk of delaying the whole execution scheme. Instead, we will tackle the problem in its full generality and allow for the possibility of idle times (even if we may end by proving that there is no idle time in the optimal solution).

The objective function is to maximize the number of load units that are processed within T time-units. Let α_i be the number of load units sent to, and processed by, worker P_i within T time-units. Owing to the linear cost model, the quantity $\frac{\sum_{i=1}^{p} \alpha_i}{T} = \rho$ does not depend on T (see Section 2.2 for a proof), and corresponds to the achieved throughput, which we aim at maximizing.

2.2   Linear Program for a Given Scenario

Given a star platform with p workers, and parameters $w_i, c_i, d_i, 1 \leq i \leq p$, how can we compute the optimal throughput? First we have to decide which workers are enrolled. Next, given the set of participating workers, we have to decide for the ordering of the initial messages. Finally we have to decide for the ordering of the return messages. Altogether, there is a finite (although exponential) number of scenarios, where a *scenario* refers to a schedule with a given set of participating workers and a fixed ordering of initial and return messages. Then, the next question is: how can we compute the throughput for a given scenario?

Without loss of generality, we can always perform all the initial communications as soon as possible. In other words, the master sends messages to the workers without interruption. If this was not the case, we would simply shift ahead some messages sent by the master, without any impact on the rest of the schedule. Obviously, we can also assume that each worker initiates its computation as soon as it has received the message from the master. Finally, we can always perform all the return communications as late as possible. In other words,

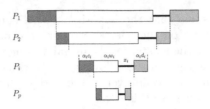

Fig. 1. LIFO strategy. Dark grey rectangles (of length $\alpha_q c_q$) represent the initial messages destined to the workers. White rectangles (of length $\alpha_q w_q$) represent the computation on the workers. Light grey rectangles (of length $\alpha_q d_q$) represent the return messages back to the master. Bold lines (of length x_q) represent the idle time of the workers.

once the master starts receiving data back from the first worker, it receives data without interruption until the end of the whole schedule. Again, if this was not the case, we would simply delay the first messages received by the master, without any impact on the rest of the schedule. Note that idle times can still occur in the schedule, but only between the end of a worker's computation and the date at which it starts sending the return message back to the master.

The simplest approach to compute the throughput ρ for a given scenario is to solve a linear program. For example, assume that we target a LIFO solution involving all processors, with the ordering $P_1, P_2, \ldots, P_p$, as outlined in Figure 1. With the notations of Section 2.1 (parameters w_i, c_i, d_i and unknowns α_i, ρ), worker P_i: (i) starts receiving its initial message at time $t_i^{\text{recv}} = \sum_{j=1}^{i-1} \alpha_j c_j$; (ii) starts execution at time $t_i^{\text{recv}} + \alpha_i c_i$; (iii) terminates execution at time $t_i^{\text{term}} = t_i^{\text{recv}} + \alpha_i c_i + \alpha_i w_i$; (iv) starts sending back its results at time $t_i^{\text{back}} = T - \sum_{j=1}^{i} \alpha_j d_j$. Here T denotes the total length of the schedule. The idle time of P_i is $x_i = t_i^{\text{back}} - t_i^{\text{term}}$, and this quantity must be nonnegative. We derive a linear equation for P_i:

$$T - \sum_{j=1}^{i} \alpha_j d_j \geq \sum_{j=1}^{i-1} \alpha_j c_j + \alpha_i c_i + \alpha_i w_i.$$

Together with the constraints $\alpha_i \geq 0$, we have assembled a linear program, whose objective function is to maximize $\rho(T) = \sum_{i=1}^{p} \alpha_i$. In passing, we check that the value of $\rho(T)$ is indeed proportional to T, and we can safely define $\rho = \rho(1)$ as mentioned before. We look for a rational solution of the linear program, with rational (not integer) values for the quantities α_i and ρ, hence we can use standard tools like Maple or MuPAD.

Obviously, this linear programming approach can be applied for any permutation of initial and return messages, not just LIFO solutions as in the above example. Note that it may well turn out that some α_i is zero in the solution returned by the linear program, which means that P_i is not actually involved in the schedule. This observation reduces the solution space: we can *a priori* assume that all processors are participating, and solve a linear program for each pair of message permutations (one for the initial messages, one for the return messages). The solution of the linear program will tell us which processors are actually involved in the optimal solution for this permutation pair.

For a given scenario, the cost of this linear programming approach may be acceptable. However, as already pointed out, there is an exponential number of scenarios. Worse, there is an exponential number of LIFO and FIFO scenarios, even though there is a single permutation to try in these cases (the ordering of the return messages is the reverse (LIFO) or the same (FIFO) as for the initial messages). The goal of Sections 3 and 4 is to determine the best LIFO and FIFO solution in polynomial time.

2.3   Counter-Examples

In Figure 2 we outline an example where not all processors participate in the optimal solution. The platform has three workers, as shown in Figure 2(a). The best throughput that can be achieved using all the three workers is obtained via the LIFO strategy represented in Figure 2(b), and is $\rho = 61/135$. However,

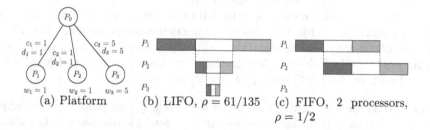

(a) Platform   (b) LIFO, $\rho = 61/135$ (c) FIFO, 2 processors,
                                                     $\rho = 1/2$

Fig. 2. The best schedule with all the three workers (shown in (b)) achieves a lower throughput than when using only the first two workers (as shown in (c))

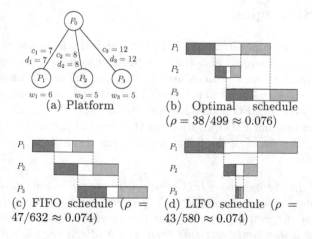

(a) Platform   (b) Optimal   schedule
                          ($\rho = 38/499 \approx 0.076$)

(c) FIFO schedule ($\rho =$ (d) LIFO schedule ($\rho =$
47/632 $\approx$ 0.074)   43/580 $\approx$ 0.074)

Fig. 3. An example where the optimal solution is neither FIFO nor LIFO

the FIFO solution which uses only the first two workers P_1 and P_2 achieves a better throughput $\rho = 1/2$. To derive these results, we have used the linear programming approach for each of the 36 possible permutation pairs. It is very surprising to see that the optimal solution does not involve all workers under a linear cost model (and to the best of our knowledge this is the first known example of such a behavior).

Next, in Figure 3, we outline an example where the best throughput is achieved using neither a FIFO nor a LIFO approach. Instead, the optimal solution uses the initial ordering (P_1, P_2, P_3) and the return ordering (P_2, P_1, P_3). Again, we have computed the throughput of all possible permutation pairs using the linear programming approach.

3   LIFO Strategies

In this section, we concentrate on LIFO strategies, where the processor that receives the first message is also the last processor that sends its results back to the master, as depicted in Figure 1. We keep the notations used in Section 2.2, namely w_i, c_i, d_i and α_i for each worker P_i.

In order to determine the optimal LIFO ordering, we need to answer the following questions: What is the subset of participating processors? What is the ordering of the initial communications? What is the idle time x_i of each participating worker? The following theorem answers these questions and provides the optimal solution for LIFO strategies (see [6] for the proof):

Theorem 1. *In the optimal LIFO solution, then: (i) all processors participate to the execution; (ii) initial messages must be sent by non-decreasing values of $c_i + d_i$; and (iii) there is no idle time, i.e. $x_i = 0$ for all i. Furthermore, the corresponding throughput can be determined in linear time $O(p)$.*

4   FIFO Strategies

In this section, we concentrate on FIFO strategies, where the processor that receives the first message is also the first processor to send its results back to the master. We keep the notations used in Sections 2.2 and 3, namely w_i, c_i, d_i and α_i for each worker P_i. The analysis of FIFO strategies is much more difficult than the analysis of LIFO strategies: we will show that not all processors are enrolled in the optimal FIFO solution. In this section, we assume that $d_i = z c_i$ for $1 \leq i \leq p$. The proof of the following theorem is long and technical, see [6] for more details:

Theorem 2. *In the optimal FIFO solution, then: (i) initial messages must be sent by non-decreasing values of $c_i + d_i$; (ii) the set of participating processors is composed of the first q processors for the previous ordering, where q can be determined in linear time; and (iii) there is no idle time, i.e. $x_i = 0$ or all i.*

Furthermore, the optimal LIFO solution and the corresponding throughput can be determined in linear time $O(p)$.

5   Simulations

In this section, we present the results of some simulations conducted with the LIFO and FIFO strategies. We cannot compare these results against the optimal schedule, since we are not able to determine the optimal solution as soon as the number of workers exceeds a few units. For instance, for a platform with 100 workers, we would need to solve $(100!)^2$ linear programs of 100 unknowns (one program for each permutation pair). Rather than computing the solution for all permutation pairs, we use the optimal FIFO algorithm as a basis for the comparisons. The algorithms tested in this section are the following: optimal FIFO solution, as determined in Section 4, called OPT-FIFO; optimal LIFO solution, as determined in Section 3, called OPT-LIFO; a FIFO heuristic using all processors, sorted by non-decreasing values of c_i (faster communicating workers first), called FIFO-INC-C; a FIFO heuristic using all processors, sorted by non-decreasing values of w_i (faster computing workers first), called FIFO-INC-W.

We present the relative performance of these heuristics on a master/worker platform with 100 workers. For these experiments, we chose $z = 0.8$, meaning that the returned data represents 80% of the input data. The performance parameters (communication and computation costs) of each worker may vary from 50% around an average value. The ratio of the average computation cost over the average communication cost (called the *w/c-ratio*) is used to distinguish between the experiments, as the behavior of the heuristics highly depends on this parameter.

Figure 4(a) presents the throughput of the different heuristics for a w/c-ratio going from 1/10 to 100. These results are normalized so that the optimal FIFO algorithm always gets a throughput of 1. We see that both OPT-FIFO and FIFO-INC-C give good results. The other heuristics (FIFO-INC-W and OPT-LIFO) perform not so well, except when the w/c-ratio is high: in this case, communications have no real impact on the schedule, and almost all schedules may achieve good performances.

In Section 4, we showed that using all processors is not always a good choice. In Figure 4(b) we plot the number of processors used by the OPT-FIFO algorithm for the previous experiments: for small values of the w/c-ratio, a very small fraction of the workers is enrolled in the optimal schedule. Finally, Figure 4(c) presents the relative performance of all heuristics when the size of the data returned is the same as the size of the input data ($z = 1$). With the exception of this new hypothesis ($z = 1$ instead of $z = 0.8$), the experimental settings are the same as for Figure 4(a). We show that for a w/c-ratio less than 10, only the OPT-FIFO algorithm gives good performance: the FIFO-INC-C heuristic is no longer able to reach a comparable throughput. We also observe that when $z = 1$ the ordering of the workers has no importance: FIFO-INC-C and FIFO-INC-W are FIFO strategies involving all the workers but in different orders, and give exactly the same results.

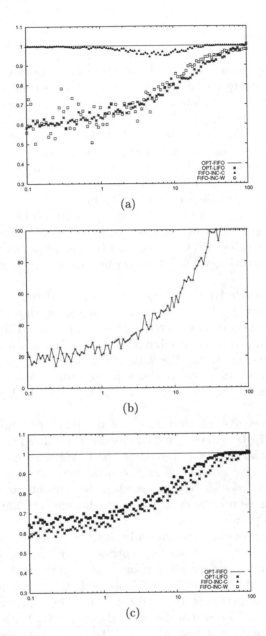

Fig. 4. (a) (a): Performance of the heuristics (optimal FIFO schedule), for different w/c-ratios. (b): Number of workers (optimal FIFO schedule), for different w/c-ratios. (c): Performance of the heuristics (optimal FIFO schedule), when $c_i = d_i$ $(z = 1)$, for different w/c-ratios.

6   Conclusion

In this paper we have dealt with divisible load scheduling on a heterogeneous master-worker platform. We have shown that including return messages from the workers after execution, although a very natural and important extension in practice, leads to considerable difficulties. These difficulties were largely unexpected, because of the simplicity of the linear model.

We have not been able to fully assess the complexity of the problem, but we have succeeded in characterizing the optimal LIFO and FIFO strategies, and in providing an experimental comparison of these strategies against simpler greedy approaches. Future work must be devoted to investigate the general case, i.e. using two arbitrary permutation orderings for sending messages from, and returning messages to, the master. This seems to be a very combinatorial and complicated optimization problem.

References

1. M. Adler, Y. Gong, and A. L. Rosenberg. Optimal sharing of bags of tasks in heterogeneous clusters. In *15th ACM Symp. on Parallelism in Algorithms and Architectures (SPAA'03)*, pages 1–10. ACM Press, 2003.
2. D. Altilar and Y. Paker. Optimal scheduling algorithms for communication constrained parallel processing. In *Euro-Par 2002*, LNCS 2400, pages 197–206. Springer Verlag, 2002.
3. G. Barlas. Collection-aware optimum sequencing of operations and closed-form solutions for the distribution of a divisible load on arbitrary processor trees. *IEEE Trans. Parallel Distributed Systems*, 9(5):429–441, 1998.
4. S. Bataineh, T. Hsiung, and T.G.Robertazzi. Closed form solutions for bus and tree networks of processors load sharing a divisible job. *IEEE Transactions on Computers*, 43(10):1184–1196, Oct. 1994.
5. O. Beaumont, H. Casanova, A. Legrand, Y. Robert, and Y. Yang. Scheduling divisible loads on star and tree networks: results and open problems. *IEEE Trans. Parallel Distributed Systems*, 16(3):207–218, 2005.
6. O. Beaumont, L. Marchal, and Y. Robert. Scheduling divisible loads with return messages on heterogeneous master-worker platforms. Research Report 2005-21, LIP, ENS Lyon, France, may 2005. Available at www.ens-lyon.fr/LIP/Pub/Rapports/RR/RR2005/RR2005-21.pdf.
7. V. Bharadwaj, D. Ghose, V. Mani, and T. Robertazzi. *Scheduling Divisible Loads in Parallel and Distributed Systems*. IEEE Computer Society Press, 1996.
8. M. Drozdowski and P. Wolniewicz. Experiments with scheduling divisible tasks in clusters of workstations. In *Proceedings of Euro-Par 2000: Parallel Processing*, LNCS 1900, pages 311–319. Springer, 2000.
9. A. L. Rosenberg. Sharing partitionable workloads in heterogeneous NOws: greedier is not better. In *Cluster Computing 2001*, pages 124–131. IEEE Computer Society Press, 2001.

A Grid Authentication System with Revocation Guarantees

Babu Sundaram and Barbara M. Chapman

Department of Computer Science,
University of Houston, Houston, TX 77204, USA
{babu, chapman}@cs.uh.edu

Abstract. Credential revocation is a critical problem in grid environments and remains unaddressed in existing grid security solutions. We present a novel grid authentication system that solves the revocation problem. It guarantees instantaneous revocation of both long-term digital identities of hosts/users and short-lived identities of user proxies. With our approach, revocation information is guaranteed to be fresh with high time-granularity. Our system employs *mediated RSA* (mRSA), adapts Boneh's notion of *semi-trusted mediators* to suit security in virtual organizations and propagates proxy revocation information as in Micali's *NOVOMODO* system. Our approach's added benefits include a configuration-free security model for end-users of the grid and fine-grained management of users' delegation capabilities.

1 Introduction and Motivation

Kohnfelder introduced the notion of a *digital certificate* [1]. In public-key encryption systems [2][3], digital certificates ascertain the identities of users, hosts and services (collectively termed as end entities). A digital certificate C is a trusted third-party's signature that validates the binding of a public key (PK) to an entity's identity (I). The trusted third-party is called a Certificate Authority (CA) and the CA uses its private key to sign end-entity certificates (EEC). Clients generate a public-private key pair and the CA signs the client's public key and includes information such as a serial number (SN), a start date (d_1) and end date (d_2) of its validity. In essence, a digital certificate $C = Sign_{CA}$(I, PK, SN, d_1, d_2). Now, any acceptor, to verify E's identity, does so by checking that E's certificate includes a valid signature from a trusted CA. Often, the life-time of a certificate C is in the order of years after which they expire. However, situations might arise that warrant immediate revocation of C before its actual expiration time. For instance, a trusted user Alice might leave her company or suspect that her private key has been compromised. Then, it is essential to immediately revoke her certificate so as to prevent acceptors from honoring stale or compromised credentials. An important design consideration in any CA implementation is handling prompt certificate revocation. Some studies [4] estimate that roughly 10% of public keys certified by a CA are revoked before they expire.

D.A. Bader et al. (Eds.): HiPC 2005, LNCS 3769, pp. 508–517, 2005.

We now stress the *revocation problem* and its importance in grids. Grids [5] are persistent infrastructures for securely sharing distributed and diverse hardware and software resources among dynamic collections of individuals, institutions and resources called virtual organizations (VO)[6].Grid security demands generality and transparency and has to ensure system, data and communication integrity. The Globus project (GT) [7][8] develops fundamental technologies to grids and provides an implementation of grid protocols and middleware. Grid Security Infrastructure (GSI) [9] is the authentication protocol of GT and provides services such as single sign-on, credential delegation and identity mapping. It requires a public key infrastructure (PKI) [10] for its operation. GSI allows grid entities to mutually authenticate using X.509 [11]certificates. It introduced the notion of *proxies*, an additional set of temporary, short-lived PKI credentials derived from user's long lived certificate to perform delegation on-behalf of the user. This eliminates the need for the users to remain online or enter passwords repeatedly whenever grid resource access is desired. User's proxies are protected using file permissions and could be compromised when someone can defeat or circumvent the security of the file system that holds the user proxy.

However, GSI provides only little support for revocation of long-term certificates (in the form of revocation lists) and no support for revoking compromised user proxies. Recent works [12] indicate hierarchical public key certification (a common grid model) is increasingly becoming the target for attackers. Hence, it has become vital to ensure revocation support in grids that can handle both long-lived digital identities and short-lived proxy identities.

In short, we make use of a variant of RSA cryptosystem [13] called mediated RSA (mRSA) as defined by Boneh and others [14] and extends the notion of *semi-trusted mediators* (SEM) to fit security in VOs. Each VO hosts a SEM-like entity to handle revocation of long-term X.509 certificates. The private key of a grid entity (E) is split (by mRSA) into two parts based on a simple 2-by-2 threshold cryptography [15]. Knowledge of a half-key cannot be used to derive the entire private key. Part of the key is held by E and the other part is held by E's SEM. So, private-key based operations (decryption / signing) can be carried out by neither E nor its SEM in isolation. Hence revocation of long-term credentials is instantaneous as the trusted SEM can simply stop exercising its part of the private key for revoked identities. We envision that the number of VOs will be much lesser than that of resources and individuals hosted by them. Our model uses GSI for delegation and resource-side user mapping. But, we add revocation capabilities for user proxies by modifying the proxy-creation process to handle validity and revocation targets as in NOVOMODO scheme [4].

The rest of the paper is organized as follows. Section 2 surveys existing efforts to solve the revocation problem in a general setting. Section 3 is a brief overview of GT's GSI. In section 4, we present a detailed explanation of our system components and protocols for grid authentication with revocation support. We discuss our future work and summarize in section 5.

2   Overview of Related Work

In a general PKI setting, many techniques exist that address revocation of digital identities. *Certificate Revocation List,* (CRL) [16] is the most popular PKI proposal for explicit revocation structures. It is a periodically generated, CA-signed list of certificates that are revoked before their intended expiry time. The acceptor of an entity E's certificate (C_E) checks that C_E is not in the latest CRL. Unfortunately, this is a very inefficient mechanism. CRLs tend to grow into unmanageable sizes with time and hence pose severe bandwidth requirements and transmission costs. In some major PKI implementations [17], CRLs form the most expensive component. Also, the long intervals between CRL distribution often result in stale revocation information. This is known as the *time granularity problem.* Further, CRLs are issuer-driven approaches and hence often fail to address recency requirements of the acceptors.

Online Certificate Status Protocol (OCSP) [18] is another PKI proposal wherein a CA replies to certificate status queries with a freshly generated signature. It is a simple request/reply protocol allowing an acceptor to query a CA for a certificate C's status. If C's identity is revoked, the CA indicates it to the querying acceptor. If the certificate is valid, it confirms it by issuing a fresh certificate. However, this model requires the CA/validation server to be available online all the time and if the validation server implementation is centralized, it becomes vulnerable to Denial of Service (DoS) attacks. Though it reduces the reply size per single status query, it poses significant loads on the CA due to the computationally expensive signature operations. This setup could easily outrun the CA resources under a heavy stream of incoming certificate status queries.

Micali proposed the *NOVOMODO Certificate Validation System* where a CA is aided by a few servers referred to as *directories* that distribute revocation information. Briefly, the CA, at the time of issuing the certificate to the client, includes 160-bit hash values indicative of the revocation information about the certificate. The CA generates two random 20-byte values X_0 and Y_0 and uses a publicly-known one-way hash function on these values. The successive hash values of X_0 are indicated as X_1, X_2, X_3 and so on. That is, $X_1 = H(X_0)$; $X_2 = H(X_1)$;. ... And, $Y_1 = H(Y_0)$. To generate a certificate with a lifetime of 1 year, the CA computes X_1 through X_{365}. Micali refers to X_{365} and Y_1 as *validity target* and *revocation target* respectively. That is, $C = \mathrm{Sign}_{CA}(U, PK, SN, d_1, d_2, \ldots, Y_1, X_{365})$. Every nth day after certificate issuance, the CA distributes the targets of all its clients to the *directories*. For revoked clients, the CA distributes their corresponding Y_0 values, the X_{365-n} values otherwise. Hence, a verifier of a certificate C on day i will query the *directories* for a target value (X or Y depending on C's validity). When a X value is returned (X_{reply}), the verifier ensures C's validity by checking $H^i(X_{reply})$ equals the X_{365} value in the certificate. When Y_0 is returned, the verifier can confirm C is revoked by checking Y_1 in the certificate equals $H(Y_0)$. NOVOMODO directory responses are concise (20-bytes) and directories cannot forge validity targets as one-way hash functions are hard to invert. Also, the computational demands on the CA are minimal as hashing is orders of magnitude cheaper to compute than signatures. However, NOVOMODO involves third-party queries and as noted in [19], its often necessary to deincentivize third party queries. And, the number of queries for a given certificate could increase dramatically as the number of querying verifiers increases, a common scenario in grid environments.

Identity-based Encryption(IBE) [20] is a public-key cryptosystem where any arbitrary string, such as a person's e-mail address or host's IP address, can act as the public key. The corresponding private keys are issued by a trusted third-party called a *Private Key Generator* (PKG). Shamir conceptualized this idea for simplified certificate management. With IBE, certification is *implicit* because decryption can happen only after getting keys from the trusted *PKG*. Implicit certification eliminates the need for third-party status queries. More details on this approach can be found in [21][22]. IBE scheme has two major drawbacks. First is the *key escrow* problem, that is, a *PKG* can decrypt messages intended for its clients. Second is the *key distribution problem* where the *PKG* has to communicate private keys to its clients via *secure channels*.

Certificate-based Encryption(CBE) was proposed by Gentry [19], where the certificate additionally serves as part of the decryption key. It eliminates the problem of key escrow by using double encryption. Certification in this model is implicit as the CA can stop issuing fresh certificates (part of decryption key) to revoked clients preventing them from decrypting any further. Gentry refined the basic CBE scheme to make use of *subset covers* to reduce the computational demands. However, CBE model requires CA to be an online entity in order to assist decryption. Further, such a CA setup becomes susceptible to DoS attacks. Also, the CA has to take part in multiple queries from various acceptors, even for the same certificate and thereby increasing the transmission costs.

The Semi-trusted Mediator (SEM) Architecture with a mRSA cryptosystem was introduced by Boneh to realize fine-grained control over security capabilities. mRSA is a simple threshold variant of RSA public key cryptosystem. With this approach, the CA initializes and distributes the mRSA keys to its clients. The threshold variant splits the private key d of an entity into two parts d_{sem} (distributed to SEM) and d_u (distributed to the client) such that $d = d_{sem} + d_u$ mod $\phi(n)$ where n is the product of two large primes p and q. Complete details of the algorithms for mRSA key generation, message encryption/decryption and signature/verification can be found in [14]. For successful decryption or signature generation, the client and SEM must co-operate and exercise their respective portions of the private key. SEMs are only *semi-trusted* because an acceptor trusts the SEM to have verified the revocation status of a client. SEMs cannot issue forged messages on behalf of revoked users (since it does not have d_u portion of the key to generate signatures). Part of our work relies upon SEM as it fits the grid model well. However, its scope disallows handling revocation of user's delegated credentials. Partly, this is because the assignment of SEMs to users is a fixed, static setup and it requires key generation support from the CA. But, user proxies are generated dynamically based on resource needs and matching frequent involvement from CA is often impractical.

3   Globus Toolkit's Grid Security Implementation

GSI is a stand-alone grid model for authentication and secure communication. GSI entities require a long-term public (RSA) key certificate - private key pair validated from a CA. GSI uses *proxies* for delegation on-behalf of a user. The user's grid identity is mapped into resource-specific user identity based on grid map files. The user's private key is encrypted with a passphrase.

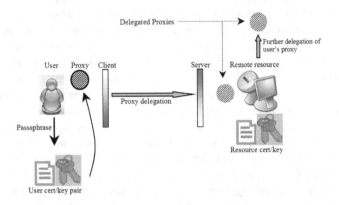

Fig. 1. Certificate-based Authentication and Delegation in GSI

To sign on to a grid resource, the user creates a proxy (with a few hours of lifetime), signs it with his private key and delegates it to the remote resource. Any resource with the delegated user proxy can now make additional resource requests on behalf of the user. To protect against malicious resources, the user can choose to restrict the proxies from any further delegation. However, there is currently no support in GSI to protect against compromised user proxies. Figure 1 illustrates the basic GSI operations. Earlier efforts [23] aimed at specifying usage restrictions as proxy certificates extensions to protect against compromised credentials. However, complete revocation capabilities were not realized. As noted earlier, grid PKI setup involves identities of various lifetimes; ranging from years for EECs to hours for user proxies. So, it is infeasible to devise a revocation scheme based solely upon any one of the above efforts. Thus, our work derives aspects from the SEM architecture as well NOVOMODO to achieve end-to-end revocation guarantees for all grid credentials.

4   Grid Authentication with Revocation Guarantees

Our model employs mRSA scheme for managing certification of users and resources. The user proxy generation process relies on NOVOMODO scheme to propagate validity status of proxies. The algorithms to generate mRSA keys and signatures are similar to SEM and are explained in the following sections.

4.1   The Architecture

Our architecture allows for instantaneous revocation capabilities and simplified credential management. Each VO's SEM and mRSA signature operations are handled by a daemon, what we call Grid Security Mediator (GSM_u and GSM_R for user's and resource's VO) that eliminates the need for CA to remain online and answer credential status queries. The GSMs generate their standard private keys, submit the public keys to their respective CAs and obtain certificates.

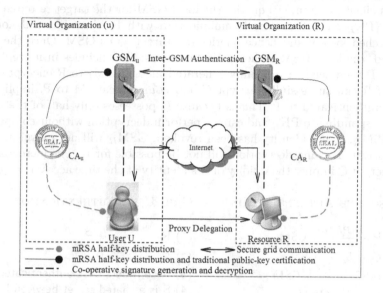

Fig. 2. Architecture of the Authentication Model

In contrast to GSI, the clients do not generate their keys themselves. But, the CAs generate and distribute a set of simple 2-by-2 threshold keys (d_u and d_{sem-u} or d_R and d_{sem-R}) based on mediated RSA to u, R and their GSMs. For each requestor, the CA generates a unique set: $\{p, q, e, d, d_{gsm}, d_u\}$ where p and q are large primes and n = pq and $\phi(n)$ = (p-1)(q-1). e is a random number prime to $\phi(n)$, that is, g.c.d($e, \phi(n)$) = 1. d is the multiplicative inverse of e modulo $\phi(n)$, that is, $d = e^{-1}$ mod $\phi(n)$. Now, (n,e) forms the public enciphering key. Now, d_{gsm} is a random integer in [1,n] and d_u = d - d_{gsm} mod $\phi(n)$. The initial key distribution does not require secure channels. Once this initial setup is complete, the client and the its SEM have to co-operate by using their respective half-keys to complete any signature or decryption operations.

4.2   Inter-GSM Authentication

A VO's GSM provides for SEM and NOVOMODO *directory* operations. By securely interacting with another VO's GSM, it also realizes freshness guarantees for user/resource credentials. For this purpose, establishing inter-GSM *trust* is vital and can be realized using traditional certificate-based handshake approach. As in classic GSI, the GSMs are configured to accept the credentials issued by each other's CAs. Optionally, it is trivial to achieve authentication between the GSMs and their clients using an optional password during the initial setup.

4.3 Protocol for Grid Resource Authentication

After the inter-GSM trust setup, whenever a user wants to use a remote resource (R), his client program (P) queries the local GSM for the target resource's public key (PK_R). For first-time communication with R, the GSM will not have PK_R cached locally and hence queries the target VO's GSM. Once the key is obtained, the local GSM caches it for future use and includes in its reply to P as well. P then sends a randomly generated message (M) to R encrypted with PK_R. If R can successfully decrypt C and communicate M to P, implicitly R stands authenticated to P. This is because R possesses only half of the private key corresponding to PK_R and cannot perform decryption without co-operation from GSM_R. If R's identity has been revoked, GSM_R will not exercise its half of R's private key and decryption will not be possible for R. The mere ability of R to decrypt C implies the validity of R's identity at the time access is requested.

RESOURCE AUTHENTICATION

(1) P $\xrightarrow{getPK(R)}$ GSM$_u$

1.1 $GSM_u \xrightarrow{R}$ GSM$_R$

1.2 $GSM_u \xleftarrow{PK_R}$ GSM$_R$

(2) P $\xleftarrow{PK_R}$ GSM_u

(3) P $\xrightarrow{C=Enc_{PK_R}(M)}$ R

(4) P $\xleftarrow{M}$ R

(5) P considers R authenticated

GRID USER AUTHENTICATION

(1) R $\xrightarrow{Enc_{PK}(S)}$ P

(2) R $\xleftarrow{Enc_s(u)}$ P

(3) R considers U as authenticated

(4) S is a shared secret between R, U

4.4 Protocol for Grid User Authentication

As indicated in [24], it is always the credential acceptor that runs the risk of accepting stale credentials and, thus, should have the ability to dictate the recency requirements for the credential on behalf of U. In our setup, R can verify that U's certificate C_U is issued by a trusted CA. R can obtain C_U from some online service or U can supply it as part of its authentication request. R now generates a secret (S) and encrypts it with the public key of U (in C_u) and sends it to U. U can get S by sorting co-operation from its local VO's GSM and this will succeed only if U's status is valid. To confirm its validity status, U sends an OK message to R encrypted with S. R can repeat the secret-key operation on this message to obtain the OK message and this implicitly confirms U's valid identity to R. If U's credentials have been revoked, GSM_u will not exercise its half of U's mRSA private key and hence U cannot decrypt to get S in the previous step. Optionally, once this protocol completes, S can be used as a shared secret between R and U for secure communication.

4.5   User Privilege Delegation Model

In our system, GSI's user proxy generation step is augmented to include support for revocation. When the proxy is created, the proxy creator makes use of NOVOMODO approach to indicate to the acceptor whether the proxy is fresh or stale. As with NOVOMODO, the proxy certificate is enhanced with the validity and revocation target values. Depending upon the intended lifetime of the proxy, the values of X_is are calculated and X and Y values are included as part of the proxy certificate extensions. This structure is ensured to comply with the proxy certificate profile specifications [25]. The functionality of a NOVOMODO *directory* is implemented as part of the GSMs and the GSM details are included as part of the proxy certificate. That is, a proxy with a lifetime of n periods (starting at time d_1, and $d_2 = d_1 + n$) contains C_{proxy}, PrivateKey$_{proxy}$ and C_u, where $C_{proxy} = \text{Sign}_U(\text{PublicKey}_{proxy}, d_1, d_2, Y1, X_n, GSM_u)$. Now, with a delegated proxy, the acceptor can query the user's GSM for revocation status. For a period i, the presence of X_{n-i} or Y_0 in GSM's reply indicates the valid or revoked status of U's proxy respectively. We argue that this modification is mandatory because, though proxies have limited lifetime, the *scale* of accessible resources with a full proxy on a grid raises serious concerns. Further, user authorization [26] on a grid is granted solely based upon a valid proxy and hence proxy compromise proves to be a serious threat.

4.6   Implementation and Discussion

Currently, our implementation of GSM is in C and uses the SEM libraries available as part of [27]. We have completed an implementation of the grid proxy creation program that adds NOVOMODO-like target values to proxy certificate extensions. Complete software information is available publicly at [28]. By using mRSA, we inherit the benefits of binding signature semantics in grids. That is, a signature's validity is equivalent to checking the public key's validity at the time the signature was generated. This could prove to be beneficial to grid accounting systems in guaranteeing non-repudiation. Also, our system is quite easy to setup and the administrative overhead involved with continued operation is trivial. For the end-users, no setup procedures or security configuration is required to use a grid. The simplified credential management in our model can aid developing simpler co-allocation tools across VOs with multiple CAs and complex trust relations. Inherently, our system eliminates *key escrow* problem because no single entity possesses the entire private key for grid users/resources and hence cannot decrypt communications. Additionally, it is trivial to achieve encrypted grid communication in our system. This is aided by the establishment of a shared secret at the end of mutual authentication protocol between grid resources and users. Caching of public keys of the resources at various GSMs helps in reduced communication between GSMs over time. Optionally, per recency requirements, a client can make the associated GSM to request a fresh copy of a chosen resource's public key.

5   Future Work and Summary

Centralized GSMs could become inter-VO communication bottleneck and liable to DoS attacks. This could be solved by replicating GSM functionalities. Replicating GSMs could complicate setting inter-GSM trust across VOs and this motivates us to devise better setup solutions. An interesting approach would be to examine minimal overhead IP security model as in [29] using low-level, host-to-host communication. To address WSRF-based [30] Globus toolkit 4 (GT4), we plan to provide an implementation of our GSM functionality as a GT4 service as well. Further, with the privilege delegation model, situations might arise where a specific resource gets compromised. So, it should be possible for the user to revoke specific delegated proxies, a problem we call *proxy subset revocation*. We plan to examine earlier approaches such as [23] to address this. It would also be interesting to study the role of online credential repositories such as MyProxy [31] in this initiative. We also envision that our model could simplify aspects of certain recent projects [32] that aim for web-based GSI support for users.

Certificate validation and revocation is universally recognized as a crucial problem. We presented an authentication system that solves the revocation issues in grid environments. The main idea is to employ mRSA approach to handle the identities of the grid users and resources. Also, user proxies are enhanced to contain revocation information using aspects of the NOVOMODO scheme. This allows for instantaneous revocation of both long-term digital identities of hosts/users and short-lived identities of user proxies. We introduced a SEM-type mediator for virtual organizations. With this approach, the users enjoy a simplified view and usage of grid security.

References

1. Kohnfelder, L. M.: Towards a Practical Public-Key Cryptosystem. B.S. Thesis, supervised by L. Adleman, MIT, May 1978.
2. Koblitz, N.: A Course in Number Theory and Cryptography. Series: Graduate Texts in Mathematics, Vol. 114, second edition, Springer-Verlag, 1994.
3. Schneier, B.: Applied Cryptography. John Wiley & Sons, second edition, 1996.
4. Micali, S.: Novomodo: Scalable Certificate Revocation and Simplified PKI Management. Proc. of 1st Annual PKI Research Workshop 2002, available at http://www.wisdom.weizmann.ac.il/ kobbi/papers.html
5. Foster, I., Kesselman, C.: The GRID: Blueprint for a new Computing Infrastructure. Morgan Kauffman Publishers, 1999.
6. Foster, I., Kesselman, C., Tuecke, S.: The Anatomy of the Grid: Enabling Scalable Virtual Organizations. International Journal of High Performance Computing Applications, 15 (3). 200-222. 2001.

7. Foster, I., Kesselman, C.: Globus: A metacomputing infrastructure toolkit. International Journal of Supercomputer Applications, Summer 1997.
8. Foster, I., Kesselman, C.: The Globus Project: A Status Report. Proc. IPPS/SPDP '98 Heterogeneous Computing Workshop, pp. 4-18, 1998.
9. Butler, R., Engert, D., Foster, I., Kesselman, C., Tuecke, S., Volmer, J., Welch, V.: A National-Scale Authentication Infrastructure. IEEE Computer, 2000.
10. Public Key Infrastructure Standards, http://csrc.nist.gov/pki/panel/warwick
11. X-509 Certificate Format, http://www.w3.org/PICS/DSig/X509_1_0.html
12. Burmester, M., Desmedt,Y.G.: Is Hierarchical Public-Key Certification the Next Target for Hackers? Communications of the ACM Vol. 47, No. 8, August 2004.
13. Rivest, R., Shamir, A., Adleman, A.: A Method for Obtaining Digital Signatures and Public-Key Cyptosystems. Communications of the ACM 21, February 1978, pages 120-126, 1978.
14. Boneh, D., Ding, X., Tsudik,G.: Fine-Grained Control of Security Capabilities. ACM Transactions on Internet Technology, Vol.4, No. 1, pages 60-82, Feb 2004.
15. Gemmel, P.: An Introduction to Threshold Cryptography. RSA Cryptobytes 2, 7.
16. X.509 Internet Public Key Infrastructure Certificate and CRL Profile , IETF RFC 2459, http://www.ietf.org/rfc/rfc2459.txt
17. Public Key Infrastructure, Final Report; MITRE Corporation; National Institute of Standards and Technology, 1994.
18. X.509 Internet Public Key Infrastructure Online Certificate Status Protocol (OCSP), IETF RFC 2560, http://www.ietf.org/rfc/rfc2560.txt
19. Gentry, C.: Certificate-based Encryption and the Certificate Revocation Problem. Cryptology ePrint Archive: Report 2003/183, 2003. http://eprint.iacr.org.
20. Shamir, A.: Identity-based Cryptosystems and Signature Schemes. In Proc. of Crypto 1984, LNCS 196, pages 47-53, Springer-Verlag, 1985.
21. Boneh, D., Franklin,M.: Identity-Based Encryption from the Weil Pairing. Proc. of Crypto 2001, LNCS 2139, pages 213-229. Springer-Verlag, 2001.
22. Lynn, B.: Authenticated Identity-Based Encryption. Cryptology ePrint Archive: Report 2002/072, 2002. http://eprint.iacr.org
23. Sundaram, B., Nebergall, C., Tuecke, S.: Policy Specification and Restricted Delegation in Globus Proxies. SuperComputing Conference 2000, Dallas, Nov 2000.
24. Rivest, R. L.: Can We Eliminate Certificate Revocation Lists? in Financial Cryptography, Rafael Hirschfield, Ed., Anguilla, British West Indies, February 1998, vol. 1465, pages 178-183, Springer Verlag 1998.
25. Tuecke, S., Engert, D., Foster, I., Thompson, M., Pearlman, L., Kesselman, C.: Internet X.509 Public Key Infrastructure Proxy Certificate Profile. IETF Draft draft-ietfpkix-proxy-06.txt, 2003.
26. Foster, I., Kesselman, C., Tsudik, G., Tuecke, S.: A Security Architecture for Computational Grids. ACM Conference on Computers and Security, 1998, 83-91.
27. The SUCSES Project, http://sconce.ics.uci.edu/sucses/
28. The HPCTools Group, Department of Computer Science, University of Houston.
29. Appenzeller, G ., Lynn. B.: Minimal Overhead IP Security using Identity-Based Encryption. http://rooster.stanford.edu/ ben/pubs
30. Web Services - Resource Framework, Specifications of the WS-Resource construct, http://www.globus.org/wsrf/specs/ws-wsrf.pdf
31. Novotny, J., Tuecke, S., Welch, V.: An Online Credential Repository for the Grid: MyProxy. Proc. of the Tenth International Symposium on High Performance Distributed Computing, pages 104-111, IEEE Press, August 2001.
32. PURSe: Portal-Based   User   Registration   Service,   http://www.gridscenter.org/solutions/purse/

Integrating a New Cluster Assignment and Scheduling Algorithm into an Experimental Retargetable Code Generation Framework

K. Vasanta Lakshmi, Deepak Sreedhar,
Easwaran Raman, and Priti Shankar

Computer Science & Automation,
Indian Institute of Science,
Bangalore - 560012, India

Abstract. This paper presents a new unified algorithm for cluster assignment and region scheduling, and its integration into an experimental retargetable code generation framework. The components of the framework are an instruction selector generator based on a recent technique, the IMPACT front end, a machine description module which uses a modification of the HMDES machine description language to include cluster information, a combined cluster allocator and an acyclic region scheduler, and a register allocator. Experiments have been carried out on the targeting of the tool to the Texas Instruments TMS320c62x architecture. We report preliminary results on a set of TI benchmarks.

1 Introduction

Building compilers for new architectures is often a bottleneck in conducting experimental research in digital signal processors. A particularly challenging problem is the construction of compiler back-ends. Current trends necessitate the design of retargetable systems where portions of the back-end are generated automatically from specifications of the instruction set architecture and the microarchitecture of the machine. Typical components of the backend are instruction selectors, schedulers, and register allocators. If the architecture is partitioned, then the compiler has the additional task of judiciously selecting the functional unit on which an operation is to be executed so that the movement of operands between clusters is minimized. This paper reports preliminary results of an attempt to integrate a combined scheduling and cluster assignment algorithm into an experimental framework using some infrastructure that is publicly available and parts that have been developed by us. The framework consists of a tree parser generator with a mechanism for handling attributes and actions so that it can output an instruction selector[11, 2]. The instruction selector traverses a sequence of expression Directed Acyclic Graphs(DAGs) representing expression trees, and can output either unscheduled code with virtual registers or a sequence of data dependency DAGs with flow dependencies. Output and anti dependency edges are added to the DDG's in a separate pass. We also propose a simple new

D.A. Bader et al. (Eds.): HiPC 2005, LNCS 3769, pp. 518–527, 2005.

integrated algorithm that performs cluster assignment as well as region scheduling together, and integrate it into the framework. A global register allocator, at
the level of a function, based on the graph colouring heuristic of Chaitin[3] is
used. We report the results of preliminary experiments on targeting the tool to
the Texas Instruments TMS320c62x architecture. We have not attempted any
hand optimization and we present the code that is completely automatically
generated(except the addition of assembler directives). We also discuss ways in
which the code can be improved in future work on this experimental tool. The
instructor selector generator and the microarchitecture description module are
described in detail in [2].

Section 2 is a brief description of the framework. Section 3 describes the instruction selector generator. Section 4 describes the architecture of the Texas
Instruments TMS320c62x Digital Signal Processor. Section 5 presents the combined cluster assignment, acyclic scheduling algorithm and register allocator.
Section 6 presents preliminary results obtained from targeting the tool to the TI
processor. Section 7 describes related work on acyclic scheduling for partitioned
architectures. Finally Section 8 concludes the paper.

2   Architecture of the Framework

The architecture of the framework is represented in figure 1. The IMPACT
frontend[19] is used to convert source code in C to low level intermediate code
Lcode. This is transformed into expression Directed Acyclic Graphs(DAGs) by a
syntax directed translation from Lcode using a YACC generated translator. The
instruction selector is generated by an Instruction Selector Generator. Specifically, the underlying regular tree grammar generates trees that correspond to intermediate code trees, and the actions correspond to the emission of code or data

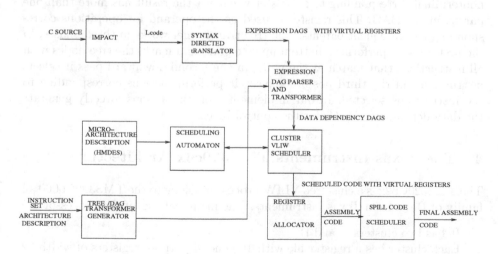

Fig. 1. The Architecture of the Code Generation Tool

dependency graphs(DDGs) as the need dictates. We have used data dependency graphs in this implementation. The tool generates a tree parser that parses intermediate code trees and generates output in the form of DDGs for unscheduled code[2]. The DDG's are at the level of superblocks[8]. This code is then cluster assigned and scheduled using the integrated algorithm described in this paper. The HMDES[20] specification is used to describe the microarchitecture. The language has been augmented to include the description of cluster information[2]. A scheduling automaton[5] is generated, which is queried by the scheduler as it attempts to construct a cluster assignment and schedule for the code. Finally the scheduled code is register allocated. Cluster allocation and scheduling are performed together and register allocation for the two clusters of the TI processors is done independently in a subsequent pass. Any spill code is scheduled in a final pass. We now briefly describe the instructor selector generator.

3   Instruction Selector Generator

The instructor selector generator(ISG) constructs a tree parser from a regular tree grammar specification of the machine instructions. The mechanism is an adaptation of LR parsing[9] to parse ambiguous tree grammars converted into context free grammars[1]. We have not yet exploited cost computations that come free with code selection, but have extended the scheme to work with DAGs so that common subexpressions are not recomputed. Attributes are used as in YACC to propagate register names,operand types and so forth as the tree parser moves up the tree selecting instructions that match and creating parts of the DDG. The ISG generates a driver and tables that encode the actions of the tree parser. Since the output of the translator of Lcode is a sequence of expression DAGs, the tree parser is modified to work on DAGs. This is done by choosing a nonterminal corresponding to a register whenever the result has more than one parent in the DAG. This register is used as the operand for operations corresponding to subsequent parents. The tree transformer works in three phases. In the first phase it performs a bottom up tree parse and marks the tree nodes with all instructions that match at the node; in the second downward pass it selects instructions; in the third postorder pass, it performs actions corresponding to the instructions selected. In our implementation the actions directly generate the data dependency DAGs for the control block.

4   The Texas Instruments TMS320c6x Architecture

The TMS320C62x is a clustered VLIW processor belong to the TMS320C6000 [6] family of DSPs from Texas Instruments. The main features of the processor are

- It has two clusters A and B
- Each cluster has a register file with 16 general purpose registers of width 32 bits. 64 bit floating point data is represented by register pairs. In addition B has a control register file.

- There is a load path and a store path to each cluster
- Each cluster has four functional units (L1, M1, S1 and D1 in A and L2, M2, S2 and D2 in B).
- There are two cross paths (1X and 2X). Cross path 1X is used by cluster B to read from A and similarly 2X by cluster A to read from B. Units L1, M1 and S1 have access to 2X and L2, M2 and S2 to 1X. L1, L2 can read either of the two source operands via the cross path while the other four can read only their second source operand.
- The data address paths DA1 and DA2 are each connected to both D units. This allows the address generated in one cluster to write into/read from a register in the other register file during a load/store.
- The processor supports predication but with a restricted set of predicate registers (A1, A2, BO, B1, B2)
- The functional unit latencies for all fixed point instructions is 1 cycle. The destination latency for multiply is 2 cycles, for branch 5 cycles, for load 4 cycles and for store 3 cycles and for other fixed point operations it is one cycle.
- Each fetch packet can contain at most 8 instructions. A fetch packet may consist of one or more execute packets, each of which contains the set of parallel operations to be executed in that cycle.
- Two instructions cannot use the same unit in a cycle. Two loads/stores cannot read from/write into the same register file in the same cycle. Two instructions cannot write to the same register in the same cycle. In any cycle there can be a maximum of four reads to the same register. Only one unit can access a cross path in a cycle.

5   Cluster Allocation, Scheduling and Register Allocation

The output of the tree parser is a sequence of DDG's one corresponding to each control block. IMPACT has the flexibility to output intermediate code at the level of basic blocks, superblocks and hyperblocks. We chose superblocks[8] for our implementation. Here each block has a single entry point but may have more than one exit point. Tail duplication[8] is a strategy used to ensure semantic correctness when superblocks are constructed. There are several cluster allocation and scheduling algorithms proposed in the literature; these are described in Section 7. Some of these[17, 10] are integrated into a single phase. Others[15, 7] use phase ordering. In terms of computational overheads of cluster scheduling, there is a wide spectrum, ranging from exponential time to linear time. We assume a "compile-and-run" environment, which renders non-polynomial time algorithms impractical. The cluster scheduling algorithm performs both cluster allocation within superblocks as well as packing of independent instructions into VLIW instructions simultaneously. The algorithm is a simple extension of a list scheduling algorithm with a dynamic priority computation strategy and is described below.

Since clusters are allocated ahead of register allocation, register allocation is carried out independently on each cluster. The unit of allocation is a web[4]

procedure 1. CB_Scheduling

{INPUT : Data Dependence Graph}

{OUTPUT : Scheduled Code With Virtual Registers}

{Terminology:

CP: Critical Path

Non_CP: Non Critical Path

$Successor(X)$: Set of instructions data dependent on X

$Predecessor(X)$: Set of instruction on which X is data dependent

$I.Lst$: latest start time for instruction I

$I.Est$: earliest start time for instruction I

$Unscheduled$: Set of unscheduled instructions

$Dependency\_edge\_latency(I, J)$: Function returns latency of edge between I and J.

$I.sched\_cluster$: Cluster in which instruction I is scheduled}

$curr\_cycle \leftarrow 0$

for Each instruction I where $Predecessor(X) = \emptyset$ **do**

  if $I.Lst = 0$ **then**

    Add to $CP\_Readylist$

  else

    Add to $Non\_CP\_Readylist$

  end if

end for

$Curr\_cycle \leftarrow 0$

while $Unscheduled \neq \emptyset$ **do**

  $CP\_selected\_set \leftarrow Select\_Max\_Subset(CP\_Readylist)$

  $NonCP\_selected\_set \leftarrow Select\_Max\_Subset(NonCP\_Readylist)$

  $Execution\_packet \leftarrow CP\_selected\_set \cup NonCP\_selected\_set$

  for Each instruction I in $Execution\_packet$ **do**

    for Each instruction J in Successor(I) **do**

      if $I.Lst \leq (curr\_cycle + Dependency\_edge\_latency(I, J))$ **then**

        Add J in $CP\_Readylist$ if all its parents are scheduled

      else

        Add J in $NonCP\_Readylist$ if all parents are scheduled

      end if

    end for

  end for

  $curr\_cycle \leftarrow curr\_cycle + 1$

end while

which is a maximal union of intersecting def-use chains. Since cluster scheduling is performed at the level of a superblock while register allocation is performed at function level, it is possible for the same virtual registers of Lcode to

procedure 2. Select_Max_Subset
{INPUT : Ready list R}
{OUTPUT : Maximum subset S of R that can be executed in parallel}
{Terminology:
$I.cluster\_weights[i]$: the priority for instruction I to be scheduled in cluster i. A higher number indicates higher priority.
$load\_balance()$: Function that returns cluster with fewer number of instructions scheduled in it.
$Preplaced(I)$: A boolean variable which is true if the target of I is a virtual register already assigned a cluster in a region previously scheduled.

$S \leftarrow \emptyset$
for Each instruction I in R **do**
    if $Preplaced(I)$ **then**
        /* The cluster in which I will be scheduled is predetermined as $pre\_placed\_cluster$*/
        $I.sched\_cluster \leftarrow pre\_placed\_cluster$
    else
        if $I.cluster\_weights[0] > I.cluster\_weights[1]$ **then**
            $I.sched\_cluster \leftarrow 0$
        else
            if $I.cluster\_weights[0] < I.cluster\_weights[1]$ **then**
                $I.sched\_cluster \leftarrow 1$
            else
                $I.sched\_cluster \leftarrow load\_balance()$
            end if
        end if
    end if
    if Resources available to execute I **then**
        $S \leftarrow S \cup I$
        for Each J in Successor(I) **do**
            /*Dynamic tuning of weights for instrcutions dependent on I*/
            $J.cluster\_weights[I.sched\_cluster] \leftarrow$
            $J.cluster\_weights[I.sched\_cluster] + 1$
            for Each K in $Predecessor(J)$ **do**
                if K in $Unscheduled$ **then**
                    /*Dynamic tuning of weights for siblings of I*/
                    $K.cluster\_weights[I.sched\_cluster] \leftarrow$
                    $K.cluster\_weights[I.sched\_cluster] + 1$
                end if
            end for
        end for
    end if
end for

be allocated to different clusters in different regions. In such a case intercluster move instructions would have to be introduced between regions. Here we ensure cluster consistency via preallocation. This may introduce some moves within a region to maintain consistency. Register allocation is carried out on each cluster independently using Chaitin's graph colouring heuristic. Since the register set in each cluster is uniform, application of the algorithm does not pose any special problems.

6   Results and Discussion

Eight TI benchmarks were compiled and the resulting code was examined for size and parallelism, and compared with the code generated by the state of the art TI compiler{TI Code Composer}. Tables 1 and 2 give the size of source code and the size of the assembly code generated by the tool and by the TI compiler for O3 optimization without software pipelining. Note that the code generated by the tool is not hand optimized and the generation is completely

Table 1. TI Compiler: With O3 Optimization and without software pipelining

Program	No. Execution Packets	No. of Instructions	Parallelism
Codebook_Search	96	119	1.239583
FIR_filter	62	77	1.241935
IIR_cascaded	46	52	1.130435
MAC_vselp	17	20	1.176471
Minimum_err	64	75	1.171875
Vector_max	29	33	1.137931
Vector_sum	12	14	1.166667
vq_MSE	18	21	1.166667

Table 2. Our Compiler: With O3 optimized lcode as IR and without software pipelining

Program	No. Execution Packets	No. of Instructions	Parallelism	lcode size	Src size
Codebook_Search	162	249	1.537037	99	57
FIR_filter	41	63	1.536585	33	16
IIR_cascaded	62	117	1.982143	59	22
MAC_vselp	30	45	1.500000	27	17
Minimum_err	121	208	2.426230	76	28
Vector_max	47	66	1.404255	35	32
Vector_sum	29	42	1.448276	24	8
vq_MSE	36	49	1.361111	25	13

automatic except for the addition of assembly directives. The table also indicates the average parallelism in a VLIW instruction generated by the compiler. Our observation is that code sizes produced automatically by the tool are considerably larger than those produced by the TI compiler with all optimizations except software pipelining turned on. There are two reasons for this. Firstly there is some improvement possible in the specifications for the multiply instruction so that compact code can be generated for short operands. At present the code generated assumes that all operands are 32 bits(as Lcode loads short operands with type 11) and therefore sometimes generates four instructions where one would suffice. This is particularly observable in the program Codebooksrch. Secondly the calling conventions of Lcode and the TI assembler for function calls have to be matched. This requires some additional move instructions. Thirdly, the assembly code generated by our tool cannot have fewer statements than the Lcode size. In many cases the Lcode size is larger than the TI assembly code size with the O3 level of optimization with no software pipelining. Fourthly some programs namely CodebookSrch, IIR Cascade and Minimum_err result in a large number of register spills, whereas there were no spills in the others. Thus some changes need to be carried out in the register allocation phase to reduce the number of spills. However, we note that the combined cluster assignment and scheduling algorithm delivers a fairly good parallelism when compared with the TI average VLIW parallelism with O3 optimization but without software pipelining. The parallelism should improve if software pipelining is integrated into the framework. Clearly, several improvements to the framework need to be incorporated before the tool is able to generate really good quality code.

7   Related Work

We briefly describe related work in cluster allocation. Pioneering work on cluster scheduling was reported in the BUG compiler[15]. The Multiflow Trace Compiler used the algorithm from the Bulldog compiler. BUG uses a two phase algorithm. The first phase traverses the dependence graph bottom up to propagate information about preplaced instructions. The second phase traverses the dependence graph top down and uses a greedy algorithm to map the instruction to the cluster that can execute it the earliest. Scheduling is done in a later pass. Capitanio et al.[14] perform scheduling before assignment. The assignment phase uses a min-cut algorithm that tries to minimize communication. Rau et al. consider code generation for VLIW processors with clustering. Partitioning of instructions is done before instruction scheduling so as to achieve load balancing in the functional units and move instructions are inserted to ensure availibility of operands in the proper cluster. However the partitioning technique is insensitive to the schedule length and there is a phase ordering problem. The unified assign and schedule algorithm[17] uses a list scheduler to perform integrated partitioning and scheduling. Desoli et al.[18] use a heuristic algorithm called *Partial Component Clustering* for the problem of partitioning DDG nodes between clusters. The basic idea is to avoid copy operations on critical paths. If the critical path length is close to the actual schedule length then this is a useful heuristic; however if there is a wide but shallow DDG then the functional units are the limiting factor and this method may not produce good code. Leupers[7] proposes a technique consisting of two interleaved phases, which are alternately invoked passing feedback to one another till all instructions are scheduled. Whereas this achieves good results the computational effort is considerable. Kailas et al.[16] propose a scheme that combines cluster assignment, register allocation and instruction scheduling into a single phase, thus eliminating problems asso&iuated with phase ordering. In a significant departure from traditional techniques, Lee et al. propose convergent scheduling[13] which is a framework for cluster assignment and scheduling using independent passes each implementing a heuristic that addresses a particular problem or constraint. This appears to be a very flexible framework. However it is not clear whether convergence is always guaranteed.

8   Conclusion

We have described the integration of a cluster assignment and scheduling algorithm into a code generation framework. Our experience indicates that such a task is feasible though some care is required to generate the specifications so that advantage is taken of special machine instructions. The code size generated by the tool needs to be reduced though the parallelism is quite good. Future work envisages including software pipelining into the tool and improving the register allocator, as at present it seems to introduce too many spills. More architectures have to be experimented on to estimate the practicality of the tool. We also intend to implement other acyclic scheduling algorithms as well so that the tool allows the choice of scheduling algorithm.

References

1. Priti Shankar, A. Gantait, A.R. Yuvaraj, M. Madhavan. A New Algorithm For Linear Regular Tree Pattern Matching, *Theoretical Computer Science 242, 2000, pp. 125-142*
2. Deepak Sreedhar, Easwaran R. Retargetable code generation for clustered embedded processors, ME Project Report. Dept of CSA, Indian Institute of Science
3. G.J. Chaitin. Register Allocation and Spilling via Graph Coloring, *Proceedings of the 1982 Symposium on Compiler Construction, June 1982, pp. 98-105*
4. Preston Briggs, Keith Cooper and Linda Torczon . Improvements to Graph Coloring Register Allocation, *ACM Transactions on Programming Languages and Systems, 16(3), May 1994, pp. 428-455*
5. T.A. Proebsting and C.W. Fraser. Detecting Pipeline Hazards Quickly. *21st ACM SIGPLAN-SIGACT Symposium on Principles of Programming Languages, Jan 1994*
6. Texas Instruments TMS320C6000 CPU and Instruction Set Reference Guide
7. R. Leupers. Instruction Scheduling for Clustered VLIW DSPs. *Proceedings of the International Conference on Parallel Architecture and Compilation Techniques. (Philadelphia,PA), October 2000.*

8. Wen-mei Hwu *et. al.* The superblock: An effective structure for VLIW and superscalar compilation. *The Journal of Supercomputing, vol. 7, pp. 229–248, Jan. 1993*

9. A.V. Aho and J.D. Ullman.*Principles of Compiler Design.*Addison-Wesley, 1977.

10. Krishnan Kailas, Kemal Ebcioglu, Ashok Agrawala. CARS: A New Code Generation Framework for Clustered ILP Processors, *7th Interbational SYmposium on High Performance Computer Architecture, pp 133-143, 2001*

11. Maya Madhavan, Priti Shankar, Siddhartha Rai, U. Ramakrishna.*Extending Graham-Glanville techniques for optimal code generation,*ACM Trans. Program. Lang. Syst. 22(6): 973-1001 (2000)

12. http://dspvillage.ti.com/docs/catalog/devtools/details.jhtml?templateId =5121&path=templatedata/cm/tooldetail/data/ccs_codegen

13. Walter Lee, Diego Puppin, Shane Swenson, Saman Amarasinghe *Convergent Scheduling. MICRO-35, November 2002 Istanbul, Turkey*

14. A. Capitanio, N. Dutt, and A. Nicolau. *Partitioned Register Files for VLIWs: A Preliminary Analysis of Tradeos. 25th International Symposium on Microarchitecture (MICRO),1992*

15. J. R. Ellis. Bulldog: A Compiler for VLIW Architectures. MIT Press, 1986.

16. K. Kailas, K. Ebcioglu, and A. K. Agrawala.*CARS: A New Code Generation Framework for Clustered ILP Processors. 7th International Symposium on High Performance Computer Architecture (HPCA)*

17. E. Ozer, S. Banerjia, and T. M. Conte. *Unified Assign and Schedule: A New Approach to Scheduling for Clustered Register File Microarchitectures. In 31st International Symposium on Microarchitecture (MICRO)*

18. G. Desoli. *Instruction Assignment for Clustered VLIW DSP Compilers: a New Approach. Technical Report HPL-98-13, Hewlett Packard Laboratories, 1998.*

19. The Impact Research Group, University of Urbana Champaign. *http://www.crhc.uiuc.edu/Impact*

20. Trimaran Compiler. *http://www.trimaran.org*

Cooperative Instruction Scheduling with Linear Scan Register Allocation

Khaing Khaing Kyi Win and Weng-Fai Wong

Department of Computer Science,
National University of Singapore,
3 Science Drive 2, Singapore 117543
khaing_khaing@alumni.nus.edu.sg, wongwf@comp.nus.edu.sg

Abstract. Linear scan register allocation is an attractive register allocation algorithm because of its simplicity and fast running time. However, it is generally felt that linear scan register allocation yields poorer code than allocation schemes based on graph coloring. In this paper, we propose a pre-pass instruction scheduling algorithm that improves on the code quality of linear scan allocators. Our implementation in the Trimaran compiler-simulator infrastructure shows that our scheduler can reduce the number of active live ranges that the linear scan allocator has to deal with. As a result, fewer spills are needed and the quality of the generated code is improved. Furthermore, compared to the default scheduling and graph-coloring allocator schemes found in the IMPACT and Elcor components of Trimaran, our implementation with our pre-pass scheduler and linear scan register allocator significantly reduced compilation times.

1 Introduction

Instruction scheduling and register allocation are one of the most important phases in compiler optimization. In compilers for machines with instruction-level parallelism, the phases of instruction scheduling and register allocation can be antagonistic. This is the well-known phase ordering problem [7] as shown in Fig. 1. One of the ways to solve that problem is to combine instruction scheduling and register allocation such that these two phases can be performed together to generate efficient code. In current optimizing compilers, a compromise consisting of a phase of instruction scheduling (pre-pass scheduling) is first performed. This is followed by register allocation and another phase of instruction scheduling (post-pass scheduling). The linear scan register allocator proposed by Poletto and Sarkar [13] is very simple and significantly faster than algorithms based on graph-coloring approaches. The performance of a linear scan register allocator is affected by the maximum number of active live intervals. If we can reduce the maximum number of active live intervals, the linear scan register allocator can generate a more efficient code by reducing the amount of spill code inserted. Thus, we propose a pre-pass local instruction scheduler which can reduce simultaneously live ranges so as to decrease the maximum number of active live

D.A. Bader et al. (Eds.): HiPC 2005, LNCS 3769, pp. 528–537, 2005.

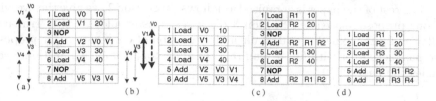

Fig. 1. Example of the phase ordering problem. (a) Sample code in a pseudo intermediate language with live ranges. (b) After instruction scheduling with live ranges. (c) Register allocation first. (d) Instruction scheduling first. Memory access operations are assumed to take two cycles while all other operations take one cycle. Comparing (a) and (b) we see the increase in the number of overlapping live ranges after instruction scheduling. If the register allocation is executed first, it would require eight cycles although only two registers is enough for register allocation. However, if instruction scheduling is done first, then although it would require only six cycles, four registers would be needed to avoid spilling. Which of these two orders is better depends upon the number of available registers and functional units.

intervals. We combined the our proposed scheduler with a linear scan register allocator and evaluated the overall performance. Some previous experimental evaluation and improvements to the linear scan register allocation can be found in [6] and [14]. Previous studies [13], [6] and [14] investigated the linear scan register allocator in isolation rather than its combination with instruction scheduling. In addition, most previous works on phase ordering problem [7], [1], [12], [10], [2] and [4] had focused on combining the instruction scheduling phase with register allocator based on graph-coloring approaches. In this paper, we focus on a *cooperative* approach that solves the phase ordering problem between instruction scheduling and linear scan register allocation. In order to evaluate the performance of combining our proposed scheduler with linear scan register allocation, we have implemented our proposed scheduler and linear scan register allocator in Trimaran. Trimaran [16] is a compiler infrastructure for supporting state of the art research in compiling for Instruction Level Parallel (ILP) architectures. The system is oriented towards EPIC (Explicitly Parallel Instruction Computing) architectures, and supports compiler research in what is typically considered to be "back end" techniques such as instruction scheduling, register allocation, and machine-dependent optimizations. In our framework, we first perform our proposed scheduler, followed by linear scan register allocation and finally the Trimaran-Elcor list scheduler. The results show that performing our proposed pre-pass scheduler can reduce the maximum active live intervals of the linear scan register allocator. This can decrease the amount of spill code inserted by the linear scan register allocator thereby increasing the quality of the generated code. Moreover, it also shows that combining our proposed pre-pass scheduler with linear scan register allocator is significantly faster than Trimaran's pre-pass scheduling, register allocation, and post-pass scheduling scheme using either the IMPACT or the region-based register allocator.

The rest of the paper is organized as follows. In Section 2, we give the overview of our proposed scheduler. Section 3 discusses register allocators in Trimaran, followed by the experimental result and discussion of cooperative our proposed pre-pass scheduler with linear scan register allocator (Section 4). Section 5 concludes the paper.

2   Overview of Our Proposed Scheduler

Our proposed scheduler is based on convergent scheduling [11]. As in the convergent scheduler, we used a weight matrix to compute the schedule time of an operation. However, while the convergent scheduler schedules the instructions at the earliest cycle, our proposed scheduler schedules the instructions in the latest cycle possible while maintaining optimal instruction scheduling length which we assumed to be the critical path of a basic block. In order to get our proposed schedule, we use a weight matrix to calculate the optimal schedule length. In particular, the weight $W_{i,t}$ of instruction i at time slot t is a value between zero and one.

Our heuristic is based on the earliest completion time (the longest path from the root node to the current node) and latest completion time (i.e., the critical path length - the longest path from the current node to a leaf node) of the dependence graph. The operation nodes exist either on the critical path (the longest path of the root node to leave node) or non-critical path of the DAG. The optimal schedule length can be assumed as the length of the critical path length if the available resources are not in conflict. Normally, we can only reorder the operation nodes which are not on the critical path length in order to get more efficient codes. This is especially in the case when there are two kinds of non-critical paths: one that starts with a node which has no dependence predecessor and ends at a node on the critical path, or a second kind that starts at a node on the critical path and ends with a node which has no dependence successor. We should schedule the operation nodes which are on the first kind of non-critical path at the latest cycle possible to reduce the simultaneously live ranges. In contrast, the second kind of non-critical paths should be scheduled at the earliest cycle possible. In [15], Chen reported that real dependence graphs of programs have more of the first kind of non-critical paths than the second. Therefore, our proposed scheduler applies as late as possible schedule. If l_e is the earliest completion time and l_l is the latest completion time, the instruction can be scheduled only in the time slots between l_e and l_l. If the instruction could not be scheduled between l_e and l_l due to insufficient parallel functional units, we increase l_l by one and reschedule again. To begin, the value of $W_{i,t}$ is initialized as follows:

$$W_{i,t} = \begin{cases} 0 & \text{if } t < l_e \text{ or } t > l_l; \\ 1/I(t) & \text{if } t \geq l_e \text{ and } t \leq l_l. \end{cases}$$

where $I(t)$ is the number of instructions that has its (l_e, l_l) crossing time t.

1. Compute the earliest completion time and latest completion time
2. Initialize the weight matrix
3. Mark the scheduling is not finished
4. **While** scheduling is not finished
5.   **While** the latest completion time is greater than zero
6.   **For** (each operation within a basic block)
7.   **If** (resource is available and weight matrix is not zero)
8.   Multiply weight matrix by 1.2 and put back into weight matrix
9.   Mark the operation with schedule
10.   Increase the number of current resources by one
11.   **If** the current resources reach the maximum resource limit
12.   Decrease the latest completion time by one
13.   Initialize the number of current resources by zero
14.   **Endif**
15.   **Endif**
16.   **Endfor**
17.   **Endwhile**
18. Mark the scheduling is finished
19. **For** (each operation within a basic block)
20.   **If** one of the operations within a basic block is not scheduled
21.   Mark the scheduling is not finished
22.   **Endif**
23. **Endfor**
24. **Endwhile**
23. Normalize the weight matrix by dividing each weight with the
    total weights for each operation
24. Choose the cycle time which has the maximum weight for each
    operation as schedule time

Fig. 2. Our proposed scheduler

We give more weight to a specific instruction to be scheduled in a given time cycle by multiplying the weight with a constant value. Then, we normalize our weights:

$$\forall i, t, \hat{W}_{i,t} \leftarrow \frac{W_{i,t}}{\sum_{t'} W_{i,t'}}$$

The schedule time for each instruction is then $\max_t\{\hat{W}_{i,t}\}$. The full algorithm of our proposed scheduler is given in Fig. 2.

3   Register Allocators in Trimaran

Global register allocation based on graph coloring was first proposed by Chaitin et al. [3]. A graph coloring register allocator iteratively builds an undirected graph called an *interference graph* that shows the overlap in live ranges. A node in an interference graph is a live range that is a candidate for register allocation and an edge connects two nodes when the corresponding live ranges overlap. The standard graph-coloring method heuristically attempts to find a k-coloring for the interference graph. A graph is k-colorable if each node can be assigned to one of k-colors such that no two adjacent nodes have the same color. If the heuristic can find a k-coloring, then k registers are sufficient to hold the content of all the live ranges. Otherwise, some candidates are chosen to be spilled, and the interference graph must be rebuilt after a spill decision is made. Another attempt is then made to obtain a k-coloring. This whole process is repeated until a k-coloring is finally obtained. In practice, the cost of graph-coloring approach can be expensive by repeatedly constructing a register interference graph until the heuristic succeeds. However, the graph-coloring based register allocators have been used in many commercial compilers to obtain significant improvements over simple register allocation heuristic. In Trimaran, there have been two global register allocators: the IMPACT register allocator [8] and region-based register allocator [9], adapted from Chow and Hennessy graph-coloring framework [5].

3.1   Linear Scan Register Allocator

Register allocation based on graph-coloring is generally considered the state-of-the-art. However, the algorithm can be computationally expensive. In light of this, Poletto and Sarkar [13] proposed an alternative algorithm for fast register allocator called linear scan register allocation. Linear scan register allocation works on the topological ordering of live ranges (also known as live intervals). Live intervals of each temporary variable are computed and assigned registers. A temporary variable with overlapping intervals can be assigned to different registers and non-overlapping intervals can be assigned to same registers. A linear scan register allocator performs the following four steps [6]:

1. sort all the instructions in topological order;
2. calculate the set of live intervals;
3. assign each temporary variable to physical register for each interval (or spill into the memory) and finally
4. rewrite the code with the obtained allocation.

Ordering of Instructions. The topological ordering of basic blocks required by the linear scan allocator is not unique. In particular, the ordering may be (1) depth-first, (2) preorder, (3) postorder, (4) breadth-first, (5) prediction, and (6) random. An experimental study of the impact of these orderings can be found in [6]. Among the different orderings, depth-first ordering was found empirically to reduce the most amount of false interference between live intervals [13,6]. However, there has been no discussion of reordering of instructions *within* a basic block. The order of instructions within a basic block impacts the allocation and the number of spill code insertions. Our proposed schedule described in Section 2 reorders instructions within a basic block to reduce simultaneously live ranges. Fig. 3(a) and (b) show the original ordering and the ordering of instruction after applying our proposed scheduler respectively.

Computation of Live Intervals. Live ranges are determined by a set of instructions within each basic block. Each live range has a start position with the first definition of the temporary and an end position with the last use of the temporary. Then, all live intervals are sorted in the order of increasing start-points so as to make the allocation more efficient. The number of live intervals with start position and end position in Fig. 3 are given in Fig. 4.

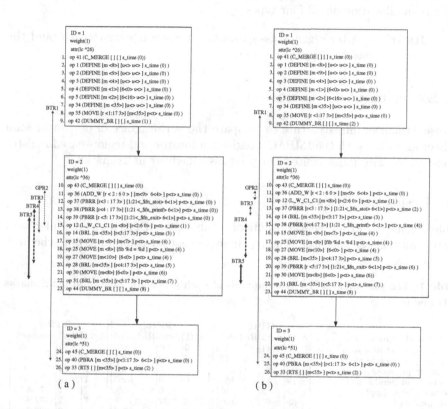

Fig. 3. Control Flow Graph (CFG) with long instructions within each basic block from Trimaran. (a) without instruction reordering. (b) with instruction reordering.

Register name	Start position	End position
BTR1	8	25
GPR2	11	15
BTR3	12	16
BTR4	13	20
BTR5	14	22

(a)

Register name	Start position	End position
BTR1	8	25
GPR2	11	12
BTR3	13	14
BTR4	15	19
BTR5	20	22

(b)

Fig. 4. A number of live intervals for the data dependent graph in Fig. 3. (a) without instruction reordering (b) with instruction reordering.

As shown in Fig. 4, without instruction reordering, BTR1, BTR3, BTR4 and BTR5 are live at the same time. However, with instruction reordering, BTR1 is only live at the same time with BTR3, BTR4 or BTR5.

Register Assignment. After sorting all live intervals by their start points, the allocation of registers to intervals can be done. In Trimaran, there are four register types : general purpose registers (GPRs), floating point registers (FPRs), branch target registers (BTRs) and predicate registers (PRs). We performed linear scan allocation on all four types.

Code Rewrite. After register assignment, the code is rewritten to bind the temporary variables to physical registers.

4 Experimental Evaluation

We use Trimaran infrastructure to compare the performance of our linear scan register allocator with the IMPACT register allocator and region-based register allocator. We also implemented our pre-pass scheduler in Trimaran.

4.1 Result and Discussion

Table 1 gives the experimental results of combining our pre-pass scheduler with linear scan register allocator. The result suggests that combining our scheduler

Table 1. The maximum active live intervals of each procedure of several benchmarks in Trimaran

Benchmarks(procedure)	GPR			FPR			BTR		
	Act1	Act2	Reduce%	Act1	Act2	Reduce%	Act1	Act2	Reduce%
181.mcf(_insert_new_arc)	16	11	31.25%	0	0	0%	1	1	0%
181.mcf(_replace_weaker_arc)	17	12	29.41%	0	0	0%	1	1	0%
181.mcf(_price_out_impl)	29	24	17.24%	0	0	0%	2	2	0%
181.mcf(_suspend_impl)	19	13	31.58%	0	0	0%	1	1	0%
181.mcf(_global_opt)	3	3	0%	0	0	0%	7	2	71.43%
101.tomcatv	69	65	5.80%	33	32	3.03%	7	2	71.43%
wc(_main)	7	7	0%	0	0	0%	3	2	33.33%
bmm (_sumup)	6	6	0%	1	1	0%	3	1	66.67%
dag	11	11	0%	0	0	0%	1	1	0%
eight	8	8	0%	0	0	0%	2	1	50.00%
example_bench(_convert_to_int)	2	2	0%	0	0	0%	3	2	33.33%
fact2	3	3	0%	0	0	0%	3	2	33.33%
fib	4	4	0%	0	0	0%	3	2	33.33%
fib_mem	6	6	0%	0	0	0%	3	2	33.33%
fir	11	11	0%	3	3	0%	3	2	33.33%
hyper	5	5	0%	0	0	0%	1	0	100.00%
ifthen	13	13	0%	0	0	0%	2	1	50.00%
mm_ double (_matmult)	11	11	0%	3	3	0%	2	1	50.00%
mm_ int	14	13	7.14%	0	0	0%	2	1	50.00%
mm	11	11	0%	3	3	0%	3	2	33.33%
nested	6	6	0%	1	1	0%	2	1	50.00%

Act1 - the number of active live intervals after Elcor pre-pass scheduler
Act2 - the number of active live intervals after our pre-pass scheduler

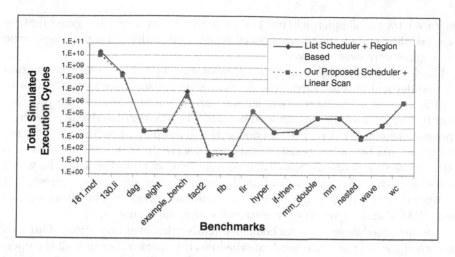

Fig. 5. Total simulated execution cycles

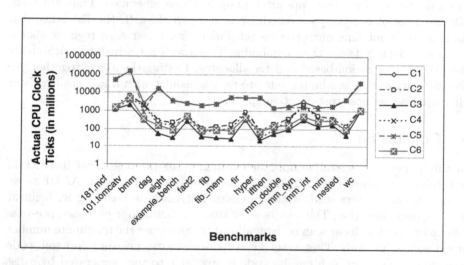

C1 - Trimaran list scheduler with IMPACT register allocator
C2 - Trimaran list scheduler with region-based register allocator
C3 - Our pre-pass scheduler with linear scan register allocator
C4 - Trimaran list scheduler with linear scan register allocator
C5 - Our pre-pass scheduler with IMPACT register allocator
C6 - Our pre-pass scheduler with region-based register allocator

Fig. 6. Actual compilation time on an 1.2 GHz AMD Athlon MP Linux system with 1 GByte RAM

with linear scan register allocator can significantly reduce the maximum active live intervals of basic block. This in turn may reduce spill code insertion and remove unnecessary dependencies. As a result, as shown in Fig. 5, the code generated by our simpler scheme performs just as well as that generated by list scheduling and graph-coloring based register allocation. In some cases, some minor gains were even achieved.

Fig. 6 shows the actual compilation time observed for the various combination of scheduler and register allocators. A linear scan register allocator attempts to find the number of live intervals which are currently active at a certain program point by visiting each lifetime interval in turn. The number of active live intervals represent the number of register needed at this point in the program. If there are insufficient number of free registers, then some active live intervals are chosen to spill and the scan proceeds. Since a linear scan register allocator scans the whole process linearly rather than repeating the process after spill code is inserted, it can operate faster than graph-coloring method based register allocators such as the IMPACT and region-based register allocators in Trimaran.

The list scheduler never unschedules already scheduled operations. Our pre-pass scheduler, on the other hand, unschedules the operations when all the operations cannot schedule within critical path length due to a shortage of functional units. However, unlike the list scheduler, our proposed scheduler does not need to recalculate l_e of successor ops after an op has been scheduled. Thus, the compilation time of our pre-pass scheduler is comparable with the list scheduler. As a result, combining our pre-pass scheduler with linear scan register allocator is significantly faster than combining Trimaran's list scheduler with either the IMPACT or region-based register allocator. Putting the above together, we therefore argue that combining our pre-pass scheduler with linear scan register allocation is the most cost-effective option.

5   Conclusion

In this paper, a cooperative approach utilizing our pre-pass local instruction scheduling and linear scan register allocation has been presented. As far as we know, this is the first study that combines instruction scheduling with linear scan register allocation. The results show that combining our proposed pre-pass scheduler with the linear scan register allocator can reduce the maximum number of active live intervals. This in turn can reduce register pressure and spill code insertion resulting in high quality code comparable to that generated by a list scheduler and a graph-coloring register allocator. Moreover, compared with the latter our scheme results in significantly lower compilation times. As a future work, we will consider the problem of how to do cooperative *global* scheduling with linear scan register allocation. Yet another approach is to fully integrate instruction scheduling with linear scan register allocation.

References

1. D.G. Bradlee, S.J. Eggers, and R.R. Henry. Integrating register allocation and instruction scheduling for RISCs. In Fourth International Conference on Architectural Support for Programming Languages and Operating Systems, pages 122-131, Santa Clara, CA, April 1991.

2. D. Berson, R. Gupta, and M.L. Soffa. Integrated instruction scheduling and register allocation techniques. In Eleventh International Workshop on Languages and Compilers for Parallel Computing, LNCS, Springer Verlag, North Carolina, Chapel Hill, August 1998.

3. G.J. Chaitin, M.A. Auslander, A.K. Chandra, J. Cocke, M.E. Hopkins, and P.W. Markstein. Register allocation via coloring. Computer Languages, 6:47–57, Jan. 1981.

4. G. Chen, and M.D. Smith. Reorganizing global schedules for register allocation. In Proceedings of the 13th international conference on Supercomputing, May 1999.

5. F. Chow, and J. Hennessy. The Priority Based Coloring Approach to Register Allocation. ACM Transactions on Programming Languages and Systems, vol. 12, 501-536, 1990.

6. E. Johansson, and K. Sagonas. Linear scan register allocation in a high performance Erlang compiler. Practical Applications of Declarative Languages: Proceedings of the PADL'2002 Symposium (Lecture Notes in Computer Science, vol. 2257). Springer: Berlin, 2002; 299-317.

7. J.R. Goodman, and W.C. Hsu. Code scheduling and register allocation in large basic blocks. In 1988 International Conference on Supercomputing, pages 442-452, Orlando, Florida, November 1988.

8. R.E. Hank. Machine Independent Register Allocation for the Impact-I C Compiler. BS Thesis, University of Illinois at Urbana-Champaign, 1990.

9. H. Kim. Region-based register allocation for EPIC architecture. Ph.D. Thesis, New York University, January 2001.

10. C. Norris, and L.L. Pollock. An experimental study of several cooperative register allocation and instruction scheduling strategies. In Proceedings of the 28th Annual International Symposium on Microarchitecture, p.169-179, November 29-December 01, 1995, Ann Arbor, Michigan, United States.

11. W. Lee, D. Puppin, S. Swanson, and S. Amarasinghe. Convergent Scheduling. In Proceedings of the 35th Annual International Symposium on Microarchitecture (MICRO), Istanbul, Turkey, November 2002.

12. S.S. Pinter. Register allocation with instruction scheduling: a new approach. In Proceedings of the SIGPLAN '93 Conference on Programming Language Design and Implementation, June 1993.

13. M. Poletto, and V. Sarkar. Linear scan register allocation. ACM Transactions on Programming Languages and Systems (TOPLAS), v.21 n.5, p.895-913, Sept. 1999.

14. K.F. Sagonas, and E. Stenman. Experimental evaluation and improvements to linear scan register allocation. Softw., Pract. Exper. 33(11): 1003-1034 (2003).

15. G. Chen. Effective Instruction Scheduling with Limited Registers. Ph.D. thesis, Harvard University, Division of Engineering and Applied Sciences, March 2001.

16. http://www.Trimaran.org

iSCSI Analysis System and Performance Improvement of iSCSI Sequential Access in High Latency Networks

Saneyasu Yamaguchi[1], Masato Oguchi[2], and Masaru Kitsuregawa[1]

[1] The University of Tokyo, 4-6-1 Komaba Meguro-Ku, Tokyo, Japan
{sane, kitsure}@tkl.iis.u-tokyo.ac.jp
[2] Ochanomizu University, 2-1-1 Otsuka Bunkyo-ku, Tokyo, Japan
oguchi@computer.org

Abstract. IP-SAN and iSCSI are expected to remedy the problems of FC-based SAN. iSCSI has a structure of multilayer protocols. A typical configuration of the protocols to realize this system is as follows: SCSI over iSCSI over TCP/IP over Ethernet. Thus, in order to improve the performance of the system, it is necessary to precisely analyze the complicated behavior of each layer. In this paper, we present an IP-SAN analysis tool that monitors each of these layers from different viewpoints. By using this analysis tool, we experimentally demonstrate that the performance of iSCSI storage access can be significantly improved by more than 60 times.

1   Introduction

The size of data processed by computer systems is increasing rapidly; thus, the large maintenance costs of storage systems have become one of the crucial issues for current computer systems. Storage consolidation using a Storage Area Network (SAN) is one of the most efficient solutions to this problem, and it has been implemented in many computer systems. However, the current-generation SAN based on FC has few demerits; for example, 1) the number of FC engineers is small, 2) the installation cost of FC-SAN is high, 3) the FC has distance limitation, and 4) the interoperability of the FC is not necessarily high. The next-generation SAN based on IP (IP-SAN) is expected to remedy these issues. The IP-SAN employs commodity technologies for a network infrastructure, including Ethernet and TCP/IP. One of the promising standard data transfer protocols of IP-SAN is iSCSI [1], which was approved by the IETF [2] in February 2003. However, the problems of low performance and high CPU utilization have been identified as the demerits of IP-SAN [3, 4, 5]. Thus, improving its performance and maintaining low CPU utilization [3, 6] are the critical issues regarding IP-SAN. In this paper, we discuss the performance issues of IP-SAN.

iSCSI is a protocol through which the SCSI protocol is transferred over TCP/IP; thus, the protocol stack of IP-SAN is "SCSI over iSCSI over TCP/IP over Ethernet." In order to improve the performance of iSCSI storage access,

D.A. Bader et al. (Eds.): HiPC 2005, LNCS 3769, pp. 538–548, 2005.

detailed information of all these layers is required because each of them may have an influence on the end-to-end performance. We propose an iSCSI analysis system and demonstrate that it can identify the causes for the decline in performance, which can be improved by resolving them.

The remainder of this paper is organized as follows. Section 2 introduces the iSCSI analysis system. Section 3 describes an actual application of this system and its role in improving the performance improvement. Section 4 mentions related work and compares them with our research. Finally, section 5 concludes this research and projects future work.

2   iSCSI Analysis System

In this section, we explain our iSCSI monitoring system. This system can monitor the internal states of each layer in an IP-SAN protocol stack. We developed an iSCSI analysis system by inserting the monitoring code into these layers. The functions of this monitoring system are as follows: 1) protocol translation (SCSI, iSCSI, and TCP/IP); 2) visualization of packet transmission with timeline; 3) monitoring behavior of the TCP flow control; 4) detection of packet loss; and 5) generation of the iSCSI storage access with a pseudo iSCSI initiator driver. An overview of our analyzing system is shown in Figure 1. We have discussed these functions in detail in the following subsections.

2.1   Protocol Translation

In this subsection, we describe the function of protocol translation in our analyzing system. With this function, the recorded iSCSI traffic data can be translated into human-readable text data. The iSCSI PDU format and an example of the translation are shown in Figure 2. In the case of "SCSI Command Read," the

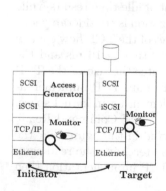

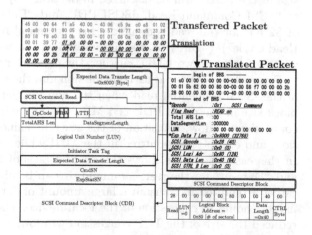

Fig. 1. Overview of the iSCSI analysis system

Fig. 2. iSCSI read command as an example of protocol translation

iSCSI PDU has a format as shown in the lower left part of Figure 2. An example of the hexadecimal dump of a transferred Ethernet packet with 100 byte data in an actual iSCSI storage access is shown in the "Transferred Packet" at the upper left part of Figure 2. The first 52 bytes written in a smaller font (bytes 1-52) construct the IP and TCP headers, and the remaining 48 bytes written in italics (bytes 53-100) construct the iSCSI PDU.

The translation function translates each of these fields into human-readable values and sections of the translated texts are shown in the upper right part of Figure 2[1]. In this case, it can be observed that the packet implies SCSI Command Read with a data length of 32 KB.

The SCSI CDB field with a size of 16 bytes is allocated in the range from bytes 33-48 in the iSCSI PDU. The fields in the range from bytes 43-48 in the PDU are padding data because the length of the SCSI CDB is 10 bytes in our experimental environment. The "Data Segment Length" and "Expected Data Transfer Length" in the iSCSI PDU are of one byte each while that of the SCSI CDB is 512 bytes. Consequently, both the "Expected Data Transfer Length" in the iSCSI header, which is 0x8000 [bytes], and the "Data Length" in the SCSI CDB, which is 0x40 [512 Bytes], are of 32 [KB]. As shown in this section, our analyzing system enables easy understanding of the transferred data using the iSCSI protocol.

2.2 Visualization of Packets Transmission

The transferred packets in the network can be visualized on a timeline with the visualization function. An example of visualized packet transmissions is shown in Figure 3. It shows the packet transmission of the iSCSI sequential read when the one-way latency time is 16 ms and the block size in the iSCSI PDU ("block size in the PDU" will be discussed in Section 3.3) is 32 KB. There are 5 cycles of "SCSI Command Read" iSCSI PDU and "Data-in" PDU with a data unit of 32 KB for the read command (this sequential read cycle is termed as "Seq. Read Cycle"). Figure 3 shows that a significant amount of idle time results while waiting for the network I/O. Further, the network utilization is considerably low.

Figure 4 shows the packet transmission and the states of the TCP flow control of the iSCSI sequential read when the one-way latency time is 16 ms and the block size in the iSCSI PDU is 4 MB. The figure indicates the size of the TCP congestion window, transition of the state machine of Linux TCP implementation, and events that occurred in the TCP while implementing in kernel space. In this case, the figure shows the iSCSI PDUs of the "SCSI Command Read" and "Data-in" for the read command. It can be observed in the figure that the target attempted to transmit a large amount of data on receiving the read command leading to congestion of the local device. The TCP implementation then reduced the size of the TCP congestion window.

---

[1] TCP/IP header and some fields in iSCSI PDU are omitted in Figure 2.

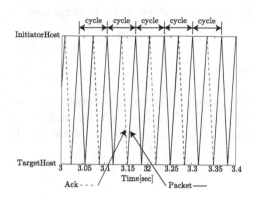

Fig. 3. Packet Transmission A

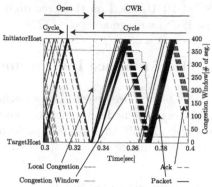

Fig. 4. Packet Transmission B

2.3 Monitoring TCP Flow Control

The TCP implementation has a flow-controlling function. In most cases, since the TCP implementation functions in the kernel space, users are not allowed to monitor its behavior. The monitoring functions of the TCP flow control in our analyzing system enables the monitoring of this behavior in user space by adding a monitoring code into the TCP implementation. This is the only function that depends on system implementation in the iSCSI analysis system. Our TCP flow control monitoring system is implemented with Linux TCP implementation. With this function, the size of the TCP congestion window, various events in the TCP implementation such as congestion detection, and the state transition of the Linux TCP can be monitored from the user space. Linux TCP is implemented as a state machine. The state transition shown in Figure 4 (transition from "Open" to "CWR") is monitored by this function. The TCP flow control has a direct influence on iSCSI performance; thus, it is important to consider this function for improving iSCSI performance.

2.4 iSCSI Access Generation

The analyzing system also has a function for iSCSI access generation. The iSCSI driver is usually installed as a SCSI HBA driver. It is then driven by a generic SCSI driver and other OS implementations. Consequently, the users cannot create the iSCSI PDU based on their requirements. In the case of Linux, since a raw device and certain driver implementations divide a single block issued by the I/O command into multiple small blocks, the users cannot issue an iSCSI read command with a large block size. This limitation does not originate from the iSCSI protocol but from the OS implementation. The iSCSI access generation function enables the iSCSI access without the OS limitation affecting it. The generator directly establishes a TCP/IP connection with iSCSI target implementation using a socket API and transmits and receives the iSCSI PDU according to the iSCSI protocol. With this generator, the users can create the iSCSI PDU based on their requirements and can measure the performance of the iSCSI storage access.

3   Performance Improvement

In this section, we present the performance evaluation of iSCSI sequential storage accesses in a long delayed network and the results of the analysis from our system. We also identify the causes of performance decline. In addition, we demonstrate how performance can be improved.

3.1   Experimental Setup

In this subsection, we describe an experiment conducted to evaluate the iSCSI performance and its environment.

We evaluated the performance of the iSCSI storage access in a heavily delayed network by performing the following experiment. The experimental system is shown in Figure 5. The iSCSI initiator and iSCSI target are constructed using PCs. A network delay emulator, which is constructed by FreeBSD dummynet [7], is placed between the initiator and the target. Further, the initiator and target establish a TCP connection over the dummynet in order that a simulated delayed network is realized between them. The initiator and dummynet are connected with a cross cable of 1 gigabit Ethernet. Further, the dummynet and target are also connected with a cross cable. Both the initiator and target are constructed by a Linux OS, and the dummynet is constructed by FreeBSD. The detailed specifications of the initiator and the target PCs are as follows: CPU Pentium 4 2.80 GHz; main memory 1 GB; OS Linux 2.4.18-3; and NIC gigabit Ethernet card Intel PRO/1000 XT Server adapter. The detailed specifications of the dummynet PC are as follows: CPU Pentium 4 1.5 GHz, main memory 128 MB, OS FreeBSD 4.5-RELEASE, and NIC Intel PRO/1000 XT Server Adapter × 2.

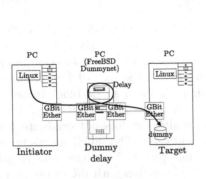

Fig. 5. Experimental setup

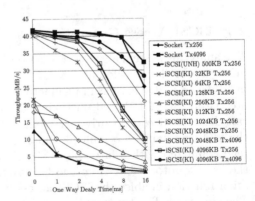

Fig. 6. Experimental result: iSCSI sequential read throughput

We employed the iSCSI implementation, which is distributed by the interoperability laboratory in the University of New Hampshire [8, 9]. This iSCSI implementation is termed as "UNH" implementation. The detailed specifications and configurations of the iSCSI implementation used for the evaluation are as follows: iSCSI initiator and target: UNH IOL Draft 18 reference implementation ver. 3; iSCSI MaxRecvDataSegmentLength, iSCSI MaxBurstLength, and iSCSI FirstBurstLength: 16777215 bytes.

The initiator establishes the iSCSI connection with the target, and a benchmark software is implemented to measure the performance. The benchmark software iterates by issuing system call `read()` to the raw iSCSI device on the initiator OS. This call is single-threaded. The size of the TCP advertised window is 2 MB. The iSCSI target runs in memory mode. It can be regarded as a storage device with exceptional performance.

3.2   Basic Performance Measurement

"Socket Tx256" and "iSCSI (UNH) 500 KB Tx256" shown in Figure 6 are obtained by the experiment described in Section 3.1. We refer to this experiment as "Exp. A." The horizontal axis represents the one-way delay time between the initiator host and the target host. The vertical axis represents the measured throughput. "0 ms" implies that the network delay was not intentionally generated by the dummynet. In this case, the network delay with the physical device is approximately 100 μs. The "iSCSI (UNH) 500 KB Tx256" shown in the figure represents the throughput of the iSCSI sequential read, while the "Socket Tx256" represents the throughput of simple socket communication by means of which the initiator and target hosts establish the TCP/IP connection and transmit data through the socket API in the same environment. This simple socket connection is referred to as "pure socket." The block size specified at system call `read()` is 500 KB, and this block size is termed as "System Call Block Size." The maximum throughput of the "pure socket" is approximately 40 [MB/s] because this value represents the performance limit of the dummynet PC.

The results obtained indicate the following: 1) The iSCSI performance is severely low although considerably high-performance socket communication is achieved in the same environment and 2) The iSCSI performance decreases as network latency increases. In the following subsections, the reasons for the performance decline of the iSCSI are discussed.

3.3   Analysis of iSCSI Access

In this subsection, we present the analysis of the iSCSI storage access, as discussed in Section 3.2. Figure 2 shows the results of the protocol translation of the iSCSI PDU in the experiment. Figure 3 shows the visualized packet transmission in the experiment.

The translation results indicate that the "SCSI Command Read" PDUs with 32-KB "PDU Block Size" were issued even though the benchmark software issued the system call `read()` with a System Call Block Size of 500 KB. With several OS implementations, the issued system calls are transmitted to the network through the block or character devices, SCSI generic driver, iSCSI driver, TCP/IP implementation, and Ethernet device driver. Such calls are rarely transmitted to the network without being modified by these drivers. In other words, the "System Call Block Size" is not always equal to the "PDU Block Size." It can also be observed that in our experimental environment, the block size of the issued system call `read()` is divided into multiple 32-KB block reads and transmitted to the target host.

The visualized figure (Figure 3) shows that when a large amount of time in the "Seq. Read Cycle" is expended in waiting for the network I/O, the network utilization is considerably low.

3.4   Discussion of Performance Decline

The analysis in Section 3.3 demonstrated that the System Call Block Size was divided into multiple small (i.e., 32 KB) "PDU Block Size"; thus, the network utilization is significantly low. The poor network utilization can be considered as the most critical reason for the decline in performance. As shown in Figure 3, the throughput of the iSCSI sequential read can be modeled as follows:

$$\frac{\text{PDU Block Size}}{4 \times \text{OneWayDelay} + \frac{\text{PDU Block Size}}{\text{LowerLayerThroughput}}} \tag{1}$$

A "Non-Idle ratio," which is the ratio of the time spent for sending data to the total "Seq. Read Cycle" time, can be modeled as follows:

$$\frac{\frac{\text{PDU Block Size}}{\text{LowerLayerThroughput}}}{4 \times \text{OneWayDelay} + \frac{\text{PDU Block Size}}{\text{LowerLayerThroughput}}} \tag{2}$$

In these models, `LowerLayerThroughput` represents the throughput of the pure socket. Consequently, the idle ratio with a 32-KB "PDU Block Size" is 84% for a one-way delay of 1 ms; 91%, 2 ms; 95%, 4 ms; 97%, 8 ms; and 98%, 16 ms. By using the analyzing system, we can identify that the reason for the performance decline is poor network utilization caused by a small block size.

The division of the block size is not only due to the limitation of the iSCSI protocol specification but also due to Linux OS implementation. As a result, we measured the performance of the essential iSCSI storage access, which is not restricted by any specific OS implementation, by using our iSCSI access generator. The experimental results are indicated in Figure 6 as "iSCSI(KI) nKB Tx256". The experiments were carried out in the same environment and termed as "Exp. B."

"iSCSI(UNH) 500 KB Tx256" in the figure represents the performance with the iSCSI sequential read access under the UNH iSCSI implementation. The System Call Block Size is 500 KB ("PDU Block Size" is 32 KB, as mentioned previously). The lines labeled as "iSCSI(KI)" represent the performance using our pseudo iSCSI initiator. Their block sizes are recorded in the labels. The "Socket Tx256" indicates the performance of the pure socket. "Tx" is mentioned in section 3.5.

The following can be obtained from the results. 1) The performance of the iSCSI increased significantly by increasing the read block size. 2) The performance obtained by the iSCSI access with a large block is not sufficiently high. In the case of 16-ms one way delay, the throughput of "iSCSI(KI) 4 MB Tx256" is 10.1 [MB/s], while that of "iSCSI(UNH) Tx256" is 0.47 [MB/s]. The performance improved more than 20 times when the block size was increased. However, the throughput of the pure socket was 25.2 [MB/s], and the performance of the iSCSI is less than half that of the socket.

3.5   Analysis Without Block Division

In this subsection, we analyze the behavior of the iSCSI storage access with a large block size.

The transitions of the throughput and size of the TCP congestion window of "iSCSI(KI) 4MB Tx256" and the "Socket Tx256" for a one-way delay of 16 ms are shown in Figure 7 and Figure 8, respectively. First, we found that the obtained throughput and the size of the TCP congestion window vary synchronously. The TCP implementation restricts the output throughput below the $\frac{\text{CongestionWindowSize}}{\text{RoundTripTime}}$. Second, Figure 7 shows that local device congestion occurs when the congestion window size is approximately 350 segments and this size cannot be exceeded for the iSCSI access sequential read. However, the congestion window size was approximately 850 segments in the case of the socket communication. The burstness of traffic progressively weakens due to self-clocking of the TCP during the pure socket communication. On the other hand, in the case of the iSCSI access, the iSCSI driver is independently synchronized using the SCSI Command Read and SCSI Response at the iSCSI layer. This results in extremely bursty traffic that is generated when the target returns the "Data-in" PDU for the "SCSI Command Read" PDU, as shown in Figure 4. As

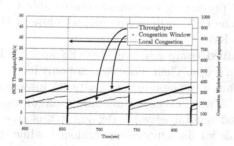

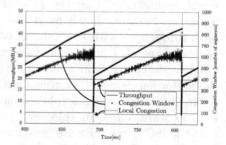

Fig. 7. Transitions of TCP Congestion Window (iSCSI)

Fig. 8. Transitions of TCP Congestion Window (Socket)

a result, the traffic burstness persists. Therefore, the iSCSI traffic patterns easily cause congestion and the throughput is restricted by the TCP implementation.

In this case, the reason for the performance decline can be attributed to the congestion in the local device and TCP flow control resulting from bursty iSCSI traffic. The local device congestion occurs when the packet descriptors are depleted in the local network interface card. This congestion can be avoided by improving the tolerance of the NIC to bursty traffic by enlarging the buffer size of the NIC device driver. The number of packet buffers in the device driver of the NICs used in this experiment environment (refer to Section 3.1) can be regulated from 80 to 4096. In "Exp. B" described in Section 3.4, they were set to the default value of 256.

The measured throughput with 4096 NIC device driver buffers are represented as lines that are labeled as "Tx4096" in Figure 6; this experiment is referred to as "Exp. C." In the labels in Figure 6, "Tx" represents the number of packets that the device driver can buffer. Further improvement in the performance was obtained by avoiding the local congestion. In the case of 4-MB block size and 16-ms one way delay, the performance improved 2.81 times. The performance in Exp. C improved 60.5 times that in Exp. A for a one-way delay of 16 ms. The performance of the pure socket, which can be considered as the performance limit of our experiment environment, was also improved by avoiding the local congestion. In Exp. C, the performance decline by adopting the iSCSI protocol was 12% below the system performance limit for a one-way delay of 16 ms, while it was 60% in Exp. B. Thus, a performance comparable to the system limit can be achieved in the iSCSI storage access.

4   Related Work

Several studies have presented the performance evaluation of IP-SAN using iSCSI [3, 4, 10, 11, 12, 13].

Sarkar et al. [4] avoided the performance evaluation of the iSCSI. In particular, this study paid attention to the CPU utilization of the iSCSI storage access since it is extremely crucial for IP-SAN. They experimented with the performances of the iSCSI storage accesses using various block sizes in a LAN environment. The work demonstrated that the TCP/IP processing consumed considerable CPU resources. Further, they showed that the CPU utilization reached 100% at the peak throughput with block size of 64 KB.

Radkov et al. [13] presented a detailed comparison of the NFS and iSCSI. Further, the comparison is very broad in scope. Their discussion not only includes the performance and CPU utilization but also the number of network messages. Both the micro- and macrobenchmarks were executed in some configurations such as warm cache or cold cache and several network delays. It was shown that the iSCSI and NFS are comparable for data-intensive workloads, while the former outperforms the latter for meta-data intensive workloads.

These studies are obtained by executing various workloads outside the IP-SAN system. Consequently, these studies do not reveal accurate behaviors inside IP-

SAN systems. Our work presents very exact behaviors inside the IP-SAN system including those in kernel space, for example TCP flow controlling. As far as we know, there is no published report discussing the performance of the iSCSI through the examination of the TCP/IP behavior, particularly the congestion window size and receive window size. This type of monitoring is a novel feature of our work. In addition, we have also identified the causes for the performance decline by employing the proposed system while the existing studies reveal the experimental results. This is also a novel feature of our work.

5 Conclusion

In this paper, we proposed and implemented an iSCSI storage access analysis system and demonstrated that the iSCSI performance can be significantly improved by detailed analysis using the proposed system and resolving the issues identified by the system. The proposed system can point out the reasons for the decline in performance. In our experiment, the performance improved more than 60 times and was comparable to the system limit performance.

In future work, our objectives are as follows: 1) to measure the performance of the iSCSI access using a real storage device; 2) to analyze not only single-threaded sequential read access but also write access, random access; and multiple access, and 3) to analyze the iSCSI storage access of some applications such as DBMS.

References

1. J. Satran et al. Internet Small Computer Systems Interface (iSCSI). http://www.ietf.org/rfc/rfc3720.txt , April 2004.
2. IETF Home Page. http://www.ietf.org/ , 2004.
3. Prasenjit Sarkar, Sandeep Uttamchandani, and Kaladhar Voruganti. Storage over IP: When Does Hardware Support help? In *Proc. FAST 2003, USENIX Conference on File and Storage Technologies*, March 2003.
4. Prasenjit Sarkar and Kaladhar Voruganti. IP Storage: The Challenge Ahead. In *Proc. of Tenth NASA Goddard Conference on Mass Storage Systems and Technologies*, April 2002.
5. Fujita Tomonori and Ogawara Masanori. Analisys fo iSCSI Target Software. In *SACSIS (Symposium on Advanced Computing Systems and Infrastructures) 2004*, April 2004. (in Japanese).
6. Jeffrey C. Mogul. Tcp offload is a dumb idea whose time has come. In *9th Workshop on Hot Topics in Operating Systems (HotOS IX)*, May 2003.
7. L. Rizzo. dummynet. http://info.iet.unipi.it/~luigi/ip_dummynet/ , 2004.
8. University of new hampshire interoperability lab. http://www.iol.unh.edu/ , 2004.
9. iSCSI reference implementation. http://www.iol.unh.edu/consortiums/iscsi/downloads.html , 2004.

10. Stephen Aiken, Dirk Grunwald, and Andy Pleszkun. A Performance Analysis of the iSCSI Protocol. In *IEEE/NASA MSST2003 Twentieth IEEE/Eleventh NASA Goddard Conference on Mass Storage Systems and Technologies*, April 2003.
11. Yingping Lu and David H. C. Du. Performance Study of iSCSI-Based Storage Subsystems. *IEEE Communications Magazine*, August 2003.
12. Wee Teck Ng, Bruce Hilly Elizabeth Shriver, Eran Gabber, and Banu Ozden. Obtaining High Performance for Storage Outsourcing. In *Proc. FAST 2002, USENIX Conference on File and Storage Technologies*, pages 145–158, January 2002.
13. Peter Radkov, Li Yin, Pawan Goyal, Prasenjit Sarkar, and Prashant Shenoy. A performance Comparison of NFS and iSCSI for IP-Networked Storage. In *Proc. FAST 2004, USENIX Conference on File and Storage Technologies*, March 2004.

Author Index

Lecture Notes in Computer Science

For information about Vols. 1–3735

please contact your bookseller or Springer

Vol. 3785: K.-K. Lau, R. Banach (Eds.), Formal Methods and Software Engineering. XIV, 496 pages. 2005.

Vol. 3784: J. Tao, T. Tan, R.W. Picard (Eds.), Affective Computing and Intelligent Interaction. XIX, 1008 pages. 2005.

Vol. 3783: S. Qing, W. Mao, J. Lopez, G. Wang (Eds.), Information and Communications Security. XIV, 492 pages. 2005.

Vol. 3781: S.Z. Li, Z. Sun, T. Tan, S. Pankanti, G. Chollet, D. Zhang (Eds.), Advances in Biometric Person Authentication. XI, 250 pages. 2005.

Vol. 3780: K. Yi (Ed.), Programming Languages and Systems. XI, 435 pages. 2005.

Vol. 3779: H. Jin, D. Reed, W. Jiang (Eds.), Network and Parallel Computing. XV, 513 pages. 2005.

Vol. 3778: C. Atkinson, C. Bunse, H.-G. Gross, C. Peper (Eds.), Component-Based Software Development for Embedded Systems. VIII, 345 pages. 2005.

Vol. 3777: O.B. Lupanov, O.M. Kasim-Zade, A.V. Chaskin, K. Steinhöfel (Eds.), Stochastic Algorithms: Foundations and Applications. VIII, 239 pages. 2005.

Vol. 3775: J. Schönwälder, J. Serrat (Eds.), Ambient Networks. XIII, 281 pages. 2005.

Vol. 3773: A. Sanfeliu, M.L. Cortés (Eds.), Progress in Pattern Recognition, Image Analysis and Applications. XX, 1094 pages. 2005.

Vol. 3772: M. Consens, G. Navarro (Eds.), String Processing and Information Retrieval. XIV, 406 pages. 2005.

Vol. 3771: J.M.T. Romijn, G.P. Smith, J. van de Pol (Eds.), Integrated Formal Methods. XI, 407 pages. 2005.

Vol. 3770: J. Akoka, S.W. Liddle, I.-Y. Song, M. Bertolotto, I. Comyn-Wattiau, W.-J. van den Heuvel, M. Kolp, J. Trujillo, C. Kop, H.C. Mayr (Eds.), Perspectives in Conceptual Modeling. XXII, 476 pages. 2005.

Vol. 3769: D.A. Bader, M. Parashar, V. Sridhar, V.K. Prasanna (Eds.), High Performance Computing – HiPC 2005. XXVIII, 550 pages. 2005.

Vol. 3768: Y.-S. Ho, H.J. Kim (Eds.), Advances in Multimedia Information Processing - PCM 2005, Part II. XXVIII, 1088 pages. 2005.

Vol. 3767: Y.-S. Ho, H.J. Kim (Eds.), Advances in Multimedia Information Processing - PCM 2005, Part I. XXVIII, 1022 pages. 2005.

Vol. 3766: N. Sebe, M.S. Lew, T.S. Huang (Eds.), Computer Vision in Human-Computer Interaction. X, 231 pages. 2005.

Vol. 3765: Y. Liu, T. Jiang, C. Zhang (Eds.), Computer Vision for Biomedical Image Applications. X, 563 pages. 2005.

Vol. 3764: S. Tixeuil, T. Herman (Eds.), Self-Stabilizing Systems. VIII, 229 pages. 2005.

Vol. 3762: R. Meersman, Z. Tari, P. Herrero (Eds.), On the Move to Meaningful Internet Systems 2005: OTM 2005 Workshops. XXXI, 1228 pages. 2005.

Vol. 3761: R. Meersman, Z. Tari (Eds.), On the Move to Meaningful Internet Systems 2005: CoopIS, DOA, and ODBASE, Part II. XXVII, 653 pages. 2005.

Vol. 3760: R. Meersman, Z. Tari (Eds.), On the Move to Meaningful Internet Systems 2005: CoopIS, DOA, and ODBASE, Part I. XXVII, 921 pages. 2005.

Vol. 3759: G. Chen, Y. Pan, M. Guo, J. Lu (Eds.), Parallel and Distributed Processing and Applications - ISPA 2005 Workshops. XIII, 669 pages. 2005.

Vol. 3758: Y. Pan, D.-x. Chen, M. Guo, J. Cao, J.J. Dongarra (Eds.), Parallel and Distributed Processing and Applications. XXIII, 1162 pages. 2005.

Vol. 3757: A. Rangarajan, B. Vemuri, A.L. Yuille (Eds.), Energy Minimization Methods in Computer Vision and Pattern Recognition. XII, 666 pages. 2005.

Vol. 3756: J. Cao, W. Nejdl, M. Xu (Eds.), Advanced Parallel Processing Technologies. XIV, 526 pages. 2005.

Vol. 3754: J. Dalmau Royo, G. Hasegawa (Eds.), Management of Multimedia Networks and Services. XII, 384 pages. 2005.

Vol. 3753: O.F. Olsen, L.M.J. Florack, A. Kuijper (Eds.), Deep Structure, Singularities, and Computer Vision. X, 259 pages. 2005.

Vol. 3752: N. Paragios, O. Faugeras, T. Chan, C. Schnörr (Eds.), Variational, Geometric, and Level Set Methods in Computer Vision. XI, 369 pages. 2005.

Vol. 3751: T. Magedanz, E.R. M. Madeira, P. Dini (Eds.), Operations and Management in IP-Based Networks. X, 213 pages. 2005.

Vol. 3750: J.S. Duncan, G. Gerig (Eds.), Medical Image Computing and Computer-Assisted Intervention – MICCAI 2005, Part II. XL, 1018 pages. 2005.

Vol. 3749: J.S. Duncan, G. Gerig (Eds.), Medical Image Computing and Computer-Assisted Intervention – MICCAI 2005, Part I. XXXIX, 942 pages. 2005.

Vol. 3748: A. Hartman, D. Kreische (Eds.), Model Driven Architecture – Foundations and Applications. IX, 349 pages. 2005.

Vol. 3747: C.A. Maziero, J.G. Silva, A.M.S. Andrade, F.M.d. Assis Silva (Eds.), Dependable Computing. XV, 267 pages. 2005.

Vol. 3746: P. Bozanis, E.N. Houstis (Eds.), Advances in Informatics. XIX, 879 pages. 2005.

Vol. 3745: J.L. Oliveira, V. Maojo, F. Martín-Sánchez, A.S. Pereira (Eds.), Biological and Medical Data Analysis. XII, 422 pages. 2005. (Subseries LNBI).

Vol. 3744: T. Magedanz, A. Karmouch, S. Pierre, I. Venieris (Eds.), Mobility Aware Technologies and Applications. XIV, 418 pages. 2005.

Vol. 3742: J. Akiyama, M. Kano, X. Tan (Eds.), Discrete and Computational Geometry. VIII, 213 pages. 2005.

Vol. 3740: T. Srikanthan, J. Xue, C.-H. Chang (Eds.), Advances in Computer Systems Architecture. XVII, 833 pages. 2005.

Vol. 3739: W. Fan, Z. Wu, J. Yang (Eds.), Advances in Web-Age Information Management. XXIV, 930 pages. 2005.

Vol. 3738: V.R. Syrotiuk, E. Chávez (Eds.), Ad-Hoc, Mobile, and Wireless Networks. XI, 360 pages. 2005.

Vol. 3737: C. Priami, E. Merelli, P. Gonzalez, A. Omicini (Eds.), Transactions on Computational Systems Biology III. VII, 169 pages. 2005. (Subseries LNBI).